ATOMS AND MOLECULES

ATOMS AND MOLECULES

An Introduction for Students of Physical Chemistry

MARTIN KARPLUS
Harvard University

RICHARD N. PORTER
University of Arkansas

W. A. Benjamin, Inc.   New York   1970

**ATOMS AND MOLECULES An Introduction for
Students of Physical Chemistry**

Copyright © 1970 by W. A. Benjamin, Inc.
All rights reserved

Standard Book Number 8053–5218-X (Clothbound Edition)

Library of Congress Catalog Card Number 72–105272

Manufactured in the United States of America
12345 K32109

W. A. Benjamin, Inc.
New York, New York 10016

To
R. I. K., C. M. P., T. M. K., J. S. P.
and their friends

PREFACE

Physical chemistry attempts to provide an understanding of chemical phenomena at a fundamental level. For a long time, most discussions were based on macroscopic relationships, such as those obtained from the laws of thermodynamics. Today, the availability of theoretical and experimental methods for the examination of atoms and molecules permits physical chemists to delve deeper and explain "macroscopic" observations in terms of "microscopic" interactions. A course in physical chemistry should follow the logic of the subject and begin the "molecular" level with the development of the quantum theory. After an understanding of atomic structure and chemical bonding is achieved, the course can progress to the interpretation of the behavior of large numbers of atoms or molecules by use of statistical mechanics and thermodynamics. Having acquired a knowledge of these basic elements of physical chemistry, the student is ready to apply them to a wide variety of chemical phenomena.

The present textbook is an introduction to the part of physical chemistry concerned with the properties of individual atoms and molecules. In the sequence outlined above, the volume would best serve as the text for the beginning quarter or semester of physical chemistry. The only prerequisites are introductory courses in calculus, chemistry, and physics. However, since the book is self-contained, it could be used equally well in the second part of a course which begins with thermodynamics. Moreover, because material beyond the scope of the usual first course is included, students can use the volume for supplementary reading. Teachers will find that the book can serve also as the basis for a more specialized course in atomic and molecular structure and spectra.

Atoms and Molecules: An Introduction for Students of Physical Chemistry originated in a lecture course recorded by one of the authors at Columbia University. We hope that some of the spirit of the lectures has persisted through the long process of rewriting, expanding, and editing. In particular, we have tried to preserve a discursive style which foregoes the succinctness of mathematical exposition to explain in words the meaning and utility of principles. This is not to say that the mathematical aspects of physical chemistry are slighted in any way; rather, they are introduced as in a lecturer's presentation which attempts to make the material plausible.

It is our intent that students should be able to read this volume and understand its contents without the need to supplement it by referring to more detailed discussions. Such an expanded treatment may make some portions of the book seem more lengthy than they are in terms of the actual required study time; that is, the teacher in his lectures and the students in their reading will be able to cover the material more rapidly than they might expect from the number of pages involved. (An indication of this is that in the original course much of Chapter 1 (40 pages) was covered in two or three lectures, Chapter 2 (60 pages) in four to six lectures, and so on. The fuller explanations require the student to read more material per lecture than he might in some other texts. However, by including the additional material we hope the result is an easier, yet more fundamental, introduction to physical chemistry.

Chapters 1 through 5 (excluding the sections in small type) contain material that can be presented in its entirety in most courses in physical chemistry. Chapters 6 and 7 provide treatments of molecular theory and of molecular spectroscopy that go beyond what needs to be covered completely in an elementary course. The instructor may consider parts of these chapters as collections of "special topics" from which he can select material to round out the course. For example, although some discussion of the spectra of diatomic molecules belongs in every physical chemistry course, the discussion of polyatomic molecules and of magnetic resonance given in Chapter 7 contains much more detail than is usually included.

The problems at the end of each chapter serve two primary purposes. Some of them are intended to amplify or complete sections of the textual material; references are made in the text to most problems of this type. Other problems provide an opportunity for the student to reinforce his understanding by working with some of the essential concepts. We have in many instances given several similar problems of the latter type, so that the instructor can vary the problem assignments from year to year without altering greatly the coverage of subject matter.

We take pleasure in acknowledging that the inspiration for the present volume can be traced back to the late nineteen fifties, when both authors were at the University of Illinois. We were impressed at that time by a modern physical chemistry course that was being developed there by Peter E. Yankwich, Aron Kuppermann, R. Linn Belford, and H. S. Gutowsky. We acknowledge also the constructive comments of Professors Bruce Berne, Thomas Dunn, Paula Getzin, Bryan Kohler, Frederick Minn, Leonard Nash, Lionel Raff, and Peter Yankwich, which assisted us in preparing the final version of the manuscript. Messrs. Sheldon Green and Steven Wofsy, who read the manuscript, made many useful suggestions and tested most of the problems. Mrs. Frankie Ward and Miss Carol Avard are due our gratitude for spending long hours preparing accurate typed versions of the manuscript in its early and late stages, respectively. We thank Mr. William Prokos for his creative advice and faithful rendition of the illustrations. We also wish to thank Prof. J. L. Whitten, Dr. R. M. Stevens, and Dr. G. G. Balint-Kurti for use of their calculations for some of the illustrations.

Much of the volume was worked out slowly over the years while one, or the other, or both authors were located at the University of Illinois, Columbia University, the University of Arkansas, Harvard University, the Joint Institute for Laboratory Astrophysics, and the Weizmann Institute. A summer on Martha's Vineyard, however, was the essential element in obtaining a completed manuscript. The opportunity for interspersing work with swimming, sailing, and other island diversions did much to make the writing a painless task. Although most readers will not be able to study the volume under such ideal conditions, we hope that this introduction to atoms and molecules will leave them, nevertheless, with a pleasant feeling for the subject.

Cambridge, Mass.
Stony Brook, N.Y.
October 1970

MARTIN KARPLUS
RICHARD N. PORTER

CONTENTS

1

The constituents of matter

A fundamental investigation of chemical phenomena begins with two basic questions: What are the elementary constituents of chemical systems? What laws govern the behavior of these constituents? Taken on face value, the first question is a somewhat ambiguous one, the answer to which changes each time the discovery of a new elementary particle is reported in *The New York Times*. In the context of physical chemistry, however, we can be satisfied with a pragmatic interpretation of the term "elementary" or "fundamental" and use it to designate the particles that must be considered for an understanding of chemical phenomena. Once these particles have been identified, the second question is concerned with the forces that exist between the particles and their motion under the influence of these forces. Knowledge of the elementary particles and their laws of motion will make it possible for us to understand chemical phenomena on a microscopic (atomic) level and ultimately to extend this understanding to the properties and transformations of macroscopic chemical systems.

To gain a first insight into the answers to these questions, we follow a quasihistorical approach; that is, we sketch the experimental and theoretical development in our knowledge of the elementary particles and their properties, but often use present understanding to modify the purely historical account. This allows us to capture the spirit of the new discoveries, while avoiding many of the blind alleys into which the early investigators were led.

1.1 CLASSICAL DETERMINISM

Let us go back to the first half of the nineteenth century, a period of great activity in the development of chemical theory. Because the concept of atomism had been shown by Dalton to be of fundamental importance for explaining the laws of chemical combination, chemists generally believed that atoms were the basic constituents of matter and that if one understood everything about the atoms, regarded as elementary particles, one would understand everything about chemistry.

Furthermore, it was believed that the laws of motion for atoms were just the same laws as had been firmly established by Newton for astronomical bodies—Newton's laws of motion of everyday mechanics. Since atoms are the basic building blocks of all matter, the nineteenth century mechanists reasoned that all phenomena (including chemistry and biology, as well as presumably the outcome of horse races and political crises) could be reduced to an exercise in differential and integral calculus.

The essence of this deterministic view is brought out by some simple and familiar examples. For an isolated particle which has a certain mass, a certain position, and a certain velocity at some initial time, it is possible from Newton's equations to predict its position and velocity exactly for all future time.

We consider a particle with velocity v directed along the x axis. If it is at the point (x_0, y_0, z_0) at time $t = 0$, and there are no forces, then according to Newton's first law the velocity remains constant and the position at some future time t is $(x_0 + vt, y_0, z_0)$.

Suppose now that the particle (initially moving in the x direction with velocity v) is in the vicinity of one or more other particles, and that these particles exert forces upon one another. If the x component of the force on the particle is F, Newton's second law tells us that the velocity of the particle changes with time at a rate proportional to the force,

$$\frac{dv}{dt} \propto F$$

and that the proportionality constant can be taken to be the reciprocal of the mass m of the particle; that is, we can write

$$F = ma \qquad (1.1)$$

where

$$a = \frac{dv}{dt} = \frac{d^2x}{dt^2}$$

is the acceleration. If we know how the force F depends upon the position of the particles, then we can integrate Eq. 1.1 to obtain the position and velocity of the particle of interest at any future time. We could then proceed, in principle, to solve Newton's equations simultaneously for all the particles in the system and thereby obtain a complete knowledge of its future behavior.

The two classical ideas—that atoms are the fundamental particles and that the motion of these fundamental particles is governed by Newton's laws—thus lead to a clear, simple, and well-defined picture of nature. As we know now, these classical hypotheses are both wrong. There are particles more elementary than atoms—namely, electrons and nuclei—which determine chemical phenomena. (We are considering here phenomena external to the nucleus, so that we need not be concerned with the details of nuclear structure.) Furthermore, electrons have such a small mass (on the order of 10^{-27} g) and their velocities can vary so rapidly over short distances [dv/dx is on the order of 10^{16} (cm/sec)/cm for electrons in atoms and molecules, or about 10^{13} times the corresponding value for an object falling slowly in the earth's gravitational field] that the extention of Newton's laws from macroscopic bodies to this microscopic realm cannot be taken for granted. In fact, Newton's laws do not apply to electrons in atoms, and a set of laws, comprising what is known as quantum mechanics, has to be obtained for a correct description of electronic behavior. Newton's laws, which are adequate for macroscopic objects (moving with velocities small compared with the speed of light), turn out to be a special case to which the equations of quantum mechanics reduce when the mass of the object becomes large and the force upon it varies slowly with position.

1.2   THE ELECTRON

The concept of the electron as a particle had its modern origin in Faraday's experiments with electrolytic cells. He discovered that the amount of a substance chemically liberated at an electrode was proportional to the amount of electricity passed through the cell. Furthermore, if several different cells were placed in series, a given quantity of electricity liberated chemically equivalent amounts of the various substances. Faraday's interpretation of these results was that equivalent weights of substances (1 g of hydrogen, 108 g of silver, 8 g of oxygen, and so on) contained equal amounts of electricity. In 1874, G. J. Stoney proposed that the natural unit of electricity should be taken as the quantity which liberates by electrolysis one atom of a univalent substance, and in 1891 he suggested the word "electron" as the name for this unit. Stoney even predicted the approximate amount of electricity represented by his electron. Faraday's experiments showed that 1000.00 C of electricity[1] when passed through a silver nitrate solution caused the deposition of 1.11797 g of silver. Since silver is univalent, Stoney's electron is the amount of electricity required to liberate one silver atom:

$$e = \left(\frac{1000.00}{1.11797} \text{ C g}^{-1}\right) \times (107.870 \text{ g mol}^{-1}) \times \left(\frac{1}{N_A} \text{ mol}\right)$$

where 107.870 is the gram equivalent weight of silver and N_A is the Avogadro number, the number of atoms per mole. Thus one finds that the charge of one mole of electrons is

$$N_A e = 9.64870 \times 10^4 \text{ C mol}^{-1}$$

From this result the charge-to-mass ratio of an ion can be calculated. Since N_A silver ions weigh 107.870 g,

$$\left(\frac{\text{charge}}{\text{mass}}\right)_{\text{Ag}} = \frac{N_A e}{M_{\text{Ag}}} = \frac{9.6487 \times 10^4 \text{ C mol}^{-1}}{107.87 \text{ g mol}^{-1}} = 8.9448 \times 10^2 \text{ C g}^{-1}$$
$$= 2.6816 \times 10^{12} \text{ esu g}^{-1}$$

Similarly, for hydrogen ions the value is

$$\left(\frac{\text{charge}}{\text{mass}}\right)_{\text{H}} = \frac{N_A e}{M_{\text{H}}} = \frac{9.6487 \times 10^4 \text{ C mol}^{-1}}{1.008 \text{ g mol}^{-1}} = 9.57 \times 10^4 \text{ C g}^{-1}$$
$$= 2.870 \times 10^{14} \text{ esu g}^{-1}$$

and so on.

Although we have introduced the symbol N_A for the Avogadro number in the above calculations we have not made use of its numerical value. If N_A is known, the charge e itself can be determined. In Stoney's time N_A could only be estimated to be of the order 10^{24} by use of simple kinetic-theory arguments. From this value for N_A, the electronic charge is calculated to be approximately 10^{-19} C; or, in electrostatic units (esu),

$$e = (10^{-19} \text{ C}) \times (3 \times 10^9 \text{ esu C}^{-1}) = 3 \times 10^{-10} \text{ esu}$$

(Stoney's actual estimate in 1874 was lower than this by a power of 10, since he used 10^{23} for N_A.)

Although these ideas were not yet supported by the direct experimental detection of charged *particles*, they helped to condition people to think in

---

[1] Units are discussed in the Appendix.

terms of "electrical atomism," that electricity, as well as matter, might be atomic in nature. This provided the scientific atmosphere which led others to design more detailed experiments and made easier the general accept-ance of their results. The search for the electron had begun.

1.2.1 Thomson's measurement of e/m The first experiment which yielded direct information about the nature of the electron was performed in 1897 by J. J. Thomson. A few years earlier, he had shown that electricity in

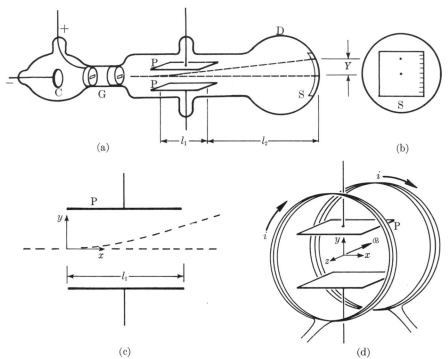

Fig. 1.1. Thomson's measurement of e/m for the electron. (Photo Courtesy of Sir George P. Thomson.)

gases was conducted by essentially the same mechanism as in electrolytic solutions, namely, by the motion of charged particles, or ions. Using a cathode tube to produce a beam of negative ions, he hoped to measure their charge-to-mass ratio e/m by deflecting the beam in electric and magnetic fields. Figure 1.1a shows a schematic drawing of his apparatus. It consists of a discharge tube D which is evacuated by means of a mercury diffusion pump, a heated cathode C which emits ions, some potential grids G to accelerate the ions, condenser plates P which make it possible to apply an electric field perpendicular to the direction of motion of the ions, and a phosphorescent screen S to permit observation of the end point of the path of the beam, as shown in Fig. 1.1b.

With the condenser circuit open (i.e., in the absence of a deflecting field), the ions pass through the discharge tube in a straight line path, forming an image on the screen shown by the lower dot in Fig. 1.1b. With the circuit closed, however, the *electric field* $\mathcal{E}$ between the condenser plates exerts a force $\mathcal{E}e$ upon each ion, accelerating it according to Newton's second law. This causes the ions to follow a curved path while they are in the region l_1, as shown in Fig. 1.1c. After leaving the condenser region, the ions have a component of velocity perpendicular to the original direction of motion toward the positive plate. They continue in the field-free region l_2 with constant velocity, forming a deflected beam which produces an image on the screen represented by the upper dot in Fig. 1.1b.

Although the beam is composed of electrons, the fields with which it interacts are macroscopic [e.g., dv/dx for the electrons in Thomson's apparatus is of the order 10^9 (cm sec^{-1}) cm^{-1}, or 10^{-7} as large as the corresponding value in atoms]. We may thus use Newton's second law to calculate the effect of the electric field $\mathcal{E}$. With the coordinate system shown in Fig. 1.1c, we see that Eq. 1.1 can be written

$$F_y = m\frac{dv_y}{dt} = \mathcal{E}e \tag{1.2}$$

where all quantities are expressed in terms of cgs-Gaussian units. Integrating Eq. 1.2, we find

$$v_y = \frac{e}{m}\mathcal{E}t \tag{1.3}$$

Here we have chosen $t = 0$ to be the time at which the ion enters the field, so that $v_y = 0$ and $y = 0$ at $t = 0$. Since

$$v_y = \frac{dy}{dt}$$

we can integrate a second time to obtain

$$y = \tfrac{1}{2}\frac{e}{m}\mathcal{E}t^2 \tag{1.4}$$

If t_1 is the time required for the ion to travel through the condenser region (l_1), the ion emerges with its position and velocity given by

$$y_1 = \tfrac{1}{2}\frac{e}{m}\mathcal{E}t_1^2$$

$$v_y = \frac{e}{m}\mathcal{E}t_1 \tag{1.5}$$

In the field-free region (l_2) there is no accelerating force and the ion moves to the screen with a constant velocity. If the time required to pass through the region l_2 is t_2, the ion with velocity v_y given by Eq. 1.3 travels a distance y_2 in the y direction, where

$$y_2 = v_y t_2 = \frac{e}{m} \, \mathcal{E} t_1 t_2 \tag{1.6}$$

The displacement Y of the deflected beam at the screen is therefore

$$Y = y_1 + y_2 = \left(\frac{e}{m}\right) \mathcal{E} \left(\tfrac{1}{2} t_1^2 + t_1 t_2\right) \tag{1.7}$$

To use Eq. 1.7 to calculate e/m, we must be able to determine all of the other quantities. We can calculate the strength of the electric field from the applied voltage and the distance between the plates, and we can measure the deflection Y. But so far we do not know the times t_1 and t_2. However, they can be obtained from the value of the x component of the velocity v_x and the dimensions of the tube. Since no forces act in the x direction, v_x is constant throughout l_1 and l_2; therefore,

$$t_1 = \frac{l_1}{v_x} \qquad t_2 = \frac{l_2}{v_x}$$

Thus, Eq. 1.7 becomes

$$Y = \left(\frac{e}{m}\right) \frac{\mathcal{E}}{v_x^2} l_1 (\tfrac{1}{2} l_1 + l_2) \tag{1.8}$$

The remaining problem is to determine v_x. Thomson accomplished this by using Helmholz coils to apply a magnetic field with induction $\mathcal{B}$,[2] in addition to the electric field $\mathcal{E}$. As indicated in Fig. 1.1d, $\mathcal{B}$ is applied in the negative z direction. By Ampere's law, the force exerted on a moving charged particle by a magnetic field is

$$F = \frac{e}{c} v \mathcal{B} \sin \theta \tag{1.9}$$

where F is perpendicular to both v and $\mathcal{B}$, and θ is the angle between v and $\mathcal{B}$. The factor $1/c$, the reciprocal of the speed of light, is required so that $\mathcal{B}$ is given in gauss (see Appendix). The measurement is made by determining the magnetic field whose force just balances that of the electric field. Since v_y and v_z are both zero under these conditions, $v = v_x$, and θ is 90° in Eq. 1.9. By the left-hand rule (for negatively charged particles) the magnetic force is in the negative y direction and we can write

---

[2] In cgs units, the magnetic field $\mathcal{H}$ (in oersteds) in a vacuum and the magnetic induction $\mathcal{B}$ (in gauss) are numerically equal, but they differ in a medium with magnetic permeability different from unity. Strictly speaking, it is $\mathcal{B}$ which enters most simply into the equation for the force on a moving charged particle (see the Appendix).

$$(F_y)_{\text{mag}} = -\frac{e}{c} v_x \mathcal{B} \tag{1.10}$$

while the force due to the electric field remains

$$(F_y)_{\text{el}} = e\mathcal{E} \tag{1.11}$$

Since $\mathcal{E}$ and $\mathcal{B}$ are adjusted to produce zero net force, we have

$$F_y = (F_y)_{\text{mag}} + (F_y)_{\text{el}} = -\frac{e}{c} v_x \mathcal{B} + e\mathcal{E} = 0 \tag{1.12}$$

Solving Eq. 1.12 for v_x, we obtain

$$v_x = \frac{\mathcal{E}c}{\mathcal{B}} \tag{1.13}$$

When this result is substituted into Eq. 1.8, e/m can be determined from the expression

$$\frac{e}{m} = \frac{Y\mathcal{E}c^2}{\mathcal{B}^2 l_1(\frac{1}{2}l_1 + l_2)} \tag{1.14}$$

in which all the quantities on the right-hand side are known. Thomson's value from this experiment was $e/m = \sim 3 \times 10^{17}$ esu g^{-1}, the modern value being 5.27274×10^{17}. This means that the magnitude of the charge-to-mass ratio for the negative ions generated in a cathode ray tube is 1840 times larger than the electrolytic value for the least massive atomic ion, hydrogen (see Section 1.2). Furthermore, Thomson's results were found to be independent of the metal used for the cathode and the nature of the residual gas in the discharge tube: Cathodes of platinum, aluminum, and iron gave essentially the same results, as did discharge tubes containing water vapor, carbon dioxide, and air. Thus, the fundamental nature of these ions appeared to be established. If one assumes that the negative ions of the discharge-tube experiment have the same charge as the univalent ions which carry current through electrolytic cells, Thomson's particle has a mass 1840 times smaller than that of the hydrogen atom. The electron was thus conjectured to be a *subatomic particle*. The direct determination of either its charge or its mass would confirm Thomson's discovery.

1.2.2 *Mass spectrometry* As is often the case, the experimental method developed by Thomson for solving a specific scientific problem was later to find widespread practical use in chemistry. Before continuing our discussion of the properties of the electron, it is appropriate at this point to describe briefly the modern mass spectrometer, which is based on the same principles as was Thomson's experiment. The mass spectrometer has become important in analyzing mixtures of many components and in determining the structures of complex organic compounds, as a means of distinguishing isotopic tracers employed to elucidate kinetic mechanisms, as an analytical tool in geochemical research, and more recently as a detector for atomic and molecular beams.

In one of its common forms, the mass spectrometer (see Fig. 1.2) consists of a sample inlet system, an ionization chamber, an analyzer tube, and a detector plate. If the substance to be analyzed is a gas, it is admitted

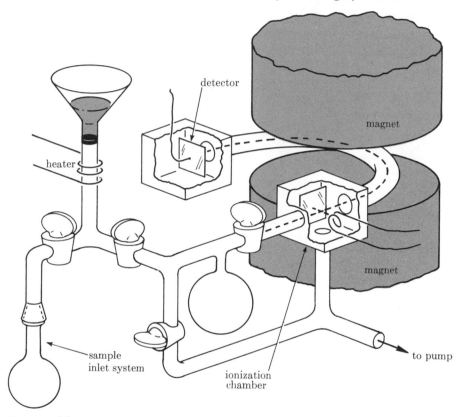

Fig. 1.2. Mass spectrometer.

directly to a sample bottle which is separated from the ionization chamber by a capillary tube. If the substance is a liquid or solid, it must first be vaporized in the inlet system by heating under reduced pressure. As the neutral molecules of the sample leak into the ionization chamber, they are bombarded with a beam of electrons accelerated through a potential of approximately 70 V. These electrons knock electrons out of the molecules, leaving the latter positively charged. The resulting positive molecular ions are then accelerated by an electric potential on the order of 1000 V and pass into the analyzer tube, which is placed between the poles of a large electromagnet to produce a uniform field perpendicular to the velocity of the ions. From Eq. 1.9 the force upon ions with charge q and velocity v in the magnetic field is

$$F = \frac{q}{c} v \, \mathcal{B}$$

According to the right-hand rule (for positively charged particles), the force on ions just entering the field is in the y direction. Since the force remains perpendicular to the direction of motion at every point (see Fig.

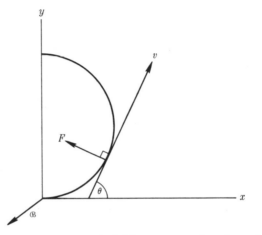

Fig. 1.3. Force exerted by a magnetic field upon a moving charged particle.

1.3), the ions follow a circular path in the field. If the ions have mass M, the radius of the path is

$$R = \frac{Mvc}{q\mathcal{B}} \tag{1.15}$$

As an instructive exercise in particle dynamics, we shall prove that the path is indeed circular and that Eq. 1.15 is the correct expression for the radius. We consider some arbitrary point along the path. The velocity v is tangential to the path and the force due to the magnetic field (which is directed along the z axis) will be at right angles to it (as shown in Fig. 1.3). If v_x and v_y are the x and y components of v, then

$$v_x = v \cos \theta$$

$$v_y = v \sin \theta \tag{1.16}$$

The force F, with magnitude $qv\mathcal{B}/c$, has components

$$F_x = \frac{q\mathcal{B}}{c} v \cos(\theta + 90°) = -\frac{q\mathcal{B}}{c} b \sin \theta = -\frac{q\mathcal{B}}{c} v_y$$

$$F_y = \frac{q\mathcal{B}}{c} v \sin(\theta + 90°) = \frac{q\mathcal{B}}{c} v \cos \theta = \frac{q\mathcal{B}}{c} v_x \tag{1.17}$$

According to Eq. 1.1 (Newton's second law),

$$M \frac{dv_x}{dt} = -\frac{q\mathcal{B}}{c} v_y \tag{1.18}$$

and

$$M \frac{dv_y}{dt} = \frac{q\mathcal{B}}{c} v_x \tag{1.19}$$

Integration of Eqs. 1.18 and 1.19 gives

$$v_x = v - \frac{q\mathcal{B}}{Mc} y \tag{1.20}$$

and

$$v_y = \frac{q\mathcal{B}}{Mc} x \tag{1.21}$$

where we have chosen $t = 0$ as the time of entering the magnetic field, so that at that instant

$$x = 0 \qquad y = 0 \qquad v_x = v \qquad v_y = 0$$

Eliminating v_y between Eqs. 1.21 and 1.18 and multiplying the result by v_x/M, we obtain

$$v_x \frac{dv_x}{dt} = \frac{1}{2} \frac{d(v_x)^2}{dt} = -\frac{q^2\mathcal{B}^2}{M^2c^2} x \frac{dx}{dt} = -\frac{q^2\mathcal{B}^2}{M^2c^2} \frac{1}{2} \frac{d(x^2)}{dt} \tag{1.22}$$

Equation 1.22 is easily integrated to give

$$v_x{}^2 = v^2 - \frac{q^2\mathcal{B}^2}{M^2c^2} x^2 \tag{1.23}$$

where the conditions at $t = 0$ are the same as for Eqs. 1.20 and 1.21. If we now eliminate v_x between Eqs. 1.23 and 1.20 and rearrange, we can write the relation between x and y as

$$x^2 + \left(y - \frac{Mvc}{q\mathcal{B}}\right)^2 = \left(\frac{Mvc}{q\mathcal{B}}\right)^2 \tag{1.24}$$

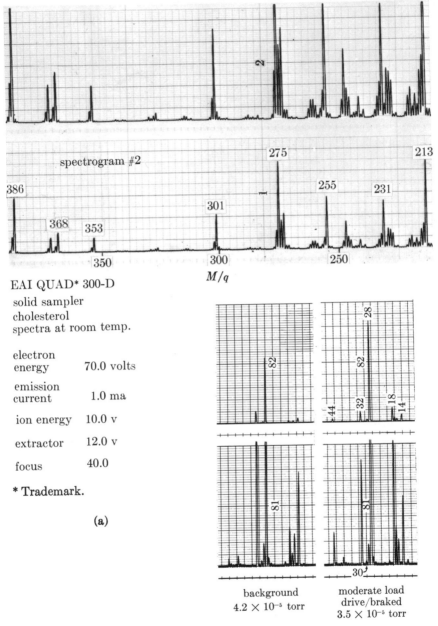

Fig. 1.4. Mass spectra. **(a)** The high-mass region of the mass spectrum of a sample of cholesterol. The sensitivity in the upper spectrum is approximately 2.7 times that of the lower spectrum. **(b)** Mass spectrum of automobile exhaust gas. The lower spectra have approximately 14 times the sensitivity of the upper spectra. [Parts (a) and (b) courtesy of Electronic Associates, Inc., Palo Alto, California.]

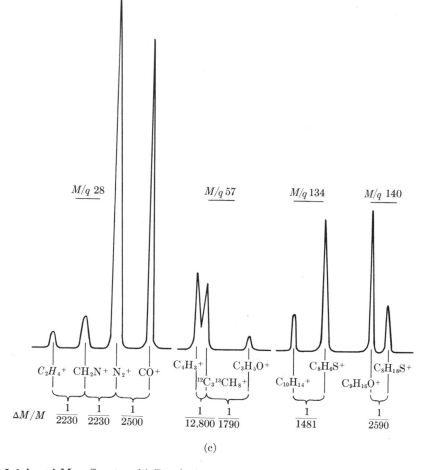

M/q 28 M/q 57 M/q 134 M/q 140

$C_2H_4{}^+$ CH_2N^+ $N_2{}^+$ CO^+ $C_4H_3{}^+$ $C_3H_5O^+$ $C_8H_6S^+$ $C_8H_{18}S^+$

                                    $^{12}C_3{}^{13}CH_8{}^+$ $C_{10}H_{14}{}^+$ $C_9H_{16}O^+$

$\Delta M/M$ $\dfrac{1}{2230}$ $\dfrac{1}{2230}$ $\dfrac{1}{2500}$ $\dfrac{1}{12{,}800}$ $\dfrac{1}{1790}$ $\dfrac{1}{1481}$ $\dfrac{1}{2590}$

(c)

Fig. 1.4. (cont.) Mass Spectra. **(c)** Resolution of several peaks with similar values of M/q. [Redrawn, with permission, from R. D. Craig and G. A. Errock, *Advances in Mass Spectrometry*, edited by J. D. Waldron (Pergamon Press, New York, 1959), p. 78.]

Equation 1.24 is the equation of a circle of radius $R = Mvc/q\mathcal{B}$ whose center is $(0, R)$, so that Eq. 1.15 is verified. By squaring Eq. 1.21 and adding the result to Eq. 1.23, we can show that the magnitude of the ion velocity is constant,

$$v_x{}^2 + v_y{}^2 = v^2 \tag{1.25}$$

This means that a magnetic field acting alone upon a moving charged particle only changes its direction of motion and does not alter its kinetic energy.

In the mass spectrometer, the velocity magnitude v is determined by the accelerating electric potential V. Equating the kinetic energy of the ion at a given point to its change in potential energy gives

$$\tfrac{1}{2}Mv^2 = qV \tag{1.26}$$

Thus, Eqs. 1.26 and 1.15 can be combined to give the path radius in terms of the electric and magnetic fields and the ratio of the mass to charge:

$$R = \left(\frac{2MV}{q}\right)^{1/2} \frac{c}{\mathcal{B}} \tag{1.27}$$

With R fixed by the dimensions of the apparatus, ions of any given M/q may be focused on the detector by varying either V or $\mathcal{B}$. It is more common to keep $\mathcal{B}$ fixed and vary V, so as to avoid the hysteresis effects obtained if $\mathcal{B}$ is swept over a large range. With V set for a certain ionic species, the current produced in the detector circuitry by the incident ions is proportional to the concentration of that species in the sample. Since the charge q is usually e or $2e$, an accurate determination of molecular masses can also be made by means of the mass spectrometer (Problems 1.5 and 1.6). Some mass spectra are shown in Fig. 1.4.

1.2.3 Millikan's measurement of e   After Thomson had discovered negative particles whose charge-to-mass ratio e/m was smaller than that of the hydrogen ion by a factor of 1840, he and others set about to determine the electronic charge itself. Several experiments were performed which gave approximate values, but in each there was an experimental or theoretical uncertainty which put the result in doubt. In 1909, R. A. Millikan, working at the University of Chicago, provided a definite measurement of e and firmly established the electron as a particle with a given charge and mass.

A schematic diagram of the experiment is shown in Fig. 1.5a. Small oil droplets are sprayed by an atomizer into the vessel and drift through a pin hole into a region between two charged condenser plates. Although the drops have radii of only a few microns, they can be illuminated by an arc lamp and observed individually by means of a short-focus telescope mounted opposite a window in the vessel, as shown. If x rays are passed through the vessel, they partially ionize the air by liberating electrons. Some of these free electrons attach themselves to the oil droplets, which thereby become negatively charged and experience a force due to the electric field.

The first part of the experiment is performed with the electric field shut off, so that a droplet of effective mass M falls under the gravitational force Mg (see Fig. 1.5b). Since air is present in the vessel, the droplet is accelerated by gravity only until it reaches a velocity at which the frictional resistance due to the viscosity of the air just balances the force Mg. The re-

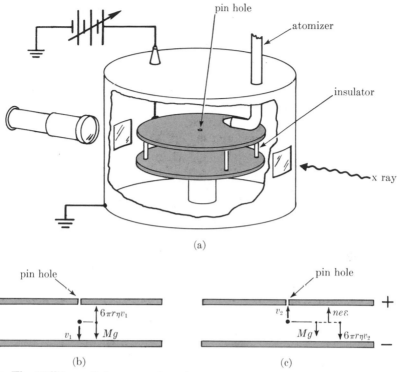

(a)

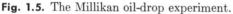

(b)   (c)

Fig. 1.5. The Millikan oil-drop experiment.

sulting terminal velocity v_1 can be related to M by use of Stokes' law, which states that a spherical object of radius r moving with velocity v through a continuous medium of viscosity η experiences a frictional force given by

$$F = 6\pi r \eta v \tag{1.28}$$

Equating the gravitational force Mg to the frictional force corresponding to v_1, we have

$$Mg = 6\pi r \eta v_1 \tag{1.29}$$

The constant velocity v_1 can be measured by recording the time required for the drop to traverse two cross hairs in the telescope. The effective mass M of the drop (corrected for the buoyancy of air) is related to its radius r by

$$M = \tfrac{4}{3}\pi r^3 (\sigma - \rho) \tag{1.30}$$

where σ and ρ are the densities of the drop and the air, respectively. By substituting Eq. 1.30 into Eq. 1.29, we can solve for the radius r:

$$r = \left(\frac{9\eta v_1}{2g(\sigma - \rho)}\right)^{1/2}$$ (1.31)

In the second part of the experiment, the electric field is turned on, so that the velocity of the drop is retarded or even reversed, depending upon its size and charge (see Fig. 1.5c). When the new terminal velocity v_2 (taken to be positive if the drop is moving toward the lower plate) is reached, the gravitational, electric, and frictional forces are balanced according to the equation

$$\tfrac{4}{3}\pi r^3(\sigma - \rho)g = 6\pi r\eta v_2 + ne\mathcal{E}$$ (1.32)

If we substitute Eq. 1.31 into Eq. 1.32 and solve for the charge ne on the drop, we obtain

$$ne = \frac{36\pi}{\mathcal{E}}\left(\frac{\eta}{2}\right)^{3/2}\frac{v_1^{1/2}(v_1 - v_2)}{[g(\sigma - \rho)]^{1/2}}$$ (1.33)

where n is equal to the number of electrons attached to the droplet. Since there is no direct way of determining n, Millikan repeated the experiment for a large number of drops, and for different charges on the same drop. The greatest common divisor of the various resulting values of ne was taken as the electronic charge e.

Although Eq. 1.33 gave reasonably consistent values for different sets of experiments, some dependence of the measured value of e upon the radius of the drop was noted. Millikan correctly attributed this discrepancy to the failure of Stokes' law when applied to very small drops. Theoretically, Stokes' law is valid only for $r \ll \eta/v\rho$ and $r \gg l$, where l is the average distance between successive collisions of the air molecules (*mean free path*). At any given time the size of the empty regions or "holes" in the air is on the order of the mean free path. Millikan's measurements showed that the first of these criteria was satisfied in his experiments, but not the second. Thus the medium in which the drop falls does not have a uniform density throughout regions large with respect to the size of the drop, as is assumed in the derivation of Stokes' law. To correct Eq. 1.28 for fluctuations in the density of the air, H. D. Arnold showed that one must divide v by a function $f(l/r)$ that goes to unity as l/r goes to zero (i.e., as the density of the air becomes uniform with respect to the drop). Assuming that the function can be approximated by a power series, and using the fact that l is inversely proportional to the pressure P at constant temperature, one can write

$$f = 1 + \frac{b}{Pr} + \frac{c}{(Pr)^2} + \cdots$$

where $b, c, \ldots$ are constants to be determined empirically. Millikan found that the first two terms in the power series provided an adequate correction to Stokes' law for his experimental conditions. The corrected form of Eq. 1.28 is

$$F = \frac{6\pi r\eta v}{1 + b/Pr} \tag{1.34}$$

If e_1 is the electronic charge calculated from Stokes' law (i.e., when Eq. 1.28 is used in the derivation of Eq. 1.33) and if e is the charge obtained if Eq. 1.34 is used for the frictional retarding force, rather than Eq. 1.28 [i.e., if η is replaced by $\eta/(1 + b/Pr)$ in Eq. 1.33], then it is easy to show (see Problem 1.4) that for a particular drop with n electrons attached and radius r,

$$(ne_1)^{2/3} = (ne)^{2/3}\left(1 + \frac{b}{Pr}\right) \tag{1.35}$$

Equation 1.35 suggests a procedure for determining the true electronic charge e. We can vary the pressure P at which we determine ne_1 and r for a given drop from Eqs. 1.33 and 1.31, respectively. Then we can plot $(ne_1)^{2/3}$ versus $1/Pr$ and obtain $(ne)^{2/3}$ as the intercept at infinite P. Alternatively, we can hold the pressure constant and make measurements on drops of various r. In a characteristically thorough investigation, Millikan and his students used both methods, making observations on a total of 58 drops over a 60 day period. His result ($e = 4.77 \times 10^{-10}$ esu) had a probable random error[3] of only 16 parts in 61,000, but contained an easily corrected systematic error which was due to his use of an inaccurate value for η. The best value for e is now

$$e = 4.80298 \times 10^{-10} \text{ esu}$$

which is reasonably close to Stoney's prediction (Section 1.2).

Millikan's result, when combined with Thomson's measurement of e/m, shows that the electron is indeed a subatomic particle whose mass is

$$m_e = 9.1091 \times 10^{-28} \text{ g}$$

1.2.4   Thomson's atom model   What does the discovery of the electron mean in terms of the structure of the atom? Since atoms have masses on the order of 10^{-22} to 10^{-24} g (10^3 to 10^5 times that of an electron), electrons can contribute very little to the total mass of an atom. Also, because atoms are known to be electrically neutral, there must be present some positive charge to neutralize that of the negative electrons. On the basis of these results and the kinetic-theory estimate that atomic radii are about 10^{-8} cm, Thomson proposed the first "modern" picture of the atom. He assumed

---

[3] The probable error is the median of the absolute deviation from the mean, which is approximately 0.675 times the standard deviation if the errors are random.

that both the mass and positive charge of an atom are provided by a fairly homogeneous "jelly" which fills the space occupied by the atom. The amount of positive charge is just sufficient to neutralize the atomic electrons, which are imbedded more or less uniformly in the jelly. Thomson's model, which can be dubbed the jelly or "currant bun" model of the atom, did not last very long, because new experiments soon showed it to be incorrect.

1.3   THE ATOMIC NUCLEUS

In 1911, Rutherford and his student Geiger (who later invented the "Geiger" radiation counter) bombarded thin foils of heavy metals (such as gold and platinum) with high energy so-called *alpha rays* (α rays); they used a scintillation counter to observe the scattered rays and hence to determine the angles through which they had been deflected (see Fig. 1.6a). The α rays, obtained from the disintegration of radon or radium, were known to consist of heavy positively charged particles (α particles). (We now know that they are helium nuclei with charge $2e$.) Thus, it was pos-

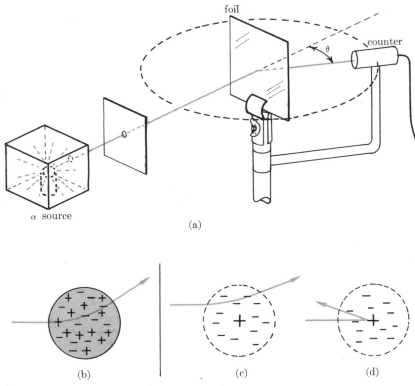

Fig. 1.6. The Rutherford scattering experiment.

sible to predict by simple classical arguments the angular distribution of α particles expected if the atoms in the foils conform to Thomson's jelly model. Since, according to this model, the atomic mass and positive charge are continuously distributed over a large volume, most of the α particles passing through the foil should be deflected through small angles, and none at all should be deflected through angles greater than 90° (see Fig. 1.6b). The striking result of the Rutherford–Geiger experiment was that deflections through all angles were observed, with a few deflection angles near 180°. The only way to explain these results is to assume that the mass and positive charge of the atom reside in a particle (the nucleus) whose dimension is very small compared with that of the atom. Calculations for point masses with charge $2e$ (α particles) colliding with a stationary point mass of charge Ze (the nucleus) show that the number of particles n_θ which are deflected through an angle θ is proportional to $Z^2/\sin^4(\theta/2)$. This dependence of n_θ upon θ, known as the Rutherford scattering formula, is found to give an accurate description of the observations. Note in particular that the scattering angles of maximum probability are in the neighborhood of 0° and that the angles close to 180° have a relatively small probability. This is physically reasonable for a target whose positive scattering centers are uniformly distributed point nuclei. Since most of the target is empty space, most of the α particles pass through far from any nucleus (see Fig. 1.6c) and therefore suffer little deflection ($\theta \sim 0°$). When an α particle happens to approach very close to one of the target nuclei, however (see Fig. 1.6d), the Coulomb repulsion of the two positive charges causes the projectile to reverse its course ($\theta \sim 180°$).

These results establish that the mass and charge of the atoms composing the foils are concentrated in a small nucleus, and thereby unequivocally invalidate the Thomson jelly model. By means of the proportionality between n_θ and Z^2 in the Rutherford formula, the experiment can also be used to determine the nuclear charge Z for various metal foils. It was found that Z was approximately half of the atomic weight; for example, the atomic weight of copper is 63.54, while $Z = 29$. Rutherford's results for Z agreed quite well with those of the pioneering x-ray experiments for determining the number of electrons in various atoms.

Although Rutherford's formula gave a good description of the observations, a very careful analysis demonstrated that it was not exact. It was found that closer agreement with experiment is obtained if the nucleus is assumed to have a finite volume. In fact, the cross-section area which the nucleus presents to the bombarding particles could be estimated to be on the order of 10^{-24} cm^2, corresponding to a radius of about 10^{-12} cm. This is to be contrasted with the kinetic-theory estimate of approximately 10^{-8} cm for the radius of the atom.

Thus, the Rutherford–Geiger experiment gives us the qualitative picture of an essentially empty atom with a heavy positive nucleus that is almost a point of 10^{-12} cm radius in the center, and with light negative electrons moving around it so as to fill the space out to a radius of about 10^{-8} cm. To obtain a quantitative interpretation of this picture of the atom is our next objective.

(Although the structure of the nucleus and the nature of its component particles is an exciting and important topic, its discussion would take us too far from our primary concern with extranuclear atomic and molecular structure. The interested reader may wish to read selected topics in References 14–16 of the Additional Reading list.)

1.4   THE BOHR ATOM

In this section we describe the development of the Bohr model for the structure of one-electron atoms.

1.4.1   Rutherford's planetary model
The first attempt to incorporate the new experimental information on atomic structure into a detailed model was made by Rutherford himself. He suggested that an atom was essentially a miniature solar system, with the electrons moving in a plane along circular paths around the nucleus by analogy with the way that the planets orbit

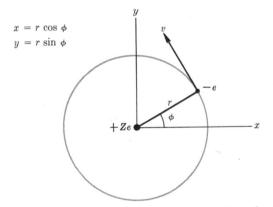

Fig. 1.7. Plane polar coordinates for the planetary model of an electron in a hydrogen atom.

the sun. (See Fig. 1.7, which introduces the plane polar coordinates r and ϕ to describe the motion of the electron.)

For a stable orbit, the attractive (centripetal) force between the two bodies—gravitational in the solar system and electrostatic in the atom—must be just balanced by the centrifugal force resulting from the planetary or electronic motion.

The electrostatic attraction between a nucleus of charge $+Ze$ and an electron of charge $-e$ gives rise to a potential energy of interaction equal to $-Ze^2/r$. Corresponding to this potential energy, there is a radial force F_r pulling the electron toward the nucleus, given by

$$F_r = -\frac{dV}{dr} = -\frac{d}{dr}\left(-\frac{Ze^2}{r}\right) = -\frac{Ze^2}{r^2}$$

If the electron is moving in a circular orbit with linear velocity v, the centrifugal force is mv^2/r; thus, the condition for stability is[4]

$$\frac{mv^2}{r} = \frac{Ze^2}{r^2} \tag{1.36}$$

The total energy E of the system is the sum of the kinetic energy T and the potential energy V of interaction; that is,

$$E = T + V = \tfrac{1}{2}mv^2 - \frac{Ze^2}{r} \tag{1.37}$$

Solving Eq. 1.36 for r or for mv^2 and substituting into Eq. 1.37, we find

$$E = -\tfrac{1}{2}mv^2 = -\frac{Ze^2}{2r} \tag{1.38}$$

We see from Eq. 1.38 that the energy of the electron in the Rutherford atom is negative. This results from our choice of the zero of energy as the energy corresponding to the electron being infinitely far from the nucleus, that is, when there is no interaction between the electron and the nucleus.

Equation 1.38 is a special case of the *virial theorem;* it states that for a system of bound particles with gravitational or electrostatic forces acting between them, the total energy is equal to half of the average potential energy or, equivalently, to the negative of the average kinetic energy. In the present case, V and T are constant for a given orbit (i.e., independent of the value of the polar angle ϕ), so that

$$E = \tfrac{1}{2}V = -T$$

In addition to the energy, the angular momentum also plays an important role in the theory of atomic structure. If ω is the angular velocity of the electron in its orbit about the nucleus, the angular momentum is

$$p_\phi = mr^2\omega \tag{1.39}$$

Since $v = r\omega$ for circular orbits, Eq. 1.39 can be used to rewrite the kinetic energy in the form

$$T = \tfrac{1}{2}mv^2 = \frac{p_\phi{}^2}{2mr^2} \tag{1.40}$$

---

[4] By using m, the electronic mass, in Eq. 1.36 we are assuming an infinite nuclear mass. (See the discussion at the end of Section 1.4.3.)

If we combine Eqs. 1.38 and 1.40, we obtain

$$r = \frac{p_\phi^2}{mZe^2} \tag{1.41}$$

This expression for r, the radius of the orbit, allows us to write the energy in terms of p_ϕ. From Eqs. 1.38 and 1.41, we have

$$E = -\frac{Ze^2}{2r} = -\frac{mZ^2e^4}{2p_\phi^2} \tag{1.42}$$

The energy of such an atomic electron could thus be calculated when either the orbit radius or the angular momentum is known.

Although the planetary model is very appealing—it is analogous to a macroscopic system which is well understood, and the purely mechanical aspects of the model are consistent with the experimental results of Rutherford and Geiger—it contains a fundamental inconsistency which has so far been ignored. Since the electron is moving in a circular orbit, there is a constant acceleration corresponding to the continuous change in the direction of its velocity. For such an accelerated charged particle, classical electrodynamics leads to an unequivocal conclusion. The accelerated electron must emit electromagnetic radiation of frequency ν equal to its circulation frequency,

$$\nu = \frac{\omega}{2\pi} \tag{1.43}$$

and of an intensity proportional to the square of the acceleration, $a^2 = \omega^4 r^2$; the proportionality factor is determined by classical electromagnetic theory to be $2e^2/3c^3$, yielding the intensity I

$$I = \frac{2e^2}{3c^3} a^2 = \frac{2e^2}{3c^3} \omega^4 r^2 \tag{1.44}$$

What would atoms be like if they were to conform to Rutherford's model and to obey classical electrodynamics? Since the atoms are continuously radiating, they are continuously losing energy. Thus, their energy continuously becomes more negative, which implies by Eqs. 1.38 and 1.42 that the radius continuously decreases until finally the electrons spiral into the nucleus and the atom ceases to exist as such. If the atom had an initial radius of 10^{-8} cm, the time required for its collapse would be on the order of 10^{-10} sec (see Problem 1.7). Thus, according to the Rutherford model, all the matter in the universe should have collapsed eons ago into a large number of neutral, massive particles on the order of nuclear dimensions (i.e., with radii about 10^{-4} times those of atoms). This would clearly result in a world nothing like the present one.

Furthermore, since such a spiraling, radiating electron has a contin-

uously decreasing radius, its "circulation" frequency must also continuously change. Equations 1.39 and 1.41 allow us to write

$$\omega = \left(\frac{Ze^2}{mr^3}\right)^{1/2} \tag{1.45a}$$

or, equivalently,

$$\omega = \frac{mZ^2e^4}{p_\phi^3} \tag{1.45b}$$

Thus, as r decreases from its initial value to zero, the frequency of the
emitted radiation, which is equal to $\omega/2\pi$ (Eq. 1.43), varies continuously
between the initial frequency and infinity. This conclusion (even if the
impossibly short atomic lifetime were temporarily ignored) was known to
be in disagreement with experimentally observed emission spectra. At the
end of the nineteenth century, scientists had observed the hydrogen spectrum emitted by stars and found it to consist of very sharp, discrete
frequencies. For sources containing molecules, the spectra are often smeared
out (see Chapter 7), but "line spectra" are characteristic of the individual
atoms which Rutherford's model attempts to describe.

From the above discussion, it is clear that the classical planetary model,
although at first sight an appealing picture of the atom, cannot be correct.
An interesting point is that Rutherford was apparently aware of the difficulties, but did not let himself be deterred by them.

What is to be done? Obviously a different approach is needed. However,
some of the features of the planetary model are so appropriate that it is
difficult to cast it aside completely. Since certain of its consequences, which
follow unequivocally from classical theory, are clearly wrong, a drastic
innovation appears to be required. Before discussing the first attempt at a
solution, which was indeed revolutionary, we shall consider in more detail
the information that was available at the time (1911–1913) concerning the
discrete line spectra of atoms. The data are important because any satisfactory new theory had to provide an explanation for them. Moreover, as
it turned out, the fact that the new theory did account quantitatively for
the hydrogen atom spectrum was instrumental in its acceptance by the
scientific community.

1.4.2   The Rydberg formula   In discussing atomic spectra, we take the
hydrogen atom as an example. This is partly because emission studies of
the sun and other stars, as well as data from discharge tubes, had provided
very detailed information about the hydrogen spectrum, and partly because it is the simplest atom.

One of the significant contributors to the experimental analysis of the
hydrogen spectrum was Balmer. In about 1885, he noted that the spacings

of a series of lines could be expressed in simple algebraic form (see Problem 1.8). If each line in the series is assigned an integer n, the frequency ν of the line is given quantitatively by the formula

$$\nu = \text{const} \left(\frac{1}{2^2} - \frac{1}{n^2} \right) \qquad n = 3, 4, 5, \ldots$$

Figure 1.8 shows the lines of the Balmer series. Qualitatively, we can see that the spacing between the lines becomes smaller with increasing n, in accordance with the formula.

In practical spectroscopy, measurements are often made in terms of the wavelength λ, rather than the frequency. One can easily be calculated from the other by the relation

$$\nu = \frac{c}{\lambda}$$

where c is the speed of light. To avoid the very large numbers obtained by multiplying by the speed of light, it is the usual practice to express the measurements in terms of the reciprocal of the wavelength. (Another reason for this is that spectroscopic data are generally more accurate than determinations of the velocity of light.) This quantity, given the symbol $\tilde{\nu}$ (nu tilde), is called the *wave number*

$$\tilde{\nu} = \frac{1}{\lambda} = \frac{\nu}{c}$$

The wave number is the number of waves per unit length; it is usually expressed in units of reciprocal centimeters (cm^{-1}) (see Section 7.1). In terms of wave numbers, the Balmer formula is

$$\tilde{\nu} = 1.09 \times 10^5 \left(\frac{1}{2^2} - \frac{1}{n^2} \right) \text{cm}^{-1}$$

where the constant was estimated by Balmer from his observations of a particular series of lines.

The hydrogen spectrum contains many lines other than the Balmer series, and a number of people did the work of identifying and classifying the lines. Important among these was Rydberg, who carried Balmer's idea further, and made more accurate and extensive measurements. Rydberg found that all of the lines in the hydrogen spectrum could be obtained from a single general formula of the form

$$\tilde{\nu} = R \left(\frac{1}{n_1^2} - \frac{1}{n_2^2} \right) \tag{1.46}$$

where R is a constant (now called the Rydberg constant) and n_1 and n_2 are both integers, with $n_1 = 1, 2, 3, 4, \ldots$ and $n_2 = n_1 + 1, n_1 + 2, \ldots$. Each n_1 value corresponds to a separate series, while the n_2 values are associated with individual lines of the series (see Fig. 1.8). Thus $n_1 = 1$, $n_2 = 2, 3, 4, \ldots$ is a series which was originally discovered by Lyman and is called the Lyman series in his honor; $n_1 = 2$, $n_2 = 3, 4, 5, \ldots$ is the Balmer series, which we have already discussed; $n_1 = 3$, $n_2 = 4, 5, 6$, $\ldots$ corresponds to a series discovered by Paschen; and so on. For all of

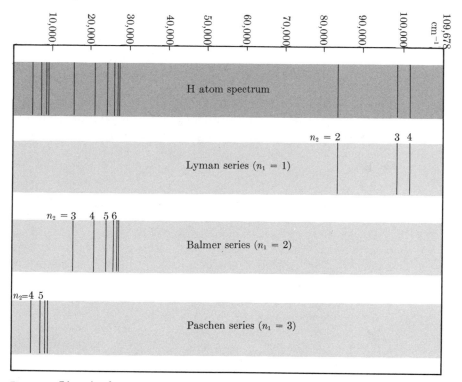

Fig. 1.8. Lines in the spectrum of atomic hydrogen.

these series, the *same* constant R applies. Its accurate present-day value is

$$R = 1.0967758 \times 10^5 \text{ cm}^{-1} \tag{1.47}$$

For one-electron atoms other than hydrogen (for example, the ions He$^+$ and Li^{2+}), the spectrum is given by a formula having the same form as Eq. 1.46, but with a different constant in place of R. Thus, we have a simple yet quantitative description of the spectra of one-electron atoms which any useful theory must be able to justify.

1.4.3   The Bohr theory of the atom   As we have already seen, the serious difficulties with Rutherford's model were the prediction of impossibly short atomic lifetimes and the prediction of continuous emission spectra for atoms. To overcome these two problems and to provide a quantitative prediction of line spectra, Bohr proposed a new theory of atomic structure in 1913. He began with the planetary model, but introduced two general assumptions which make atoms differ markedly from macroscopic systems. The first assumption is that an atom *can* exist for a long time without radiating in certain states with discrete energies; the states are called *stationary states*. Secondly, Bohr assumed that under certain conditions transitions between these stationary states do occur and that these transitions are accompanied by the emission or absorption of radiation. In essence, these two assumptions state that the classical deductions from the planetary model do not apply to atoms. Although this may seem to us an obvious way out of the difficulties we have discussed, in the context of the times, Bohr's ideas constituted a revolutionary hypothesis. They gave rise to violent discussions and disagreements and were finally accepted only because Bohr was able to draw from his assumptions a number of conclusions which agreed quantitatively with the experimental observations on one-electron atoms.

Bohr's first assumption, that certain discrete energy states of atoms are stable, allows us to number the possible energies in increasing order starting with the one of lowest (most negative) value: E_1, E_2, E_3, . . . , and so on. According to Bohr's second assumption, if an atom in one of the particular stationary states, say, E_1, is placed in a beam of light, it may absorb enough energy from the radiation to make the transition to a stationary state of higher energy, say, E_2. Conversely, an atom in an *excited state*, say, E_2, may emit enough energy in the form of radiation to make a transition to a stationary state of lower energy, E_1. If this mechanism is to provide information about the frequency spectrum of the atom, a quantitative relation between the energy and frequency of the radiation is required. Planck had already used such a relationship to explain the observed variation with wavelength of the intensity of radiation in equilibrium inside a hollow cavity, often referred to as "blackbody" radiation. He had found that the experimental data could be fitted if the energy ΔE absorbed or emitted by the "oscillators" making up the walls of the blackbody is assumed to be directly proportional to the frequency of the radiation; that is,

$$\Delta E = h\nu$$

where the constant of proportionality h (Planck's constant) has the value 6.6256×10^{-27} erg sec. Planck's relation suggests that radiation is quantized; that is, it comes in energy bundles $h\nu$, called *quanta* (one *quantum*) of energy. Bohr assumed that Planck's result applied to atoms with ΔE equal to the energy between two states; that is, the frequency of radiation emitted or absorbed by an atom as a result of a transition between the states E_{n_2} and E_{n_1} is determined by the *frequency rule*

$$\nu = \frac{E_{n_2} - E_{n_1}}{h} \quad \text{or} \quad \bar{\nu} = \frac{E_{n_2} - E_{n_1}}{hc} \tag{1.48}$$

The integer n_2 is chosen as the larger of the two indices so that ν is always positive, whether the process is absorption or emission.

For the Bohr frequency rule to give the experimentally observed frequencies, it must yield the same results as the Rydberg formula. Comparison of Eqs. 1.46 and 1.48 suggests that the energies of the n_1th and n_2th stationary states are

$$E_{n_1} = -\frac{A}{n_1^2} \qquad E_{n_2} = -\frac{A}{n_2^2} \tag{1.49}$$

where A is a constant equal to Rhc. If the stationary-state energies are proportional to n^{-2} and the planetary model is applicable, Eq. 1.42 can be used to solve for the stationary-state radii and angular momenta and thereby to show that they must be proportional to n^2 and n, respectively. Letting a_0 and b_0 be the proportionality constants, we have

$$r = n^2 a_0 \qquad\qquad p_\phi = n b_0 \tag{1.50}$$

The energy, radius, and angular momentum of a hydrogen atom are seen by the Bohr hypothesis to be restricted to a set of discrete values determined by the integer n. Therefore, we say that E, r, and p_ϕ are "quantized," and refer to n as the *quantum number*. A picture of the stationary orbits would look like Fig. 1.9; the smallest orbit $(n = 1)$ with energy E_1

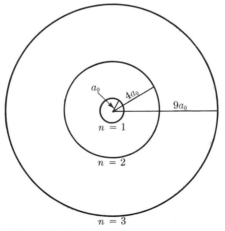

Fig. 1.9. Stationary orbits of the Bohr atom.

$(= -Rhc)$ would have a radius a_0, the next orbit $(n = 2)$ with energy E_2 $(= -\frac{1}{4}Rhc)$ would have a radius $4a_0$, and so on.

If a determination of any one of the proportionality constants of Eqs. 1.49 or 1.50 were made independently of the experimental spectral data, the Rydberg constant R could be calculated theoretically. Bohr accomplished this very important step by introducing the *correspondence principle* which bears his name. The essence of Bohr's correspondence principle is that in the limit of large quantum numbers, when the orbits become macroscopic, classical behavior must be approached; that is, the frequency of radiation for transitions between neighboring states, as given by the frequency rule (Eq. 1.48) should coincide with the classical frequency, as given by Eqs. 1.43 and 1.45. Also the spacing of stationary energy values (Eq. 1.49) becomes so small that they can be regarded as practically continuous. To examine the mathematical consequences of the correspondence principle, we substitute for p_ϕ in Eq. 1.42 by Eq. 1.50 and obtain

$$E_n = -\frac{mZ^2e^4}{2b_0^2n^2} \tag{1.51}$$

Introducing this expression into the Bohr frequency rule (Eq. 1.48) and setting $n_2 = n_1 + 1$, we obtain

$$\nu_{n_1,n_1+1} = \frac{mZ^2e^4}{2b_0^2h}\left(\frac{1}{n_1^2} - \frac{1}{(n_1+1)^2}\right) = \frac{mZ^2e^4}{2b_0^2h}\left(\frac{2n_1+1}{n_1^2(n_1+1)^2}\right) \tag{1.52}$$

According to the correspondence principle, as n_1 becomes large, the right-hand side of Eq. 1.52 must become equal to the classical frequency. For very large n_1, we can neglect unity in both the numerator and the denominator of the last term in Eq. 1.52,

$$\frac{2n_1+1}{n_1^2(n_1+1)^2} \simeq \frac{2}{n_1^3} \qquad (n_1 \to \infty)$$

Thus, we can write

$$\nu_{n_1,n_1+1} \simeq \frac{mZ^2e^4}{2b_0^2h}\left(\frac{2}{n_1^3}\right) \qquad (n_1 \to \infty)$$

If we now use Eq. 1.50 to reintroduce the angular momentum p_ϕ, the frequency ν_{n_1,n_1+1} at large n_1 becomes

$$\nu_{n_1,n_1+1} \xrightarrow[n_1\to\infty]{} \frac{mZ^2e^4b_0}{p_\phi^3h}$$

Equating this limiting expression to the classical frequency given by Eqs. 1.43 and 1.45b, we obtain

$$\frac{mZ^2e^4b_0}{p_\phi^3 h} = \nu_{\text{class}} = \frac{mZ^2e^4}{2\pi p_\phi^3} \tag{1.53}$$

Equation 1.53 shows that in order to satisfy the correspondence principle, the constant b_0 must be set equal to $h/2\pi$. Thus, from Eq. 1.50,

$$p_\phi = \frac{nh}{2\pi} \qquad n = 1, 2, \ldots \tag{1.54}$$

that is, the angular momentum is restricted to integer multiples of $h/2\pi$. Since the combination $h/2\pi$ often appears in quantum-mechanical formulas, it is usually written as the special symbol $\hbar$ (h bar), where $\hbar$ is equal to 1.05450×10^{-27} erg sec. In terms of $\hbar$, the angular momentum of the nth stationary state is $p_\phi = n\hbar$. For this reason, $\hbar$ is sometimes called the *quantum of angular momentum*.

The orbit radius for the nth stationary state can be calculated by substituting Eq. 1.54 into Eq. 1.41; the result is

$$r = \frac{n^2\hbar^2}{mZe^2} \tag{1.55}$$

Comparison of Eq. 1.55 with the first of Eqs. 1.50 shows that for the hydrogen atom ($Z = 1$), the smallest radius corresponding to the stationary state of lowest energy is

$$a_0 = \frac{\hbar^2}{me^2} \tag{1.56}$$

This constant is called the *Bohr radius* and is numerically equal to 0.529167×10^{-8} cm. The value of the Bohr radius is a very reasonable result, since kinetic-theory data have indicated that atoms have radii of about 10^{-8} cm.

Returning to the question of the energy levels, we substitute Eqs. 1.55 and 1.56 into Eq. 1.42 and obtain

$$E_n = -\frac{Z^2me^4}{2\hbar^2}\left(\frac{1}{n^2}\right) = -\frac{Z^2e^2}{2a_0}\left(\frac{1}{n^2}\right) \qquad n = 1, 2, 3, \ldots \tag{1.57}$$

The stationary state of lowest (most negative) energy, which is often called the *ground state*, is the state for which $n = 1$. For hydrogen ($Z = 1$), the ground-state energy is -2.18×10^{-11} erg. Because this is such a small number, it is convenient to introduce another unit of energy, namely, the electron volt (eV). This unit is the amount of kinetic energy that an electron gains when it falls through a potential of one volt. A simple conversion of units shows that 1 eV = $(4.80298 \times 10^{-10}$ esu$) \times (1/299.7925$ statvolt$) = 1.60210 \times 10^{-12}$ erg. Thus, in terms of electron volts, the ground-state energy of hydrogen is

$$E_1 = -13.6 \text{ eV}$$

This means that 13.6 eV of energy is required to take an electron from its most stable stationary state in the hydrogen atom ($n = 1$) and excite it to a state with $n \to \infty$. Since this final state is one for which $r(= n^2 a_0)$ is infinity, the potential energy of interaction between the electron and the nucleus is zero. The electron is now free of the nucleus and the atom is said to have been *ionized*. The energy supplied for the ionization process is

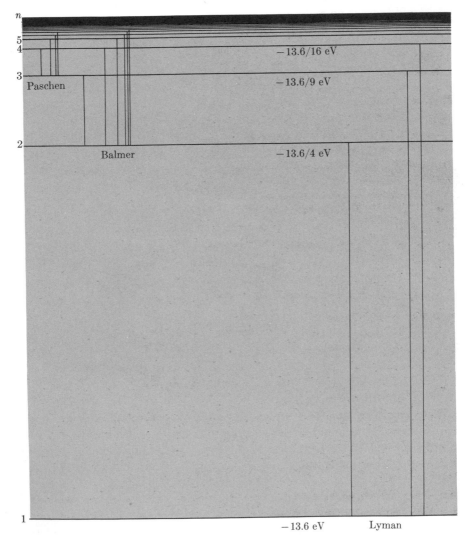

Fig. 1.10. Energy levels of the hydrogen atom.

called the *ionization potential* (IP). For hydrogen, the Bohr theory thus yields IP $= 13.6$ eV.

Figure 1.10 shows the results obtained for the hydrogen energy levels by the Bohr theory. Between $n = 1$ and $n \to \infty$ (ionization), there is an infinite set of discrete energy levels. The first few of these ($E_2 = -13.6/4$ eV, $E_3 = -13.6/9$ eV, $E_4 = -13.6/16$ eV) are shown by lines in Fig. 1.10. As we noted before, the energy levels become more and more closely spaced with increasing n, until a virtual continuum is reached at very large n. For still higher energies (i.e., an ionized hydrogen atom), Eq. 1.57 does not apply and there is no restriction on the allowed energies.

We can utilize the energy-level formula for the calculation of the Rydberg constant. Combining Eqs. 1.57 with the frequency rule (Eq. 1.48), we have

$$\bar{\nu} = \frac{1}{hc}\left(\frac{Z^2 m e^4}{2\hbar^2}\right)\left(\frac{1}{n_1{}^2} - \frac{1}{n_2{}^2}\right) \tag{1.58}$$

Setting $Z = 1$ for hydrogen and using the known values for the constants m, $\hbar$, c, and e, we obtain

$$R = \frac{1}{hc}\left(\frac{m e^4}{2\hbar^2}\right) = 1.097 \times 10^5 \text{ cm}^{-1} \tag{1.59}$$

and

$$\bar{\nu} = 1.097 \times 10^5 \left(\frac{1}{n_1{}^2} - \frac{1}{n_2{}^2}\right)$$

in agreement with the experimental value given in Eq. 1.47. It is now clear that the lines of the emission spectrum in each of the series correspond to transitions from various excited levels n_2 to a final state specified by the quantum number n_1. Thus, the Lyman series ($n_1 = 1$) consists of all transitions to the ground state ($E_2 \to E_1$, $E_3 \to E_1$, $E_4 \to E_1$, etc.), while the Balmer series ($n_1 = 2$) consists of transitions to the first excited state ($E_3 \to E_2$, $E_4 \to E_2$, $E_5 \to E_2$, etc.).

The Bohr theory so far appears very satisfactory, since it is able to explain the sizes of atoms and the hydrogen spectrum quantitatively. Furthermore, it is found that Eq. 1.58 applies as well to the other one-electron atoms (He^+, Li^{2+}, etc.) if the value of Z corresponding to the nuclear charge is used; that is, $Z = 2$ for He^+, $Z = 3$ for Li^{2+}, and so on.

1.4.4 Correction for finite nuclear mass Since the Bohr theory is so successful for one-electron atoms, a small but significant refinement is worthy of discussion. The need for the refinement becomes apparent experimentally when it is discovered that the frequencies of the lines in the helium ion spectrum are not given exactly by multiplying the hydrogen frequencies by 4, as Eq. 1.58 suggests. The source of the discrepancy is our assumption that the nucleus is stationary and that only the motion of the electron about the nucleus contributes to the energy of the system. This would be strictly true only if the nucleus had infinite mass. In actual fact, the nucleus of the hydrogen atom has about 2000 times the electron mass, while for helium the ratio is about 6000 or 8000, depending upon the isotope. Since both the nucleus and the electron have finite mass, they move about their common center of mass, the nucleus sweeping out a tiny circle while the electron sweeps out a circle with a radius several thousand times larger.

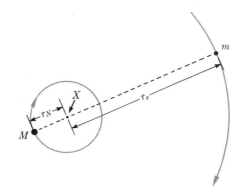

Fig. 1.11. Center of mass of a one-electron atom with finite nuclear mass. The proton and the electron rotate about point X.

To obtain the Rydberg constant for a system with a nucleus of finite mass, we must rederive relations similar to Eqs. 1.36, 1.39, and 1.40, from which the Bohr formula was shown to follow. Consider an electron of mass m and a nucleus of mass M rotating about their center of mass (Fig. 1.11). The distances of the electron r_e and of the nucleus r_N from the center of mass are related by

$$Mr_N = mr_e \qquad (1.60)$$

and the distance between the particles is

$$r = r_N + r_e \qquad (1.61)$$

From Eqs. 1.60 and 1.61, we obtain

$$r_e = \frac{M}{M+m}r \qquad r_N = \frac{m}{M+m}r \qquad (1.62)$$

The total angular momentum p_ϕ is the sum of the angular momenta of the two particles; that is,

$$p_\phi = Mv_Nr_N + mv_er_e = Mr_N{}^2\omega + mr_e{}^2\omega \tag{1.63}$$

or, making use of Eq. 1.62,

$$p_\phi = \left(\frac{Mm}{M+m}\right)r^2\omega \tag{1.64}$$

The angular momentum formula for the two-particle system can thus be reduced to the same form as for a single particle (Eq. 1.39), except that m is replaced by the *reduced mass* μ defined by

$$\mu = \frac{Mm}{M+m} \tag{1.65}$$

Since M is much larger than m, the reduced mass μ is very close to m, so that the Bohr formula is a very good approximation.

Similarly, the total kinetic energy is

$$T = \tfrac{1}{2}(Mv_N{}^2 + mv_e{}^2) = \tfrac{1}{2}(Mr_N{}^2 + mr_e{}^2)\omega^2 \tag{1.66}$$

or, from Eqs. 1.62 and 1.64,

$$T = \tfrac{1}{2}\,\mu r^2\omega^2 = \frac{p_\phi{}^2}{2\mu r^2} \tag{1.67}$$

which is the analog of Eq. 1.40. The centrifugal force on the electron is

$$F = \frac{mv_e{}^2}{r_e} = mr_e\omega^2 \tag{1.68}$$

Making use of Eqs. 1.62 and 1.64 to eliminate r_e and v_e, we obtain

$$F = \mu r\omega^2 = \frac{p_\phi{}^2}{\mu r^3} \tag{1.69}$$

In this way, all of the equations needed for finding the energy E for the planetary model (see Problem 1.12) are shown to be identical to those obtained under the assumption of infinite nuclear mass, except that the electron mass m is replaced by μ as defined in Eq. 1.65.

By replacing m in Eqs. 1.58 and 1.59 by the appropriate value of μ, we can calculate the Rydberg constants for different nuclear masses. If we let R_∞ denote the Rydberg constant for infinite nuclear mass, we have the previous result

$$R_\infty = \frac{me^4}{2hc\hbar^2}$$

while for the hydrogen atom ($\mu = \mu_{\mathrm{H}}$) the Rydberg constant R_{H} is

$$R_{\mathrm{H}} = \frac{\mu_{\mathrm{H}}e^4}{2hc\hbar^2}$$

The accurate value of R_∞ is 1.0973731×10^5 cm^{-1}, which is seen to differ somewhat from the experimental value for R_H given in Eq. 1.47. The ratio R_∞/R_H is calculated from the electron mass m and the proton mass m_p by the relation

$$\frac{R_\infty}{R_\mathrm{H}} = \frac{m}{\mu_\mathrm{H}} = \frac{m + m_p}{m_p}$$

This means that the assumption of an infinite mass for the nucleus leads to a calculated Rydberg constant which is slightly larger than the one obtained if the finite mass is properly taken into account. Using the values $m = 9.1091 \times 10^{-28}$ g and $m_p = 1.67252 \times 10^{-24}$ g, we find that the ratio R_∞/R_H is 1.00054 and that the calculated value of R_H is

$$R_\mathrm{H} = \frac{m_p}{m + m_p} R_\infty = \frac{1.0973731 \times 10^5}{1.00054} = 1.09678 \times 10^5 \text{ cm}^{-1}$$

which is in excellent agreement with the experimental value. Correspondingly, the experimentally determined ratio $R_\mathrm{H}/R_\mathrm{He}$ turns out to be the predicted ratio of the reduced masses ($\mu_\mathrm{H}/\mu_\mathrm{He}$). Since spectroscopic measurements are very accurate, it is possible to use ratios of Rydberg constants to obtain refined values of relative nuclear masses. Such measurements played a significant role in the discovery of isotopes.

The excitement created by the brilliant successes of Bohr's theory soon spread throughout the scientific world. It rapidly became clear that a significant advance had been made in the understanding of atoms and the interactions among material particles of which they are composed. The price paid was high, however. Classical theory, which had long been accepted as universal, was relegated to the role of a special case, applicable only for macroscopic systems in the limit of large quantum numbers. (The spirit of this scientific revolution is described in Reference 2 of the Additional Reading list.)

ADDITIONAL READING

[1] R. A. MILLIKAN, *The Electron* (University of Chicago Press, Chicago, 1963); this is a facsimile of the original 1917 edition, with an introduction by J. W. M. DuMond.

[2] RUTH MOORE, *Niels Bohr* (Knopf, New York, 1967).

[3] G. P. THOMSON, *The Inspiration of Science* (Oxford University Press, London, 1961).

[4] G. K. T. Conn and H. D. Turner, *The Evolution of the Nuclear Atom* (American Elsevier, New York, 1965).

[5] *Classical Scientific Papers; Physics*, Introduction by S. WRIGHT (American Elsevier, New York, 1964).

[6] *Nobel Lectures; Physics 1922–1941* (Elsevier, Amsterdam, 1965).

[7] R. W. KISER, *Introduction to Mass Spectrometry and Its Applications* (Prentice-Hall, Englewood Cliffs, N.J., 1965).

[8] F. W. McLAFFERTY, *Interpretation of Mass Spectra* (Benjamin, New York, 1966).

[9] W. KAUZMANN, *Quantum Chemistry* (Academic, New York, 1957), Chap. 9.

[10] M. BORN, *Atomic Physics* (Blackie and Son, Glasgow, 1945).

[11] F. K. RICHTMEYER, E. H. KENNARD, and T. LAURITSEN, *Introduction to Modern Physics* (McGraw-Hill, New York, 1955), 5th ed.

[12] U. FANO and L. FANO, *Basic Physics of Atoms and Molecules* (Wiley, New York, 1959).

[13] F. L. FRIEDMAN and L. SARTORY, *The Classical Atom* (Addison-Wesley, Reading, Mass., 1965).

[14] L. KERWIN, *Atomic Physics* (Holt, Rinehart, and Winston, New York, 1963).

[15] M. R. WEHR and J. A. RICHARDS, JR., *Physics of the Atom* (Addison-Wesley, Reading, Mass., 1960).

[16] D. E. CARO, J. A. McDONELL, and B. M. SPICER, *Introduction to Atomic and Nuclear Physics* (Aldine, Chicago, 1962).

PROBLEMS

1.1 A Thomson experiment to measure e/m is made using a tube like that shown in Fig. 1.1a, with $l_1 = 3$ and $l_2 = 20$ cm. When an electric field of 1 statvolt cm^{-1} is applied across the plates and a magnetic induction of 8 G is used, the electrons arrive at the screen without deviating from a straight-line path. When the magnetic induction is turned off, a deviation of 2.44 cm from the straight-line path is observed.

Calculate e/m from these results.

1.2 A modification of Thomson's experiment with cathode rays is made by placing two sets of condenser plates in the tube, as shown in Fig. 1.12, and eliminating the magnetic field.

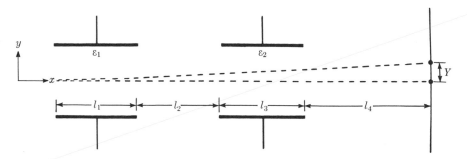

Fig. 1.12.

(a) Find Y in terms of ε_1, ε_2, l_1, l_2, l_3, l_4, v_x, and e/m.

(b) Can this apparatus be used to determine e/m? Explain your answer.

1.3 Consider a Millikan oil-drop experiment in which (a) charged drops are allowed to fall in a gravitational field and then (b) charged drops rise in the presence of an electric field that overbalances the gravitational field. From the data given below determine the elementary unit of charge.

Distance between plates = 0.7135 cm

Potential between plates = 879 V (300 V = 1 statvolt)

Distance through which drop lifted or fell = 0.145 cm

Density of oil = 0.893 g cm^{-3}

Density of air = 0.00117 g cm^{-3}

Viscosity of air = 0.001817 poises (dyn sec cm^{-2})

Time for free fall = 34.2 sec

Times for rise with field = 10.3, 28.6, 10.5, 4.4, 6.1, 28.9 sec

1.4 Using Eq. 1.34 for the frictional force on an oil drop of radius r with n electrons attached, find an expression for ne in terms of the terminal velocity v_1 of the drop when the electric field is turned off and the terminal velocity v_2 of the drop in an electric field ε. Comparing your result with Eq. 1.33, and show that Eq. 1.35 holds.

1.5 A mass spectrometer similar to the one shown in Fig. 1.2 has a constant magnetic field such that the CF_2^+ ion is focused at the collector when the accelerating voltage drop V is 1234.16 V. What is the mass of a singly charged positive ion which is focused in the same mass spectrometer when $V = 2123.65$ V? If the ion is known to be derived from a hydrocarbon, what is its identity?

1.6 A certain 180° mass spectrometer analyzer tube (see Fig. 1.2) is constructed in such a way that an ion path radius of 15 cm is required for focusing an ion beam at the detector. A voltage drop of 10 statvolt and a magnetic induction of 2230 G are required to focus an ion with unit positive charge. Determine the mass of this ion.

1.7 A Rutherford atom initially with radius a_0 radiates classically until the electron collapses into the nucleus. Find expressions for r and ω as functions of time and calculate the lifetime of the atom. (*Hint:* Use Eq. 1.44, which gives the rate of energy dissipation into radiation dE/dt, to set up a differential equation for r.)

1.8 Balmer originally analyzed the hydrogen spectrum in terms of wavelengths, and found that λ is given by

$$\lambda = C\left(\frac{n^2}{n^2 - 4}\right)$$

where C is a constant and n is an integer greater than 2. Show that this formula is equivalent to Eq. 1.46 with $n_1 = 2$ and $n_2 = n$. Find the value of C in terms of the Rydberg constant R.

1.9 Calculate the kinetic energy in ergs and the magnitude of the velocity in cm sec^{-1} for electrons in the hydrogen atom Bohr orbits with $n = 1$ and with $n = 4$.

1.10 Use the results of Problem 1.9 to determine the number of times per second the electron completes a circuit about the nucleus in each of the hydrogen atom Bohr orbits with $n = 1$ and $n = 2$. How do these frequencies compare with the frequency of the radiation absorbed in going from level $n = 1$ to $n = 2$?

1.11 Calculate the frequencies of radiation emitted in the transitions ($n = 101 \rightarrow n = 100$) and ($n = 102 \rightarrow n = 100$) for the hydrogen atom. Compare these frequencies with the classical radiation frequency for the orbit $n = 100$. What is the significance of your result?

1.12 Derive an expression for the energy E_n of the nth level of a Bohr atom with a nucleus of mass M using Eqs. 1.60–1.69. By what factor is the result related to Eq. 1.57?

1.13 Calculate the ground-state radius, shortest transition wavelength, and the classical lifetime of the ground state (see Problem 1.7) for positronium (an atomic system consisting of an electron and a positron).

1.14 From handbook values of the masses of the earth and moon, the mean distance between them, and the gravitation constant G, calculate the value of n for the Bohr model of the earth-moon "atom." Is this result meaningful? Explain. [For data, see "earth" and "solar system" in *Handbook of Chemistry and Physics* (The Chemical Rubber Company).]

1.15 Use the Bohr formula to compute the ionization potential of Li^{2+} (in volts) and the wave number, frequency, and wavelength of each of the first three lines of the Balmer series of Li^{2+}. Compare the radius of the orbit of the first excited state of the hydrogen atom with that of a uranium ($+91$) ion in the first excited state.

1.16 H. C. Urey, F. G. Brickwedde, and G. M. Murphy, *Phys. Rev.* **40,** 464 (1932) reported very faint lines accompanying the Balmer lines of the hydrogen spectrum. They attributed the faint lines to the presence of deuterium in the sample since they appeared at the calculated wavelengths for a hydrogen atom of mass 2. What is the displacement of the deuterium wavelength from the hydrogen wavelength for the first Balmer line ($n = 3 \rightarrow n = 2$)?

1.17 Sequential analysis of oligopeptides (condensation polymers of a few amino acids) can be made by means of their mass spectra after reduction to β-polyamines with lithium aluminum deuteride, since rupture of one of the C—C bonds adjacent to the NH group (α-cleavage) is the principal fragmentation mechanism [K. Biemann and W. Vetter, *Biochem. Biophys. Res. Commun.* **3,** 587 (1960)]. Consider a tripeptide composed of three amino acids, each with the formula

$$NH_2CHCO_2H,$$
$$|$$
$$R_i$$

where R_i is given by the following table.

amino acid	R_i	M_i
alanine (Ala)	CH_3	15
valine (Val)	$CH(CH_3)_2$	43
leucine (Leu)	$CH_2CH(CH_3)_2$	57

where M_i is the mass of R_i in amu. The esterified, N-acylated tripeptide thus has the formula

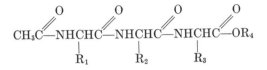

After reducing twice with LiAlD₄, one obtains

Given the fact that the parent ion can easily abstract a hydrogen atom from another molecule to increase its mass by 1 amu and that the C—C bonds indicated by dashes in the last formula above are the bonds most likely to break, show that the principal peaks in the mass spectrum of the reduced tripeptide are at the M/q values

281	$221 - R_1$
265	$218 - R_3$
237	$62 + R_3$
223	$59 + R_1$

If the observed peaks are at $M/q = 281, 265, 237, 223, 175, 164, 116, 105$ find the order of the amino acids in the tripeptide.

1.18 An oligopeptide known to be composed of four amino acids was treated by the procedure given in Problem 1.17. The resulting mass spectrum had major peaks at $M/q = 340, 324, 282, 223, 220, 175, 164, 132, 116, 117$. Determine the masses in amu of the R groups in order from the acylated to the esterified end of the chain. What is the sequence of amino acids? [For formulas of amino acids, see, e.g., J. D. Roberts and M. C. Caserio, *Basic Principles of Organic Chemistry* (W. A. Benjamin, New York, 1965).]

1.19 The mass spectra of pure CH_3OH, pure CH_3CH_2OH, and an unknown mixture of the two compounds have peaks at the following M/q, with intensities (in arbitrary units) as indicated. The intensities of the pure compounds have been normalized to 100.00 for the most intense (base) peak. The sensitivities of the pure compounds are the ratios of the respective base peaks to the sample pressure in 10^{-3} Torr.

M/q	CH_3OH	CH_3CH_2OH	mixture
31	100.00	100.00	288.97
32	68.03	1.14	50.46
45	$\cdots$	37.33	81.55
sensitivity	8.76	17.98	

Assuming a linear response of the ion detector at each M/q, calculate the partial pressures P_i of the components in the mixture sample and their mole fractions $x_i = P_i/(P_1 + P_2)$. (*Hint:* First use the intensity at $M/q = 45$ to find the partial pressure of CH_3CH_2OH. Then use $M/q = 32$ to find the partial pressure of CH_3OH by difference. The intensity at $M/q = 31$ can be used as a check.)

For the following problems a programmable desk calculator or digital computer will be useful.

C1.1 From the Rydberg formula for the hydrogen atom (i.e., with $R = R_H$) calculate the first 20 transitions of the Lyman, Balmer, Paschen, and Brackett series in units of Å, cm^{-1}, microns (μ) or nanometers, Hz (cps), eV, and a. u. (e^2/a_0).

C1.2 An ion, thought to be hydrogenic (i.e., to have a single extranuclear electron), displays spectral lines at wavelengths (in units of 10 microns) given below. (The data have been corrected to the nonrelativistic limit.) Determine the value of RZ^2 that minimizes the sum of the square of the differences between the experimental wave numbers and those predicted by the Rydberg formula. Compare the experimental spectrum with the least-squares prediction. Is the ion hydrogenic? What ion is it (i.e., what is the value of Z)? Are the data sufficiently precise to allow determination of the mass of nucleus? Are the data sufficiently precise to allow you to distinguish between the naturally occuring isotopes if you know the sample to be isotopically pure?

33.752	13.403	8.618	6.392	5.091
28.478	13.065	8.552	6.373	5.084
27.002	12.915	8.517	6.361	5.080
26.369	12.835	8.495	6.352	5.075
26.038	12.778	8.482	6.347	5.073
25.842	12.757	8.473	6.344	5.072
25.716	12.735	8.467	6.342	5.070
25.629	12.721	8.462	6.341	5.068

C1.3 Calculate the relative number of α particles deflected in the Rutherford–Geiger foil experiment through angles $20° \leq \theta \leq 180°$ in steps of $2°$. Make a polar plot of your result; that is, plot the curve $\rho = \sin^{-4}(\theta/2)$ in polar coordinates. At what angles besides $0°$ do local maxima occur?

C1.4 A portion of Millikan's original raw data for the fall times t_1 in the gravitational field alone and rise times t_2 in the gravitational plus electric field for a single oil drop is given. Calculate ne for each observation (neglecting deviations from Stokes' law) and find the average greatest common divisor e. Estimate the probable error and the root mean square deviation of your result [see H. Margenau and G. M. Murphy, *The Mathematics of Physics and Chemistry* (Van Nostrand, Princeton, N.J., 1956), p. 510].

t_1 (sec)	t_2 (sec)	t_1 (sec)	t_2 (sec)
11.848	80.708	11.904	29.236
11.890	22.366	11.870	137.308
11.908	22.390	11.952	34.638
11.904	22.368	11.846	22.104
11.882	140.565	11.912	22.268
11.906	79.600	11.910	500.1
11.838	34.748	11.918	19.704
11.816	34.762	11.870	19.668
11.776	34.846	11.888	77.630
11.840	29.286	11.894	77.806
		11.878	42.302

plate distance = 16 mm   air density = 1.1871×10^{-3} g cm^{-3}
fall distance = 10.21 mm   oil density = 0.9199 g cm^{-3}
average potential = 16.96   air viscosity = 1.824×10^{-4} poise
                statvolts

C1.5   Construct a table of $\nu_{n,n+1}$ and ν_{class} for orbit n of hydrogen for $1 \leq n \leq 500$. Above what value of n is the classical description adequate?

C1.6   Calculate the magnitude of the electron velocity v in cm sec^{-1} for the hydrogenic ions with $2 \leq Z \leq 150$, assuming the Bohr model with infinite nuclear mass. For what range of Z is the velocity greater than that of light? Above what values of Z do you expect relativistic effects to be important?

C1.7   The following M/q peaks, intensities, and sensitivities (ratios of intensity of most intense peak to sample pressure in 10^{-3} Torr) were found in the mass spectra for the pure butyl alcohol isomers and an unknown mixture. Write down the system of linear equations for the partial pressures of the four components in the mixture and solve them. Calculate the mole fraction of each component (see Problem 1.19).

M/q	n butyl alcohol	sec-butyl alcohol	tert-butyl alcohol	isobutyl alcohol	mixture
31	100.00	20.31	35.53	63.10	368.6
43	61.36	9.83	14.45	100.00	292.7
45	6.59	100.00	0.59	5.03	322.6
59	0.26	17.78	100.00	4.98	301.5
sensitivity	11.51	26.98	20.93	12.05	

(*Hint:* Calculate the pressure per unit intensity at each M/q for each component. Let this array be M_{ij}, where $i = 1, \ldots, 4$ denotes the M/q value and $j = 1, \ldots,$ 4 denotes the component. Then

$$M_{11}P_1 + M_{12}P_2 + M_{13}P_3 + M_{14}P_4 = I_1$$

$$\begin{matrix} \cdot & \cdot & \cdot & \cdot & \cdot \\ \cdot & \cdot & \cdot & \cdot & \cdot \\ \cdot & \cdot & \cdot & \cdot & \cdot \end{matrix}$$

$$M_{41}P_1 + M_{42}P_2 + M_{43}P_3 + M_{44}P_4 = I_4$$

where I_i is the intensity of the mixture at the ith M/q value.)

Ans. P_1 (*n* butyl alcohol) $= 11.18 \times 10^{-3}$ Torr.

2

The quantum theory

Since the Bohr model was so successful in providing an explanation of the properties of one-electron atoms, scientists tried to apply it to more complicated atoms and to molecules. However, even for the simplest many-electron system—the helium atom, composed of a nucleus and two electrons—Bohr's theory was a failure. Many modifications were introduced into the theory to take account of the interaction between the electrons, but no adjustment of the equations was able to fit the helium spectrum. It became more evident only that this simple and beautiful theory, which was in such excellent quantitative agreement with the experimental results for hydrogen, could not be extended to complex atoms. Efforts to understand the chemical bond on the basis of the Bohr theory were similarly unsuccessful. There appeared to be no way to calculate the binding energy of even the simplest neutral molecule, H_2. This dashed the hope that the new understanding of atomic structure might be employed to gain insight into chemical problems.

In addition to these limitations of the Bohr theory,

objections were raised to its rather arbitrary use of quantum restrictions to modify an essentially classical model. A philosophically acceptable theory would have to be based upon more fundamental hypotheses from which the quantization of energy, angular momentum, and other properties followed in a logical manner.

The need for a theory that gave results identical to the Bohr model for hydrogen, but was generally applicable to atoms and molecules, was clearly evident. As often happens in scientific developments, the key to the new theory came from experiments in a different area, rather than from the further study of atomic and molecular problems.

2.1   THE NATURE OF RADIATION AND MATTER

It had been assumed that matter, which is composed of particles like atoms and molecules (or more basically, electrons and nuclei), is distinct from radiation, which is the transmission of energy by wave motion. However, experiments which were done in the period 1887–1927 demonstrated that the boundary between matter with its particle-like behavior, and radiation with its wave-like behavior, is not as rigid as had been supposed. In fact, the observations implied that both matter and radiation can behave as if they are composed of waves or as if they are composed of particles, the behavior manifested depending on the nature of the experiment. Attempts to understand these results led to a new concept of radiation and matter that was essential to the development of the quantum theory. To clarify what is involved, it is helpful to consider a few of the important experiments. However, before doing so we review briefly the wave theory of radiation.

2.2   WAVE THEORY

In the simplest case, radiation can be represented by a function of the coordinate x and the time t of the form

$$a(x, t) = a_0 \cos\left[2\pi \left(\frac{x}{\lambda} - \nu t \right) \right] \tag{2.1}$$

Figures 2.1a and 2.1b show plots of $a(x, 0)$ and $a(0, t)$. The maximum value of the function, a_0, is called the *amplitude*. For $t = 0$, the function attains the value a_0 when x/λ is an integer or zero; that is, for $x = \ldots , -3\lambda,$ $-2\lambda, -\lambda, 0, \lambda, 2\lambda, 3\lambda, \ldots$. Thus λ, the distance between successive maxima, is called the *wavelength*. At the point $x = 0$, the function $a(0, t)$ exhibits a maximum at the times when νt is an integer or zero; that is, at $t = \ldots , -3/\nu, -2/\nu, -1/\nu, 0, 1/\nu, 2/\nu, 3/\nu, \ldots$. The time interval, $1/\nu$ between successive maxima at a fixed point in space is called the *period*

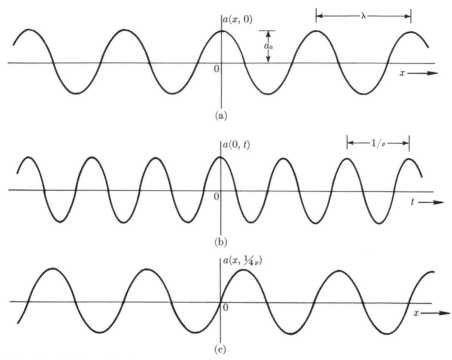

Fig. 2.1. Motion of a plane wave.

τ; the number of maxima per unit time which pass a fixed point is equal to the reciprocal of τ and is called the *frequency* ν. Since the radiation described by Eq. 2.1 has the single frequency ν, it is said to be *monochromatic;* if there were a range of frequencies, it would be *polychromatic* radiation.

To examine the motion of the wave in more detail, we consider it at a time $t_1 = (4\nu)^{-1}$. The wave has the form

$$a(x, t_1) = a_0 \cos\left(2\pi \frac{x}{\lambda} - \frac{\pi}{2}\right) = a_0 \sin\left(\frac{2\pi x}{\lambda}\right)$$

shown in Fig. 2.1c. Comparison with Fig. 2.1a demonstrates that the wave has moved a distance $\lambda/4$ to the right in the time $(4\nu)^{-1}$, so that the velocity in the positive x direction is

$$v = \frac{\Delta x}{\Delta t} = \frac{\lambda/4}{(4\nu)^{-1}} = \lambda\nu \tag{2.2}$$

We refer to $a(x, t)$ as a plane monochromatic wave moving in the x direction with velocity $\lambda\nu$; the term "plane wave" is used to describe the fact that (for a given value of x and t) the disturbance is uniform throughout a plane, the yz plane in the present case.

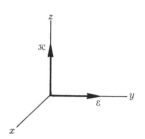

Fig. 2.2. Directions of electric and magnetic fields associated with a y-polarized electromagnetic wave traveling in the x direction.

If electromagnetic radiation is represented by the plane wave $a(x, t)$, the physical quantity corresponding to $a(x, t)$ is the electric field $\mathcal{E}$. It is known from electromagnetic theory that the electric field is oriented perpendicular to the direction of motion of the wave; that is, for the plane wave $a(x, t)$ moving in the x direction, the electric field direction must be in the yz plane. For *polarized* radiation, the electric field is oriented in a particular direction in the yz plane; for example, if the radiation is y polarized, the field is directed along the y axis (Fig. 2.2) and we can write

$$\mathcal{E}(x, t) = \mathcal{E}_y{}^0 \cos\left[2\pi\left(\frac{x}{\lambda} - \nu t\right)\right] \tag{2.3}$$

Associated with the electric field of such a wave, there is also present a magnetic field $\mathcal{H}$. It is known to be oriented perpendicular to the plane of motion and, furthermore, to be perpendicular to the electric field and in phase with it; that is, for the plane-polarized electric wave given in Eq. 2.3, the magnetic wave associated with it is

$$\mathcal{H}(x, t) = \mathcal{H}_z{}^0 \cos\left[2\pi\left(\frac{x}{\lambda} - \nu t\right)\right]$$

and the amplitude $\mathcal{H}_z{}^0$ is equal to $\mathcal{E}_y{}^0$ (Fig. 2.3).

The *energy density* (i.e., the energy per unit volume) corresponding to a plane wave is proportional to the square of its amplitude (see Section 2.10 for a discussion of this point in relation to the wave motion of a harmonic oscillator). For electromagnetic radiation, the energy density $U(x, t)$ is given by

$$U(x, t) = \frac{1}{4\pi} [\mathcal{E}(x, t)]^2$$

$$= \frac{(\mathcal{E}_y{}^0)^2}{4\pi} \left\{ \cos\left[2\pi\left(\frac{x}{\lambda} - \nu t\right)\right] \right\}^2 \tag{2.4}$$

We can use this result to obtain an expression for the *intensity* I, which is the average rate of flow of energy per unit area past a given point in space.

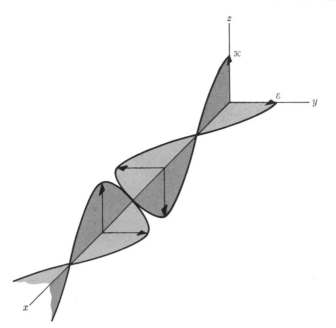

Fig. 2.3. Sinusoidal dependence of the electric and magnetic fields upon the displacement x at fixed time t for a y-polarized electromagnetic wave moving in the x direction.

At time $t = 0$, the part of the wave that will move past the point $x = 0$ during the time interval $t = 0$ to $t = t_1$ lies between $x = -\lambda \nu t_1$ and $x = 0$. Thus, the intensity can be determined by finding the total energy in the volume defined by the unit cross-section area A and the length $x = \lambda \nu t_1$ and dividing the result by the time interval t_1 (Fig. 2.4). Since we know the energy density from Eq. 2.4, the energy in this region as a function of t_1 is found by integrating $U(x, 0)$ over x to give

$$U(t_1) = \frac{(\mathcal{E}_y{}^0)^2}{4\pi} \int_{-\lambda \nu t_1}^{0} \cos^2 \left(\frac{2\pi x}{\lambda} \right) dx$$

$$= \frac{(\mathcal{E}_y{}^0)^2}{8\pi} \lambda \nu t_1 + \frac{(\mathcal{E}_y{}^0)^2}{32\pi^2} \lambda \sin(4\pi \nu t_1)$$

Since the intensity I is defined as the *average* rate of energy flow, we assume that t_1 corresponds to a large time ($t_1 \to \infty$) and divide $U(t_1)$ by the time t_1; that is, the second term does not contribute and

$$I = \lim_{t_1 \to \infty} \frac{1}{t_1} U(t_1) = \frac{(\mathcal{E}_y{}^0)^2}{8\pi} \lambda \nu \qquad (2.5)$$

Thus, the intensity I is proportional to $(\mathcal{E}_y{}^0)^2$ and to $\lambda \nu$, which is the velocity of the wave; for electromagnetic radiation in a vacuum, the velocity is equal 'o c, the velocity of light.

central, or zero-order, maximum; namely, that the difference in path length d of the two rays must equal

$$d = r_2 - r_1 = n\lambda \qquad n = 1, 2, 3, \ldots$$

If $\lambda \ll s \ll l$, as is usually the case, it can be shown (Problem 2.1) that the first-order maximum $(n = 1)$ appears at

$$D = \lambda \frac{l}{s} \qquad (2.7a)$$

By similar arguments, the first minimum (destructive interference) occurs at

$$D = \tfrac{1}{2}\lambda \frac{l}{s} \qquad (2.7b)$$

The *refractive index* μ of a medium is defined as the ratio of the velocity of light in the medium to the velocity in empty space, c. Since the frequency is known to remain constant, the change in velocity (Eq. 2.2) results from a change in wavelength. If the light emanating from one of the slits of Fig. 2.5 is made to pass through a medium of different index of refraction from that of the other, the waves from the two sources will propagate with different velocities and wavelengths. Consequently, the effective path lengths to a screen are different for the two slits, and a change in diffraction pattern is the result. An interferometer (Fig. 2.6) is a device that is based upon this principle. It measures the displacement of the maxima of the diffraction pattern at the focal plane FF' and hence the difference in refractive index between the two media in cells C_I and C_{II} (Problem 2.21). Since the refractive index of a gas mixture or liquid solution depends upon the concentration of the components, rapid and very accurate chemical analyses can often be made by interferometric techniques; for example, the

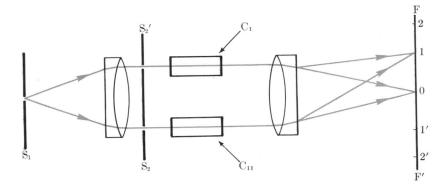

Fig. 2.6. Schematic diagram of a simple interferometer. [Adapted from N. Bauer and S. Z. Lewin, *Technique of Organic Chemistry*, A. Weissberger, Ed. (Interscience, New York, 1960), 3rd ed., Vol. I, Part II.]

technique has been used to determine trace quantities of excess D_2O in water and to analyze blood serum for protein content. Also, interferometry has come to be an important method for the determination of molecular spectra (Section 7.5).

2.2.1   Complex wave forms   Although we have introduced real wave forms in the earlier part of this section, for many applications it is more convenient to work with waves expressed in the *complex* form

$$a(x, t) = a_0 \left\{ \cos \left[2\pi \left(\frac{x}{\lambda} - vt \right) \right] \pm i \sin \left[2\pi \left(\frac{x}{\lambda} - vt \right) \right] \right\} \quad (2.8)$$

where i is the *imaginary* number $\sqrt{-1}$. We can write this function more compactly; to see how to do this, we compare the series expansions

$$e^x = 1 + x + \frac{1}{2!} x^2 + \cdots + \frac{1}{n!} x^n + \cdots \quad (2.9)$$

$$\sin x = x - \frac{1}{3!} x^3 + \frac{1}{5!} x^5 - \cdots + (-1)^n \frac{1}{(2n+1)!} x^{2n+1} + \cdots \quad (2.10)$$

$$\cos x = 1 - \frac{1}{2!} x^2 + \frac{1}{4!} x^4 - \cdots + (-1)^n \frac{1}{(2n)!} x^{2n} + \cdots \quad (2.11)$$

From Eqs. 2.10 and 2.11 and the definition of i, we have

$$\cos x + i \sin x = 1 + ix - \frac{1}{2!} x^2 - \frac{i}{3!} x^3 + \frac{1}{4!} x^4 + \frac{i}{5!} x^5 + \cdots$$
$$= 1 + ix + \frac{1}{2!} (ix)^2 + \frac{1}{3!} (ix)^3 + \cdots + \frac{1}{n!} (ix)^n + \cdots \quad (2.12)$$

Comparison of the last line of Eq. 2.12 with Eq. 2.9 shows that

$$\cos x + i \sin x = e^{ix} \quad (2.13)$$

which is the *Euler identity*. With the aid of Eq. 2.13 we can write Eq. 2.8 in the form

$$a(x, t) = a_0 \exp \left[\pm 2\pi i \left(\frac{x}{\lambda} - vt \right) \right] \quad (2.14)$$

where exp (x) is used for e^x. To obtain equations by which the real wave forms can be converted into linear combinations of the complex forms, we use Eq. 2.13 and its *conjugate* (obtained by replacing i by $-i$),

$$\cos x - i \sin x = e^{-ix} \quad (2.15)$$

From Eqs. 2.13 and 2.15, we obtain

$$\cos x = \frac{1}{2} (e^{ix} + e^{-ix}) \quad (2.16)$$

$$\sin x = \frac{1}{2i} (e^{ix} - e^{-ix}) \quad (2.17)$$

Thus, the real wave form given in Eq. 2.1 can be written

$$a(x, t) = \frac{a_0}{2} \left\{ \exp\left[2\pi i \left(\frac{x}{\lambda} - \nu t \right) \right] + \exp\left[-2\pi i \left(\frac{x}{\lambda} - \nu t \right) \right] \right\} \quad (2.18)$$

2.3   THE PARTICLE-LIKE CHARACTER OF RADIATION

Although the wave nature of radiation is well established by experiments of the type discussed in Section 2.2, Bohr's model for radiation by the hydrogen atom provides a hint that radiant energy can have a particle-like character as well. Since radiation was assumed by Bohr to be emitted or absorbed in indivisible units or quanta, $h\nu$, it is reasonable to suppose that these quanta might behave like particles. To attempt to deduce some properties of the radiation particles, called *photons*, we make use of the theory of relativity. If a particle has (rest) mass m and velocity v, special relativity theory gives the momentum p as

$$p = \frac{mv}{(1 - v^2/c^2)^{1/2}} \quad (2.19)$$

Since the velocity of a particle of radiation is the velocity of light c, the denominator of Eq. 2.19 is zero. Thus, in order to have a finite momentum, photons must have *zero* mass. Even with this assumption, Eq. 2.19 yields zero over zero, so that it is of no use for finding the momentum. It can be shown that the energy E of a relativistic particle of mass m moving freely in space with momentum p is

$$E = (m^2 c^4 + p^2 c^2)^{1/2} \quad (2.20)$$

For a photon, $\epsilon = h\nu$ and $m = 0$, so that Eq. 2.20 yields

$$p = \frac{E}{c} = \frac{h\nu}{c} \quad (2.21)$$

This formula for the photon momentum characterizes one of the particle aspects of radiation.

We now examine some experimental results which cannot be understood in terms of the wave theory.

2.3.1   The photoelectric effect   If ultraviolet light with wavelength in the range 2000–4000 Å (2.5×10^4 to 5×10^4 cm^{-1}) is allowed to impinge upon the clean surface of a metal (such as sodium) in an evacuated vessel, electrons are liberated from the metal. The photoelectrons can be collected by a plate and a circuit set up to measure the current (Fig. 2.7). The photoelectric current, which is the amount of charge arriving at the plate per unit time, is proportional to the rate of liberation of electrons from the metal

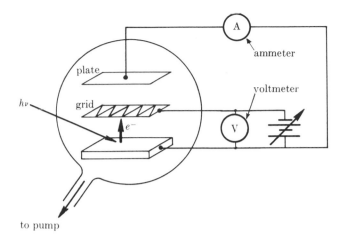

Fig. 2.7. Schematic diagram of apparatus for photoelectric measurements.

surface; that is, if Δn_e is the number of free electrons produced in the time interval Δt and i is the current,

$$\frac{\Delta n_e}{\Delta t} = \frac{i}{e}$$

To determine the velocity with which the photoelectrons travel, a potential is applied to a grid mounted between the metal surface and the plate (Fig. 2.7). The potential creates an electric field which decelerates the photoelectrons. If the potential difference between the grid and the emitting metal is increased, one reaches a value above which the electrons are stopped before they reach the plate and the current ceases to flow. This potential is called the *stopping voltage* V_s. At the stopping voltage, the initial kinetic energy of the photoelectrons liberated from the metal by the light has all been converted to potential energy, so that we can write

$$\tfrac{1}{2}mv^2 = eV_s \tag{2.22}$$

Thus, by measuring i and V_s we know the number of electrons produced per second and their maximum kinetic energy as functions of the intensity and frequency of the incident light.

An analysis of the photoelectric effect by means of the classical wave theory demonstrates that the energy of the radiation should be absorbed continuously by the electrons in the metal. After an electron has absorbed an amount of energy in excess of its binding energy eV_0, it may be ejected from the surface. Figure 2.8 shows a simplified plot of the potential energy of the photoelectron. The adjustable potential V_s is used to stop electrons whose energy exceeds eV_0 by eV_s or less. Since the intensity I of the

Angstrom longer than λ. The dependence of the scattered wavelength λ' upon the angle θ between the primary and scattered beams is found to be

$$\lambda' = \lambda + k \sin^2\left(\frac{\theta}{2}\right) \tag{2.25}$$

where k is a constant. An increase of wavelength can be predicted from classical wave theory, since an electron in the sample will be accelerated by the impinging radiation and will therefore emit waves with longer wavelengths (the Doppler effect). The classical theory does not correctly explain the observations, however, since (a) the Doppler shift is proportional to the wavelength of the primary radiation and (b) the Doppler shift increases with the electron velocity and therefore should increase with time, since the electrons are accelerated continuously while they absorb energy during the irradiation. Neither of these predictions is corroborated by the experimental results. As was the case for the photoelectric effect, the observations can be explained quantitatively by the photon theory of radiation and the laws of conservation of energy and momentum for particles.

According to Eq. 2.21, the incident photon with wavelength λ and frequency $\nu = c/\lambda$ has momentum $h\nu/c$. Correspondingly, the scattered photon, which has a longer wavelength λ', and therefore a lower frequency $\nu' = c/\lambda'$, has a lower momentum $h\nu'/c$. Since ν is in the x-ray region ($\lambda \sim 1$–10Å), the energy ($h\nu \sim 1000$ eV) is so much greater than the binding energy of the electrons (~ 10 eV) that to a first approximation the latter be neglected. Thus, the electron is ejected in the direction ϕ with a momentum mv which is calculable from an energy and momentum balance for the process (see Fig. 2.10). The classical equations of conservation of energy and of the two components of the linear momentum are

$$h\nu = h\nu' + \tfrac{1}{2}mv^2 \qquad \text{(energy)}$$

$$\frac{h\nu}{c} = \frac{h\nu'}{c}\cos\theta + mv\cos\phi \quad \text{(x component of momentum)} \tag{2.26}$$

$$0 = \frac{h\nu'}{c}\sin\theta - mv\sin\phi \quad \text{(y component of momentum)}$$

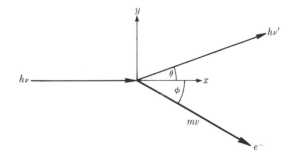

Fig. 2.10. The Compton effect.

Eliminating v and ϕ from these equations, introducing λ by the definition $\lambda = c/\nu$, and making the approximation that $\lambda\lambda' \simeq \lambda^2$, we find that (Problem 2.7)

$$\Delta\lambda = \lambda' - \lambda = 2\frac{h}{mc}\sin^2\left(\frac{\theta}{2}\right) \tag{2.27}$$

in agreement with Eq. 2.25; for λ in Angstroms, this gives

$$\Delta\lambda = 0.0485 \sin^2(\theta/2)$$

If the ejected electron is treated relativistically (i.e., if Eq. 2.20 is used for the total energy and the kinetic energy is obtained by subtracting the rest energy mc^2), one can derive Eq. 2.27 without resorting to the approximation that $\lambda' \simeq \lambda$ (see Problem 2.8). The maximum shift is seen to occur for $\theta = \pi$, where $\Delta\lambda = 0.0485$ Å.

The photon mechanism may be submitted to an independent check by using gamma rays (γ rays) of energy $\sim 10^6$ eV and making observations of both the scattered photon and the Compton electron by means of modern scintillation counters. W. G. Cross and N. F. Ramsey, *Phys. Rev.* **80,** 929 (1950) found that the angles ϕ and θ for an electron and a photon which are simultaneously detected are within $\pm 1°$ of those required by the conservation laws, Eq. 2.26.

Our discussion of the photoelectric and Compton effects shows that the particle viewpoint and Newtonian mechanics lead to a simple and quantitatively correct interpretation of these experiments, and that predictions based upon the classical wave theory are wrong. However, we have seen that for certain other phenomena, namely, diffraction and interference, the wave theory works perfectly well, while the particle theory is inadequate. We conclude that the character exhibited by radiation, whether wave-like or particle-like, depends upon the type of experiment that is done. If the interaction of radiation with matter produces a measurable change in the matter, such as the ejection of an electron, the phenomenon appears to require the photon theory for its interpretation. If the interaction produces a measurable change in the spatial distribution of the radiation, such as diffraction at a slit, but produces no measurable change in the matter, the wave theory must be invoked. These results suggest that we need a synthesis of the two points of view which takes into account the nature of the experiment being analyzed; that is, the measuring process itself must be included in the theory.

2.4 THE WAVE NATURE OF MATTER

The conclusions that we have just reached about the dual nature of radiation apply equally to matter. The classical viewpoint had been that matter is composed of particles whose measurable positions and momenta were governed by Newton's equations of motion. It was discovered, however, that experiments could be performed that are satisfactorily interpreted only if a wave-like character is ascribed to electrons and even to atoms as a whole.

2.4.1 De Broglie's hypothesis   The duality of matter was suggested on the basis of theoretical considerations prior to its experimental verification. For a photon, the momentum and wavelength are simply related; by substituting into Eq. 2.21 with $\nu = c/\lambda$, one obtains $p = h/\lambda$. In 1924, de Broglie reasoned that a similar relation should hold for material particles, since the same relativistic equations of motion apply to photons and to particles of nonzero rest mass. Thus an electron with momentum p should have associated with it a wave whose wavelength is given by

$$\lambda = \frac{h}{p} \tag{2.28}$$

If there are matter waves, why had they not been observed before? Assuming that Eq. 2.28 is correct, we can determine the conditions under which the wave-like character of an electron could be observed; that is, we calculate the wavelength λ for an electron accelerated through a potential V. The kinetic energy of the electron is $E = \frac{1}{2}mv^2 = eV$ and the momentum is $p = mv = (2meV)^{1/2}$. Substituting for p in Eq. 2.28, we find that the wavelength is

$$\lambda = \frac{h}{p} = \frac{h}{(2meV)^{1/2}} = 7.0830 \times 10^{-9} V^{-1/2} \text{ cm} \tag{2.29}$$

where V is given in statvolts. For a potential of 1 statvolt (300 V), which is a reasonable potential for experiments with electrons, the de Broglie wavelength is on the order of 10^{-8} cm, or 1 Å. As we saw in the discussion of diffraction (Section 2.2), if the wave characteristics are to be experimentally manifested, arrangements must be made for detecting differences on the order of λ between the lengths of alternate "paths." This can be accomplished only if certain dimensions of the apparatus are on the order of magnitude of the wavelength λ. With $\lambda \sim 1$ Å, it is immediately clear why diffraction of electrons is not observed in an ordinary apparatus whose significant dimensions are macroscopic; for example, in cathode-ray tubes and similar devices which we have discussed, only particle-like behavior is exhibited. To make an experimental test of de Broglie's hypothesis, we must have gratings or slits with spacings on the order of 10^{-8} cm, that is, on the order of atomic diameters. Although gratings with such small spacings would be impossible to manufacture, suitable natural gratings are furnished by the regular array of atoms in a crystal.

2.4.2 Diffraction of electrons   C. Davisson and L. H. Germer in 1927 made
the first successful study of diffraction patterns for electrons reflected from
a crystal. The experimental arrangement is shown in Fig. 2.11a. A beam of
electrons which have been accelerated through a potential difference V is
directed upon a single crystal of nickel at an angle normal to the crystal
surface. The scattered electrons are detected by means of a movable·col-
lector which measures the electron current as a function of the angle θ. A
voltage drop of 0.9 V between the target and collector serves to stop
electrons which have lost more than 10% of their kinetic energy, and thus
restricts the measurements to elastically scattered electrons. The crystal
of nickel is composed of a regular array of atoms (see Fig. 2.11b), each of

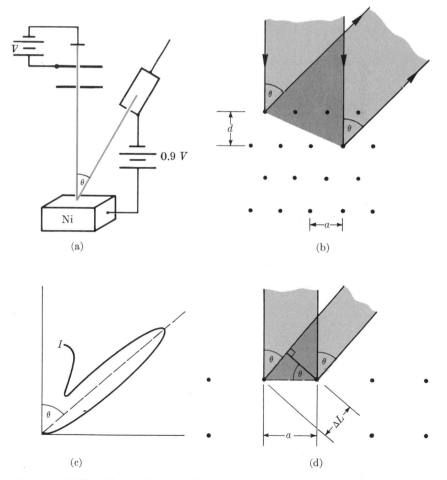

Fig. 2.11. Diffraction of electrons from a nickel crystal.

which can scatter the incident electrons in all directions. By the wave theory (Section 2.2), if λ is the wavelength of a wave associated with the electron beam, constructive interference occurs for angles at which the waves scattered by different atoms have paths differing in length by $n\lambda$, where n is an integer; if the path lengths differ by an amount $n\lambda/2$ (when n is odd), destructive interference occurs. Davisson and Germer obtained plots of the electron current I versus θ similar to that in Fig. 2.11c. In this plot polar coordinates are used, so that the distance from the origin of a point on the intensity curve is proportional to I, and the angle with the vertical axis is θ. The appearance of a maximum in the I-versus-θ plot clearly shows that the electron beam is scattered preferentially at a certain angle. To see whether de Broglie's hypothesis (Eqs. 2.28 and 2.29) can account quantitatively for the result, we begin by analyzing the expected diffraction pattern in a manner similar to that used earlier in the chapter for diffraction at two slits. In the case of x-ray beams, the reflections from all the planes of the crystal (Fig. 2.11b) would have to be taken into account. For electron beams with the low velocities used by Davisson and Germer ($V \sim$ 10–400 V), however, the penetration of the electrons into the crystal is very slight. Consequently, the assumption that the diffraction occurs solely as a result of reflections from a single plane (the surface plane) is a good approximation. By looking at Fig. 2.11d we see that the difference in path length ΔL for waves scattered at angle θ by atoms spaced a distance a apart is $a \sin \theta$. If this distance is an integral number of wavelengths, the waves are in phase and an intensity maximum is observed at θ. Thus, the law for single-plane diffraction is

$$n\lambda = a \sin \theta \qquad (2.30)$$

with the strong first-order peak corresponding to $n = 1$. Davisson and Germer found that an electron beam accelerated through a potential $V = 0.18$ statvolt had a maximum scattering angle at $\theta = 50°7'$. Since the spacing between the nickel atoms exposed to the beam is $a = 2.15$ Å (as determined by x rays), the wavelength λ calculated from Eq. 2.30 is 1.65 Å. According to Eq. 2.29, the de Broglie wavelength associated with electrons accelerated through 0.18 statvolt is

$$\lambda = \frac{7.0830 \times 10^{-9} \text{ cm}}{(0.18)^{1/2}} = 1.67 \text{ Å}$$

in good agreement with Davisson and Germer's result.

G. P. Thomson, the son of the discoverer of the electron, also studied electron diffraction patterns. His experiments differed from those of Davisson and Germer in that his targets were very thin films of metals such as gold, aluminum, and platinum, and the electron energies were sufficient to

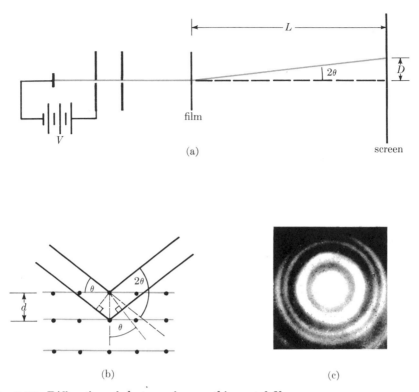

Fig. 2.12. Diffraction of electrons from a thin metal film.

penetrate the films. As shown in Fig. 2.12a, the beam of electrons passes through the metal film and is observed on a screen or photographic plate. In this case the film is composed not of a single crystal, but of many microscopic crystals randomly oriented with respect to the electron beam. The angle of incidence of the beam with respect to a crystal plane in the film thus can have any value between 0° and 90°. Since the target is penetrated (in contrast to the Davisson–Germer experiment), we must consider the reflections from layers of atoms deep in the microcrystals. Reflection of the electrons from two parallel planes is illustrated in Fig. 2.12b. It is evident from the figure that the beam which emerges from a crystal with the orientation shown is deflected through an angle 2θ from the direction of incidence. If this deflection angle is one of maximum intensity, the difference in path length between the rays reflected from the different planes must be an integral multiple of the wavelength. From Fig. 2.12b it is seen that for two adjacent planes this difference is $2d \sin \theta$; thus, the intensity maxima occur for angles which satisfy Bragg's law, namely,

(see Eq. 1.39). Elimination of r from Eq. 2.35 leads directly to the quantization rule for angular momentum,

$$p_\phi = \frac{nh}{2\pi} \qquad (2.36)$$

which was derived previously by a correspondence principle argument (Section 1.4.3). The remainder of the calculation for the hydrogen atom planetary model could now be completed in the same way as before to yield energy levels and spectral frequencies identical with those obtained from the Bohr theory.

2.5 LOCALIZED WAVES AND THE UNCERTAINTY PRINCIPLE

The wave-particle duality of both matter and radiation faces us with the task of finding the best general description for microscopic phenomena. An ideal description would apply to all types of experiments, so that one would not have to decide beforehand whether to use a theory based upon the wave or the particle viewpoint. Such a unified formulation, known as wave mechanics, has been achieved. It is based upon the assumption that there are waves associated with matter and radiation, and that the waves furnish information about the positions of the particles; that is, if the *wave* amplitude is large in a certain region of space, the "particle" of matter or of radiation is likely to be found in that region. Suppose, for example, that as a result of a measurement an electron is known to be localized in a certain region, say at the origin ($x = 0$). From the wave assumption, we would expect that the wave associated with the electron has a large amplitude in the region where the electron was observed and a small amplitude everywhere else (see Fig. 2.14a). By contrast, if we had not measured the position of the electron, it could be anywhere along the x axis. Since we know nothing about the electron position in this case, we could use a sine wave (Fig. 2.14b) which has nonzero amplitudes in all parts of space ($-\infty < x < \infty$). Thus the measurement (or lack of it) of the position of a particle determines the form of the wave associated with it; conversely, knowledge of the wave form allows us to predict the probable position of the particle.

From the above, it is clear that to be able to associate a wave with localized particles, we must have a method for constructing localized waves. In our discussion of sine waves (Section 2.2) it was pointed out that the principle of superposition provides a means of constructing arbitrarily shaped wave forms. For a particle localized at the origin ($x = 0$), we must add together a number of sine waves of the form of Eq. 2.1 which are in phase near $x = 0$ and out of phase elsewhere on the x axis; that is, by constructive interference a large amplitude is produced in the region of the origin and by destructive interference the amplitude elsewhere is small. Such a wave would look similar to the one already shown in Fig. 2.14a.

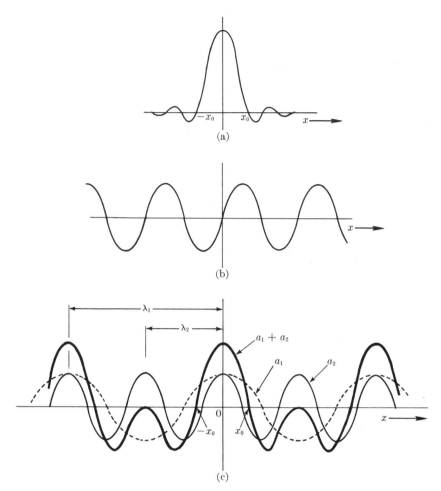

Fig. 2.14. Construction of a localized wave by superposition of plane waves.

We suppose that at time $t = 0$ a measurement was made of the position of the electron and it was found to lie somewhere between $-x_0$ and x_0. Setting $t = 0$ in Eq. 2.1, we obtain

$$a_1(x, 0) = a_0 \cos\left(\frac{2\pi\, x}{\lambda_1}\right) \tag{2.37}$$

where we have introduced the subscript 1 to indicate that $a_1(x, 0)$ is the wave with wavelength $\lambda = \lambda_1$. Its form, which is shown in Fig. 2.14c, is similar near $x = 0$ to the form of the localized wave we seek, but the amplitude does not damp out for $x < -x_0$ or $x > x_0$. Adding to $a_1(x, 0)$ a second wave $a_2(x, 0)$ with amplitude a_0 and wavelength λ_2,

$$a_2(x, 0) = a_0 \cos\left(\frac{2\pi x}{\lambda_2}\right) \tag{2.38}$$

we choose the wavelengths λ_1 and λ_2 such that the waves $a_1(x, 0)$ and $a_2(x, 0)$ cancel exactly at x_0 and $-x_0$. As we have seen in Eq. 2.6, destructive interference at x_0 requires that the two waves be out of phase by π at that point; that is, for

$$A(x_0, 0) = a_1(x_0, 0) + a_2(x_0, 0)$$

$$= a_0\left(\cos\frac{2\pi x_0}{\lambda_1} + \cos\frac{2\pi x_0}{\lambda_2}\right) = 0$$

we must have (taking $\lambda_1 > \lambda_2$)

$$\frac{x_0}{\lambda_2} = \frac{x_0}{\lambda_1} + \frac{1}{2}$$

or

$$x_0\left(\frac{1}{\lambda_2} - \frac{1}{\lambda_1}\right) = \frac{1}{2} \tag{2.39}$$

If we add together only these two waves, there are points along the x axis other than $x = 0$ at which constructive interference occurs. Hence we obtain a series of *beats*, each representing a possible region of localization of the particle (Fig. 2.14c). Thus, although the waves cancel at $x = \pm x_0$, localization between $-x_0 < x < x_0$ has not been achieved. However, by adding together a very large (infinite) number of sine waves of the form $a_0 \cos(2\pi x/\lambda)$ with wavelengths in the range $\lambda_2 \leq \lambda \leq \lambda_1$ (i.e., replacing the sum by an integral over the range λ_2 to λ_1), a completely localized wave can be produced (Problem C2.2).

If the wave amplitude is restricted to the interval $(-x_0, x_0)$, the uncertainty Δx in position is $2x_0$; that is,

$$\Delta x = 2x_0$$

Substituting into Eq. 2.39, we have

$$\Delta x\left(\frac{\lambda_1 - \lambda_2}{\lambda_1\lambda_2}\right) = 1 \tag{2.40}$$

This relationship corresponds to the result that, to localize the wave representing a particle in a region Δx, sine waves whose wavelengths are spread over the range $\lambda_2 \leq \lambda \leq \lambda_1$ must be used. Since $\bar{\nu} = 1/\lambda$, this is equivalent to a wave-number range of $1/\lambda_1 \leq \bar{\nu} \leq 1/\lambda_2$. Denoting this range by $\Delta\bar{\nu}$,

$$\Delta\bar{\nu} \equiv \frac{1}{\lambda_2} - \frac{1}{\lambda_1} = \frac{\lambda_1 - \lambda_2}{\lambda_1\lambda_2} \qquad (2.41)$$

we obtain

$$\Delta x \, \Delta\bar{\nu} = 1 \qquad (2.42)$$

Destructive interference at x_0 can also be brought about by use of a *larger* range for λ than the one employed in deriving Eq. 2.42; for example, λ_1 and λ_2 could have been chosen so that the two waves were out of phase by any odd multiple of π at $x = x_0$. Thus, the value of $\Delta\bar{\nu}$, which is related to Δx in Eq. 2.42, represents the *minimum* spread in $\bar{\nu}$ necessary to localize the wave; this fact can be expressed by writing

$$\Delta x \, \Delta\bar{\nu} \geq 1 \qquad (2.43)$$

It should be pointed out that the relation in Eq. 2.43 depends on the definition of Δx and $\Delta\bar{\nu}$. As an alternative, we could add together waves of the form $a(\bar{\nu}) \cos 2\pi\bar{\nu}x$, where $a(\bar{\nu})$ is proportional to $\exp(-\alpha\bar{\nu}^2)$ with α a constant, and $\bar{\nu}$ is allowed to range from $-\infty$ to $+\infty$. The resulting wave form is proportional to $\exp(-\pi^2x^2/\alpha)$, a "Gaussian" function centered at $x = 0$. In this case, even though both $\bar{\nu}$ and x have infinite ranges, one may define $\Delta\bar{\nu}$ and Δx as the *root-mean-square deviations*

$$\Delta\bar{\nu} \equiv [\langle(\bar{\nu} - \langle\bar{\nu}\rangle)^2\rangle]^{1/2}$$
$$\Delta x \equiv [\langle(x - \langle x\rangle)^2\rangle]^{1/2}$$

where $\langle \; \rangle$ indicates *average value*; thus, $\langle x \rangle$ is the average value of x. The uncertainty product for the Gaussian wave form can be shown to be

$$\Delta x \, \Delta\bar{\nu} = \frac{1}{4\pi^2}$$

This value differs from Eq. 2.43, so that both should be viewed as giving the order of magnitude for the minimum of $\Delta x \, \Delta\bar{\nu}$, rather than an exact lower limit.

Equation 2.43, which relates the wave number range $\Delta\bar{\nu}$ of the superposed waves required to produce a localized wave or *wave packet* and the region Δx, is a purely classical result. Only the association of the wave with the position of a particle is nonclassical. Making use of the nonclassical de Broglie relation,

$$p = \frac{h}{\lambda} = h\bar{\nu}$$

we see that a spread $\Delta \bar{\nu}$ in wave number corresponds to a spread $\Delta p = h \Delta \bar{\nu}$ in the momentum. Substituting into Eq. 2.43, we obtain

$$\Delta x \, \Delta p \geq h = 6.6 \times 10^{-27} \text{ erg sec} \tag{2.44}$$

We see that the representation of a particle by a localized wave introduces a limit on the accuracy of a simultaneous measurement of its position and momentum. If at $t = 0$ we measure x with an uncertainty limit Δx, then our knowledge of p at $t = 0$ can be no better than the uncertainty Δp, where

$$\Delta p \geq \frac{h}{\Delta x}$$

If one wishes to know the position *exactly* (i.e., $\Delta x = 0$), one must forego any knowledge whatsoever of the momentum (i.e., $\Delta p \to \infty$). Conversely, an exact measurement of the momentum is made at the expense of a complete uncertainty in the position.

Suppose that instead of localizing the particle in space at a particular time as we have done in the above analysis, we look for the particle at a certain point, say $x = 0$, and record the time interval Δt during which the particle can be found at that point. To build up a wave localized in the time interval Δt, we use sine waves of the form $a(0, t) = a_0 \cos 2\pi\nu t$ with a range of values for the frequency ν. The superposition procedure, which is identical to the one employed above for space localization (see Problem 2.12), leads to the result

$$\Delta \nu \, \Delta t \geq 1$$

With the relation $E = h\nu$, we have

$$\Delta E \, \Delta t \geq h$$

This implies that a system must exist for a long time if we wish to determine its energy with great accuracy, in agreement with Bohr's concept of discrete energy levels associated with stationary (long-lived) states. Moreover, for unstable systems (such as dissociating molecules) that exist for short times, the energy levels are imprecisely defined ($\Delta E \geq h/\Delta t$, where Δt is the lifetime) and a diffuse spectrum is observed(see Section 7.7).

The principle that one cannot simultaneously measure both the coordinate and the momentum exactly is called the *indeterminacy principle* or the Heisenberg *uncertainty principle*, after W. Heisenberg, who first demonstrated its existence and pointed out its fundamental importance. The uncertainty principle shows us why classical theory is not applicable to phenomena on a microscopic (i.e., atomic or subatomic) scale. As we saw in Chapter 1, Newton's equation of motion for a particle (Eq. 1.1) assumes that one can measure exactly the coordinates and momentum of the particle at a certain time, say, t_0. Once these "initial conditions" are specified, the coordinates and momentum of the particle for any future time ($t > t_0$) can

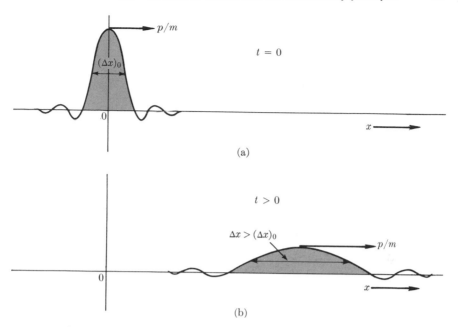

Fig. 2.15. Spreading of a wave packet with time.

be predicted exactly by integrating Eq. 1.1 from t_0 to t. Correspondingly, one can integrate from t_0 to some time $t < t_0$ and obtain the complete past history of the particle as well. The uncertainty principle denies the possibility of doing this, no matter how precise our measuring devices may be. Suppose we have made a measurement that shows the electron to be somewhere between $-x_0$ and x_0 at $t = 0$ (Fig. 2.15a) and we wish to determine where it will be at some later time. Because of the uncertainty in the momentum, $\Delta p \geq h/\Delta x$, we do not know how far the particle will move in the given time interval. As time progresses, the uncertainty in position increases; after an interval t, the uncertainty is $\Delta x + (\Delta p/m)t$ and the wave associated with the particle has spread out over a large range (Fig. 2.15b). The fundamental concept of classical mechanics that a particle is known to follow a unique path, which is completely determined by the initial conditions and the forces, is thus rendered untenable by the uncertainty principle.

That Newtonian mechanics is applicable to the macroscopic world, in spite of the limitations set by the uncertainty principle, is a consequence of the fact that Planck's constant h is very small relative to macroscopic quantities. From Eq. 2.44 the simultaneous uncertainties in measurements of position and momentum can be as small as 10^{-13} cgs units. For planets or billiard balls, such uncertainties are far below the possible accuracy of measurement. Consequently, the uncertainty principle imposes no practical

mathematical machinery. We wish to find a quantitative method for determining the functional form of the wave for a particle in a variety of experimental situations. Having obtained the wave form, we need to define clearly how it is related to the probability concept, which permits us to extract physically meaningful information. We would, for example, like to be able to calculate the wave associated with the electron in a hydrogen atom and to use it to obtain the probability of finding the electron at certain positions.

The principle of *superposition* provides the clue as to how we should proceed. We have noted that arbitrary wave forms can be built up from sine waves. Thus, if we can obtain an equation which is obeyed by sine waves, we may be able to generalize it for the problem of finding arbitrary wave forms. Certain differential equations have the property that the sum of two or more solutions is also a solution; that is, their solutions obey the principle of superposition. We begin, therefore, by finding an equation of this type (called a linear homogeneous differential equation) which is obeyed by the simple sine waves representing nonlocalized free particles in the absence of potential energy (see Section 2.2). The function given in Eq. 2.1 is such a wave with velocity $\lambda \nu$. However, as we saw from the discussion of the de Broglie picture of the hydrogen atom, we are primarily interested not in traveling waves of this type, but rather in the stationary waves which are associated with stationary states. To form a stationary wave, we superpose two traveling waves which are moving in opposite directions; that is, if we add the waves

$$a_1(x, t) = a_0 \cos\left[2\pi\left(\frac{x}{\lambda} - \nu t\right)\right] \qquad a_2(x, t) = a_0 \cos\left[2\pi\left(\frac{x}{\lambda} + \nu t\right)\right]$$

we obtain

$$A(x, t) = a_1(x, t) + a_2(x, t) = 2a_0 \cos\left(2\pi\frac{x}{\lambda}\right)\cos\left(2\pi\nu t\right) \qquad (2.51)$$

The wave $A(x, t)$ is seen to be factored into coordinate-dependent and time-dependent parts. Inspection of the right-hand side of Eq. 2.51 shows that the *magnitude* of the amplitude at a given point fluctuates with time, but the positions of the maxima and minima remain fixed in space (see Fig. 2.16). Since we are not interested in the time dependence of the amplitude for most applications (for the time-dependent treatment, see Section 3.11), we consider only the time-independent part of Eq. 2.51,

$$A(x) = a_0 \cos\left(2\pi\frac{x}{\lambda}\right) \qquad (2.52)$$

To obtain the differential equation satisfied by $A(x)$, we differentiate twice with respect to x,

$$\frac{d^2A(x)}{dx^2} = -a_0\left(\frac{2\pi}{\lambda}\right)^2\cos\left(2\pi\frac{x}{\lambda}\right) = -\left(\frac{2\pi}{\lambda}\right)^2 A(x) \tag{2.53}$$

which is the classical time-independent wave equation. Equation 2.53 could be solved immediately for $A(x)$ if we did not already know its form. To transform Eq. 2.53 into a quantum-mechanical equation, we proceed as with the uncertainty principle and introduce the momentum p for the wavelength by means of the de Broglie relation $(1/\lambda = p/h)$; thus,

$$\frac{d^2A(x)}{dx^2} = -\frac{p^2}{\hbar^2}A(x)$$

where $\hbar = h/2\pi$. When a wave amplitude (or *wave function*) such as $A(x)$ is used to describe a de Broglie wave, it is usually given the symbol $\psi(x)$. For de Broglie waves associated with a free particle, we have, therefore,

$$\frac{d^2\psi(x)}{dx^2} = -\frac{p^2}{\hbar^2}\psi(x) \tag{2.54}$$

Since the energy E of a free particle is entirely kinetic energy (i.e., there are no forces acting on the particle so that there is no potential energy), we can write

$$E = \tfrac{1}{2}mv^2 = \frac{p^2}{2m} \tag{2.55}$$

or

$$p^2 = 2mE$$

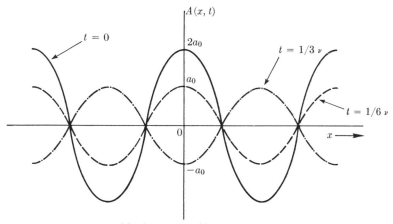

$$A(x, t) = 2a_0 \cos\left[(2\pi x)/\lambda\right] \cos\left(2\pi\nu t\right)$$

Fig. 2.16. Standing wave constructed from two plane waves moving in opposite directions.

Substituting for p^2 in Eq. 2.54 by means of Eq. 2.55, we obtain

$$\frac{d^2\psi(x)}{dx^2} = -\frac{2mE}{\hbar^2}\psi(x)$$

or, rearranging,

$$-\frac{\hbar^2}{2m}\frac{d^2\psi(x)}{dx^2} = E\psi(x) \tag{2.56}$$

Equation 2.56 is called the *Schrödinger amplitude equation* for a free particle, because Schrödinger first formulated quantum mechanics in terms of such an equation and its more generally applicable form (see Section 2.9). If we were given Eq. 2.56 without any knowledge of its solution, there would be two quantities that we would have to determine: the wave function ψ and the energy E. How this is done is most easily demonstrated by solving some examples.

2.6.1 Solution of the Schrödinger equation for a free particle   The first system that we consider is that of a free particle with no potential energy. This is a trivial example, since we expect to obtain the wave form from which Eq. 2.56 was derived; however, working out the solution will help us to see how Eq. 2.56 is used. We begin by introducing the constant k, defined by

$$k^2 = \frac{2mE}{\hbar^2}$$

in terms of which Eq. 2.56 can be written

$$\frac{d^2\psi(x)}{dx^2} = -k^2\psi(x) \tag{2.57}$$

A standard and widely used method for solving differential equations can be applied to Eq. 2.57. If we do not know the form of the solution, we assume that it can be written as a power series

$$\psi(x) = a_0 + a_1x + a_2x^2 + a_3x^3 + a_4x^4 + a_5x^5 + \cdots \tag{2.58}$$

and attempt to determine the coefficients a_0, a_1, a_2 by means of the differential equation. Differentiating Eq. 2.58, we obtain

$$\frac{d\psi}{dx} = a_1 + 2a_2x + 3a_3x^2 + 4a_4x^3 + 5a_5x^4 + \cdots$$

and

$$\frac{d^2\psi}{dx^2} = 2a_2 + 6a_3x + 12a_4x^2 + 20a_5x^3 + \cdots \tag{2.59}$$

Substituting Eqs. 2.58 and 2.59 into Eq. 2.57, we find

$$2a_2 + 6a_3x + 12a_4x^2 + 20a_5x^3 + \cdots$$
$$+ k^2(a_0 + a_1x + a_2x^2 + a_3x^3 + \cdots) = 0$$

or, collecting the coefficients of each power of x,

$$(2a_2 + k^2a_0) + (6a_3 + k^2a_1)x + (12a_4 + k^2a_2)x^2$$
$$+ (20a_5 + k^2a_3)x^3 + \cdots = 0 \qquad (2.60)$$

If the left-hand side of Eq. 2.60 is to be zero for every value of x, the co-efficients of all powers of x must vanish; that is,

$$x^0: 2a_2 + k^2a_0 = 0 \qquad x^1: 6a_3 + k^2a_1 = 0$$
$$x^2: 12a_4 + k^2a_2 = 0 \qquad x^3: 20a_5 + k^2a_3 = 0$$

or, in general, for the coefficient of x^{n-2},

$$n(n-1)a_n + k^2a_{n-2} = 0 \qquad (2.61)$$

Equation 2.61 provides a relationship called a *recursion formula*, between the coefficients a_n and a_{n-2}; that is, if a_{n-2} is known, a_n is given by

$$a_n = \left(-\frac{k^2}{n(n-1)}\right)a_{n-2} \qquad (2.62)$$

Since $a_{n-2}, a_{n-4}, \ldots$ must also obey Eq. 2.62, we can write

$$a_n = \left(-\frac{k^2}{n(n-1)}\right)a_{n-2} = \left(-\frac{k^2}{n(n-1)}\right)\left(-\frac{k^2}{(n-2)(n-3)}\right)a_{n-4}$$
$$= \left(-\frac{k^2}{n(n-1)}\right)\left(-\frac{k^2}{(n-2)(n-3)}\right)\left(-\frac{k^2}{(n-4)(n-5)}\right)a_{n-6}$$
$$= \cdots$$

Thus, we can express all of the coefficients in the power series for $\psi(x)$ (Eq. 2.58) in terms of the first two, a_0 and a_1; that is, if a_0 is given, we obtain $a_2, a_4, a_6, \ldots$ by repeated use of Eq. 2.62; similarly, if a_1 is given, we obtain $a_3, a_5, a_7, \ldots$. Grouping even and odd power terms, we find

$$\psi(x) = a_0\left[1 + \left(-\frac{k^2}{2\cdot 1}\right)x^2 + \left(-\frac{k^2}{4\cdot 3}\right)\left(-\frac{k^2}{2\cdot 1}\right)x^4 + \cdots\right]$$
$$+ a_1\left[x + \left(-\frac{k^2}{3\cdot 2}\right)x^3 + \left(-\frac{k^2}{5\cdot 4}\right)\left(-\frac{k^2}{3\cdot 2}\right)x^5 + \cdots\right]$$
$$= a_0\left[1 - \frac{(kx)^2}{2!} + \frac{(kx)^4}{4!} - \frac{(kx)^6}{6!} + \cdots\right] \qquad (2.63)$$
$$+ a_1'\left[kx - \frac{(kx)^3}{3!} + \frac{(kx)^5}{5!} - \frac{(kx)^7}{7!} + \cdots\right]$$

where we have replaced the constant a_1 by $a_1' = a_1/k$ in the second series. Examination of the two power series appearing in Eq. 2.63 shows that they are just the Maclaurin expansions of $\cos kx$ and $\sin kx$, respectively. Thus, we can write

$$\psi(x) = a_0 \cos kx + a_1' \sin kx \qquad (2.64)$$

That is, by expanding $\psi(x)$ in a power series and determining a recursion relation for the coefficients, we have been able to demonstrate that the general solution is a sum of the known functions $\cos kx$ and $\sin kx$. Readers who have met Eq. 2.57 previously will have been able to write the solution directly. By studying the example in detail, however, we have shown how one can proceed if the solution is *not* known. Additional examples appear later in this chapter (the harmonic oscillator) and in Chapter 3 (the hydrogen atom).

Equation 2.64 for $\psi(x)$ contains the constants a_0 and a_1', which are the two arbitrary constants of integration expected for a second-order differential equation. To evaluate these constants, restrictions on the solutions that are not contained in the differential equation itself must be introduced. These restrictions arise from the physical nature of the problem under discussion. One of the constants, a_0 or a_1', could be determined from a knowledge of the value of $\psi(x)$ at some point along the x axis. For example, if we choose the origin of our coordinate system so that the wave function has a maximum at $x = 0$, then a_1' must be zero, and the wave function has the form $a_0 \cos kx$. To see this, let us impose upon $\psi(x)$ the condition for a maximum at $x = 0$; that is, we require that the first derivative of ψ vanish at $x = 0$. Differentiation of Eq. 2.64 gives

$$\frac{d\psi}{dx} = -ka_0 \sin kx + ka_1' \cos kx$$

If we now set x equal to zero, we obtain the requirement that

$$\left.\frac{d\psi}{dx}\right|_{x=0} = -ka_0 \sin (0) + ka_1' \cos (0) = ka_1' \qquad (2.65)$$

For a maximum in ψ at $x = 0$,

$$\left.\frac{d\psi}{dx}\right|_{x=0} = ka_1' = 0$$

and the constant a_1' is zero, so that $\psi(x)$ has the form $a_0 \cos kx$. If instead, we had required that ψ have a node (i.e., that it cross the axis) at $x = 0$, we would have obtained

$$\psi(0) = a_0 \cos (0) + a_1' \sin (0) = a_0 = 0 \qquad (2.66)$$

which leads to the result that the constant a_0 is zero; consequently, the wave function in this case has the form $a_1' \sin kx$. Such information about the value of the wave function or its derivative at a certain point is called a *boundary condition*. As we have just seen, imposing boundary conditions allows us to specify the wave function more completely than is possible by means of the Schrödinger equation alone.

For a free particle, we have no knowledge of its position, so that arbitrary relative values of a_0 and a_1' are possible. Moreover, the boundary conditions impose no restrictions upon the value of k and the energy given by

$$E = \frac{\hbar^2 k^2}{2m} \qquad (2.67)$$

can assume any value. The energy of a free particle is therefore continuous, rather than quantized as in the case of the Bohr atom or the particle in a box discussed below.

2.6.2 Particle in a one-dimensional box

Another simple, but very important application of Eq. 2.56 is to a particle of mass m in the presence of the potential-energy function shown in Fig. 2.17a. The function forms a one-dimensional potential-energy "box" of length l that is bounded at $x = 0$ and at $x = l$ by infinitely high potential-energy walls; inside the box the potential energy is zero. This means that the particle (an electron, for example) is confined to the box by the potential walls; hence, the term *particle-in-a-box* is often used to describe this problem. Since the particle must be between $x = 0$ and $x = l$, the wave function associated with the position of the particle is zero outside the box; that is, $\psi(x) = 0$ for $x < 0$ and $x > l$. This reduces the problem to that of determining the form of the wave function inside the box. Because the potential energy is zero inside the box ($0 \leq x \leq l$), Eq. 2.56 is applicable in this region and the general solution for $\psi(x)$ is given by Eq. 2.63, with $k^2 = 2mE/\hbar^2$ as before.

We now apply the boundary conditions appropriate for this problem to obtain the solution ψ in a more definite form; that is, we wish to evaluate a_0, a_1', and k (or a_0, a_1', and E). We know that $\psi(x)$ is zero for $x < 0$ and $x > l$ because the infinite potential energy denies the particle access to these regions. Furthermore, since $\psi(x)$ is related to the probability distribution for the particle, it is physically reasonable that we require $\psi(x)$ to be a continuous function of x. This continuity condition, which is a general one in quantum mechanics, requires in the present case that the wave function obey the boundary conditions

$$\psi(0) = \psi(l) = 0 \qquad (2.68)$$

Application of Eq. 2.68 to the wave function given by Eq. 2.64 leads to

$$\psi(0) = a_0 \cos (0) + a_1' \sin (0) = a_0 = 0 \qquad (2.69)$$

and

$$\psi(l) = a_1' \sin kl = 0 \qquad (2.70)$$

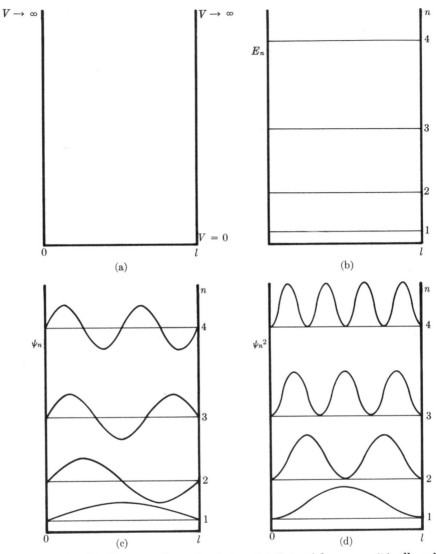

Fig. 2.17. Particle in a one-dimensional box. **(a)** Potential energy; **(b)** allowed energy levels; **(c)** wave functions ψ_n; **(d)** probability densities ψ_n^2.

Equation 2.69 is satisfied by setting a_0 equal to zero; that is, the cosine term cannot contribute to the particle-in-a-box solution and the wave function takes the form

$$\psi(x) = a_1' \sin kx \qquad (2.71)$$

To satisfy Eq. 2.70, we could choose $a_1' = 0$. However, the result would be a wave function which is zero everywhere; that is, we would have no particle at all in the box. The alternative is to require that $\sin kl = 0$, which occurs for

$$kl = \pi, 2\pi, 3\pi, \ldots$$

Thus, k is limited to the discrete values

$$k = \frac{\pi}{l}, \frac{2\pi}{l}, \frac{3\pi}{l}, \ldots$$

or, in general,

$$k = \frac{n\pi}{l} \qquad n = 1, 2, 3, \ldots \tag{2.72}$$

The value $k = 0$ has been excluded because its insertion into Eq. 2.71 yields a wave function that is zero for all values of x.

The requirement that the particle be confined to the box [i.e., that $\psi(x) = 0$ for $x < 0$ and $x > l$] and the continuity condition [i.e., that $\psi(0) = \psi(l) = 0$] have determined the constant a_0 and restricted the values of k. From Eqs. 2.71 and 2.72, we see that the wave function is

$$\psi_n(x) = a_1' \sin \frac{n\pi x}{l} \qquad 0 \le x \le l; \, n = 1, 2, 3, \ldots$$
$$\psi_n(x) = 0 \qquad\qquad x < 0, \, x > l \tag{2.73}$$

where the subscript n on the left-hand side indicates that the function depends upon the value assigned to the integer n. The constant a_1', which could depend upon n, must still be evaluated to completely characterize the wave function. Before doing so (Section 2.6.3), we note an important implication of Eq. 2.72. Since the constant k is related to the energy E by Eq. 2.67, we can substitute from Eqs. 2.72 into 2.67 to obtain

$$E_n = \frac{n^2 h^2}{8ml^2} \qquad n = 1, 2, 3, \ldots \tag{2.74}$$

where the subscript n indicates the dependence of E upon the value assigned to this integer. Thus E, as well as k, is limited to certain discrete values; that is, the energy of the particle-in-a-box is *quantized*. The integer n which specifies the wave function, or state, and the associated energy is called the *quantum number*. It is important to note that, in contrast to the method used by Bohr in treating the hydrogen atom, the discreteness of the energy levels has not been introduced by a specific postulate. Instead, the quantization is a direct consequence of the application of the appropriate boundary conditions to the solution of the Schrödinger equation. The stationary-state wave functions and energy values that satisfy the boundary conditions are often referred to as *eigenfunctions* and *eigenvalues*, respectively.

It is of interest to ascertain whether the results for the problem of the particle in a one-dimensional box are consistent with the uncertainty principle. Since we know that the particle must be somewhere inside the box, the uncertainty in its position Δx is just equal to the size of the box; that is, $\Delta x = l$. According to Eq. 2.44, the uncertainty limit on the momentum of such a particle is $\Delta p \geq h/l$. If we take the minimum momentum of the particle to be on the order of the uncertainty itself (that is, if we take $p_{min} \simeq h/l$) then the minimum kinetic energy, which equals the total energy in this case, is

$$E_{min} \simeq \frac{p_{min}^2}{2m} \simeq \frac{h^2}{2ml^2} \tag{2.75}$$

From Eq. 2.74, the lowest allowed state for a particle confined to a box of length l is obtained by setting $n = 1$,

$$E_1 = \frac{h^2}{8ml^2} \tag{2.76}$$

Thus, the two results for the minimum energy are of the same order of magnitude. As we have already mentioned, Eq. 2.44 is only approximate; in the present case, if the minimum uncertainty product were written as $h/2$ instead of h, the exact energy would have been obtained. The important result is that the form of Eq. 2.75 is the same as Eq. 2.76; that is, the minimum energy, whether from the uncertainty principle or the Schrödinger equation, varies inversely with the mass of the particle and the square of the size of the box. The qualitative implications of this result are quite general in quantum theory: Confining a particle to a smaller region (e.g., reducing the value of l) leads to an increase in momentum and kinetic energy. Like quantization, the uncertainty principle and its consequences follow naturally from the Schrödinger equation and the continuity condition.

2.6.3 *Probability density; normalization*

To complete the determination of the particle-in-a-box wave function, we must evaluate the constant a_1' in Eq. 2.73. Since a_1' is arbitrary as far as the restrictions imposed by the Schrödinger equation and the boundary conditions are concerned, we seek an additional condition which the wave function must satisfy. This new condition is intimately connected with the probability interpretation of the wave function ψ. We have seen already that it is possible to relate the amplitude of ψ in a certain region to the probability of finding the particle in that region: The probability is large where ψ is large and small where ψ is small. However, the form of Eq. 2.73 shows that ψ inside the box is both positive and negative, whereas a probability is always a positive quantity; the probability can be zero in some regions (as it is outside the box) but it can never be negative. This rules out a simple proportionality between ψ and the probability.

The correct relation is suggested by our previous discussion of electromagnetic waves. Classical radiation theory shows that the energy per unit volume is proportional to the square of the wave amplitude (see Eq. 2.4). Since the radiant energy is equal to the photon energy $h\nu$ times the number of photons, the number density (i.e., the number per unit volume) of photons in the neighborhood of a point in space is proportional to the square of the amplitude of the electromagnetic wave at that point. Extension of this concept to material particles suggests that their number density should be proportional to the square of the wave function. Alternatively, if the wave function ψ is to be associated with a single particle, the square, ψ^2, is to be interpreted quantitatively as the *probability per unit volume* or the *probability density* for the particle. Use of ψ^2, rather than ψ itself, clearly limits the probability to positive values, as long as one is using real functions; for complex wave functions, the probability expression has to be generalized (see Section 3.8).

For the particle-in-a-box, we have from Eq. 2.73 (writing A for a_1')

$$\psi_n{}^2(x) = A^2 \sin^2\left(\frac{n\pi x}{l}\right)$$

where in this one-dimensional problem the probability density $\psi_n{}^2(x)$ is defined as the probability per unit length of finding the particle at x. Equivalently, the function

$$\psi_n{}^2(x)dx = A^2 \sin^2\left(\frac{n\pi x}{l}\right) dx \tag{2.77}$$

is the probability of finding the particle in the range x to $x + dx$. Integration of $\psi_n{}^2(x)dx$ over all values of x must give unity since the particle must be *somewhere;* that is,

$$\int_{-\infty}^{+\infty} \psi_n{}^2(x)dx = 1 \tag{2.78}$$

Applying this condition to the particle-in-a-box, we see that the constant A has to be chosen so as to satisfy Eq. 2.78. As the wave function is nonzero only between 0 and l, we can limit the integration interval to this range and write

$$\int_0^l \psi_n{}^2(x)dx = A^2 \int_0^l \sin^2\left(\frac{n\pi x}{l}\right) dx = 1 \tag{2.79}$$

Evaluation of the integral in Eq. 2.79 gives

$$\int_0^l \sin^2\left(\frac{n\pi x}{l}\right) dx = \frac{l}{n\pi} \int_0^{n\pi} \sin^2 y \, dy = \frac{l}{n\pi} \int_0^{n\pi} \tfrac{1}{2}(1 - \cos 2y)dy$$

$$= \frac{l}{n\pi}\left(\frac{n\pi}{2}\right) = \frac{l}{2} \tag{2.80}$$

Consequently, Eq. 2.78 will be satisfied if $A = (2/l)^{1/2}$. The complete form of the wave function for this system is thus

$$\psi_n(x) = \left(\frac{2}{l}\right)^{1/2} \sin\left(\frac{n\pi x}{l}\right) \qquad 0 \leq x \leq l \tag{2.81}$$

$$\psi_n(x) = 0 \qquad x < 0 \text{ or } x > l$$

The process of evaluating the constant A by means of Eq. 2.78 is referred to as *normalization* of the wave function, and A is called the normalization constant.

The interpretation of ψ^2 as the probability density is not an obvious one, even though it can be rationalized in the manner that we have used. However, the fundamental probability assumption, originally introduced by M. Born, has been shown to be correct by the agreement between calculated and experimental results for a large number of different systems. The pragmatic qualities of physical theory again come into play: The probability density assumption, like other assumptions upon which the quantum theory is based, is adopted because it gives valid results.

Having completed the solution of the particle-in-a-box problem, we consider the energies, wave functions, and probability densities for the first few states. In Fig. 2.17b we have plotted the energy levels for $n = 1, 2, 3$, and 4. Plots of the functions ψ_n and ψ_n^2 versus x for each of these states are shown in Figs. 2.17c and 2.17d.

It is evident that, as n increases and the energy goes up (quadratically in n), the wave functions oscillate with higher frequencies and shorter wavelengths. In fact, λ is simply $2l/n$; inserting this result into the de Broglie formula, we have

$$p = \frac{h}{\lambda} = \frac{nh}{2l} \qquad E = \frac{p^2}{2m} = \frac{n^2 h^2}{8ml^2}$$

in agreement with Eq. 2.74. The exact equivalence of the energy determined by the de Broglie relation and the Schrödinger equation holds only when the wavelength is constant, as it is for the present case and in similar problems with constant or zero potential energy. However, a more qualitative correspondence between wavelength and momentum or kinetic energy holds generally. One way of looking at this is in terms of the number of nodes (points at which the wave function is zero, other than the boundary points) in the wave function. We see from Fig. 2.17c that the ground (lowest energy) state ($n = 1$) is nodeless, the first excited state ($n = 2$) has one node, the second ($n = 3$) has two nodes, and so on; that is, for the state associated with the quantum number n, there are $n - 1$ nodes. Thus, the energy of the system increases with the number of nodes, a high energy corresponding to a large number of nodes.

From Fig. 2.17d it is clear that the probability of finding the particle in an interval x to $x + dx$ is not uniform throughout the box and changes as a function of the state of the system. For the ground state $(n = 1)$ the probability has a maximum in the center $(x = l/2)$, while for the first excited state $(n = 2)$ the probability is zero at $x = l/2$ and is a maximum at $l/4$ and $3l/4$; and so on. As n increases, these regions of high probability become more numerous, until the classical limit of equal probability throughout the box is approached, except for the high-frequency periodic fluctuations. When n is very large, the spacing between the nodes is smaller than the shortest measurable distance, so that quantum effects disappear in accordance with the correspondence principle.

Although the particle-in-a-box is a very simple problem and may seem far removed from atoms and molecules, it serves as a valid, though approximate, model for a number of real systems. One application is to the theory of metals, for which the particle-in-a-box treatment is called the *free-electron* model. For conjugated molecules, a similar free-electron approximation can be used (Section 7.11.3). Also, the behavior of monatomic gas atoms can be determined by assuming that they are restricted to a potential box. We shall discuss some of these applications in subsequent chapters of this book. They are mentioned here only to emphasize that model problems play a very important role in the quantum theory. Since many of the systems of interest are too complicated for exact calculations, the qualitative information that can be obtained from simplified models is extremely useful.

2.7   THE CALCULATION OF AVERAGE VALUES

From the probability interpretation of the wave function, we know that $\psi_n^2(x)dx$ is the probability of finding the particle in the range x to $x + dx$. To clarify the significance of such a probability, we consider a large number of identical particle-in-a-box systems, every one of which is in the state $n = 2$. We divide the region $0 \leq x \leq l$ into N segments Δx_1, Δx_2, . . . , Δx_N of equal width, the midpoints of the segments being x_1, x_2, . . . , x_N, respectively (see Fig. 2.18a). We determine the segment of the x axis in which each of the particles is located by measuring its position. If we were to plot as a fraction the number of times S_i a particle is found in the ith segment, relative to the total number of measurements M, we might find the result shown in Fig. 2.18b. Taking the limit as the segments $\Delta x_i \rightarrow 0$ and as the number of segments and the number of systems measured go to infinity. we expect to obtain the curve $\psi_2^2(x)$ shown in Fig. 2.17d.

The square of the wave function, which gives the probability distribution, can be used to evaluate the average value of the particle coordinate or some function of the coordinate. For the case discussed above where we

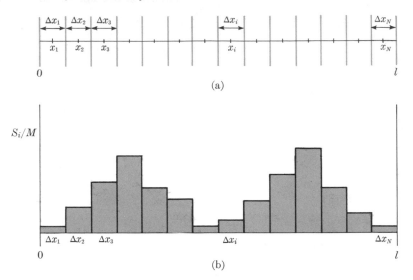

Fig. 2.18. Determination of the average position of a particle.

divide the range of x into finite intervals Δx, the average value of x for the set of M measurements is

$$\langle x \rangle = \frac{S_1}{M} x_1 + \frac{S_2}{M} x_2 + \cdots + \frac{S_N}{M} x_N = \sum_{i=1}^{N} \frac{S_i}{M} x_i \qquad (2.82)$$

From the interpretation of $\psi_2{}^2$ as the probability density, the fraction of the time that the particle is expected to be found in the ith segment is

$$\frac{S_i}{M} = \psi_2{}^2(x_i)\Delta x_i \qquad (2.83)$$

Equations 2.82 and 2.83 give the calculated quantum-mechanical average or expectation value $\langle x \rangle$ as

$$\langle x \rangle = \sum_{i=1}^{N} \psi_2{}^2(x_i)x_i\Delta x_i \qquad (2.84)$$

This value for $\langle x \rangle$ is only approximate, since the segments are of finite width. If we again take the limit as the segments become narrower ($\Delta x \to 0$) and larger in number ($N \to \infty$), we obtain for a probability density ψ^2

$$\langle x \rangle = \lim_{\substack{\Delta x \to 0 \\ N \to \infty}} \sum_{i=1}^{N} \psi^2(x_i)x_i\Delta x_i = \int_{-\infty}^{+\infty} \psi^2(x)x \, dx$$

or, as it is often written (for the reason given below),

$$\langle x \rangle = \int_{-\infty}^{+\infty} \psi(x) x \, \psi(x) \, dx \tag{2.85}$$

A similar argument shows that the average or expectation value of x^2 is

$$\langle x^2 \rangle = \int_{-\infty}^{+\infty} \psi(x) x^2 \, \psi(x) \, dx \tag{2.86}$$

Generalizing this result to a function $f(x)$, we have

$$\langle f(x) \rangle = \int_{-\infty}^{+\infty} \psi(x) f(x) \psi(x) \, dx \tag{2.87}$$

In addition to the position of the particle, we may wish to obtain information about its momentum and energy. For a system in a given stationary state, the energy is known exactly from the solution of the Schrödinger equation, as we have seen from the particle-in-a-box problem. However, as suggested by the uncertainty principle, the momentum, like the position, is usually not completely specified, and average or expectation values are of interest. From the introduction of the wave equation for the free particle by means of the de Broglie relation, we have (Eq. 2.54)

$$-\hbar^2 \frac{d^2\psi(x)}{dx^2} = p^2 \psi(x)$$

that is, differentiation of the free-particle wave function twice with respect to x and multiplication by $-\hbar^2$ yields the square of the momentum times the wave function. Thus, the operation which can be symbolized by $-\hbar^2 \, d^2/dx^2$ corresponds in some way to p^2. Moreover, for the momentum itself (i.e., the square root of p^2), the result suggests that the operation corresponding to the square root of $-\hbar^2 \, d^2/dx^2$ [i.e., $i\hbar \, d/dx$ or $(\hbar/i) \, d/dx$, where $i = \sqrt{-1}$] may be used. Although such reasoning does not provide a derivation, comparison of calculations with experimental results has shown that the association of the *differential operator* $(\hbar/i)d/dx$ with the momentum yields correct results in the quantum theory. In particular, the average value of the momentum expected from measurements on a large number of identical systems, each with the wave function $\psi(x)$, is

$$\langle p \rangle = \int_{-\infty}^{+\infty} \psi(x) \left(\frac{\hbar}{i} \frac{d}{dx} \right) \psi(x) \, dx = \int_{-\infty}^{+\infty} \psi(x) \frac{\hbar}{i} \frac{d\psi}{dx} \, dx \tag{2.88}$$

and for measurements of p^2,

$$\langle p^2 \rangle = \int_{-\infty}^{+\infty} \psi(x) \left(\frac{\hbar}{i} \frac{d}{dx} \right)^2 \psi(x) \, dx = \int_{-\infty}^{+\infty} \psi(x) \left(-\hbar^2 \frac{d^2\psi(x)}{dx^2} \right) dx \tag{2.89}$$

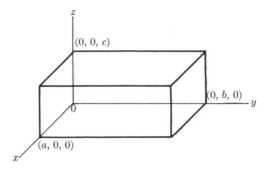

Fig. 2.19. Particle in a three-dimensional box.

As a simple illustration, Eq. 2.94 can be applied to the motion of a particle in a rectangular box of sides a, b, and c (see Fig. 2.19). The potential energy is zero inside the box and infinite outside, so that the probability of finding the particle outside is zero. Hence, the wave function $\psi(x, y, z)$ is zero except for the region

$$0 \leq x \leq a$$
$$0 \leq y \leq b$$
$$0 \leq z \leq c$$

The procedure for finding ψ inside this potential box corresponds to that for the one-dimensional problem; that is, one determines the general solution of the Schrödinger equation and applies the boundary condition that the wave function must be zero at the edge of the box. The resulting normalized wave functions and the associated values of the energy E are

$$\psi_{n_x,n_y,n_z}(x, y, z) = \left(\frac{2}{a}\right)^{1/2} \left(\frac{2}{b}\right)^{1/2} \left(\frac{2}{c}\right)^{1/2} \sin\left(\frac{n_x\pi x}{a}\right)$$
$$\times \sin\left(\frac{n_y\pi y}{b}\right) \sin\left(\frac{n_z\pi z}{c}\right) \tag{2.98}$$

and

$$E_{n_x,n_y,n_z} = \frac{h^2}{8ma^2} n_x^2 + \frac{h^2}{8mb^2} n_y^2 + \frac{h^2}{8mc^2} n_z^2 \tag{2.99}$$

where $n_x = 1, 2, \ldots$, $n_y = 1, 2, \ldots$, $n_z = 1, 2, \ldots$. Equations 2.98 and 2.99 can be verified by substituting into Eqs. 2.94 and 2.95.

For the three-dimensional problem, three quantum numbers are necessary to describe the state of the system. In the present case, each quantum number is associated with the motion along one coordinate: n_x with x, n_y with y, and n_z with z. Thus, the energy is the *sum* of kinetic energies for the three directions and the wave function is the *product* of wave functions for the three directions. The result for each individual coordinate—x, y, or z —is just that obtained previously in the one-dimensional box problem (Eq. 2.81). The relation between quantum numbers and the number of nodes is the same as for one dimension; that is, there are $n_x - 1$ nodal planes perpendicular to the x axis, $n_y - 1$ nodal planes perpendicular to the y axis, and $n_z - 1$ nodal planes perpendicular to the z axis (exclusive of the boundaries of the box). In accordance with the general quantum-mechanical result, Eq. 2.99 shows that the energy E_{n_x,n_y,n_z} increases with the number of these nodal planes.

Although many of the results for the three-dimensional box problem correspond to those for one dimension, the extra dimensionality manifests itself in an important way if the box possesses geometric symmetry. In the one-dimensional problem, the state of the system could be specified by giving the value of the quantum number or the energy since there is a one-to-one correspondence between the two. In the three-dimensional case specialized to a cubic potential box with $c = b = a$, the state (i.e., the wave function) is not determined by specifying the energy, except in a few instances. To illustrate, we write the equation for the energy of a particle in the cubic box as

$$E_{n_x,n_y,n_z} = \frac{h^2}{8ma^2} (n_x^2 + n_y^2 + n_z^2) \tag{2.100}$$

For the ground state, $n_x = n_y = n_z = 1$, the energy is

$$E_{1,1,1} = \frac{3h^2}{8ma^2} \tag{2.101}$$

For the first excited state, the set of quantum numbers $(n_x, n_y, n_z) = (2, 1, 1)$ leads to the result

$$E_{2,1,1} = \frac{6h^2}{8ma^2} \tag{2.102}$$

However, the same energy is also given by the set of quantum numbers $(1, 2, 1)$ and $(1, 1, 2)$; that is,

$$E_{2,1,1} = E_{1,2,1} = E_{1,1,2} \tag{2.103}$$

In other words, three different wave functions (i.e., three different states of the system) correspond to the same energy. One of the functions, $\psi_{2,1,1}(x, y, z)$, has a node along the x axis, the second, $\psi_{1,2,1}(x, y, z)$, has a node along the y axis, and the third, $\psi_{1,1,2}(x, y, z)$, has a node along the z axis. This situation is described by saying that *the energy level is threefold degenerate*, or that the degeneracy g is equal to 3. The ground-state energy, which corresponds to only one wave function (i.e., the one for which $n_x = n_y = n_z = 1$), is said to be *nondegenerate*, or to have a degeneracy $g = 1$.

Other nondegenerate states also exist; for example, the energy level $27h^2/8ma^2$ corresponds to the unique state for which $n_x = n_y = n_z = 3$. Table 2.1 lists the degeneracies of the lowest energy states of a particle in a cubic box of side a.

Table 2.1

State n_x n_y n_z	Energy [units of $h^2/(8ma^2)$]	Degeneracy g
1 1 1	3	1
2 1 1	6	
1 2 1	6	3
1 1 2	6	
1 2 2	9	
2 1 2	9	3
2 2 1	9	
3 1 1	11	
1 3 1	11	3
1 1 3	11	
2 2 2	12	1
1 2 3	14	
1 3 2	14	
2 1 3	14	6
3 1 2	14	
2 3 1	14	
3 2 1	14	
3 2 2	17	
2 3 2	17	3
2 2 3	17	
4 1 1	18	
1 4 1	18	3
1 1 4	18	

The source of the degeneracy in the present system is the fact that the three dimensions of the potential box are equal. Thus, there is no physical difference between the three directions in space, and a node along one coordinate direction is equivalent to a node along another axis. The fact that one is called the x axis and another the y axis cannot make any difference in the energy associated with the wave function, the definition of coordinate axes being arbitrary. If the box dimensions are different ($a \neq b \neq c$), each direction is unique and no degeneracy is expected, in general.

The three-dimensional box is often used as a model for the motion of ideal gas molecules in a closed vessel. In that case, the concept of degeneracy plays an important role. Degeneracies also arise for many other systems, including all atoms and the molecules for which there is some spatial symmetry.

2.9 THE WAVE EQUATION FOR A PARTICLE WITH POTENTIAL ENERGY

The Schrödinger equation that we have been using up to this point is applicable only to problems in which the potential energy is zero everywhere or zero except at the walls of a box, so that the potential energy enters the problem only by way of the boundary conditions. In atoms and molecules, which are the primary concern of chemistry, the potential energy of the electrons is a continuous function of the coordinates. To apply quantum mechanics to such problems, we must generalize Eq. 2.94 so that a variable potential energy $V(x, y, z)$ is included.

Since Eq. 2.94 was obtained for cases in which the energy E is entirely kinetic energy T, the equation can be written

$$-\frac{\hbar^2}{2m}\left(\frac{\partial^2 \psi}{\partial x^2} + \frac{\partial^2 \psi}{\partial y^2} + \frac{\partial^2 \psi}{\partial z^2}\right) = T\psi \tag{2.104}$$

If a potential energy V is present the total energy E is the sum of the kinetic and potential energies, and we can write

$$T = E - V \tag{2.105}$$

Introducing this expression for T in Eq. 2.104, we have

$$-\frac{\hbar^2}{2m}\left(\frac{\partial^2 \psi}{\partial x^2} + \frac{\partial^2 \psi}{\partial y^2} + \frac{\partial^2 \psi}{\partial z^2}\right) = (E - V)\psi$$

or, rearranging,

$$-\frac{\hbar^2}{2m}\left(\frac{\partial^2 \psi}{\partial x^2} + \frac{\partial^2 \psi}{\partial y^2} + \frac{\partial^2 \psi}{\partial z^2}\right) + V(x, y, z)\psi = E\psi \tag{2.106}$$

Equation 2.106 is the Schrödinger equation for a particle of mass m in the presence of a potential $V(x, y, z)$. Although we have not "derived" this equation, we shall assume that it is correct because its application to a large number of problems, including those of atomic and molecular structure, has yielded results in agreement with experiment. The question of the validity Eq. 2.106 for the interactions in atomic nuclei has not been settled, but this does not affect its use for the extranuclear chemical problems that form the subject matter of this course.

The first term on the left-hand side of Eq. 2.106 is associated with the kinetic energy of the particle, and the second term with the potential energy—in fact, to find the average value of the potential energy, we make use of Eq. 2.87 generalized to three dimensions (see Eq. 2.96) and write

$$\langle V \rangle = \int_{-\infty}^{+\infty} \int_{-\infty}^{+\infty} \int_{-\infty}^{+\infty} \psi(x, y, z) \, V(x, y, z) \, \psi(x, y, z) \, dx \, dy \, dz \quad (2.107)$$

Correspondingly, the average value of the kinetic energy is

$$\langle T \rangle = \int_{-\infty}^{+\infty} \int_{-\infty}^{+\infty} \int_{-\infty}^{+\infty} \psi(x, y, z) \left[-\frac{\hbar^2}{2m} \left(\frac{\partial^2 \psi}{\partial x^2} + \frac{\partial^2 \psi}{\partial y^2} + \frac{\partial^2 \psi}{\partial z^2} \right) \right] dx \, dy \, dz$$

$$(2.108)$$

Although the total energy of a system in a stationary state is well defined, in quantum mechanics only average values can be determined for its component parts, the kinetic energy and the potential energy.

2.10 HARMONIC OSCILLATOR

To illustrate the Schrödinger equation for a system with potential energy, we consider a one-dimensional problem. A particle of mass m is subject to a force F which is proportional to the displacement from the origin; that is,

$$-\frac{dV}{dx} = F = -kx \quad (2.109)$$

where the proportionality constant k is called the *force constant*. If we take the zero of potential energy to be at the origin $x = 0$ and integrate Eq. 2.109, we obtain the potential energy at x,

$$V(x) = \int_0^x dV = k \int_0^x x \, dx = \tfrac{1}{2} kx^2 \quad (2.110)$$

A system with this form of potential energy, which corresponds to the Hooke's-law force (Eq. 2.109), is called a *harmonic oscillator*. Such an oscillator plays an important role in the quantum theory; among other applications, it serves as a model for the vibrational motion of a diatomic molecule.

Before solving the Schrödinger equation, we introduce some useful concepts by considering the classical motion of the harmonic oscillator. From Newton's second law we have

$$F = m \frac{d^2x}{dt^2} = -kx$$

or

$$\frac{d^2x}{dt^2} = -\frac{k}{m} x \quad (2.111)$$

which is a differential equation for the particle trajectory $x = x(t)$. The form of the equation is the same as Eq. 2.57 for $\psi(x)$. Thus, the solution to Eq. 2.111 is

$$x(t) = A \sin\left[\left(\frac{k}{m}\right)^{1/2} t\right] + B \cos\left[\left(\frac{k}{m}\right)^{1/2} t\right]$$

If we assume that $x = 0$ at $t = 0$, B is zero and we obtain

$$x(t) = x_0 \sin\left[\left(\frac{k}{m}\right)^{1/2} t\right] \tag{2.112}$$

where the constant A has been replaced by x_0, which is the maximum displacement of the particle from the origin. Equation 2.112 shows that, according to classical mechanics, the position of the particle oscillates between $-x_0$ and x_0 in a sinusoidal manner. The period of the oscillation (see Section 2.2) is

$$\tau = 2\pi \left(\frac{m}{k}\right)^{1/2} \tag{2.113}$$

and the classical frequency ν_0, which is defined as the reciprocal of τ, is

$$\nu_0 = \frac{1}{2\pi} \left(\frac{k}{m}\right)^{1/2} \tag{2.114}$$

From Eq. 2.112 the velocity v of the classical harmonic oscillator is calculated to be

$$v(t) = \frac{dx}{dt} = \left(\frac{k}{m}\right)^{1/2} x_0 \cos\left[\left(\frac{k}{m}\right)^{1/2} t\right]$$

The velocity therefore has a sinusoidal time dependence like that of the coordinate, but the two are out of phase by $\pi/2$. The total energy, given by the sum of the kinetic and potential energies, is

$$\begin{aligned}
E &= \frac{1}{2} mv^2 + \frac{1}{2} kx^2 \\
&= \frac{1}{2} m \left(\frac{k}{m}\right) x_0^2 \cos^2\left[\left(\frac{k}{m}\right)^{1/2} t\right] \\
&\quad + \frac{1}{2} kx_0^2 \sin^2\left[\left(\frac{k}{m}\right)^{1/2} t\right] \\
&= \frac{1}{2} kx_0^2
\end{aligned} \tag{2.115}$$

It is seen that the energy, which is independent of time, is proportional to the square of the maximum displacement, called the *amplitude* of the oscillator.

Turning to the quantum-mechanical description of the harmonic oscillator, we substitute the potential $V = \frac{1}{2} kx^2$ into the one-dimensional form of the Schrödinger equation with a potential (Eq. 2.106), and obtain

$$-\frac{\hbar^2}{2m}\frac{d^2\psi(x)}{dx^2} + \frac{1}{2} kx^2\psi(x) = E\psi(x) \tag{2.116}$$

Equation 2.116 can be solved by a series-expansion technique analogous to that used for the free-particle problem (see Section 2.6.1). The details of the procedure are given in standard texts on quantum mechanics (see Reference 13 of the Additional Reading list). Since it is rather involved, we concern ourselves only with some of the results. We consider the simple Gaussian function

$$\psi(x) = A\, e^{-\alpha^2 x^2/2} \tag{2.117}$$

with A and α constants. The second derivative of this function is

$$\frac{d^2\psi(x)}{dx^2} = -\alpha^2(1 - \alpha^2 x^2)\psi(x) \tag{2.118}$$

which can be compared with Eq. 2.116 in the rearranged form

$$\frac{d^2\psi(x)}{dx^2} = -\frac{2m}{\hbar^2}(E - \frac{1}{2} kx^2)\psi(x) \tag{2.119}$$

It is clear that the function $\psi(x)$ given by Eq. 2.117 satisfies Eq. 2.119 if the constant α is chosen such that

$$\alpha^2 = \frac{(mk)^{1/2}}{\hbar} \tag{2.120a}$$

and if the energy E is restricted to the value

$$E = \frac{\alpha^2\hbar^2}{2m} = \frac{h}{4\pi}\left(\frac{k}{m}\right)^{1/2} \tag{2.120b}$$

Substitution of Eq. 2.114 into Eq. 2.120b gives

$$E = \frac{1}{2} h\nu_0 \tag{2.121}$$

Thus, $\psi(x)$ is a stationary-state solution for the harmonic oscillator with energy $\frac{1}{2} h\nu_0$. To determine the constant A, the normalization condition is applied as for the particle-in-a-box problem (Eq. 2.78); there results the normalized function

$$\psi(x) = \left(\frac{\alpha^2}{\pi}\right)^{1/4} e^{-\alpha^2 x^2/2} \qquad \alpha^2 = \frac{(mk)^{1/2}}{\hbar} \tag{2.122}$$

The function $\psi(x)$ is plotted in Fig. 2.20a; it has its maximum at the minimum point of the potential and drops rapidly as x increases in magnitude in both the positive and negative directions. We note that $\psi(x)$ is

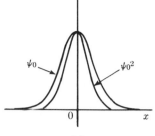

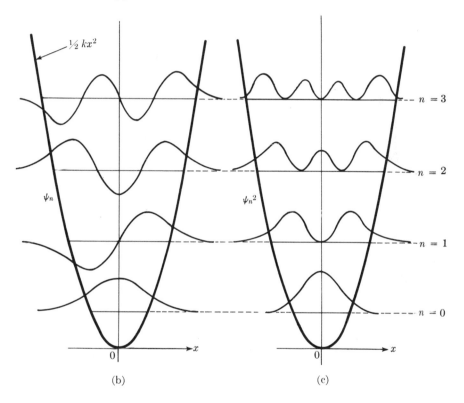

Fig. 2.20. Harmonic oscillator. **(a)** Ground-state wave function ψ_0 and probability density $\psi_0{}^2$; **(b)** potential-energy function $\frac{1}{2}kx^2$, wave functions ψ_n, and energy levels E_n; **(c)** probability densities $\psi_n{}^2$ and energy levels E_n.

a nodeless function. In correspondence with our previous discussion of the relation between nodes in the wave function and the system energy, this suggests—as is indeed the case—that $\psi(x)$ is the ground-state wave function; that is, its energy $\frac{1}{2}h\nu_0$ is the lowest energy. The existence of a minimum energy greater than zero is in accordance with the uncertainty principle.

Higher states of the harmonic oscillator can be found by multiplying the Gaussian exp $(-\alpha^2 x^2/2)$ by a polynomial in x, the coefficients of the polynomial and the associated energy being determined by the Schrödinger equation (see Problem 2.22). For the lowest states, the functions $\psi(x)$ shown in Fig. 2.20b are obtained. Their energies are given by the simple formula

$$E_n = (n + \tfrac{1}{2}) h\nu_0 \qquad n = 0, 1, 2, \ldots$$

It is seen that the allowed energies for this case are equally spaced, the constant interval being $h\nu_0$. This contrasts with the particle-in-a-box result, where the energy level separation increases with the quantum number n.

SUMMARY

In this chapter we have provided the foundation for the development of physical chemistry. We have introduced the basic ideas of the quantum theory in terms of the Schrödinger equation and its boundary conditions. They determine the wave functions and the allowed stationary-state energies. From the wave function, the probability distribution for the particle coordinates is obtained. The operators associated with the variables of interest (coordinates, momenta, etc.) can then be used to calculate their average or expectation values.

In the following chapters the quantum theory will be used to gain an understanding of atomic and molecular structure, beginning with the simplest problem, that of the hydrogen atom. The additional postulates required to complete the theory, such as intrinsic spin and the Pauli exclusion principle, will be discussed as the need for them arises.

ADDITIONAL READING

[1] *Classical Scientific Papers; Physics*, Introduction by S. Wright (American Elsevier, New York, 1964).

[2] *Nobel Lectures; Physics 1932–1941* (Elsevier, Amsterdam, 1965).

[3] C. A. COULSON, *Waves*, 7th ed. (Interscience, New York, 1958).

[4] W. E. WILLIAMS, *Applications of Interferometry* (Wiley, New York, 1960).

[5] NIELS BOHR, *Atomic Physics and Human Knowledge* (Wiley, New York, 1958).

[6] C. W. SHERWIN, *Introduction to Quantum Mechanics* (Henry Holt, New York, 1959).

[7] D. BOHM, *Quantum Theory* (Prentice-Hall, Englewood Cliffs, N.J., 1951).

[8] R. D. STUART, *Fourier Analysis* (Wiley, New York, 1961).

[9] M. BORN, *Atomic Physics* (Blackie and Son, Glasgow, 1945).

[10] F. K. RICHTMEYER, E. H. KENNARD, and T. LAURITSEN, *Introduction to Modern Physics* (McGraw-Hill, New York, 1955), 5th ed.

[11] U. FANO and L. FANO, *Basic Physics of Atoms and Molecules* (Wiley, New York, 1959).

[12] E. GOLDWASSER, *Optics, Waves, Atoms, and Nuclei* (Benjamin, New York, 1965).

[13] L. PAULING and E. B. WILSON, JR., *Introduction to Quantum Mechanics* (McGraw-Hill, New York, 1935).

[14] J. C. D. Brand and J. C. Speakman, *Molecular Structure* (Arnold, London, 1964), Chap. 9.

PROBLEMS

2.1 Show that the first minimum and maximum in the diffraction pattern for two slits (Fig. 2.5) occur at $d = \lambda l/2s$ and $\lambda l/s$, respectively, when $\lambda \ll s \ll l$.

2.2 The electric field intensity ε of an electromagnetic wave of velocity c obeys the wave equation

$$\frac{\partial^2 \varepsilon}{\partial x^2} = \frac{1}{c^2} \frac{\partial^2 \varepsilon}{\partial t^2}$$

Show that the function $\varepsilon(x, t)$ given by Eq. 2.3 is a solution to this wave equation.

2.3 Calculate the energy and momentum in cgs units of a photon with each of the following wavelengths:

(a) 10 A   (x ray)
(b) 2000 A (ultraviolet)
(c) 6000 Å (visible)
(d) 10^5 Å   (infrared)
(e) 100 cm (microwave)
(f) 10 m   (radio wave)

2.4 Calculate the de Broglie wavelength for

(a) an electron with a kinetic energy of 200 eV
(b) a proton with a kinetic energy of 10^5 eV
(c) a 300 g pie thrown with a velocity of 1 m sec^{-1}
(d) the earth moving in its orbit about the sun
(e) an electron in a Bohr orbit corresponding to $n = 1$ and 100

2.5 When the surface of a piece of copper is irradiated with light of wavelength $\lambda = 2537$ Å from a mercury arc, the value of the potential necessary to stop the emitted electrons is found to be 0.24 V. What is the maximum wavelength which will eject an electron from a copper surface?

2.6 When a clean tungsten surface is illuminated by light of wavelength 2000 Å, a potential of 1.68 V is required to stop the emitted electrons; when the wavelength of light is 1500 Å, a potential of 3.74 V is required. From these data determine Planck's constant (in erg sec) and the potential barrier to emission (in eV) for tungsten.

2.7 From the equations of conservation of energy and momentum, show that the shift in wavelength for the Compton effect is given by Eq. 2.27.

2.8 Derive Eq. 2.27 from a relativistic treatment of the initial and final energy of the electron.

2.9 Determine the maximum scattering angle of 75 V electrons from a nickel crystal in the Davisson–Germer experiment.

2.10 What is the percentage error introduced by ignoring the effect of special relativity in calculating the de Broglie wavelength of a 10 kV electron?

2.11 Assume that the diameter of an atomic nucleus is 10^{-12} cm; use the uncertainty principle to compute approximately the least kinetic energy that
(a) a proton
(b) an electron
within the nucleus can have.
(c) If the binding energy per nucleon is 5×10^6 eV, what is the physical significance of your results for parts (a) and (b)?

2.12 By arguments analogous to those leading to Eq. 2.43, find the range $\Delta \nu$ of frequencies required for superposing waves of the form $\cos 2\pi \nu t$ to obtain a wave form localized in the time interval Δt. Show that for matter waves your result leads to the uncertainty relation $\Delta E \, \Delta t \geqslant h$.

2.13 When an electron in an atom is excited into a level above the ground state, it remains in the excited state for an average length of time on the order of 10^{-8} sec before returning to the ground state. If the radiation emitted has a wavelength $\lambda = 5000$ Å, use the uncertainty principle to compute the minimum width of the spectral line.

2.14 Assuming that an electron in a one-dimensional box of length $l = 2$ Å obeys the Bohr frequency rule, find
(a) the wavelength λ (in Å)
(b) the wave number $\bar{\nu}$ (in cm^{-1})
of the radiation emitted for a transition from the level $n + 1$ to n.
(c) What is the longest wavelength in the spectrum of this system?

2.15 Find the average (expectation) values of x^2 and p^2 for a particle in the state $n = 3$ of a one-dimensional box of length l.

2.16 Calculate the uncertainty product $\Delta x \, \Delta p$ for a particle-in-a-one-dimensional box in state n if both Δx and Δp are defined as root-mean-square deviations; that is,

$$\Delta x \equiv [\langle (x - \langle x \rangle)^2 \rangle]^{1/2}$$
$$\Delta p \equiv [\langle (p - \langle p \rangle)^2 \rangle]^{1/2}$$

Compare your result with Eq. 2.90.

2.17 Show that

$$\int_0^l \psi_1(x)\psi_2(x) \, dx = 0$$

for a particle in a one-dimensional box of length l. When an integral of the product of two functions vanishes, the functions are said to be *orthogonal*. In quantum mechanics, any two functions corresponding to different energy levels of a system are orthogonal.

2.18 For a particle in a one-dimensional box of length l, calculate the probability that the particle will be found within a distance of $0.001l$ on either side of the point $x = l/3$ for the states $n = 1$, $n = 2$, and $n = 3$.

2.19 (a) Show that the wave function given in Eq. 2.98 for the three-dimensional box is a solution of the Schrödinger equation ($0 < x < a$, $0 < y < b$, $0 < z < c$) when the appropriate energy value is used.
(b) Show that the function is normalized.
(c) For the case in which $a = b = c = 1$ Å, find the probability that an electron will be found in a volume element whose dimensions are $\Delta x = \Delta y = \Delta z = 0.001$ Å and whose center is at $x = 0.2$ Å, $y = 0.3$ Å, $z = 0.5$ Å for the two states:

n_x	n_y	n_z
2	1	1
1	1	2

2.20 For a particle in a three-dimensional box of sides a, b, and c, with $a \neq b = c$ (see Fig. 2.19), make a table of n_x, n_y, n_z, the energies, and the degeneracies of the levels in which the quantum numbers range from 0 to 5.

2.21 The two identical sample cells C_I and C_{II} of an interferometer like that of Fig. 2.6 initially contain air ($\mu = 1.0002926$) at 0°C and 760 Torr. As pure methane ($\mu = 1.00044$) is allowed to slowly displace the air in C_{II}, the interference pattern observed with sodium D light ($\lambda = 5893$ Å) moves past a reference line in the focal plane of the eyepiece. After all the air in C_{II} has been displaced by methane and the sample brought again to 0°C and 760 Torr, the pattern has moved a distance equal to 25.31 times the distance between successive maxima. When the experiment is repeated, with nitric oxide displacing the methane, the pattern moves in the opposite direction. When the nitric oxide has completely displaced the methane and has been brought to 0°C and 760 Torr, the pattern has moved a distance equal to 24.64 times the distance between successive maxima. Use these data and the fact that maxima are observed when the difference between the optical path (μx distance) through the air and the optical path through the sample is an integral multiple of the wavelength to determine the refractive index of nitric oxide at 0°C and 760 Torr.

Ans. 1.000297.

2.22 By substituting the function

$$\psi(x) = (a + bx^2) \exp(-cx^2)$$

into Eq. 2.119 and performing the differentiation, find the relations that must be satisfied by the constants a, b, and c if ψ is to be a normalized solution of the Schrödinger equation for the harmonic oscillator. What is the energy in units of $h\nu_0$ corresponding to this function? How many nodes, maxima, and minima does the function have?

2.23 It is often convenient to describe the superposition of a sine and cosine wave of the same frequency in terms of the *phase* of the resultant wave. For example, the sum

$$A \cos \nu + B \sin \nu$$

can be transformed into the expression

$$C \cos (\nu - \delta)$$

where the angle δ is called the *phase angle*, and the resultant wave is described as being "out of phase with cos ν by δ radians." Find δ for $B/A = 0$, ∞, 1, -1, $1/\sqrt{3}$. [*Hint:* Use the identity $\cos(\alpha + \beta) = \cos \alpha \cos \beta - \sin \alpha \sin \beta$.]

2.24 (a) For a classical harmonic oscillator of mass m, force constant k, and amplitude x_0, find the time average value of the kinetic energy and the potential energy by averaging over one period. Compare your result with the total energy.
(b) For a quantum harmonic oscillator with the same mass and force constant in its ground state, find $\langle T \rangle$ and $\langle V \rangle$ and compare with the total energy.

The answers obtained here are another example of the virial theorem, in this case for a harmonic potential; compare with the virial theorem result obtained in Section 1.4.

3

The structure of one-electron atoms and ions

In Chapter 2 the mathematical apparatus of the quantum theory was introduced. As a first step in its application to the study of the interactions between electrons and nuclei, we again consider the hydrogen atom and compare the results obtained from the quantum theory with those from the Bohr model. Whereas the Bohr theory could be applied to systems composed of only one electron moving in the field of a single nucleus, the quantum theory allows us to treat both simple and complex atoms and molecules. In Chapter 4, the hydrogen-atom energy levels and wave functions, appropriately modified to take account of the electron-electron repulsion, are used as the basis for a general examination of the structure of many-electron atoms.

103

For the following problems a programmable desk calculator or digital computer will be useful.

C2.1   Evaluate the function $f_m(x)$ between $x = 0$ and $x = \pi$ in steps of $\pi/100$, where

$$f_m(x) = \sum_{n=1}^{m} a_n \sin nx$$

for $m = 5$ and for $m = 50$ with a_n given by

(a) $a_n = \dfrac{2}{n\pi}(1 - \cos n\pi)$

(b) $a_n = -\dfrac{2}{n\pi}\cos n\pi$

(c) $a_n = \dfrac{8}{n^2\pi^2}\sin\dfrac{n\pi}{2}$

Plot your results, and sketch in the function $f(x)$ to which $f_m(x)$ seems to be converging in each of the three cases as $m \to \infty$.

C2.2   A function $f(x)$ which is symmetric about $x = 0$ can be represented by the *Fourier integral*

$$f(x) = (2\pi)^{1/2}\int_{-\infty}^{+\infty} g(\bar{\nu} - \bar{\nu}_0)\cos[2\pi(\bar{\nu} - \bar{\nu}_0)x]\,d\bar{\nu}$$

where the weighting function $g(\bar{\nu})$ is given by

$$g(\bar{\nu} - \bar{\nu}_0) = (2\pi)^{-1/2}\int_{-\infty}^{+\infty} f(x)\cos[2\pi(\bar{\nu} - \bar{\nu}_0)x]\,dx$$

(a) Find the analytical expression for $g(\bar{\nu} - \bar{\nu}_0)$ which gives $f(x)$ the form

$$\begin{aligned} f(x) &= \tfrac{1}{2}a & -a < x < a \\ &= 0 & |x| > a \end{aligned}$$

(b) Find the limit of $g(\bar{\nu} - \bar{\nu}_0)$ as $\bar{\nu}$ goes to $\bar{\nu}_0$ and the smallest value of ϵ such that $g(\epsilon) = 0$. If the range of $\bar{\nu}$ is defined as $\Delta\bar{\nu} = 2\epsilon$, what is the uncertainty product $\Delta\bar{\nu}\,\Delta x$?

(c) Approximate the Fourier integral by a sum over m equally spaced discrete values of $\bar{\nu}$ between $\bar{\nu}_0 - \epsilon$ and $\bar{\nu}_0 + \epsilon$. [*Caution:* Use the limiting form for $g(0)$.] Thus calculate

$$f_m(x) = \sum_{i=1}^{m} g(\bar{\nu}_i - \bar{\nu}_0)\cos[2\pi(\bar{\nu}_i - \bar{\nu}_0)x]$$

for $m = 10$, 100, and 1000. Plot $f_{10}(x)$, $f_{100}(x)$, $f_{1000}(x)$, and $f(x)$ on the same graph.

(d) For each of the values of m in part (c), calculate the root-mean-square deviation

$$\Delta x_m \equiv [\langle (x - \langle x \rangle)^2 \rangle]^{1/2}$$

by integrating Eqs. 2.85 and 2.86 (trapezoidal rule), using the normalized wave function

$$\psi_m(x) = f_m(x) \Big/ \int_{-L}^{+L} [f_m(x)]^2 \, dx$$

where $L = 5a$. Compare your results with Δx_∞ [calculated from $f(x)$]. Hence calculate the uncertainty products $\Delta \bar{\nu} \, \Delta x_m$ for $m = 10, 100, 1000$, and ∞.

3.1 THE SCHRÖDINGER EQUATION FOR ONE-ELECTRON ATOMS

To apply the Schrödinger equation (Eq. 2.106) to the hydrogen atom, we need to know the form of the potential energy $V(x, y, z)$. As in the planetary model (Chapter 1), the potential energy of an electron in the electrostatic field of a nucleus of charge Ze at a distance r is

$$V = - \frac{Ze^2}{r} \tag{3.1}$$

Since Eq. 2.106 is expressed in Cartesian coordinates, we rewrite Eq. 3.1 in the form

$$V(x, y, z) = \frac{-Ze^2}{(x^2 + y^2 + z^2)^{1/2}} \tag{3.2}$$

Thus, for a one-electron atom or ion with a stationary nucleus of charge Ze, the Schrödinger equation is

$$-\frac{\hbar^2}{2m} \left(\frac{\partial^2 \psi}{\partial x^2} + \frac{\partial^2 \psi}{\partial y^2} + \frac{\partial^2 \psi}{\partial z^2} \right) - \frac{Ze^2}{(x^2 + y^2 + z^2)^{1/2}} \psi = E\psi \tag{3.3}$$

This form for the Schrödinger equation, a partial differential equation in the variables x, y, and z, is quite cumbersome. In fact, it cannot be simply solved for the function $\psi(x, y, z)$ in Cartesian coordinates because there is no way to separate the variables; that is, the solution cannot be expressed

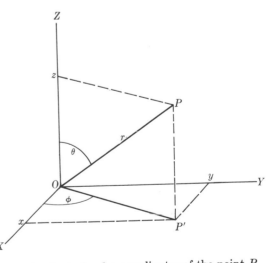

Fig. 3.1. Cartesian and spherical polar coordinates of the point P.

as a simple product function $\psi_x(x)\,\psi_y(y)\,\psi_z(z)$, analogous to that used in the three-dimensional particle-in-a-box problem Eq. 2.98.

To simplify the solution of a differential equation such as Eq. 3.3, it is often helpful to transform it to a different coordinate system. Since the potential energy given in Eq. 3.1 is spherically symmetric (i.e., it depends only on the distance r from the nucleus and not on the orientation angles), it is appropriate to introduce spherical polar coordinates. The relations between the Cartesian coordinates x, y, and z and the spherical polar coordinates r, θ, and ϕ are illustrated in Fig. 3.1. Let P be a point in space whose position is described by the projections x, y, and z of the line segment $\overline{OP}$ upon the x, y, and z axes, respectively. The position of the point P may also be given by specifying the length r of the line segment $\overline{OP}$, the angle θ which $\overline{OP}$ makes with the z axis, and the angle ϕ which the projection $\overline{OP'}$ of $\overline{OP}$ onto the xy plane makes with the x axis. From Fig. 3.1, it is evident that

$$z = \overline{OP}\cos\theta$$
$$\overline{OP'} = \overline{OP}\sin\theta \tag{3.4a}$$

and that

$$x = \overline{OP'}\cos\phi = \overline{OP}\sin\theta\cos\phi$$
$$y = \overline{OP'}\sin\phi = \overline{OP}\sin\theta\sin\phi \tag{3.4b}$$

Upon substitution of $\overline{OP} = r$ in Eq. 3.4, one obtains the equations which define the transformation from Cartesian coordinates to spherical polar coordinates,

$$x = r\sin\theta\cos\phi$$
$$y = r\sin\theta\sin\phi \tag{3.5}$$
$$z = r\cos\theta$$

Solving these equations for r, θ, and ϕ, we find the inverse transformation

$$r = (x^2 + y^2 + z^2)^{1/2}$$
$$\theta = \arccos\frac{z}{(x^2 + y^2 + z^2)^{1/2}} \tag{3.6}$$
$$\phi = \arctan\frac{y}{x}$$

The student should be aware that, although the definitions of θ and ϕ given in Eq. 3.6 conform to those commonly used in chemistry and physics, the angles are often interchanged in mathematics textbooks.

Having defined the spherical polar coordinates, we introduce them into the Schrödinger equation (Eq. 3.3). Although the potential-energy term takes the simple form given by Eq. 3.1, the kinetic-energy term becomes complicated. To see how the transformation of the kinetic-energy term is carried out, we begin by transforming the derivative $\partial\psi/\partial x$. According to the "chain rule" for differentiation, we have

$$\frac{\partial\psi}{\partial x} = \left(\frac{\partial r}{\partial x}\right)_{y,z}\left(\frac{\partial\psi}{\partial r}\right)_{\theta,\phi} + \left(\frac{\partial\theta}{\partial x}\right)_{y,z}\left(\frac{\partial\psi}{\partial\theta}\right)_{r,\phi} + \left(\frac{\partial\phi}{\partial x}\right)_{y,z}\left(\frac{\partial\psi}{\partial\phi}\right)_{r,\theta} \quad (3.7)$$

Each term on the right-hand side of Eq. 3.7 is seen to be a product of two factors; e.g., in the first term the left-most factor $(\partial r/\partial x)$ is a partial derivative in which the remaining *Cartesian* coordinates y, z are held constant, while in the right-most factor $(\partial\psi/\partial r)$ the remaining *polar* coordinates θ, ϕ are held constant. We can evaluate the left-most factors directly from Eq. 3.6 and insert them into Eq. 3.7 to obtain

$$\frac{\partial\psi}{\partial x} = \sin\theta\cos\phi\left(\frac{\partial\psi}{\partial r}\right)_{\theta,\phi} + \frac{1}{r}\cos\theta\cos\phi\left(\frac{\partial\psi}{\partial\theta}\right)_{r,\phi} - \frac{\sin\phi}{r\sin\theta}\left(\frac{\partial\psi}{\partial\phi}\right)_{r,\theta} \quad (3.8)$$

The next step is to find the form of $\partial^2\psi/\partial x^2$ and then to repeat the calculation for $\partial^2\psi/\partial y^2$ and $\partial^2\psi/\partial z^2$. Without going through the details, which are straightforward but lengthy,[1] we give the resulting Schrödinger equation for the hydrogen atom in spherical polar coordinates:

$$-\frac{\hbar^2}{2m}\left[\frac{1}{r^2}\frac{\partial}{\partial r}\left(r^2\frac{\partial\psi}{\partial r}\right) + \frac{1}{r^2\sin\theta}\frac{\partial}{\partial\theta}\left(\sin\theta\frac{\partial\psi}{\partial\theta}\right) + \frac{1}{r^2\sin^2\theta}\frac{\partial^2\psi}{\partial\phi^2}\right]$$
$$-\frac{Ze^2}{r}\psi = E\psi \quad (3.9)$$

where ψ is now considered to be a function of r, θ, and ϕ. We see in Eq. 3.9 that the derivatives with respect to x, y, z have been replaced by derivatives with respect to r, θ, ϕ.

In seeking solutions to Eq. 3.9, it is instructive to recall the results for the other three-dimensional problem that we have discussed, namely, the particle in a three-dimensional box. In Chapter 2 it was noted that the solution to this problem was the product of three solutions to the one-dimensional-box problem; that is, the wave function given in Eq. 2.98 has the factored form

$$\psi(x, y, z) = X(x)\,Y(y)\,Z(z) \quad (3.10)$$

where $X(x)$ is a function of x alone, $Y(y)$ of y alone, and $Z(z)$ of z alone. It turns out that the solution to Eq. 3.9 can also be written as a product of three functions, each one of which is a function of a single polar coordinate; thus the coordinate transformation has led to the desired result of making possible a separation of variables. We write the solution to Eq. 3.9 as

$$\psi(r, \theta, \phi) = R(r)\,\Theta(\theta)\Phi(\phi) \quad (3.11)$$

---

[1] The most direct route is not always the easiest one; this point is illustrated in Problems 3.1 and 3.2.

where $R(r)$ depends only on r, $\Theta(\theta)$ only on θ, and $\Phi(\phi)$ only on ϕ. In contrast to the particle-in-a-box problem, where $X(x)$, $Y(y)$, and $Z(z)$ have the same functional form, we expect the three functions $R(r)$, $\Theta(\theta)$, and $\Phi(\phi)$ to have different dependences on their respective variables because, unlike x, y, and z, each of the three polar coordinates has a unique character. To find the form of the functions is the object of the following sections.

3.2 THE GROUND STATE OF ONE-ELECTRON ATOMS

We begin by looking for the solution corresponding to the ground state of the hydrogen atom, in other words, the wave function corresponding to the lowest possible energy. Recalling the general relationship between the energy and the number of nodes, we assume that ground-state solution has no nodes. To eliminate the possibility of nodes in the angular functions Θ and Φ, we choose these functions to be constant.[2] Since the wave function ψ is taken to be independent of θ and ϕ, it is spherically symmetric and the derivatives with respect to θ and ϕ are zero; that is,

$$\frac{\partial \psi}{\partial \theta} = 0 \qquad \frac{\partial \psi}{\partial \phi} = 0$$

This means that there are no angular contributions to the kinetic-energy term. Writing $\psi(r, \theta, \phi) = R(r)$, we have

$$-\frac{\hbar^2}{2m}\left[\frac{1}{r^2}\frac{\partial}{\partial r}\left(r^2\frac{\partial R}{\partial r}\right) + \frac{1}{r^2 \sin\theta}\frac{\partial}{\partial\theta}\left(\sin\theta\frac{\partial R}{\partial\theta}\right)\right.$$
$$\left. + \frac{1}{r^2\sin^2\theta}\frac{\partial^2 R}{\partial\phi^2}\right] = -\frac{\hbar^2}{2m}\left(\frac{d^2R}{dr^2} + \frac{2}{r}\frac{dR}{dr}\right) \quad (3.12)$$

since

$$\frac{\partial R}{\partial \theta} = 0 \qquad \frac{\partial R}{\partial \phi} = 0$$

On the right-hand side of Eq. 3.12, we have written total derivatives since R depends only on r, and have performed the indicated differentiation. Substituting into Eq. 3.9, we obtain the *radial Schrödinger equation*

$$-\frac{\hbar^2}{2m}\left(\frac{d^2R}{dr^2} + \frac{2}{r}\frac{dR}{dr}\right) - \frac{Ze^2}{r}R = ER \quad (3.13)$$

[2] It is always possible to try such a simplifying assumption; if it is correct, the resulting function satisfies the Schrödinger equation and is a solution; if it were incorrect, no function satisfying the Schrödinger equation would be found and a more complicated form for the solution would have to be tried. This type of trial-and-error procedure for solving partial differential equations is often used in real problems, though for the present case the general solutions can be found directly by a modification of the series expansion technique used in our discussion of the free particle and the particle-in-a-box (Sections 2.6.1 and 2.6.2).

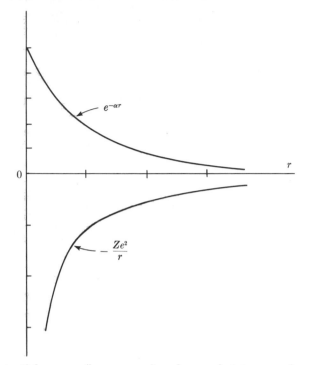

Fig. 3.2. Potential energy (lower curve) and ground-state wave function (upper curve) for the electron in a hydrogen atom.

Since the squares of the solutions to Eq. 3.13 give probability distributions for the position of the electron with respect to the nucleus (see Chapter 2), the state of lowest total energy should have a wave function that is large in regions where the potential energy is large and negative. However, the distribution of the electron cannot be confined to too small a volume, because this would require a high energy, in accordance with the uncertainty principle. A plot of the potential energy as a function of r (Fig. 3.2) shows that the region of lowest potential energy is near the origin, which suggests that the ground-state wave function $R(r)$ should have a large amplitude at small r and should decrease rapidly as r increases. To find a solution of this general form, we look at Eq. 3.13 for values of r so large that the potential-energy term, which varies inversely with r, can be neglected. At such distances, the term $(2/r)\, dR/dr$, with its $1/r$ dependence, is also expected to be small relative to the remaining terms. Writing the approximate equation for large r, called the *asymptotic* equation, we have

$$-\frac{\hbar^2}{2m}\frac{d^2R}{dr^2} = ER \qquad \text{(large } r\text{)} \tag{3.14}$$

or

$$\frac{d^2R}{dr^2} = \alpha^2 R \qquad \text{(large } r) \qquad (3.15)$$

where $\alpha^2 = -2mE/\hbar^2$. Equation 3.15 has a form similar to Eq. 2.57 for the free particle, differing only in that the minus sign appearing in front of k^2 in Eq. 2.57 has been incorporated into the constant α^2 here. The reason for this choice is that, from the Bohr theory of the hydrogen atom, we know that the ground state and all other bound states have a negative energy relative to the ionization limit. For $E < 0$, the quantity $-2mE/\hbar^2$ is positive and $\alpha = (-2mE/\hbar^2)^{1/2}$ is a positive real number. Equation 3.15 is known to have solutions of the form $Ae^{+\alpha r}$ and $Be^{-\alpha r}$, as can be verified by substitution or demonstrated by the series-expansion method (Problem 3.3).

Of the two exponential solutions, $Ae^{+\alpha r}$ increases as r increases. The probability density ($\psi^2 = A^2 e^{+2\alpha r}$) associated with this function has a physically unacceptable form in that the probability of finding the electron at a distance r from the nucleus increases without bound as r increases; this is in contradiction to the expectation mentioned above for the form of the ground-state wave function. By contrast, $Be^{-\alpha r}$ is of the appropriate form, being large near the nucleus and decreasing exponentially as r increases. Thus, $Be^{-\alpha r}$ is reasonably behaved for a bound-state solution and satisfies the asymptotic equation (Eq. 3.15) valid for large r. To determine whether $Be^{-\alpha r}$ is actually a solution to the complete Schrödinger equation, we substitute $R(r) = Be^{-\alpha r}$ into Eq. 3.13 and obtain after rearranging

$$\alpha^2 Be^{-\alpha r} + \frac{2}{r}(-\alpha Be^{-\alpha r}) + \frac{2m}{\hbar^2}\left(\frac{Ze^2}{r} + E\right)Be^{-\alpha r} = 0$$

Multiplication of every term by $B^{-1}e^{\alpha r}$ yields

$$\left(\alpha^2 + \frac{2mE}{\hbar^2}\right) + \frac{1}{r}\left(\frac{2mZe^2}{\hbar^2} - 2\alpha\right) = 0 \qquad (3.16)$$

If Eq. 3.16 is satisfied for all values of r, then the asymptotic (large r) form for $R(r)$ is a solution to Eq. 3.13. The only way that Eq. 3.16 can hold for all values of r is for each of the terms in parentheses to be separately equal to zero. This may be seen as follows: If either term is zero, the equation itself states that the other is zero; on the other hand, if neither is zero, Eq. 3.16 may be solved for r. Suppose the solution is $r = r_1$; then for $r \neq r_1$ Eq. 3.16 does not hold, and we cannot assert that it holds for all r.

Therefore, if and only if

$$\alpha^2 + \frac{2mE}{\hbar^2} = 0 \qquad (3.17)$$

and

$$\frac{2mZe^2}{\hbar^2} - 2\alpha = 0 \qquad (3.18)$$

is $R(r) = Be^{-\alpha r}$ a solution to the Schrödinger equation for the hydrogen atom. Equation 3.17 is the definition of α^2 used in the asymptotic equation (Eq. 3.15). It relates α to the energy E but does not determine either of them. However, Eq. 3.18 can be solved for α; it yields

$$\alpha = \frac{Zme^2}{\hbar^2} = \frac{Z}{a_0} \qquad (3.19)$$

where a_0 is the Bohr radius (Eq. 1.56). When this value for α is inserted into Eq. 3.17, E is found to be

$$E = -\frac{Z^2\hbar^2}{2ma_0^2} = -\frac{\hbar^2}{me^2}\frac{Z^2e^2}{2a_0^2} = -\frac{Z^2e^2}{2a_0} \qquad (3.20)$$

The wave function that we have guessed is shown to be a solution if α and E are given by Eqs. 3.19 and 3.20, respectively. As in the particle-in-a-box and in the harmonic oscillator, the quantized energy thus follows without any additional assumptions from the Schrödinger equation and the boundary condition on the wave function, in this case, that $R(r) \rightarrow 0$ as $r \rightarrow \infty$. Moreover, the energy given by Eq. 3.20 is identical with the Bohr-theory result for the ground state, which we have already seen to be in agreement with experiment.

Substituting for α from Eq. 3.19, the ground-state wave function ψ_1 for a one-electron atom with nuclear charge Ze is found to be

$$\psi_1 = N \exp\left(-\frac{Zr}{a_0}\right) \qquad (3.21)$$

where N is a normalizing constant yet to be determined. The criterion for determining the value of N is that the probability of finding the electron somewhere in space be unity; that is, since the square of the wave function is the probability density, we require that

$$\int_{\substack{\text{all}\\\text{space}}} \psi_1^2 \, dv = N^2 \int \exp\left(-\frac{2Zr}{a_0}\right) dv = 1 \qquad (3.22)$$

where dv is the differential volume element. As stated in Section 2.6.3, a wave function that satisfies Eq. 3.22 is said to be normalized and N is called the normalizing factor. Since the wave function is given as a function of r, the integral of Eq. 3.22 can be evaluated most easily in spherical polar coordinates. This involves finding the volume element dv, which is simply

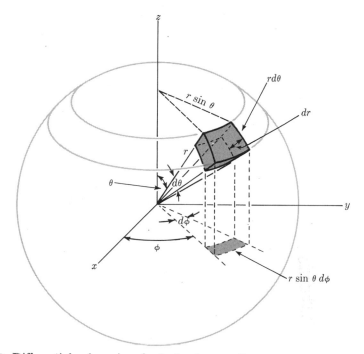

Fig. 3.3. Differential volume in spherical polar coordinates; see Eq. 3.23.

$dx\ dy\ dz$ in Cartesian coordinates, in terms of the differential quantities dr, $d\theta$, and $d\phi$. Such a differential volume located at the point r, θ, and ϕ is shown in Fig. 3.3. Since dr, $d\theta$, and $d\phi$ are differential quantities, the curvature of the sides of the "cube" can be neglected in calculating the volume. Multiplying the lengths of the three sides together, we obtain

$$dv = r^2 \sin\theta\ dr\ d\theta\ d\phi \qquad (3.23)$$

We also need the limits of the variables r, θ, and ϕ which correspond to an integration over all space. Since r is a distance, it is always positive and goes from 0 to ∞. For a given r, θ going from 0 to π generates an arc from $z = r$ to $z = -r$. If this arc is now swept through ϕ values from 0 to 2π, it covers the entire surface of the sphere. Thus,

$$
\begin{aligned}
0 &\leq r \leq \infty \\
0 &\leq \theta \leq \pi \\
0 &\leq \phi \leq 2\pi
\end{aligned}
\qquad (3.24)
$$

Introducing the explicit volume element and the integration limits, we write the normalization integral (Eq. 3.22) for the ground-state wave function as

$$\int_{\substack{\text{all}\\ \text{space}}} \psi_1^2 \, dv = N^2 \int_0^{2\pi} d\phi \int_0^{\pi} \sin \theta \, d\theta \int_0^{\infty} \exp \left(-\frac{2Zr}{a_0} \right) r^2 \, dr = 1 \qquad (3.25)$$

Since ψ_1 is independent of θ and ϕ, the integrals over the angles can be done immediately to give the result 4π. Thus, Eq. 3.25 becomes

$$4\pi N^2 \int_0^{\infty} \exp \left(-\frac{2Zr}{a_0} \right) r^2 \, dr = 1$$

The remaining radial integral is of the form (Problem 3.4)

$$\int_0^{\infty} e^{-\beta r} r^n \, dr = \frac{n!}{\beta^{n+1}} \qquad (3.26)$$

Here $n = 2$ and $\beta = 2Z/a_0$, so that

$$4\pi N^2 \frac{2!}{(2Z/a_0)^3} = \frac{\pi N^2 a_0^3}{Z^3} = 1 \qquad (3.27)$$

Solving for N, we find

$$N = \left(\frac{Z^3}{\pi a_0^3} \right)^{1/2} \qquad (3.28)$$

The normalized wave function is therefore

$$\psi_1 = \left(\frac{Z^3}{\pi a_0^3} \right)^{1/2} \exp \left(-\frac{Zr}{a_0} \right) \qquad (3.29)$$

By the above discussion we have shown that the introduction of a nodeless function, independent of angle, leads to a solution of the Schrödinger equation for the hydrogen atom and one-electron ions with nuclear charge Ze. That this solution (Eq. 3.29) corresponds to the ground state is suggested by its nodeless character. Proof of this fact would require a more general treatment (see Reference 2 of the Additional Reading list).

3.2.1 Correction for finite proton mass   For very accurate results, the electron mass m that has been used in these calculations of the energy and wave function should be replaced by the reduced mass μ of the nucleus-electron system, as defined in Eq. 1.65. The reason for employing μ rather than m is the same as for the planetary model; that is, the electron and nucleus both move about the common center of mass (see Fig. 1.11), while the motion of the center of mass itself contributes nothing to the internal energy of the atom (see Problem 3.17). In atoms with more than one electron, however, the methods required for taking into account the finite nuclear mass become much more complicated. Since the nuclei of the more complex atoms are several times more massive than the proton, the assumption of a stationary nucleus and the use of the electronic mass m in the Schrödinger equation is an even better approximation than in the present case. In molecules, the error introduced by using the electronic mass is also quite small. For these reasons, we follow the common practice of omitting the nuclear-mass correction in our calculations of electronic energies and wave functions for atoms and molecules.

3.3 GROUND-STATE WAVE FUNCTION AND PROBABILITY

Let us now examine the spatial distribution of an electron in the ground state of the hydrogen atom ($Z = 1$). Figures 3.4a and 3.4b show plots of $\psi_1(r)$ and $\psi_1^2(r)$, respectively. The latter gives the probability per unit volume of finding the electron at a point in space at a distance r from the nucleus. The point with the maximum probability density is seen to be the origin, with the probability density decreasing exponentially with r. Thus, the electron is somewhat "smeared out" over the atom, rather than being confined to a Bohr orbit at a well-defined distance from the nucleus. To make a better comparison with the Bohr theory, we divide the space around the nucleus into a set of concentric spherical shells spaced an infinitesimal distance dr apart. Now let us ask what is the probability of finding the electron between the shell with radius r and the shell with radius $r + dr$. The answer is the probability per unit volume, $\psi_1^2(r)$, multiplied by the infinitesimal volume between the spheres, which equals $4\pi r^2 dr$. The function $P_1(r)$,

$$P_1(r) = 4\pi r^2 \psi_1^2(r)$$

is a probability distribution (*radial distribution function*) for the variable r; it is the probability per unit distance of observing the electron at a distance r from the origin, irrespective of the angles θ and ϕ. A plot of this function for the ground state of hydrogen is shown in Fig. 3.4c. At $r = 0$, the probability distribution for r is zero, since the volume element shrinks to zero at this point. As r increases, the volume between the concentric spheres of spacing dr increases; this volume increase is enough to compensate at first for the exponential decrease of $\psi_1^2(r)$, so that the probability distribution for r rises. At large r, the exponentially decreasing function $\psi_1^2(r)$ dominates, and the radial distribution function approaches zero asymptotically. The most probable value for r corresponds to the point for which the distribution $P_1(r)$ has its maximum. It can be found by differentiating $P_1(r)$ and setting the derivative equal to zero; for $Z = 1$,

$$\frac{d}{dr}\left[r^2 \exp\left(-\frac{2r}{a_0}\right)\right] = \left(2r - \frac{2r^2}{a_0}\right)\exp\left(-\frac{2r}{a_0}\right) = 0 \qquad (3.30)$$

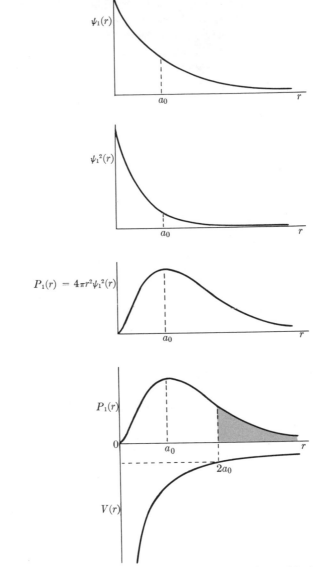

Fig. 3.4. Ground-state of the hydrogen atom: **(a)** wave function $\psi_1(r)$, **(b)** probability density $\psi_1^2(r)$, **(c)** radial distribution $P_1(r)$, and **(d)** classically forbidden region.

The solution to Eq. 3.30 is seen to be $r = a_0$. Thus, the most probable value of r is equal to the Bohr radius. This is the quantum-theory analog to the Bohr picture of the ground state, in which the electron orbits the nucleus at the fixed distance a_0. Although the most probable result of a measurement is a_0, any value between zero and infinity can be obtained since the probability distribution is nonzero for all $r > 0$. The "size" of atoms is therefore not well defined by the quantum-theory model, but depends somewhat on the nature of the interaction upon which a measurement of the size is based. One possible definition of size is in terms of the maximum of $P(r)$, which for the hydrogen atom is just a_0. An alternative possibility is to use the average or expectation value of r. By Eq. 2.87, $\langle r \rangle$ for a hydrogen-like atom or ion is

$$\langle r \rangle = \int_{\substack{\text{all} \\ \text{space}}} \psi_1 r \, \psi_1 dv = 4\pi \left(\frac{Z^3}{\pi a_0^3} \right) \int_0^\infty \exp\left(-\frac{2Zr}{a_0} \right) r^3 \, dr$$

$$\tag{3.31}$$

$$= 4\pi \left(\frac{Z^3}{\pi a_0^3} \right) \left(\frac{3!}{(2Z/a_0)^4} \right) = \frac{3}{2} \frac{a_0}{Z}$$

for hydrogen ($Z = 1$), $\langle r \rangle = \frac{3}{2} a_0$. Thus, the average value of r is somewhat larger than the most probable value. Another useful limit is that obtained by finding the value $r_{(0.9)}$ such that there is a 90% probability that the radial coordinate is less than or equal to $r_{(0.9)}$. To determine $r_{(0.9)}$, we consider the radial integral

$$4\pi \left(\frac{Z^3}{\pi a_0^3} \right) \int_0^{r_{(0.9)}} \exp\left(-\frac{2Zr}{a_0} \right) r^2 \, dr = 0.9$$

Performing the integration and solving for $r_{(0.9)}$, one finds $r_{(0.9)} \simeq 2.6a_0$.

For most purposes, an effective atomic radius in this range (a_0 to $2.5a_0$) is suitable. Thus, the size is on the order of 10^{-8} cm. The experimental facts that atoms have diameters of about 1 Å and that various kinds of measurements often give somewhat different values are both compatible with the quantum-theory results. Also, in contrast to the Bohr theory, the new picture does not violate the uncertainty principle, since the uncertainties Δr and Δp_r are no longer both zero.

3.4 TUNNEL EFFECT

A characteristic and very important quantum-mechanical consequence of the smeared out electron distribution is the so-called *tunnel effect*. Classically, a particle has a *turning point* where the potential energy becomes equal to the total energy. Since the kinetic energy and therefore the velocity are equal to zero at such a point, the classical particle is expected to be turned around or *reflected* by the potential barrier. For an electron in the hydrogen-atom ground state, such a classical turning point occurs where

$$V(r) = -\frac{e^2}{r} = E = -\frac{e^2}{2a_0}$$

that is, at $r = 2a_0$. Figure 3.4d shows, however, that the probability distribution $P(r)$ has a nonzero value for $r > 2a_0$; that is, the electron has access to a region (represented by shading in the diagram) which is forbidden by classical theory. Such penetration or *tunneling* into or through potential energy barriers is typical of quantum-theory results. The tunneling concept has been used as a model for radioactive decay of nuclei and for explaining some anomalously high chemical reaction rates. Tunneling is also important in solid-state physics and has been put to practical use in the design of certain electronic devices (e.g., tunnel diodes).

The possibility of tunneling in quantum mechanics illustrates an important point about the theory. If the electron had a value of $r > 2a_0$, then its kinetic energy would have to be *negative* to satisfy the condition $E = T + V$, with $V > E$. A negative kinetic energy is absurd, and in fact does not have to be introduced to explain tunneling if sufficient care is used. In quantum mechanics, it is meaningful to say that a particle is in a certain region only if a measurement of the particle position has been made. This measurement introduces an uncertainty in the particle momentum and a corresponding uncertainty in the particle kinetic energy. It can be shown that a measurement which proves that the particle is in the classically forbidden region introduces an uncertainty in the kinetic energy sufficiently large to compensate for the negative value required by conservation of energy. Thus, consideration of the measuring process and its relation to the phenomenon leads to consistent results in quantum mechanics.

3.5 SPHERICAL EXCITED STATES OF ONE-ELECTRON ATOMS

When we sought the ground-state wave function for hydrogen, we began by assuming that the angular functions Θ and Φ in Eq. 3.11 were constant, so that there would be no angular nodes. This resulted in Eq. 3.13, which was satisfied by the radial function $R(r)$. In Section 3.3 a nodeless solution to Eq. 3.13 was found and was taken to be the ground state. We now seek the wave function for the first excited state. It is reasonable to suppose that this solution has a single node. If we assume this node to be in the radial function R rather than the angular functions Θ and Φ, we can proceed in much the same way as before; that is, we take Θ and Φ to be constant and solve Eq. 3.13 for $R(r)$, subject to the requirement that $R(r)$ cross the zero axis for some value of r. As for the ground state, we expect that the wave function approaches zero asymptotically at large r, since we are dealing with a bound state, which is stabilized by a large negative potential energy.

An appropriate function with such properties is a first-degree polynomial in r multiplied by an exponential function, similar to the one used for the ground state; that is,

$$\psi_2(r) = N_2(b - r)e^{-ar} \tag{3.32}$$

The factor $(b - r)$ assures us of a radial node at $r = b$, while the exponential function provides for the appropriate asymptotic behavior at large r. The constants a and b and the energy E are evaluated in the same way as for the ground state: The function given in Eq. 3.32 is substituted for $R(r)$ in Eq. 3.13, the coefficients of the different powers of r are collected, and they are separately set equal to zero because Eq. 3.13 must hold for all r. The results obtained (see Problem 3.5) are

$$a = \frac{Z}{2a_0}$$

$$b = \frac{2a_0}{Z}$$

$$E_2 = -\frac{1}{4}\frac{Z^2e^2}{2a_0}$$

and the normalized wave function can be written

$$\psi_2(r) = \frac{1}{4}\left(\frac{Z^3}{2\pi a_0^3}\right)^{1/2}\left(2 - \frac{Zr}{a_0}\right)\exp\left(-\frac{Zr}{2a_0}\right) \tag{3.33}$$

The same procedure can be repeated for states with 2, 3, 4, . . . nodes in the radial function $R(r)$ and with constant angular functions. In each case the wave function is an exponential function of r, multiplied by a polynomial in r whose degree is the number of nodes. Specifically, if $n - 1$ is the number of nodes, the wave functions for spherically symmetric states (no θ or ϕ dependence) can be written

$$\psi_n(r) = N_n\left(\sum_{i=0}^{n-1} b_i r^i\right)e^{-ar} \tag{3.34}$$

where n is an integer which is called the *principal quantum number;* it can take on the values

$$n = 1, 2, 3, \ldots$$

and is associated with a function that has $n - 1$ nodes, all of which are radial for the spherically symmetric solutions. For the nodeless ground state, $n = 1$.

When the general radial wave function (Eq. 3.34) is substituted into Eq. 3.13, which is valid as long as Θ and Φ are constants, it is found that one can choose the b_i and a so that $\psi_n(r)$ satisfies the equation for all values of r, if the energy E has a certain value E_n that depends on the principle quantum number n. For $n = 1$ and $n = 2$, we have seen that $E_1 = -Z^2e^2/2a_0$ and $E_2 = -\frac{1}{4}Z^2e^2/2a_0$, respectively. In the general case, it can be shown that

$$E_n = -\frac{1}{n^2}\frac{Z^2e^2}{2a_0} \qquad n = 1, 2, 3, \ldots \tag{3.35}$$

Comparison of Eqs. 3.35 and 1.57 demonstrates that for spherically symmetric states of the hydrogen atom, the energy levels (and therefore the Rydberg constant) predicted by the Bohr model and the quantum theory are identical. The spherically symmetric wave functions are discussed further below (see Section 3.10).

3.6 FUNCTIONS WITHOUT SPHERICAL SYMMETRY

Let us now look at the hydrogen-atom states in which the wave function is not spherically symmetric. The wave function is of the general form given in Eq. 3.11 and satisfies the complete Schrödinger equation (Eq. 3.9). The possibility now arises for nodes in the angular functions $\Theta(\theta)$ and $\Phi(\phi)$, since they are no longer constant for states that are not spherically symmetric. As the simplest functions of this type, we consider those with one angular node. They have the form $x\,f(r)$, $y\,f(r)$, $z\,f(r)$ or $r\sin\theta\cos\phi\,f(r)$, $r\sin\theta\sin\phi\,f(r)$, $r\cos\theta\,f(r)$, with $f(r)$ to be determined by substitution into the Schrödinger equation. These three functions are equivalent except for their orientation, and each has one angular node; for example, for any value of r, the function $z\,f(r)$ is zero for $\cos\theta = 0$ (i.e., $\theta = \pi/2$).

To see if these functions satisfy the Schrödinger equation, we substitute them into Eq. 3.9. Considering $x\,f(r) = r\sin\theta\cos\phi\,f(r)$ as an example, we have for the angular derivatives,

$$\frac{\partial^2[r\sin\theta\cos\phi\,f(r)]}{\partial\phi^2} = r\sin\theta\,f(r)\,\frac{\partial^2(\cos\phi)}{\partial\phi^2}$$
$$= -r\sin\theta\cos\phi\,f(r)$$

and

$$\frac{1}{\sin\theta}\frac{\partial}{\partial\theta}\sin\theta\frac{\partial}{\partial\theta}[r\sin\theta\cos\phi\,f(r)]$$
$$= r\cos\phi\,f(r)\,\frac{1}{\sin\theta}\frac{\partial}{\partial\theta}\left(\sin\theta\frac{\partial\sin\theta}{\partial\theta}\right)$$
$$= r\cos\phi\,f(r)\left(\frac{\cos^2\theta - \sin^2\theta}{\sin\theta}\right)$$

The radial derivative is

$$
\frac{1}{r^2}\frac{\partial}{\partial r}\left(r^2\frac{\partial}{\partial r}\left[r\sin\theta\cos\phi\,f(r)\right]\right)
$$
$$
= \sin\theta\cos\phi\,\frac{1}{r^2}\frac{\partial}{\partial r}\left(r^2\frac{\partial}{\partial r}\left[rf(r)\right]\right)
$$
$$
= \sin\theta\cos\phi\left(r\frac{d^2f}{dr^2}+4\frac{df}{dr}+\frac{2}{r}f(r)\right)
$$

Introducing these results into Eq. 3.9, we obtain

$$
-\frac{\hbar^2}{2m}\left(\frac{d^2f}{dr^2}+\frac{4}{r}\frac{df}{dr}+\frac{2}{r^2}f(r)+\frac{\cos^2\theta-\sin^2\theta}{r^2\sin^2\theta}f(r)\right.
$$
$$
\left.-\frac{1}{r^2\sin^2\theta}f(r)\right)-\frac{Ze^2}{r}f(r)=Ef(r)\quad(3.36)
$$

where we have divided[3] both sides by $r\sin\theta\cos\phi$. Combining the angular terms, we see that they cancel the term $(2/r^2)f(r)$, so that Eq. 3.36 reduces to the radial equation

$$
-\frac{\hbar^2}{2m}\left(\frac{d^2f}{dr^2}+\frac{4}{r}\frac{df}{dr}\right)-\frac{Ze^2}{r}f(r)=Ef(r)\quad(3.37)
$$

The identical equation is obtained from the function $y\,f(r)$ and $z\,f(r)$, verifying the assumption that the radial part $f(r)$ is the same for all three functions (see Problem 3.6). The form of Eq. 3.37 is very similar to the previous radial equation (Eq. 3.13), the only difference being that the factor $2/r$ in front of the first-derivative term is now $4/r$. This suggests that we try the same exponential-times-polynomial form for $f(r)$ as was used for $R(r)$. Furthermore, since we are interested in the lowest state with one angular node, we assume a simple nodeless exponential; that is, $f(r)=e^{-ar}$ with a to be determined from Eq. 3.37. The result of setting the coefficient of each power of r equal to zero is

$$
a=\frac{Z}{2a_0}\qquad E=-\frac{1}{4}\frac{Z^2e^2}{2a_0}
$$

Evaluation of the normalization constant yields

---

[3] Strictly speaking, one can divide by $r\sin\theta\cos\phi$ only if θ and ϕ have values for which $\sin\theta\cos\phi \neq 0$. Thus, for $\theta=0,\pi$ or for $\phi=\pi/2,3\pi/2$, Eq. 3.9 is satisfied by any function $f(r)$, while for all values of θ and ϕ other than these, $f(r)$ must satisfy Eq. 3.37. The requirement that the wave function be continuous (otherwise, there would be unacceptable discontinuities in the probability density) means that $f(r)$ must satisfy Eq. 3.37 for all values of θ and ϕ. The result is therefore the same as if we divide both sides of the Schrödinger equation by $r\sin\theta\cos\phi$, without regard to the values of θ and ϕ.

where the quantum number m is a positive or negative integer or zero. The sign of m determines whether the z component of angular momentum points in the positive or negative z direction. In contrast to the Bohr-theory rule for p_ϕ, however, the quantum number m is distinct from the principal quantum number n. As we shall see later, m may be measured by observing the spectrum of the atom in a magnetic field; for this reason, m is called the *magnetic quantum number*. Since M_z is one component of the angular momentum M, the absolute value of M_z cannot exceed M; that is, the inequality $|m| \leq [l(l+1)]^{1/2}$ must hold. With the additional requirement that both m and l are integers, the allowed values of m are seen to be

$$m = 0, \pm 1, \pm 2, \ldots, \pm l$$

Specified in this way, the absolute value of the quantum number m, that is, $|m|$, turns out to be the number of nodal planes of $\psi(r, \theta, \phi)$ that are perpendicular to the xy plane.

Summarizing the above discussion, we conclude that the quantum numbers n, l, and m specify the hydrogen-atom wave functions, with allowed values

$$
\begin{aligned}
n &= 1, 2, 3, 4, \ldots \\
l &= 0, 1, 2, 3, \ldots, n-1 \\
m &= 0, \pm 1, \pm 2, \pm 3, \ldots, \pm l
\end{aligned}
\tag{3.42}
$$

The principal quantum number n determines the total energy,

$$E_n = -\frac{1}{n^2}\left(\frac{Z^2 e^2}{2a_0}\right)$$

the azimuthal quantum number l determines the square of the total angular momentum

$$M^2 = l(l+1)\hbar^2$$

and the magnetic quantum number m determines the z component of the angular momentum

$$M_z = m\hbar$$

Since the energy of the hydrogen atom is given by Eq. 3.35, states with different l and m but with the same n are degenerate. For a given n, Eq. 3.42 shows that there are n different possible values of l. In turn, for each value of l, Eq. 3.42 shows that $2l + 1$ different values of m are possible. The total number of states with the same energy (that is, the degeneracy) is

$$g = \sum_{l=0}^{n-1}(2l+1) = 1 + 3 + 5 + \cdots + 2n - 1 = \frac{(1 + 2n - 1)n}{2} = n^2$$

$$\tag{3.43}$$

It should be emphasized again that this n^2-fold degeneracy for the hydrogen atom or hydrogen-like atoms is due to the fact that $V(r) = -Ze^2/r$. A change in the form of the potential energy to a non-Coulombic interaction makes the energy depend on both n and l. If in addition the spherical symmetry of the potential is removed by applying an external magnetic field, the energy depends on m as well.

3.8 WAVE FUNCTIONS FOR ONE-ELECTRON ATOMS

In terms of the quantum numbers n, l, and m, the wave functions for a one-electron atom with nuclear charge Ze can be specified as

$$\psi_{nlm}(r,\ \theta,\ \phi) = R_{nl}(r)\,\Theta_{lm}(\theta)\Phi_m(\phi) \tag{3.44}$$

Here the quantum numbers associated with each of the factors R, Θ, and Φ are the ones required to determine the form of that function uniquely. Table 3.1 lists the quantum numbers and the corresponding functions of all the states ψ_{nlm} with $n = 1$, 2, and 3. The first five functions in the table are the ones that we have considered above. Comparison of the functions in Table 3.1 for various n, l, m shows that the R factors depend on both n and l, that the Θ factors depend on l and m, and that the Φ factors depend on m alone.

The wave functions given in Table 3.1 are all normalized; that is,

$$\iiint\limits_{\text{all space}} \psi_{nlm}^2\ (r,\ \theta,\ \phi)\ r^2\ dr\ \sin\theta\ d\theta\ d\phi = 1$$

the required normalization constants being given at the bottom of the table. Another property of the functions in the table is that they are *orthogonal*; that is,

$$\iiint\limits_{\text{all space}} \psi_{nlm}\,\psi_{n'l'm'}\ r^2\ dr\ \sin\theta\ d\theta\ d\phi = 0 \tag{3.45}$$

for one or more of the quantum numbers $n'l'm'$ not equal to nlm; that is, the integral in Eq. 3.45 is zero for two hydrogen-atom wave functions as long as the two functions are different. The orthogonality of different wave functions in quantum mechanics is a general property. The term "orthogonal" is used in analogy to the orthogonality of two vectors $\mathbf{a}$, $\mathbf{b}$ which point at $90°$ so that their scalar product $(a_x b_x + a_y b_y + a_z b_z)$ is zero. To illustrate the orthogonality, we consider the integral

$$\iiint\limits_{\text{all space}} \psi_{100}(r,\ \theta,\ \phi)\psi_{210}(r,\ \theta,\ \phi)\ r^2\ dr\ \sin\theta\ d\theta\ d\phi$$

$$= N_1 N_2 \int_0^\infty \exp\left(-\frac{Zr}{a_0}\right)\frac{Zr}{a_0}\exp\left(-\frac{Zr}{2a_0}\right)r^2\ dr \int_0^\pi \cos\theta\ \sin\theta\ d\theta \int_0^{2\pi} d\phi$$

Table 3.1. *States of one-electron atoms*

n	l	$|m|$	Spectroscopic designation	E_n in units of $e^2/2a_0$	g	$\psi_{n,l,m}(r,\theta,\phi)$
1	0	0	$1s$	-1	1	$N_1 \exp(-Zr/a_0)$
2	0	0	$2s$	$-\tfrac{1}{4}$	4	$N_2(2 - Zr/a_0)\exp(-Zr/2a_0)$
2	1	0	$2p_z$	$-\tfrac{1}{4}$		$N_2(Zr/a_0)\exp(-Zr/2a_0)\cos\theta$
2	1	1, cos	$2p_x$	$-\tfrac{1}{4}$		$N_2(Zr/a_0)\exp(-Zr/2a_0)\sin\theta\cos\phi$
2	1	1, sin	$2p_y$	$-\tfrac{1}{4}$		$N_2(Zr/a_0)\exp(-Zr/2a_0)\sin\theta\sin\phi$
3	0	0	$3s$	$-\tfrac{1}{9}$	9	$N_3[27 - 18(Zr/a_0) + 2(Zr/a_0)^2]\exp(-Zr/3a_0)$
3	1	0	$3p_z$	$-\tfrac{1}{9}$		$N_3\sqrt{6}\,(6 - Zr/a_0)(Zr/a_0)\exp(-Zr/3a_0)\cos\theta$
3	1	1, cos	$3p_x$	$-\tfrac{1}{9}$		$N_3\sqrt{6}\,(6 - Zr/a_0)(Zr/a_0)\exp(-Zr/3a_0)\sin\theta\cos\phi$
3	1	1, sin	$3p_y$	$-\tfrac{1}{9}$		$N_3\sqrt{6}\,(6 - Zr/a_0)(Zr/a_0)\exp(-Zr/3a_0)\sin\theta\cos\phi$
3	2	0	$3d_{3z^2-r^2}$	$-\tfrac{1}{9}$		$N_3\sqrt{1/2}\,(Zr/a_0)^2\exp(-Zr/3a_0)(3\cos^2\theta - 1)$
3	2	1, cos	$3d_{zx}$	$-\tfrac{1}{9}$		$N_3\sqrt{6}(Zr/a_0)^2\exp(-Zr/3a_0)\sin\theta\cos\theta\cos\phi$
3	2	1, sin	$3d_{zy}$	$-\tfrac{1}{9}$		$N_3\sqrt{6}(Zr/a_0)^2\exp(-Zr/3a_0)\sin\theta\cos\theta\sin\phi$
3	2	2, cos	$3d_{x^2-y^2}$	$-\tfrac{1}{9}$		$N_3\sqrt{3/2}(Zr/a_0)^2\exp(-Zr/3a_0)\sin^2\theta\cos 2\phi$
3	2	2, sin	$3d_{zy}$	$-\tfrac{1}{9}$		$N_3\sqrt{3/2}(Zr/a_0)^2\exp(-Zr/3a_0)\sin^2\theta\sin 2\phi$

$$N_1 = \left(\frac{Z^3}{\pi a_0^3}\right)^{1/2}, \quad N_2 = \frac{1}{4}\left(\frac{Z^3}{2\pi a_0^3}\right)^{1/2}, \quad N_3 = \frac{1}{81}\left(\frac{Z^3}{3\pi a_0^3}\right)^{1/2}$$

where we have used Eq. 3.24 for the ranges of the polar coordinates. Although the r and ϕ integrals are not zero for this case, the orthogonality condition is satisfied by the θ integral since

$$\int_0^\pi \cos\theta \sin\theta \, d\theta = \int_{-1}^{+1} x \, dx = 0$$

Before looking at the electron distributions corresponding to the various states of the hydrogen atom, a word should be said about the ϕ dependence of the functions given in Table 3.1. When the variables of Eq. 3.9 are separated and the equation for $\Phi_m(\phi)$ is solved, one possible set of normalized functions is

$$\Phi_0 = (2\pi)^{-1/2}$$
$$\Phi_{|m|\cos} = \pi^{-1/2} \cos|m|\phi \qquad (3.46)$$
$$\Phi_{|m|\sin} = \pi^{-1/2} \sin|m|\phi$$

where $m = 0, \pm1, \pm2, \ldots$. These are just the $\Phi_m(\phi)$ functions in Table 3.1. They depend only on $|m|$ (the absolute value of m), and can be distinguished by their cosine or sine dependence upon ϕ. However, there exists an alternative set of solutions, that are directly related to the sign as well as the magnitude of m. These can be written in the form

$$\Phi_m(\phi) = (2\pi)^{-1/2} (\cos m\phi + i \sin m\phi) = (2\pi)^{-1/2} e^{im\phi} \qquad (3.47)$$

where $m = 0, \pm1, \pm2, \ldots$; thus, Eq. 3.47 is valid for m positive, negative, or zero. Clearly for $m \neq 0$, the function $\Phi_m(\phi)$, and therefore $\psi_{nlm}(r, \theta, \phi)$, is complex when Eq. 3.47 is used. This means that the relationship between the coordinate probability distribution and the wave function must be generalized from that given in Section 2.6.3, since a physically meaningful probability must be real, as well as positive and normalized The correct generalization for a complex wave function is that the probability $P(r, \theta, \phi)$ of finding a particle in the volume element $r^2 \, dr \sin\theta \, d\theta \, d\phi$ is the square of the absolute value of the wave function, namely,

$$P(r, \theta, \phi) = \psi^*(r, \theta, \phi)\psi(r, \theta, \phi) = |\psi(r, \theta, \phi)|^2 \qquad (3.48)$$

where $\psi^*(r, \theta, \phi)$ is the complex conjugate of $\psi(r, \theta, \phi)$, obtained by replacing i wherever it appears in $\psi(r, \theta, \phi)$ by $-i$. Thus, for the particular function $\Phi_m(\phi)$ given in Eq. 3.47, we have

$$\Phi_m^*(\phi) = (2\pi)^{-1/2}(\cos m\phi - i \sin m\phi) = (2\pi)^{-1/2}e^{-im\phi}$$

and

$$|\Phi_m(\phi)|^2 = \Phi_m^*(\phi)\Phi_m(\phi) = \frac{1}{2\pi}(\cos^2 m\phi + \sin^2 m\phi)$$
$$= \frac{1}{2\pi} e^{-im\phi}e^{im\phi} = \frac{1}{2\pi}$$

The complex functions $\Phi_m(\phi)$ and the sign of m are important primarily when the electron is acted upon by an external magnetic field. For most problems in the absence of magnetic fields, we may use the real functions of Eq. 3.46. They are related to the complex solutions in a simple way, as can be seen from Eq. 3.47. Taking $\Phi_m(\phi)$ and $\Phi_{-m}(\phi)$ with $m > 0$ and forming the real linear combinations

$$\frac{1}{\sqrt{2}}[\Phi_m(\phi) + \Phi_{-m}(\phi)] = \pi^{-1/2}\cos m\phi \qquad m > 0$$

and .

$$\frac{1}{i\sqrt{2}}[\Phi_m(\phi) - \Phi_{-m}(\phi)] = \pi^{-1/2}\sin m\phi \qquad m > 0$$

we obtain the functions in Eq. 3.46.

Table 3.1 lists the real functions because they are particularly convenient in chemical applications. This is a consequence of the fact that the real angular parts behave like Cartesian coordinates (e.g., like x, y, and z for $n = 2, l = 1$) or simple functions of Cartesian coordinates, so that they are well suited for describing the directional properties of chemical bonds. This property of the functions, which is discussed further below, permits us to distinguish the real wave functions with a given n and l by symbols such as $x, y, z, xy, x^2 - y^2$, etc., as shown in column 4 of Table 3.1. The first symbol in that column is the principal quantum number n, as can be seen by comparing it with column 1. The second symbol indicates the quantum number l by means of the following correspondence

$l = 0$	1	2	3	4	5
$\updownarrow$	$\updownarrow$	$\updownarrow$	$\updownarrow$	$\updownarrow$	$\updownarrow$
s	p	d	f	g	h

which arose from observations and interpretations of the one-electron spectra of atoms prior to the advent of quantum mechanics; the letters s, p, d, and f are the first letters of the adjectives "sharp," "principal," "diffuse," and "fundamental," respectively; letters corresponding to higher values of l follow in alphabetical order. Thus, for example, a $3p_x$ function, which is often called a $3p_x$ orbital, is a wave function with $n = 3, l = 1$, and $|m| = 1$; the θ and ϕ dependence is $\sin\theta\cos\phi$, corresponding to the coordinate x. This type of designation for atomic functions, as given in column 4 of Table 3.1, is the one most often used in chemistry.

3.9   THE VIRIAL THEOREM FOR THE HYDROGEN ATOM

In our discussion of the classical Rutherford planetary model for the hydrogen atom (Chapter 1), we mentioned that it obeys the virial theorem for a system of bound particles with electrostatic interaction potentials, namely,

$$E = \tfrac{1}{2}\overline{V} = -\overline{T}$$

where, in the general classical case, $\overline{V}$ and $\overline{T}$ are time averages of the potential and kinetic energies, respectively. In the quantum theory, the virial theorem for the hydrogen atom takes the analogous form

$$E_n = \tfrac{1}{2}\langle V \rangle_{nlm} = - \langle T \rangle_{nlm}$$

where $\langle V \rangle_{nlm}$ and $\langle T \rangle_{nlm}$ are now the quantum-mechanical average values

$$\langle V \rangle_{nlm} = \left\langle - \frac{Ze^2}{r} \right\rangle_{nlm}$$

$$= -\int_0^\infty \int_0^{2\pi} \int_0^\pi \psi_{nlm}(r, \theta, \phi) \frac{Ze^2}{r} \psi_{nlm}(r, \theta, \phi) r^2 \sin\theta \, dr \, d\theta \, d\phi$$

$$\langle T \rangle_{nlm} = \left\langle \frac{p^2}{2m} \right\rangle_{nlm}$$

$$= -\int_0^\infty \int_0^{2\pi} \int_0^\pi \psi_{nlm}(r, \theta, \phi) \frac{\hbar^2}{2m} \left(\frac{\partial^2}{\partial x^2} + \frac{\partial^2}{\partial y^2} + \frac{\partial^2}{\partial z^2} \right) \psi_{nlm}(r, \theta, \phi)$$

$$r^2 \sin\theta \, dr \, d\theta \, d\phi$$

As a simple example, we calculate $\langle V \rangle_{100}$ and $\langle T \rangle_{100}$ from the ground-state wave function $\psi_{100} = (Z^3/\pi a_0^3)^{1/2} \exp(-Zr/a_0)$.

$$\langle V \rangle_{100} = - \left(\frac{Z^3}{\pi a_0^3} \right) (Ze^2) \int_0^{2\pi} d\phi \int_0^\pi \sin\theta \, d\theta \int_0^\infty \exp\left(-\frac{2Zr}{a_0} \right) r \, dr = - \frac{Z^2 e^2}{a_0}$$

$$\langle T \rangle_{100} = - \left(\frac{\hbar^2}{2m} \right) \left(\frac{Z^3}{\pi a_0^3} \right) \int_0^{2\pi} d\phi \int_0^\pi \sin\theta \, d\theta \int_0^\infty \exp\left(-\frac{Zr}{a_0} \right)$$

$$\times \left[\frac{1}{r^2} \frac{d}{dr} r^2 \frac{d}{dr} \exp\left(-\frac{Zr}{a_0} \right) \right] r^2 \, dr$$

$$= - \left(\frac{\hbar^2}{2m} \right) \left(\frac{Z^3}{\pi a_0^3} \right) (4\pi) \left(\frac{Z}{a_0} \right) \int_0^\infty \left(\frac{Z}{a_0} r^2 - 2r \right) \exp\left(-\frac{2Zr}{a_0} \right) dr = \frac{Z^2 e^2}{2a_0}$$

Thus, $\langle V \rangle_{100}$ and $\langle T \rangle_{100}$ are seen to satisfy the virial theorem.

For a general state nlm, we obtain

$$\langle V \rangle_{nlm} = 2E_n = - \frac{Z^2 e^2}{n^2 a_0}$$

$$\langle T \rangle_{nlm} = -E_n = \frac{Z^2 e^2}{2n^2 a_0}$$

3.10 GEOMETRIC DETAILS OF HYDROGEN-LIKE ORBITALS

An advantage of the Bohr model, with its fixed orbits for the various states of the electron, was its pictorial simplicity. We have seen that in the quantum-mechanical description of the atom, the orbits are replaced with the probability distribution functions for the coordinates of the electron, such as the one shown in Fig. 3.4b for the ground state of hydrogen. An electron in an atom is thus represented by a probability "cloud" whose density, as given by the square of the wave function, varies from point to point. Different states of the electron (corresponding to different wave functions) have distributions with different shapes and density patterns. Although the concept of a classical orbit for an electron in an atom is an imprecise one in quantum mechanics, plots of the radial factor $R(r)$ and angular factor $\Theta\Phi$ can be used to give pictorial representations of the electronic wave functions for the various states. These one-electron wave functions are called atomic orbital functions, or more often *atomic orbitals*, by analogy with the simpler Bohr picture in which the orbits are well defined. In Figs. 3.5–3.12, we illustrate the behavior of the functions listed in Table 3.1.

The radial functions $R_{nl}(r)$ are plotted in Fig. 3.5a. These functions have been normalized so that

$$\int_0^\infty [R_{nl}(r)]^2 r^2 \, dr = 1$$

Note that the number of radial nodes is $n - l - 1$ and that only the s functions ($l = 0$) have nonzero values at the origin.

The squares of the radial functions are plotted in Fig. 3.5b. These plots give the electron density as a function of distance from the nucleus measured along a given direction (at fixed θ and ϕ). It is seen that the s orbitals have a fairly large density at the nucleus, while the p and d orbitals have zero probability that the electron is at the nucleus. We shall make use of this result in Chapter 4 in the discussion of shielding in many-electron atoms.

The plots of $r^2 R_{nl}^2$ in Fig. 3.5c represent the radial distribution functions; that is, they give the probability per unit length that the electron will be found a distance r from the nucleus, regardless of direction. The number of maxima in these functions is $n - l$. The single maxima for the $1s$, $2p$, and $3d$ orbitals fall at a_0/Z, $4a_0/Z$, and $9a_0/Z$, respectively; they give a rough estimate of the extension of the electron clouds for these states. As the principal quantum number n increases, the clouds become more expanded, since the most probable distance is proportional to n^2. As the nuclear charge Z increases, the clouds are more contracted, the most probable distance being proportional to $1/Z$. This relation of atomic size to n and Z is the same as that predicted by the Bohr model, except that the diffuseness of the electron cloud makes the concept of size somewhat less precise in the quantum-mechanical theory.

The angular distribution of the electron can be shown by polar plots of the function $\Theta\Phi$ in one or more planes through the origin. In the polar plots shown in Figs. 3.6–3.8, the distance measured along a given direction from the origin to the curve is proportional to the value of the function $\Theta\Phi$; that is, for fixed r it is the relative value of the orbital in the particular θ, ϕ directions. For example, in Fig. 3.6a the distance from the origin to the curve is constant, since the s orbital has no angular dependence. In Fig. 3.6b, the distance is $|\cos\theta|$; the result is two circular lobes tangential to the horizontal axis, the upper lobe corresponding to the positive values of $\cos\theta$ and the lower lobe corresponding to negative values. Since p_z is independent of ϕ, the orbital is symmetric about the z axis and the polar plot for the zx plane is the same as that for zy. The functions p_x and p_y have angular plots of the same shape as that of p_z, except that they are rotated so that the centers of the two lobes lie on the x and y axes, respectively. Note that the angular parts of the p functions have a single nodal plane, which is the expected result since $l = 1$ for these functions.

Polar plots of the five d orbitals are shown in Fig. 3.7 for various planes. The $d_{3z^2-r^2}$ function displays large positive lobes along the z axis and small negative lobes in the xy plane. Since the function is symmetric about the z axis, polar plots of $d_{3z^2-r^2}$ in the zx and yz planes are identical. The polar plots of the d_{zx} and d_{yz} functions have four lobes of alternating sign at right angles to one another in the zx and yz planes, respectively. For these functions, lobes of the same sign lie along axes that make 45° angles with the coordinate axes. For any fixed value of θ greater than zero and less than $\pi/2$, polar plots of d_{zx} and d_{yz} in a plane parallel to the xy plane consist of two circular lobes along the x and y axes, respectively, as shown in the right-hand figures of Figs. 3.7b and 3.7c. According to the right-hand figure of Fig. 3.7d, the function $d_{x^2-y^2}$ has a polar plot in the xy plane that is identical in form to that of d_{zx} in the zx plane, with the positive and negative lobes of the former oriented along the x and y axes, respectively. A cross section of a lobe is obtained by setting $\phi = 0$ and plotting $d_{x^2-y^2}$ in the zx plane (see the left-hand figure of Fig. 3.7d). Except for $d_{3z^2-r^2}$, the d functions shown in Fig. 3.7 also have identical shapes but different orientations. Since $l = 2$, there are two nodal surfaces in the angular plot of each d function. For $d_{x^2-y^2}$, d_{xy}, d_{yz}, and d_{zx}, the nodal surfaces are planes intersecting each other at right angles; for $d_{3z^2-r^2}$, the nodal surfaces are right circular cones.

The relation of $d_{3z^2-r^2}$ to the other d functions is not immediately apparent from the shape of its angular plot. However, let us consider two alternative angular functions which we call $d_{z^2-x^2}$ and $d_{z^2-y^2}$; their angular dependence is given by

$$d_{z^2-x^2} \sim \cos^2\theta - \sin^2\theta \cos^2\phi$$
$$d_{z^2-y^2} \sim \cos^2\theta - \sin^2\theta \sin^2\phi \qquad (3.49)$$

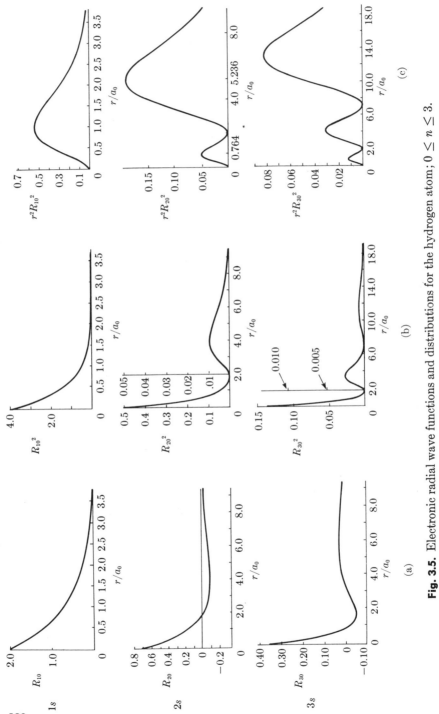

Fig. 3.5. Electronic radial wave functions and distributions for the hydrogen atom; $0 \leq n \leq 3$.

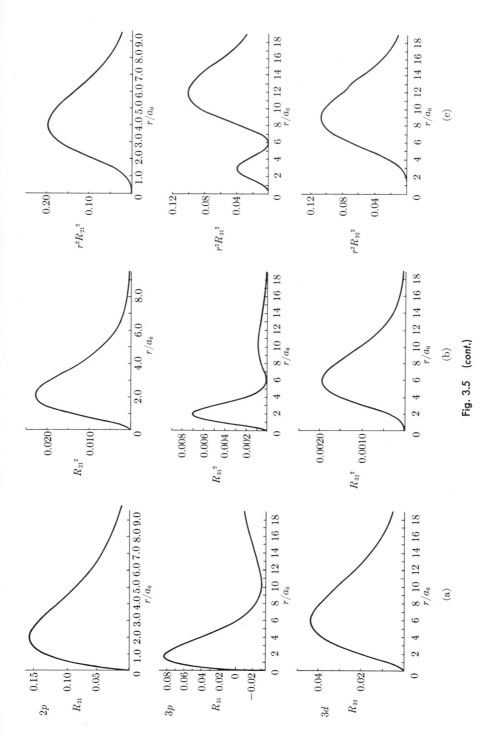

Fig. 3.5 (cont.)

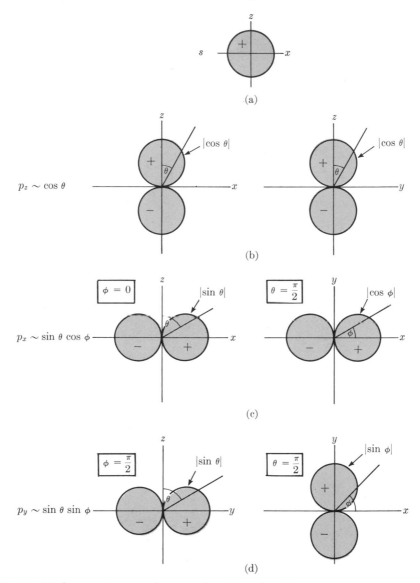

Fig. 3.6. Hydrogen-atom angular wave functions; $l = 0, 1$.

Polar plots of these functions (see Problem 3.7) in the zx, zy, and xy planes show that they have the same shape as the $d_{x^2-y^2}$, d_{xy}, d_{yz}, and d_{zx} functions, but with the lobes oriented along different directions. The $d_{z^2-x^2}$ function has positive lobes along the z axis and negative lobes along the x axis; the

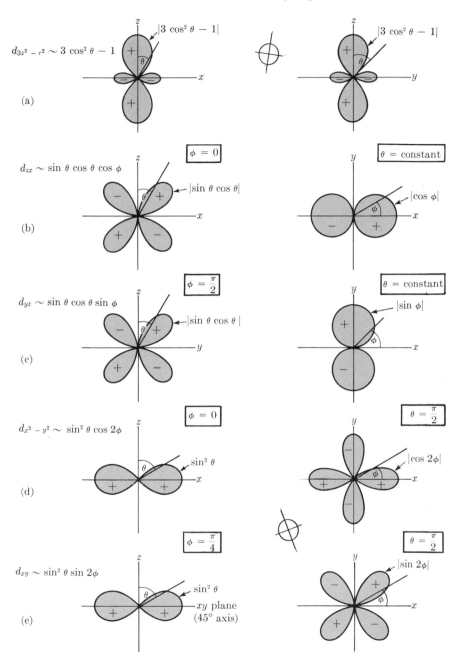

Fig. 3.7. Hydrogen-atom angular wave functions; $l = 2$.

$f_{z(5z^2 - 3r^2)} \sim \cos\theta \, (5\cos^2\theta - 3)$

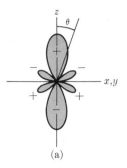

(a)

$f_{x(5z^2 - r^2)} \sim \sin\theta \, (5\cos^2\theta - 1) \cos\phi$

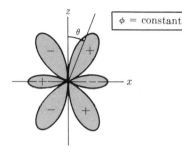

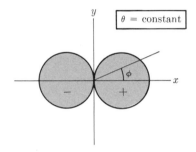

(b)

$f_{y(5z^2 - r^2)} \sim \sin\theta \, (5\cos^2\theta - 1) \sin\phi$

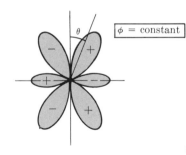

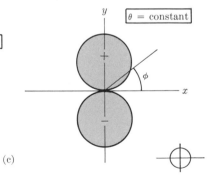

(c)

$f_{z(x^2 - y^2)} \sim \sin^2\theta \cos\theta \cos 2\phi$

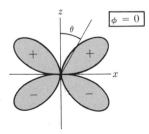

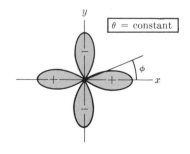

(d)

$f_{zxy} \sim \sin^2 \theta \cos \theta \sin 2\phi$

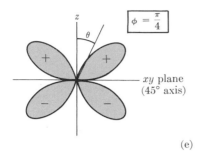

$$\boxed{\phi = \frac{\pi}{4}}$$

xy plane
(45° axis)

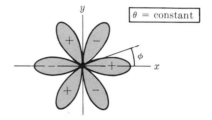

$$\boxed{\theta = \text{constant}}$$

(e)

$f_{x(x^2 - 3y^2)} \sim \sin^3 \theta \ (\cos^3 \phi - 3 \sin^2 \phi \cos \phi)$

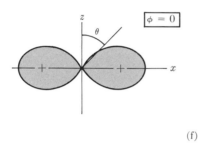

$$\boxed{\phi = 0}$$

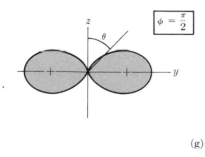

$$\boxed{\theta = \text{constant}}$$

(f)

$f_{y(y^3 - 3x^2)} \sim \sin^3 \theta \ (\sin^3 \phi - 3 \sin \phi \cos^2 \theta)$

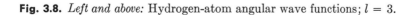

$$\boxed{\phi = \frac{\pi}{2}}$$

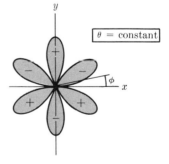

$$\boxed{\theta = \text{constant}}$$

(g)

Fig. 3.8. *Left and above:* Hydrogen-atom angular wave functions; $l = 3$.

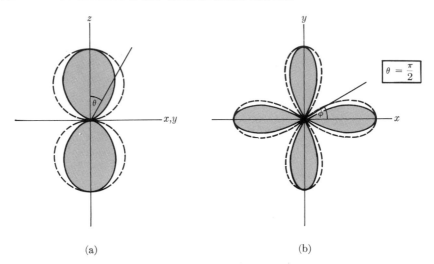

Fig. 3.9. Comparison of $|\Theta\Phi|$ and $\Theta^2\Phi^2$ for **(a)** p_z and **(b)** $d_{x^2-y^2}$ wave functions.

$d_{z^2-y^2}$ has positive lobes along the z axis and negative lobes along the y axis. If we take certain linear combinations of these two functions, we find

$$d_{z^2-y^2} + d_{z^2-x^2} \sim 2\cos^2\theta - \sin^2\theta = 3\cos^2\theta - 1 \sim d_{3z^2-r^2}$$
$$d_{z^2-y^2} - d_{z^2-x^2} \sim (\sin^2\theta)(\cos^2\phi - \sin^2\phi) = \sin^2\theta\cos 2\phi \sim d_{x^2-y^2}$$

Thus, we have shown $d_{3z^2-r^2}$ to be the sum of two functions with the characteristic shape of the other d functions. The particular linear combination involved, which results in the cylindrically symmetrical shape of $d_{3z^2-r^2}$, is a consequence of the requirement that all five d orbitals be orthogonal to one another (Problem 3.16); however, the choice of the z axis as the axis of symmetry for the nonequivalent function is simply one of mathematical convenience.

Plots of the higher nl orbitals become increasingly complicated. We give the polar diagrams for f orbitals in Fig. 3.8, because of their importance for the rare-earth elements.

Since the probability density is equal to the square of the wave function, polar plots of the function $\Theta^2\Phi^2$ are also of interest. In these plots (see Fig. 3.9) all of the lobes are, of course, positive. Comparison of the polar plot of $\Theta^2\Phi^2$ with that of $|\Theta\Phi|$ for the p_z orbital (Fig. 3.9a) shows that the general shapes are similar, but that the lobes of $\Theta^2\Phi^2$ are no longer circular. A corresponding result is found for the $d_{x^2-y^2}$ orbitals; for both p_z and $d_{x^2-y^2}$ the $\Theta^2\Phi^2$ lobes are narrower than are those of $|\Theta\Phi|$.

In our discussion of complex wave functions with Φ having the form of Eq. 3.47, we showed that the probability density (Eq. 3.48) is independent of ϕ, that is, symmetric about the z axis. Thus the functions

$$np_1 = \frac{1}{\sqrt{2}} (np_x + inp_y) = R_{n1}(r) \frac{\sqrt{3}}{2} \sin \theta \, (2\pi)^{-1/2} e^{i\phi}$$

$$np_{-1} = \frac{1}{\sqrt{2}} (np_x - inp_y) = R_{n1}(r) \frac{\sqrt{3}}{2} \sin \theta \, (2\pi)^{-1/2} e^{-i\phi}$$

both have the angular probability distribution

$$|\Theta\Phi|^2 = \frac{3}{8\pi} \sin^2 \theta$$

Polar plots of $|\Theta\Phi|^2$ for p_x, p_y, and $p_{\pm 1}$ appear in Fig. 3.10.

In the polar plots just discussed, r was, in effect, taken to be constant. If we wish to represent the probability density for an orbital $\psi_{nlm}{}^2(r,\theta,\phi)$ at various points in space, we must consider both the radial and angular

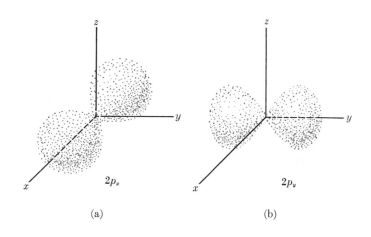

(a)   (b)

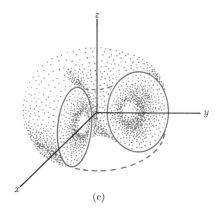

(c)

Fig. 3.10. Polar plots of $|\Theta\Phi|^2$ for **(a)** p_x, **(b)** p_y, and **(c)** $p_{\pm 1}$.

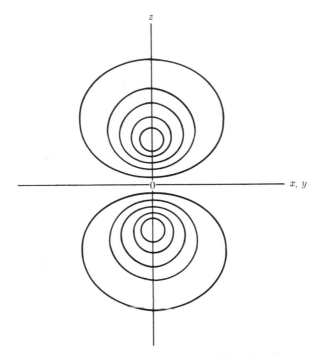

Fig. 3.11. Contour diagram of the $2p_z$ orbital probability density.

dependence. One method is to choose various planes through the origin and make a contour diagram showing lines of constant ψ^2 in each plane. Such a diagram for the probability density of the $2p_z$ orbital in planes containing the z axis is shown in Fig. 3.11. In connection with such contour diagrams, mention should be made of a common method of pictorial representation of both the size and shape of an atomic orbital by sketching the *90% boundary surface*. This surface is the contour of constant probability density within which the probability of finding the electron is 0.9. Such a sketch for the $2p_z$ orbital is shown in Fig. 3.12.

As we have already pointed out, the wave functions, ψ_{nlm}, for the various states of the hydrogen atom are sometimes called atomic orbitals. To preserve the pictorial semiclassical terminology of the older Bohr model, in which an electron in a particular state was said to "occupy the nth Bohr orbit," we can in quantum mechanics characterize an electronic state of the hydrogen atom by saying that the electron "occupies the 2s orbital." By this we mean that the quantum numbers $n = 2$, $l = 0$, $m = 0$ specify the electronic state of the atom and that the corresponding 2s wave function provides information about its probability distribution and other measurable properties.

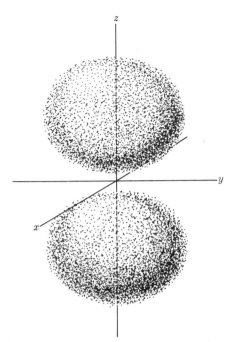

Fig. 3.12. The 90% boundary surface for the $2p_z$ orbital.

3.11 ENERGY LEVELS AND SPECTRUM OF THE HYDROGEN ATOM

Once the various possible states of the hydrogen atom and the corresponding energy levels are found by solving the Schrödinger equation, the Bohr frequency rule (Eq. 1.48) can still be used to determine the wave number of the spectral line corresponding to transitions from one level to another. The result is the same as for the Bohr model, in agreement with Eqs. 1.46 and 1.47. An assumption implicit in the derivation of Eq. 1.46 is that transitions can take place between states with any two values of n. A quantum-mechanical calculation of transition probabilities shows that this assumption is justified. However, it is found that only those transitions are allowed for which the azimuthal quantum number l changes by plus or minus one. This means that the Lyman series, corresponding to transitions to the $1s$ ground state (Fig. 1.10), must originate from the np states, while transitions to the $2p$ state must originate from the ns and nd states, and so on. The *selection rules* for the most common spectral transitions due to the interactions between the electron and the dipole electric field of the radiation are

$$n_1 - n_2 = \Delta n \text{ arbitrary}$$
$$l_1 - l_2 = \Delta l = \pm 1 \tag{3.50}$$
$$m_1 - m_2 = \Delta m = 0, \pm 1$$

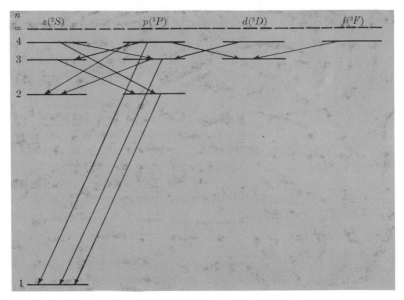

Fig. 3.13. States of the hydrogen atom and allowed transitions.

where the symbols n_1, l_1, m_1 correspond to the initial state and the symbols n_2, l_2, m_2 to the final state of the transition.

Figure 3.13 shows the states and allowed transitions for the hydrogen-atom spectrum for $n = 0$, 1, 2, 3, 4. Corresponding to $n = 1$, only an s ($l = 0$) state exists; for $n = 2$, we have s and p ($l = 1$) states; for $n = 3$, we have s, p, and d ($l = 2$) states; for $n = 4$, we have s, p, d, and f ($l = 3$) states, and so on.

3.11.1 *Time-dependent theory and selection rules* To determine the selection rules and the intensities of the spectral lines, we calculate the probability of finding the system in one state (say, the excited state) at time t, given that the system was in another state (say, the ground state) at time zero. Such a time-dependent problem requires that we introduce the *time-dependent* form of the Schrödinger equation, which is the quantum-mechanical analog of the classical equation of motion. When we introduced the *time-independent* Schrödinger equation for stationary states (Section 2.6), we used the relation between the energy of a free particle ($E = p^2/2m$) and the second derivative with respect to the coordinate x of the *standing* wave associated with the particle (Eq. 2.52),

$$A(x) = a_0 \cos\left(2\pi \frac{x}{\lambda}\right) = a_0 \cos\left(\frac{xp}{\hbar}\right)$$

If a *traveling* wave is used instead of the standing wave, a corresponding procedure allows us to obtain the time-dependent Schrödinger equation. This is what one would expect, since the equation of motion of a system describes the time dependence of the amplitude of the wave associated with it. For our present purposes, it is most convenient to make use of the complex exponential form of the wave, introduced in Section 2.2.1, namely,

$$A(x, t) = a_0 \exp\left[2\pi i\left(\frac{x}{\lambda} - \nu t\right)\right] \tag{3.51}$$

Separately differentiating $A(x, t)$ twice with respect to x and once with respect to t, we have

$$\frac{\partial^2 A(x, t)}{\partial x^2} = -\left(\frac{2\pi}{\lambda}\right)^2 A(x, t) \tag{3.52}$$

$$\frac{\partial A(x, t)}{\partial t} = -2\pi i \nu\, A(x, t) \tag{3.53}$$

Use of the de Broglie relation (Eq. 2.28) and the Bohr frequency rule (Eq. 1.48) to replace λ and ν in Eqs. 3.52 and 3.53 by the related quantities p and E gives

$$\frac{\partial^2 A(x, t)}{\partial x^2} = -\frac{p^2}{\hbar^2} A(x, t) \tag{3.54}$$

$$\frac{\partial A(x, t)}{\partial t} = -\frac{iE}{\hbar} A(x, t) \tag{3.55}$$

As in Section 2.6, we assume that because Eq. 3.51 corresponds to a free particle we can use the relation between the energy and momentum; setting $p^2 = 2mE$ in Eq. 3.54 and eliminating E between Eqs. 3.54 and 3.55, we obtain

$$-\frac{\hbar^2}{2m} \frac{\partial^2 A(x, t)}{\partial x^2} = i\hbar\, \frac{\partial A(x, t)}{\partial t}$$

or, introducing the usual symbol $\Psi(x, t)$ for the time-dependent wave function in place of $A(x, t)$,

$$-\frac{\hbar^2}{2m} \frac{\partial^2 \Psi(x, t)}{\partial x^2} = i\hbar\, \frac{\partial \Psi(x, t)}{\partial t} \tag{3.56}$$

Equation 3.56 is the one-dimensional time-dependent Schrödinger equation for a free particle (i.e., a particle with constant potential energy). To obtain the equation for a particle in the presence of a potential energy $V(x, t)$ that depends upon the coordinate and the time, we follow Section 2.9; that is, we assume that Eqs. 3.54 and 3.55 remain correct when there are forces acting on the particle. Substituting by Eq. 2.105 for p^2 in Eq. 3.54,

$$p^2 = 2m[E - V(x, t)]$$

and eliminating E between Eqs. 3.54 and 3.55, we find

$$-\frac{\hbar^2}{2m} \frac{\partial^2 \Psi(x, t)}{\partial x^2} + V(x, t)\, \Psi(x, t) = i\hbar\, \frac{\partial \Psi(x, t)}{\partial t} \tag{3.57}$$

Equation 3.57 is the one-dimensional time-dependent Schrödinger equation for particle of mass m with a potential energy $V(x, t)$.

Comparing Eq. 3.57 with the corresponding one-dimensional time-independent Schrödinger equation,

$$-\frac{\hbar^2}{2m}\frac{d^2}{dx^2}\Psi(x) + V(x)\,\Psi(x) = E\Psi(x) \tag{3.58}$$

we note some important similarities and differences. On the left-hand side of both equations are the operators associated with the kinetic energy

$$-\frac{\hbar^2}{2m}\frac{d^2}{dx^2}$$

and with the potential energy of the particle. However, the energy E which appears on the right-hand side of Eq. 3.58 is replaced in Eq. 3.57 by $i\hbar$ times the time derivative (i.e., by the operator $i\hbar\;\partial/\partial t$). This suggests that the operator $i\hbar(\partial/\partial t)$ is to be associated with the total energy of the system in the same way that the operator

$$\frac{\hbar}{i}\frac{\partial}{\partial x}$$

is associated with the momentum (see Section 2.7).

Another difference is that the potential energy $V(x, t)$ in Eq. 3.57 is a function of both the coordinate and the time, while the potential energy $V(x)$ in Eq. 3.58 depends on the coordinate alone. For an isolated system (e.g., a harmonic oscillator, $V = \frac{1}{2}kx^2$), V is independent of the time $[V = V(x)]$. In this case both equations can be used to determine the wave function; Eq. 3.58 can describe only a stationary state of the system (i.e., a state in which the probability density is independent of the time), while Eq. 3.57 can also describe the development of the system with time. If an external time-dependent field is applied (e.g., if the system is exposed to electromagnetic radiation) the potential V is a function of the time $[V = V(x, t)]$ and only Eq. 3.57 is applicable; that is, in the presence of a time-dependent potential, the system cannot remain in a single stationary (time-independent) state for an extended period.

For a three-dimensional system (e.g., the hydrogen atom), the time-dependent Schrödinger equation is (Problem 3.18)

$$-\frac{\hbar^2}{2m}\left(\frac{\partial^2\Psi}{\partial x^2} + \frac{\partial^2\Psi}{\partial y^2} + \frac{\partial^2\Psi}{\partial z^2}\right) + V(x, y, z, t)\,\Psi(x, y, z, t)$$
$$= i\hbar\frac{\partial}{\partial t}\Psi(x, y, z, t) \tag{3.59}$$

Applying Eq. 3.59 to an isolated hydrogen atom, we obtain

$$-\frac{\hbar^2}{2m}\left(\frac{\partial^2\Psi}{\partial x^2} + \frac{\partial^2\Psi}{\partial y^2} + \frac{\partial^2\Psi}{\partial z^2}\right) + V(r)\Psi = i\hbar\frac{\partial\Psi}{\partial t} \tag{3.60}$$

where $\Psi = \Psi(r, \theta, \phi, t)$. Since V $(= -e^2/r)$ is not a function of time in this case, we assume, by analogy with the separation used in Section 3.1, that we can write

$$\Psi(r, \theta, \phi, t) = \psi_{nlm}(r, \theta, \phi) \ \phi(t) \qquad (3.61)$$

where ψ_{nlm} is one of the known solutions of the time-independent Schrödinger equation (Eq. 3.9) and $\phi(t)$ is an unknown function of the time. Substituting Eq. 3.61 into Eq. 3.60, we have

$$\left[-\frac{\hbar^2}{2m}\left(\frac{\partial^2 \psi}{\partial x^2} + \frac{\partial^2 \psi}{\partial y^2} + \frac{\partial^2 \psi}{\partial z^2}\right) + V(r)\psi_{nlm}(r, \theta, \phi)\right]\phi(t) = i\hbar\psi_{nlm}(r, \theta, \phi)\ \frac{\partial \phi(t)}{\partial t}$$

In deriving this equation, we have used the fact that the kinetic energy involves only coordinate derivatives and therefore does not operate on $\phi(t)$, and that correspondingly $\partial/\partial t$ does not operate on $\psi_{nlm}(r, \theta, \phi)$. From the result for the stationary-state hydrogen atom (Eq. 3.9), the factor in square brackets on the left-hand side is equal to $E_n\psi_{nlm}(r, \theta, \phi)$, so that

$$E_n \ \psi_{nlm}(r, \theta, \phi) \ \phi(t) = i\hbar \ \psi_{nlm}(r, \theta, \phi) \ \frac{\partial \phi(t)}{\partial t}$$

or, on division of both sides by $\psi_{nlm}(r, \theta, \phi)$,

$$i\hbar \ \frac{d\phi(t)}{dt} = E_n\phi(t) \qquad (3.62)$$

Equation 3.62 is a first-order differential equation which has the known solution

$$\phi(t) = A \ \exp \ (E_n t/i\hbar) = A \ \exp \ (-iE_n t/\hbar) \qquad (3.63)$$

as can be demonstrated by substituting Eq. 3.63 into Eq. 3.62. Choosing A equal to unity, so that the entire wave function is normalized at each instant of time, we have

$$\Psi_{nlm}(r, \theta, \phi, t) = \psi_{nlm}(r, \theta, \phi) \ \exp(-iE_n t/\hbar) \qquad (3.64)$$

that is, the complete time-dependent solution for the stationary state nlm of the hydrogen atom is the time-independent spatial function multiplied by an oscillating, complex, time-dependent exponential factor. To obtain the probability distribution corresponding to the function (Eq. 3.64), we multiply it by its complex conjugate (see Sections 2.2.1 and 3.8) and find

$$
\begin{aligned}
P_{nlm}(r, \theta, \phi, t) &= \Psi_{nlm}{}^*(r, \theta, \phi, t)\Psi_{nlm}(r, \theta, \phi, t) \\
&= \psi_{nlm}{}^*(r, \theta, \phi) \ \exp(iE_n t/\hbar)\psi_{nlm}(r, \theta, \phi) \\
&\qquad \times \exp(-iE_n t/\hbar) \\
&= \psi_{nlm}{}^*(r, \theta, \phi)\psi_{nlm}(r, \theta, \phi) = |\psi_{nlm}(r, \theta, \phi)|^2
\end{aligned}
\qquad (3.65)
$$

Equation 3.65 shows that, although the wave function itself depends on the time, the probability distribution obtained from it does not. The average value of a measurable quantity such as a coordinate or a momentum is likewise independent of the time, since

$$\begin{aligned}
\langle x \rangle &= \int \Psi_{nlm}^*(r, \theta, \phi, t) x \Psi_{nlm}(r, \theta, \phi, t)\, dv \\
&= \int \psi_{nlm}^*(r, \theta, \phi) \exp(iE_n t/\hbar)\, x \psi_{nlm}(r, \theta, \phi) \exp(-iE_n t/\hbar)\, dv \\
&= \exp(iE_n t/\hbar) \exp(-iE_n t/\hbar) \int \psi_{nlm}^*(r, \theta, \phi) x \psi_{nlm}(r, \theta, \phi)\, dv \\
&= \int \psi_{nlm}^*(r, \theta, \phi) x \psi_{nlm}(r, \theta, \phi)\, dv
\end{aligned}$$

and

$$\begin{aligned}
\langle p_z \rangle &= \int \Psi_{nlm}^*(r, \theta, \phi, t) \left(\frac{\hbar}{i} \frac{\partial}{\partial x} \right) \Psi_{nlm}(r, \theta, \phi, t)\, dv \\[1em]
&= \exp(iE_n t/\hbar) \exp(-iE_n t/\hbar) \int \psi_{nlm}^*(r, \theta, \phi) \left(\frac{\hbar}{i} \frac{\partial}{\partial x} \right) \psi_{nlm}(r, \theta, \phi)\, dv \\[1em]
&= \int \psi_{nlm}^*(r, \theta, \phi) \left(\frac{h}{i} \frac{\partial}{\partial x} \right) \psi_{nlm}(r, \theta, \phi)\, dv
\end{aligned}$$

We have thus justified our description of $\psi_{nlm}(r, \theta, \phi)$ as a stationary state (i.e., a state whose properties are independent of time).

For a hydrogen-like atom interacting with radiation, the potential-energy function includes, in addition to the attraction of the electron to the positive nucleus $(-Ze^2/r)$, terms representing the interaction of the electron with the electromagnetic wave. The most important of these is the *electric dipole* term. For a wave with the electric field oriented in the x direction, the electric dipole term is $-ex\mathcal{E}_x$. Writing the electric field of a plane wave moving in the z direction in the form given by Eq. 2.3, $\mathcal{E}_x = \mathcal{E}_x{}^0 \cos [2\pi(z/\lambda - \nu t)]$, we have

$$V(x, y, z, t) = -\frac{Ze^2}{r} - ex\mathcal{E}_x{}^0 \cos \left[2\pi \left(\frac{z}{\lambda} - \nu t \right) \right] \tag{3.66}$$

The solution of Eq. 3.59 with $V(x, y, z, t)$ a function of time as in Eq. 3.66 is no longer given by the stationary states described by Eq. 3.64. Instead, the time-dependent part of the potential due to the presence of the radiation field produces a coupling between the different stationary states that can result in transitions from one state to another. To determine the rate of such transitions, we assume that at a certain time (say, $t = 0$) when the radiation field is turned on, we know that the atom is in one of its stationary states, say, ψ_{n_1}; thus

$$\Psi_{n_1}(r, \theta, \phi, t = 0) = \psi_{n_1}(r, \theta, \phi) \tag{3.67}$$

since $\exp(-iE_{n_1}t/\hbar) = 1$ for $t = 0$ (we use the single quantum number n_1 to represent $n_1 l_1 m_1$ in the following discussion to simplify the writing). If the radiation frequency ν obeys the Bohr frequency rule

$$\nu = \frac{E_{n_2} - E_{n_1}}{h} \tag{3.68}$$

then at times $t > 0$, there is a certain probability that the atom will be in the state n_2 with wave function $\Psi_{n_2}(r, \theta, \phi, t)$. Considering only the two states n_1 and n_2 that are coupled by the radiation, we make use of the superposition principle to write the complete wave function in the form

$$\Psi(r, \theta, \phi, t) = a_{n_1}(t)\Psi_{n_1}(r, \theta, \phi, t) + a_{n_2}(t)\Psi_{n_2}(r, \theta, \phi, t) \quad (3.69)$$

where the coefficients $a_{n_1}(t)$ and $a_{n_2}(t)$ represent the contribution[4] of the functions Ψ_{n_1} and Ψ_{n_2}, respectively, to the total wave function Ψ at time t. Correspondingly, the absolute value squared of the coefficients gives the probabilities of being in the state n_1 and n_2, respectively, at time t; that is,

$$P_{n_1}(t) = a_{n_1}{}^*(t)a_{n_1}(t) \qquad P_{n_2}(t) = a_{n_2}{}^*(t)a_{n_2}(t) \quad (3.70)$$

Since Eq. 3.67 is assumed to hold at $t = 0$, we see that $a_{n_1}(0) = 1$ and $a_{n_2}(0) = 0$; thus

$$P_{n_1}(0) = 1 \qquad P_{n_2}(0) = 0$$

as required by the assumption that the system is initially in the state n_1.

From the above discussion it is clear that to determine the probability of transition from the state n_1 to n_2 we need to determine the coefficients $a_{n_1}(t)$ and $a_{n_2}(t)$ as functions of time by solving the time-dependent Schrödinger equation (Eq. 3.59) with V given by Eq. 3.66. Substituting for Ψ by Eq. 3.69, we have

$$
\begin{aligned}
& a_{n_1}(t)\left[-\frac{\hbar^2}{2m}\left(\frac{\partial^2 \Psi_{n_1}}{\partial x^2} + \frac{\partial^2 \Psi_{n_1}}{\partial y^2} + \frac{\partial^2 \Psi_{n_1}}{\partial z^2} \right) - \frac{Ze^2}{r}\Psi_{n_1} \right] \\
& + a_{n_2}(t)\left[-\frac{\hbar^2}{2m}\left(\frac{\partial^2 \Psi_{n_2}}{\partial x^2} + \frac{\partial^2 \Psi_{n_2}}{\partial y^2} + \frac{\partial^2 \Psi_{n_2}}{\partial z^2} \right) - \frac{Ze^2}{r}\Psi_{n_2} \right] \\
& - e x \mathcal{E}_x{}^0 \cos\left[2\pi\left(\frac{z}{\lambda} - \nu t\right) \right][a_{n_1}(t)\Psi_{n_1} + a_{n_2}(t)\Psi_{n_2}] \\
& = i\hbar \frac{\partial}{\partial t}[a_{n_1}(t)\Psi_{n_1} + a_{n_2}(t)\Psi_{n_2}]
\end{aligned}
\quad (3.71)
$$

Since Ψ_{n_1} and Ψ_{n_2} both satisfy the time-independent Schrödinger equation, we have

$$
\begin{aligned}
-\frac{\hbar^2}{2m}\left(\frac{\partial^2 \Psi_{n_1}}{\partial x^2} + \frac{\partial^2 \Psi_{n_1}}{\partial y^2} + \frac{\partial^2 \Psi_{n_1}}{\partial z^2} \right) - \frac{Ze^2}{r}\Psi_{n_1} = E_{n_1}\Psi_{n_1} \\
-\frac{\hbar^2}{2m}\left(\frac{\partial^2 \Psi_{n_2}}{\partial x^2} + \frac{\partial^2 \Psi_{n_2}}{\partial y^2} + \frac{\partial^2 \Psi_{n_2}}{\partial z^2} \right) - \frac{Ze^2}{r}\Psi_{n_2} = E_{n_2}\Psi_{n_2}
\end{aligned}
\quad (3.72)
$$

[4] Other states can also be mixed in, but if Eq. 3.68 is obeyed by the radiation frequency, only n_1 and n_2 are important.

To carry out the differentiation on the right-hand side of Eq. 3.71, we introduce the form of Ψ_{n_1} and Ψ_{n_2} (Eq. 3.64) and write

$$\frac{\partial}{\partial t}[a_{n_1}(t)\Psi_{n_1}] = \Psi_{n_1}\frac{da_{n_1}(t)}{dt} + a_{n_1}(t)\left(-\frac{i}{\hbar}E_{n_1}\right)\Psi_{n_1}$$

$$\frac{\partial}{\partial t}[a_{n_2}(t)\Psi_{n_2}] = \Psi_{n_2}\frac{da_{n_2}(t)}{dt} + a_{n_2}(t)\left(-\frac{i}{\hbar}E_{n_2}\right)\Psi_{n_2}$$

(3.73)

Substituting Eqs. 3.72 and 3.73 into Eq. 3.71 and canceling terms, we obtain Problem 3.20)

$$-ex\mathcal{E}_x{}^0\cos\left[2\pi\left(\frac{z}{\lambda} - \nu t\right)\right][a_{n_1}(t)\Psi_{n_1} + a_{n_2}(t)\Psi_{n_2}]$$

$$= i\hbar\left(\Psi_{n_1}\frac{da_{n_1}(t)}{dt} + \Psi_{n_2}\frac{da_{n_2}(t)}{dt}\right)$$

(3.74)

To find the differential equation for $a_{n_2}(t)$, we multiply both sides of Eq. 3.74 by $\Psi_{n_2}{}^*$ and integrate over all space; that is,

$$-e\mathcal{E}_x{}^0\left\{a_{n_1}(t)\int\Psi_{n_2}{}^*\, x\cos\left[2\pi\left(\frac{z}{\lambda} - \nu t\right)\right]\Psi_{n_1}\, dv\right.$$

$$\left. +a_{n_2}(t)\int\Psi_{n_2}{}^*\, x\cos\left[2\pi\left(\frac{z}{\lambda} - \nu t\right)\right]\Psi_{n_2}\, dv\right\}$$

$$= i\hbar\left(\frac{da_{n_1}(t)}{dt}\int\Psi_{n_2}{}^*\,\Psi_{n_1}\, dv + \frac{da_{n_2}(t)}{dt}\int\Psi_{n_2}{}^*\,\Psi_{n_2}\, dv\right)$$

(3.75)

The functions Ψ_{n_1} and Ψ_{n_2} are normalized and orthogonal, so that

$$\int\Psi_{n_2}{}^*\Psi_{n_2}\, dv = 1$$
$$\int\Psi_{n_2}{}^*\Psi_{n_1}\, dv = 0$$

(3.76)

Furthermore, the range of z that contributes significantly to the integrals is on the order of the diameter of the atom ($\sim 10^{-8}$ cm), while the shortest wavelength λ for a hydrogen-atom transition (corresponding to ionization of the ground state) is a factor of about 10^3 greater ($\sim 10^{-5}$ cm). Thus the fraction z/λ can be neglected in the first and second terms on the left-hand side of Eq. 3.75 and the integrals in these terms can be approximated by

$$\int\Psi_{n_2}{}^*\, x\cos\left[2\pi\left(\frac{z}{\lambda} - \nu t\right)\right]\Psi_{n_1}\, dv$$

$$\simeq \cos(2\pi\nu t)\int\Psi_{n_2}{}^*\, x\Psi_{n_1}\, dv$$

and

$$\int\Psi_{n_2}{}^*\, x\cos\left[2\pi\left(\frac{z}{\lambda} - \nu t\right)\right]\Psi_{n_2}\, dv$$

$$\simeq \cos(2\pi\nu t)\int\Psi_{n_2}{}^*\, x\Psi_n\, dv$$

respectively. The first integral above, which involves both the initial and the final state, is called a *transition integral* for reasons demonstrated below; the second integral is the average value of x in the state n_2 which can be shown to be zero (Problem 3.8).

Substituting these results and Eq. 3.76 into Eq. 3.75, we obtain

$$i\hbar \frac{da_{n_2}(t)}{dt} = -e\mathcal{E}_x^0 a_{n_1}(t) \cos(2\pi\nu t) \int \Psi_{n_2}{}^* \, x \, \Psi_{n_1} \, dv$$

To find the initial transition rate, we consider times near zero, so that the following approximations hold:

$$a_{n_1} \simeq 1 \qquad \cos(2\pi\nu t) \simeq 1$$

$$\exp\left(-\frac{iE_{n_1}t}{\hbar}\right) \simeq 1 \qquad\qquad \exp\left(-\frac{iE_{n_2}t}{\hbar}\right) \simeq 1$$

$$\Psi_{n_1} \simeq \psi_{n_1} \qquad\qquad \Psi_{n_2}^* \simeq \psi_{n_2}$$

Then

$$i\hbar \frac{da_2(t)}{dt} \simeq -e\mathcal{E}_x^0 \int \psi_{n_2}{}^* \, x \, \psi_{n_1} \, dv \tag{3.77}$$

Since the right-hand side of the equation is independent of time, we can integrate directly to obtain

$$a_{n_2}(t) \simeq -\frac{e\mathcal{E}_x^0 t}{i\hbar} \int \psi_{n_2}{}^* \, x \, \psi_{n_1} \, dv \tag{3.78}$$

Substituting this solution for $a_{n_2}(t)$ into Eq. 3.70, we see that the probability of finding the atom in the state Ψ_{n_2} at time t is

$$P_{n_2}(t) \simeq \frac{e^2(\mathcal{E}_x^0)^2}{\hbar^2} \left| \int \psi_{n_2}{}^* \, x \, \psi_{n_1} \, dv \right|^2 t^2 \tag{3.79}$$

The intensity $I_{n_1 \to n_2}$ of the transition is proportional to the transition rate; that is, the transition probability per unit time. Differentiating Eq. 3.79 to find the rate of change of the probability, we have

$$I_{n_1 \to n_2} \propto \frac{d}{dt} P_{n_2}(t) \simeq \frac{2e^2(\mathcal{E}_x^0)^2 t}{\hbar^2} \left| \int \psi_{n_2}{}^* \, x \, \psi_{n_1} \, dv \right|^2 \tag{3.80}$$

Equation 3.80 demonstrates that the intensity is proportional to the square of the absolute value of the transition integral,

$$I_{n_1 \to n_2} \propto \left| \int \psi_{n_2}{}^* x \psi_{n_1} \, dv \right|^2 \tag{3.81}$$

The choice of x as the direction of the radiation field was of course arbitrary, so that in general all three transition integrals (called electric–dipole transition integrals)

$$\int \psi_{n_2}{}^* \, x \, \psi_{n_1} \, dv$$

$$\int \psi_{n_2}{}^* \, y \, \psi_{n_1} \, dv \qquad\qquad (3.82)$$

$$\int \psi_{n_2}{}^* \, z \, \psi_{n_1} \, dv$$

must be considered. If one or more of the integrals of Eq. 3.82 is nonzero for a given transition, the corresponding line appears in the spectrum and the transition is said to be "allowed." Transitions which are not allowed by the electric-dipole interaction are said to be "forbidden." Evaluation of the integrals in Eq. 3.82 for specific n_1 and n_2 or, writing out all the quantum numbers, (n_1, l_1, m_1) and (n_2, l_2, m_2), yields the selection rules stated in Eq. 3.50 (Problem 3.9).

For some cases in which all three of the integrals of Eq. 3.82 are zero, the interaction of the electron with the magnetic field associated with the electromagnetic wave leads to a nonzero, but fainter, intensity for the transition. In addition to this *magnetic dipole* interaction, the variation of the electric field over the extension of the atom can give rise to very faint intensities of the electric dipole forbidden transitions due to the *electric quadrupole* interaction (see Reference 8 of the Additional Reading list).

The assumption that the radiation frequency ν corresponds exactly to a Bohr transition frequency is more nearly applicable to absorption of radiation from a precisely tuned laser than to the more usual case of absorption from radiation composed of a broad band of frequencies, such as "white" light. Apart from constant factors, the main difference between the more general treatment, which takes account of the frequency distribution, and the one we have outlined here is that dP_{n_2}/dt is found to be independent of t in the former, rather than proportional to t, as in Eq. 3.80. The important result for our purposes, namely, that the intensity is proportional to the square of the absolute value of the electric dipole transition integral, remains unchanged.

3.12 ANGULAR MOMENTUM, MAGNETIC MOMENT, AND SPIN

In the discussion of the quantum numbers l and m (Section 3.7) it was pointed out that the magnitude of the square of the electronic angular momentum vector in the hydrogen atom is given by

$$M^2 = l(l + 1)\hbar^2 \qquad l = 0, 1, 2, \ldots, n - 1 \qquad (3.83)$$

and that the component of the angular momentum along the z axis is

$$M_z = m\hbar \qquad m = 0, \pm1, \pm2, \ldots, \pm l \qquad (3.84)$$

Classically, the angular momentum vector could be oriented in any direction, so that the component M_z could take any value between M and $-M$; the maximum (M) and the minimum $(-M)$ values would correspond to the electronic angular momentum pointing directly along the positive or negative z axis, respectively. In quantum theory, not only is M^2 limited to the discrete values given in Eq. 3.83, but the directions in which the vector is allowed to point are limited to those which correspond to a value of M_z given by Eq. 3.84. This directional limitation upon the angular momentum is sometimes called *space quantization*. In Fig. 3.14 the allowed directions of M are shown for p orbitals ($l = 1$) and d orbitals ($l = 2$), where in this quantitative consideration of the angular momentum, we use the complex form for the ϕ-dependent part of the wave function (Eq. 3.47). In Fig. 3.14a, the angular momentum vector $\mathbf{M}$ of length $\sqrt{2}\hbar$ must lie in either of the two conical surfaces ($m = \pm 1$) or in the xy plane ($m = 0$). Similarly, in Fig. 3.14b the vector $\mathbf{M}$ of length $\sqrt{6}\hbar$ is limited to the surfaces of four cones ($m = \pm 2, \pm 1$) or the xy plane ($m = 0$). In neither case is the vector allowed to point along the z axis, for to specify exactly the direction of the angular momentum implies that the electronic motion is confined to a plane perpendicular to that direction, which, as we have seen, violates the uncertainty principle. The choice of the z axis as the axis of quantization is arbitrary, except that it leads to the simplest formulas in terms of the spherical polar coordinate system.

To demonstrate the existence of space quantization, we need an experiment which can measure the z component of angular momentum. The simplest procedure is to place the atom in a magnetic field and to observe

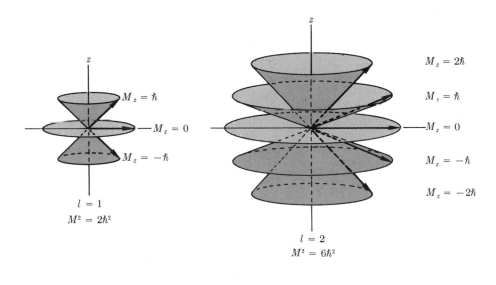

(a)

(b)

Fig. 3.14. Space quantization for $l = 1, 2$.

the resulting changes in the energy levels. According to the laws of electricity and magnetism, a moving particle with a charge q and a vector angular momentum $\mathbf{M}$ has a vector magnetic dipole moment $\mathbf{\mu}_{\text{mag}}$ proportional to $\mathbf{M}$ and in the same direction; that is,

$$\mathbf{\mu}_{\text{mag}} = \gamma \mathbf{M} \tag{3.85}$$

The constant γ, called the *magnetogyric ratio*, is given by

$$\gamma = \frac{q}{2m_q c} \tag{3.86}$$

where m_q is the mass of the particle, and c is the velocity of light (see the Appendix). If a magnetic field is present, the magnetic dipole moment associated with the motion of the particle interacts with the field in a way analogous to the coupling between an electric dipole moment and an electric field; that is, the energy of the moving charge is increased by an amount equal to the product of the magnetic induction $\mathcal{B}$ and the component of the magnetic moment in the direction of $\mathcal{B}$. Choosing the coordinate system so that $\mathcal{B}$ is in the z direction, we find that the magnetic energy is

$$\Delta E_{\text{mag}} = -(\mu_{\text{mag}})_z \mathcal{B} = -\gamma M_z \mathcal{B} \tag{3.87}$$

For a hydrogen atom in a magnetic field, Eqs. 3.84, 3.86, and 3.87 and the fact that $q = -e$ show that[5]

$$\Delta E_{\text{mag}} = m\,\frac{e\hbar}{2m_e c}\,\mathcal{B} \tag{3.88}$$

The constant $e\hbar/2m_e c$ is called the *Bohr magneton* μ_B, and has the value

$$\mu_B = \frac{e\hbar}{2m_e c} = 9.2732 \times 10^{-21}\ \text{erg G}^{-1}$$

in cgs–mixed Gaussian units.

According to Eq. 3.88, the presence of a magnetic field removes the degeneracy with respect to the quantum number m. For example, if the hydrogen atom is in a $2p$ state ($l = 1$), the possible energy levels are

$$E(2p_0) = -\frac{1}{4}\frac{e^2}{2a_0}$$

$$E(2p_{+1}) = -\frac{1}{4}\frac{e^2}{2a_0} + \mu_B \mathcal{B} \tag{3.89}$$

$$E(2p_{-1}) = -\frac{1}{4}\frac{e^2}{2a_0} - \mu_B \mathcal{B}$$

---

[5] The symbol m_e is used here to designate the electronic mass (instead of the usual m) to distinguish it from the magnetic quantum number.

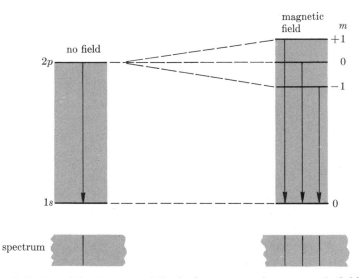

Fig. 3.15. Splitting of the $2p$ state of the hydrogen atom in a magnetic field; spin is neglected.

where the subscript on the symbol $2p$ indicates the quantum number m corresponding to the complex orbitals ($m = 0$, ± 1). Thus, the $2p_0$ state is unchanged in energy and the $2p_{+1}$ and $2p_{-1}$ states are shifted by equal amounts in opposite directions (Fig. 3.15). If the atom is in an s state, m is necessarily zero and no change of energy is expected in a magnetic field. We can therefore test the predictions of Eq. 3.88 by observing the $2p \leftrightarrow 1s$ transition in the presence of a magnetic field. The selection rules for the possible transitions (Eq. 3.50) show that $2p_0 \leftrightarrow 1s$, $2p_{+1} \leftrightarrow 1s$, $2p_{-1} \leftrightarrow 1s$ are all allowed, so that three equally spaced lines are expected. In a very strong magnetic field, such a triplet of lines is indeed observed. It is referred to as a Zeeman triplet, after Zeeman, who first studied the splitting of spectral lines in magnetic fields before the advent of the quantum theory (1896); the limiting Zeeman behavior of the lines in high magnetic fields is called the Paschen–Back effect, after F. Paschen and E. Back, who published their study in 1912. It should be noted that the actual splitting of the energy levels even for large magnetic induction values is very small; for example, for $\mathcal{B} = 5 \times 10^4$ G, the shift of the $2p_{+1}$ and $2p_{-1}$ states is 4.64×10^{-16} erg $= 2.9 \times 10^{-4}$ eV, as compared with the $1s$, $2p$ energy difference of 10.2 eV.

3.12.1 Electron spin When the Zeeman triplet is examined under high resolution, the two outer lines are found to be split into two lines each, giving five lines in all. This indicates that the expression for the splitting of the energy levels, Eq. 3.89, is incomplete. Observations of corresponding anomalies in the spectra of many-electron atoms (e.g., the presence of two lines in the sodium spectrum, where there should be only one) suggested to Goudsmit and Uhlenbeck in 1925 that the electron has an intrinsic magnetic moment independent of its orbital motion; that is, even an atomic electron in an s state, which has zero orbital angular momentum ($l = 0$), has such a magnetic moment. If we associate this magnetic moment with an intrinsic angular momentum or *spin* (e.g., in classical terms, it might be associated with an electron of finite size spinning about its own axis), we can treat it analogously to the orbital angular momentum. We associate the quantum number s with the magnitude of the total spin angular momentum S and the quantum number m_s with its z component S_z, such that

$$S^2 = s(s + 1)\hbar^2$$

and

$$S_z = m_s\hbar \qquad m_s = -s, -s + 1, \ldots, +s \qquad (3.90)$$

Thus the quantum numbers s and m_s are the spin analogs to l and m for orbital angular momentum (Eqs. 3.83 and 3.84). However, in contrast to the orbital angular momentum, the spin quantum number s has been found to have only a single value, $s = \frac{1}{2}$, and the square of the spin angular momentum is

$$S^2 = s(s + 1)\hbar^2 = \tfrac{3}{4}\hbar^2$$

The spin-component quantum number m_s can have the values $\pm\frac{1}{2}$ by Eq. 3.90 and

$$S_z = \pm\tfrac{1}{2}\hbar$$

Thus, there are only two possible orientations of the spin angular momentum (see Fig. 3.16).

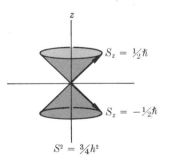

Fig. 3.16. Space quantization of spin angular momentum.

In correspondence with Eq. 3.85 we can write the z component of the intrinsic magnetic moment as

$$(\mu_s)_z = \gamma_s S_z = m_s \gamma_s \hbar \tag{3.91}$$

where γ_s is the magnetogyric ratio for the spin angular momentum. Examination of the splitting of spectral lines in magnetic fields, as well as other evidence, has shown that the value of γ_s is almost exactly twice the magnetogyric ratio for the orbital motion of the electron; that is,

$$\gamma_s = -\frac{e}{m_e c} \tag{3.92}$$

Returning to the experiment of a one-electron atom in a very strong magnetic field, we consider the modification of the energy levels and the allowed transitions due to the presence of electron spin. For the z axis chosen along the magnetic field direction, we can combine Eqs. 3.87, 3.88, 3.91, and 3.92 to obtain

$$\Delta E_{\mathrm{mag}} = m \frac{e\hbar}{2m_e c} \mathcal{B} + m_s \frac{e\hbar}{m_e c} \mathcal{B}$$

$$= \frac{e\hbar}{2m_e c} \mathcal{B} (m + 2m_s) = \mu_B \mathcal{B}(m + 2m_s) \tag{3.93}$$

If we again concern ourselves with the $2p$ and $1s$ states, we have for $2p$,

$$m = 0, \pm 1 \qquad m_s = \pm\tfrac{1}{2}$$

and for $1s$,

$$m = 0 \qquad m_s = \pm\tfrac{1}{2}$$

The resulting energy levels and spectrum in a very strong field are shown in Fig. 3.17b. Each level is labeled by the quantum numbers m and m_s to which it corresponds. On comparing with Fig. 3.15, we note that for $l = 1$ there are now five levels (instead of only three as before), one of the levels being doubly degenerate; the latter corresponds to the states $m = 1$, $m_s = -\tfrac{1}{2}$ and $m = -1$, $m_s = +\tfrac{1}{2}$. For both of these states the energy is not displaced by the magnetic field, according to Eq. 3.93. The $1s$ state, which was unaffected by the magnetic field in the absence of electron spin, is now split into two levels, corresponding to the two possible spin orientations.

To determine the allowed transitions from the energy-level diagram, we need to supplement the selection rules given in Eq. 3.50 by the one for the quantum number m_s. For the present case of electric dipole induced transitions, the rule is

$$\Delta m_s = 0 \tag{3.94}$$

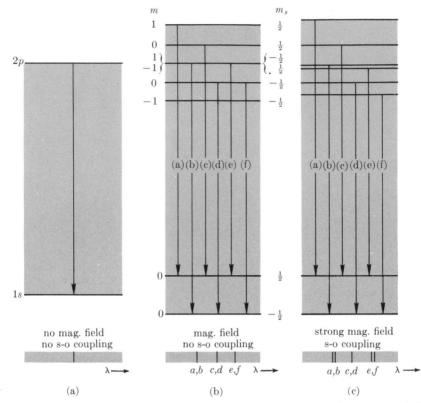

Fig. 3.17. Energy levels, transitions, and spectrum for the $1s$ and $2p$ states of the hydrogen atom. **(a)** No magnetic field or spin-orbit coupling; **(b)** magnetic field only; **(c)** strong magnetic field, effects of spin-orbit coupling included.

that is, the electron-spin orientation does not change. Making use of Eqs. 3.50 and 3.94, we find that there are six allowed transitions ($\Delta l = -1$)

(a) $m = +1$, $m_s = +\frac{1}{2} \rightarrow m = 0$, $m_s = +\frac{1}{2}$ ($\Delta m = -1$)
(b) $m = +1$, $m_s = -\frac{1}{2} \rightarrow m = 0$, $m_s = -\frac{1}{2}$ ($\Delta m = -1$)
(c) $m = 0$, $m_s = +\frac{1}{2} \rightarrow m = 0$, $m_s = +\frac{1}{2}$ ($\Delta m = 0$)
(d) $m = 0$, $m_s = -\frac{1}{2} \rightarrow m = 0$, $m_s = -\frac{1}{2}$ ($\Delta m = 0$)
(e) $m = -1$, $m_s = +\frac{1}{2} \rightarrow m = 0$, $m_s = +\frac{1}{2}$ ($\Delta m = +1$)
(f) $m = -1$, $m_s = -\frac{1}{2} \rightarrow m = 0$, $m_s = -\frac{1}{2}$ ($\Delta m = +1$)

These transitions are shown by the labels (a)–(f) in the diagram. Since the spacing of the five $2p$ levels is the same, $(e\hbar/2m_ec)B$, and that of the s levels is just twice that of the $2p$ levels, only three distinct frequencies are expected from the six transitions. As can be verified by substitution into Eq. 3.93 and application of the Bohr frequency rule, these are

$$\tilde{\nu} = \frac{E_{2p} - E_{1s}}{hc} + \frac{\mu_B}{hc} \mathscr{B} \quad \Delta m = -1 \quad \text{(a), (b)}$$

$$\tilde{\nu} = \frac{E_{2p} - E_{1s}}{hc} \qquad\qquad \Delta m = 0 \qquad \text{(c), (d)} \qquad (3.95)$$

$$\tilde{\nu} = \frac{E_{2p} - E_{1s}}{hc} - \frac{\mu_B}{hc} \mathscr{B} \quad \Delta m = +1 \quad \text{(e), (f)}$$

Thus, from the one-electron theory including the electron spin, the results appear to be a normal Zeeman triplet identical to that obtained previously without spin. Although this is qualitatively correct, there is a quantitative disagreement with experiment, since as we have pointed out the two outer lines of Eq. 3.95 are both resolvable into two lines.

3.12.2   Spin-orbit coupling

The additional splitting can be understood in the framework of the relativistic treatment of the hydrogen atom, given by P. A. M. Dirac in 1928. One of the correction terms is due to an important interaction, called *spin-orbit coupling*, which arises from the interaction between the magnetic moments associated with the orbital and spin angular momenta of the electron; that is, the magnetic dipole moment due to the orbital motion creates a magnetic field which interacts with the electron spin magnetic moment. For the very strong magnetic field being considered here, the primary effect of the spin-orbit coupling is to shift all the levels with $m m_s > 0$ upward slightly and those with $m m_s < 0$ downward slightly. Also, the remaining degeneracy in the $2p$ state is lifted; that is, the levels $m = -1$, $m_s = \frac{1}{2}$ and $m = 1$, $m_s = -\frac{1}{2}$ have slightly different energies. The transitions are correspondingly shifted (see Fig. 3.17c). Thus, transitions (a) and (b), which have the same frequency without spin-orbit coupling, now have slightly different frequencies; also transitions (e) and (f) have slightly different frequencies. Consequently, the so-called normal Zeeman pattern of three lines is seen to consist of five lines resulting from the splitting of each of the outer lines into two closely spaced lines (anomalous Zeeman effect). This is in agreement with experimental observations.

As the external magnetic field is decreased in magnitude and the spin-orbit coupling becomes relatively more important, changes in the spectrum occur. For a discussion of this behavior, which can be calculated in complete detail by quantum mechanics, the interested reader should look at texts on atomic spectra. Also, there are some very small additional effects (e.g., the Lamb shift discovered in 1947) that have been observed by refined spectroscopic techniques (e.g., radio-frequency spectroscopy) and

designations s, p, d, f, g, ... A left superscript to the L symbol gives the spin degeneracy $2S + 1$ of the atom which is $2s + 1 = 2$ for a one-electron system, and a right subscript gives the value of the quantum number $J = L + S$, $L + S - 1$, . . . , $|L - S|$ (see Chapter 4) for the total angular momentum, which is the sum of the orbital and spin angular momenta (i.e., $\mathbf{J} = \mathbf{M} + \mathbf{S}$). The principal quantum number n of the occupied orbital is sometimes written to the left of the term symbol. Thus, the term symbol has the form

$$n^{2S+1}L_J$$

With these definitions, the ground state of the hydrogen atom (corresponding to $n = 1$, $L = l = 0$, $S = s = \frac{1}{2}$, and $J = \frac{1}{2}$) has the term symbol $1^2S_{1/2}$. The two $2p$ states ($n = 2$, $L = 1$, $S = \frac{1}{2}$, $J = L \pm S$) have the term symbols $2\,^2P_{3/2}$ and $2\,^2P_{1/2}$. Terms such as these with the same values of n, L, and S, but with different values of J, differ slightly in energy because of spin-orbit coupling. In Fig. 3.13, which shows the energy levels and allowed transitions for the hydrogen atom (ignoring spin-orbit coupling), the term symbols are given at the top of each column.

SUMMARY

In this chapter we have seen how the Schrödinger equation can be applied to one-electron atoms and ions. The quantum numbers n, l, and m were introduced to specify the Schrödinger-equation solutions, called orbitals, and were seen to be related to the numbers of various types of nodes in the orbitals. The geometry of the orbitals, which is important for the interpretation of the chemical bond, was discussed and illustrated in some detail. It was shown how the quantum-mechanical energy levels and selection rules determine the spectrum of the hydrogen atom. The effects of a magnetic field upon the energy levels were found to give rise to a splitting of the spectral lines (Zeeman effect). The concepts of spin and intrinsic magnetic moment were introduced, and the effect of spin-orbit interaction upon energy levels was discussed. Finally, term symbols for one-electron atoms and ions were defined.

In the next chapter, we see how the ideas and terminology introduced here for the hydrogen atom can be extended to understand the structure of many-electron atoms.

ADDITIONAL READING

[1] H. B. GRAY, *Electrons and Chemical Bonding* (Benjamin, New York, 1965), Chap. 1.

[2] L. PAULING and E. B. WILSON, JR., *Introduction to Quantum Mechanics* (McGraw-Hill, New York, 1935), Chap. 5.

[3] F. L. Pilar, *Elementary Quantum Chemistry* (McGraw-Hill, New York, 1968), Chap. 7.

[4] H. G. Kuhn, *Atomic Spectra* (Academic Press, New York, 1962).

[5] R. M. Hochstrasser, *Behavior of Electrons in Atoms* (Benjamin, New York, 1964).

[6] W. Heitler, *Elementary Wave Mechanics* (Clarendon Press, Oxford, 1956), 2nd ed.

[7] J. W. Linnett, *Wave Mechanics and Valency* (Methuen, London, 1960).

[8] D. Bohm, *Quantum Theory* (Prentice-Hall, New York, 1951).

PROBLEMS

3.1 The cylindrical coordinates ρ, ϕ, z are defined by the relations

$$x = \rho \cos \phi$$
$$y = \rho \sin \phi$$
$$z = z$$

Use the appropriate chain rule analogous to Eq. 3.7 to find expressions for $\partial\psi/\partial x$, $\partial\psi/\partial y$, $\partial\psi/\partial z$, where ψ is an arbitrary function of ρ, ϕ, z. Hence derive an expression for

$$\frac{\partial^2\psi}{\partial x^2} + \frac{\partial^2\psi}{\partial y^2} + \frac{\partial^2\psi}{\partial z^2}$$

in terms of

$$\frac{\partial^2\psi}{\partial\rho^2}, \quad \frac{\partial\psi}{\partial\rho}, \quad \frac{\partial^2\psi}{\partial\phi^2}, \quad \text{and} \quad \frac{\partial^2\psi}{\partial z^2}.$$

$$Ans. \quad \frac{\partial^2\psi}{\partial\rho^2} + \frac{1}{\rho}\frac{\partial\psi}{\partial\rho} + \frac{1}{\rho^2}\frac{\partial^2\psi}{\partial\phi^2} + \frac{\partial^2\psi}{\partial z^2}$$

3.2 Show that the spherical polar coordinates r, θ, ϕ can be defined in terms of the cylindrical coordinates ρ, ϕ, z by the equations

$$\rho = r \sin \theta$$
$$\phi = \phi$$
$$z = r \cos \theta$$

Use the appropriate chain rule analogous to Eq. 3.7 to find $\partial^2\psi/\partial\rho^2$, $\partial\psi/\partial\rho$, $\partial^2\psi/\partial\phi^2$, and $\partial^2\psi/\partial z^2$, where ψ is an arbitrary function of r, θ, ϕ. Hence use the results of Problem 3.1 to derive an expression for

$$\frac{\partial^2\psi}{\partial x^2} + \frac{\partial^2\psi}{\partial y^2} + \frac{\partial^2\psi}{\partial z^2}$$

in terms of

$$\frac{\partial^2\psi}{\partial r^2}, \quad \frac{\partial\psi}{\partial r}, \quad \frac{\partial^2\psi}{\partial\theta^2}, \quad \frac{\partial\psi}{\partial\theta}, \quad \text{and} \quad \frac{\partial^2\psi}{\partial\phi^2}.$$

Insert the resulting expression into Eq. 3.3 and compare with Eq. 3.9.

3.3 Following the series-expansion method used in Section 2.6.1 for solving Eq. 2.57, show that solutions to Eq. 3.15 are of the form $e^{\pm ar}$.

3.4 The integral in Eq. 3.26 appears frequently in atomic and molecular problems. A simple derivative makes use of Leibniz' theorem for differentiation of a definite integral, namely,

$$\frac{d}{d\beta} \int_a^b f(\beta,r)\, dr = \int_a^b \frac{\partial}{\partial\beta} f(\beta,r)\, dr$$

Thus

$$\int_0^\infty e^{-\beta r} r\, dr = - \int_0^\infty \frac{\partial}{\partial\beta} e^{-\beta r}\, dr = - \frac{d}{d\beta} \int_0^\infty e^{-\beta r}\, dr$$

$$= - \frac{d}{d\beta} \left(\frac{1}{\beta}\right) = \frac{1}{\beta^2}$$

Generalize this result to show that

$$\int_0^\infty e^{-\beta r} r^n\, dr = \frac{n!}{\beta^{n+1}}$$

(*Hint:* Take the nth derivative of $\int_0^\infty e^{-\beta r}\, dr$ with respect to β.)

3.5 Find the values of a and b which make the function given in Eq. 3.32 a solution to Eq. 3.13. Determine the value of the energy to which this function corresponds.

3.6 Show that Eq. 3.37 is obtained for $f(r)$ when $xf(r)$, $yf(r)$, or $zf(r)$ are substituted for ψ in Eq. 3.9.

3.7 Sketch polar plots of the functions $d_{z^2-x^2}$ and $d_{z^2-y^2}$ given in Eq. 3.49, and compare with the $d_{x^2-y^2}$, d_{xy}, d_{yz}, and d_{zz} plots given in Fig. 3.7.

3.8 Find the average values of x and y for an electron in *any* state (nlm) of the hydrogen atom. (*Hint:* Do the integrations over ϕ first.) What can you say about the average value of z without doing the integration?

3.9 Evaluate the integrals of Eq. 3.82 for each of the following pairs of hydrogen-atom states. From your results, determine which transitions are allowed, and compare with the selection rules given in Eq. 3.50.

n	l	m	n'	l'	m'
1	0	0	2	0	0
1	0	0	2	1	1
1	0	0	2	1	-1
1	0	0	2	1	0
2	0	0	3	2	1

3.10 Calculate the probability of finding a hydrogen $1s$ electron in the classically forbidden region (that is, with a value of r greater than $2a_0$).

3.11 Determine the probability of finding a hydrogen $1s$ electron in
(a) the volume bounded by $r = 1.1a_0$, $r = 1.105a_0$, $\theta = 0.2\pi$, $\theta = 0.201\pi$, $\phi = 0.6\pi$, $\phi = 0.601\pi$
(b) the volume bounded by $r = 1.1a_0$, $r = 1.105a_0$
(c) the volume lying within the sphere $r = 0.005a_0$
[*Hint:* In parts (a) and (b) you may assume that the wave functions are constant throughout the small volume considered.]

3.12 Determine the probability of finding a hydrogen $2p_z$ electron in each of the three volumes described in Problem 3.11.

3.13 Calculate the splitting ($\Delta\nu$ in cm^{-1}) of a hydrogen $1s$ energy level in a magnetic field with magnetic induction $\mathfrak{B} = 1000$ G.

3.14 Calculate in cm sec^{-1} the velocity of a $1s$ electron in an atom with nuclear charge Z. What value of Z would make the $1s$ electron travel at the velocity of light? Comment upon the implications of this result for the likelihood of producing an element with atomic number of about 140. What effect has been ignored in the calculation which limits the physical validity of your answer?

3.15 Sketch spherical polar plots of the seven f orbitals, given the projections shown in Fig. 3.8 and the d-orbital spherical polar plots in Fig. 3.10 as a guide.

3.16 Show that the hydrogen wave functions $3d_{3z^2-r^2}$, $3d_{zz}$, $3d_{yz}$, $3d_{x^2-y^2}$, and $3d_{xy}$ are mutually orthogonal. To which of these functions are $3d_{z^2-x^2}$ and $3d_{z^2-y^2}$ (Eq. 3.49) orthogonal, if any?

3.17 If (x_e, y_e, z_e) are the coordinates of the electron of mass m and (x_n, y_n, z_n) are the coordinates of the proton of mass M in a hydrogen atom, the Schrödinger equation for the two-particle system is

$$-\frac{\hbar^2}{2M}\left(\frac{\partial^2}{\partial x_n{}^2} + \frac{\partial^2}{\partial y_n{}^2} + \frac{\partial^2}{\partial z_n{}^2}\right)\psi - \frac{\hbar^2}{2m}\left(\frac{\partial^2}{\partial x_e{}^2} + \frac{\partial^2}{\partial y_e{}^2} + \frac{\partial^2}{\partial z_e{}^2}\right)\psi - \frac{Ze^2}{r}\psi = E\psi$$

where

$$r = [(x_n - x_e)^2 + (y_n - y_e)^2 + (z_n - z_e)^2]^{1/2}$$

ψ is a function $\psi(x_n, y_n, z_n; x_e, y_e, z_e)$

The coordinates of the center of mass (x_c, y_c, z_c) and the coordinates of the electron relative to those of the nucleus (x, y, z) are related to the electron and nucleus coordinates by

$$x_c = \frac{m}{M+m}x_e + \frac{M}{M+m}x_n$$

$$y_c = \frac{m}{M+m}y_e + \frac{M}{M+m}y_n$$

$$z_c = \frac{m}{M+m}z_e + \frac{M}{M+m}z_n$$

$$x = x_n - x_e$$

$$y = y_n - y_e$$

$$z = z_n - z_e$$

(a) Show from the chain rule that

$$\frac{\partial \psi}{\partial x_n} = \frac{\partial x_c}{\partial x_n}\frac{\partial \psi}{\partial x_c} + \frac{\partial x}{\partial x_n}\frac{\partial \psi}{\partial x} = \frac{M}{M+m}\frac{\partial \psi}{\partial x_c} + \frac{\partial \psi}{\partial x}$$

and

$$\frac{\partial \psi}{\partial x_e} = \frac{\partial x_c}{\partial x_e}\frac{\partial \psi}{\partial x_c} + \frac{\partial x}{\partial x_e}\frac{\partial \psi}{\partial x} = \frac{m}{M+m}\frac{\partial \psi}{\partial x_c} - \frac{\partial \psi}{\partial x}$$

(b) Show that

$$\frac{\partial^2 \psi}{\partial x_n^2} = \left(\frac{M}{M+m}\right)^2 \frac{\partial^2 \psi}{\partial x_c^2} + \frac{2M}{M+m}\frac{\partial^2 \psi}{\partial x_c \partial x} + \frac{\partial^2 \psi}{\partial x^2}$$

$$\frac{\partial^2 \psi}{\partial x_e^2} = \left(\frac{m}{M+m}\right)^2 \frac{\partial^2 \psi}{\partial x_c^2} - \frac{2m}{M+m}\frac{\partial^2 \psi}{\partial x_c \partial x} + \frac{\partial^2 \psi}{\partial x^2}$$

(c) Use your results for part (b) and analogous expressions for $\partial^2 \psi/\partial y_n^2$, $\partial^2 \psi/\partial y_e^2$, $\partial^2 \psi/\partial z_n^2$, and $\partial^2 \psi/\partial z_e^2$ to show that the Schrödinger equation for the hydrogen atom in terms of (x_c, y_c, z_c) and (x, y, z) is

$$-\frac{\hbar^2}{2(M+m)}\left(\frac{\partial^2}{\partial x_c^2} + \frac{\partial^2}{\partial y_c^2} + \frac{\partial^2}{\partial z_c^2}\right)\psi - \frac{\hbar^2}{2\mu}\left(\frac{\partial^2}{\partial x^2} + \frac{\partial^2}{\partial y^2} + \frac{\partial^2}{\partial z^2}\right)\psi - \frac{Ze^2}{r}\psi = E\psi$$

where

$$\mu = \frac{mM}{m+M}$$

$$r = (x^2 + y^2 + z^2)^{1/2}$$

3.18   A three-dimensional analog of the complex plane wave in Eq. 3.51 is

$$A(x, y, z, t) = a_0 \exp\left[2\pi i\left(\frac{x}{\lambda_x} + \frac{y}{\lambda_y} + \frac{z}{\lambda_z} - \nu t\right)\right]$$

For a de Broglie wave, λ_x, λ_y, λ_z are related to the momenta p_x, p_y, p_z by

$$\lambda_x = \frac{h}{p_x}$$

$$\lambda_y = \frac{h}{p_y}$$

$$\lambda_z = \frac{h}{p_z}$$

and ν is related to the energy E by the Bohr frequency rule

$$\nu = \frac{E}{h}$$

Use the three-dimensional wave and the above relations to obtain the time-dependent Schrödinger equation in three dimensions by the method used to obtain Eq. 3.56.

3.19 Show that if $\Psi(r, \theta, \phi, t)$, $\Psi_{n_1}(r, \theta, \phi, t)$, and $\Psi_{n_2}(r, \theta, \phi, t)$ are normalized for all values of t, then the coefficients $a_{n_1}(t)$ and $a_{n_2}(t)$ in Eq. 3.69 obey the equation

$$|a_{n_1}(t)|^2 + |a_{n_2}(t)|^2 = 1$$

3.20 Derive Eq. 3.74 from Eqs. 3.71, 3.72, and 3.73.

For the following problems a programmable desk calculator or digital computer will be useful.

C3.1 Plot as a function of r on the same graph the probability per unit volume of finding a hydrogen $2p_x$ electron in the following directions:
(a) the x axis
(b) in the xy plane at 45° to the x axis
(c) in the xz plane at 45° to the x axis

C3.2 Plot the probability of finding an electron at a distance less than r ($0 \leq r \leq 2a_0$) from the origin for the following hydrogen-atom states: $1s$, $2s$, $2p_x$, $2p_{+1}$.

C3.3 (a) Integrate the radial Schrödinger equation for spherically symmetric states of the hydrogen atom (Eq. 3.13) numerically, using a Runge–Kutta or other technique, after reducing the second-order differential equation to two coupled first-order equations by defining $y_1 = R(r)$, $y_2 = R'(r)$, and using the boundary conditions $y_1(0) = 1$, $y_2(0) = -1$, and $E = -0.5$ (set $\hbar = a_0 = e = m = 1$). [See, e.g., D. D. McCracken and W. S. Dorn, *Numerical Methods and Fortran Programming* (Wiley, New York, 1964), Chap. 10.] Compare your result with e^{-r}. Choose $E = -0.6$ and attempt to find values of $y_1(0)$ and $y_2(0)$ that give finite y_1 at large r.

(b) Find by trial and error a value of E (three significant figures) other than -0.5 that gives finite y_1 at large r when $y_1(0)$ and $y_2(0)$ are appropriately chosen. Compare your results with the energies predicted by the Bohr formula. To what value of n does your wave function correspond?

C3.4   The general formula for the radial part of the hydrogen-atom wave function is

$$R_{nl}(r) = -N_{nl}e^{-\rho/2}\rho^l L_{n+l}^{2l+1}(\rho)$$

where $\rho = 2Zr/na_0$, the normalization constant N_{nl} is given by

$$N_{nl} = \left(\frac{2Z}{na_0}\right)^{3/2}\left(\frac{(n-l-1)!}{2n[(n+l)!]^3}\right)^{1/2}$$

and the *associated Laguerre polynomial* $L_{n+l}^{2l+1}(\rho)$ has the form

$$L_{n+l}^{2l+1}(\rho) = [(n+l)!]^2 \sum_{p=0}^{n-l-1}(-1)^{p+1}\frac{\rho^p}{p!(n-l-p-1)!(2l+p+1)!} .$$

Write a program to calculate $R_{nl}(r)$ for $0 \le r \le 10a_0$ in steps of $0.05a_0$, with Z, n, and l to be supplied as parameters. Use the program to calculate $R_{4l}(r)$ for this range of r with $l = 0, 1, 2, 3$ and $Z = 1$. Plot R_{4l}, $R_{4l}{}^2$, and $r^2R_{4l}{}^2$ as functions of r.

C3.5   The θ-dependent part of the hydrogen-atom wave function is given by the general formula

$$\Theta_{lm}(\theta) = \left(\frac{(2l+1)(l-|m|)!}{2(l+|m|)!}\right)^{1/2}P_l^{|m|}(u); \qquad u = \cos\theta$$

where the *associated Legendre function* $P_l^{|m|}(u)$ has the form

$$P_l^{|m|}(u) = 2^{-l}l!(l+|m|)!(1-u^2)^{|m|/2}$$

$$\times \sum_{p=|m|}^{l}(-1)^{l-p}\frac{(1+u)^{p-|m|}(1-u)^{l-p}}{p!(l+|m|-p)!(l-p)!(p-|m|)!}$$

and $u = \cos\theta$. Write a program to calculate $\Theta_{lm}(\theta)$ for $0 \le \theta \le \pi$ in steps of $\pi/64$, with l and $|m|$ arbitrary integers. Calculate $\Theta_{lm}(\theta)$ for $l = 3$, $|m| = 0, 1, 2, 3$ and plot the results in polar coordinates; compare the plot of Θ_{30} with Fig. 3.8a.

4

Many-electron atoms

The periodicity in the properties of the chemical elements was established by purely chemical evidence long before any clear understanding of atomic structure had been achieved. It was noted later that the electronic spectra of atoms followed the same periodicity; that is, elements of the same group have corresponding sets of spectral lines. This suggested that there exist certain similarities in the electronic structure of atoms in the same group. Application of the quantum theory to many-electron atoms should thus provide an understanding of the periodic table.

The principles of the quantum theory for complex atoms, and for molecules, are the same as those already described for the hydrogen atom. However, the electrostatic interaction of the electrons with one another, as well as with the nucleus, makes accurate calculations of the wave functions and energy levels for many-electron atoms much more difficult. Nevertheless, it is possible to develop qualitative concepts for understanding atomic structure and atomic spectra. This is done by introducing suitable approximations for the wave function and energies of the atomic electrons. These approximations are very important in chemical theory because they also play a fundamental role in the description of molecules (Chapter 5).

4.1 GROUND STATE OF THE HELIUM ATOM

Most of the ideas that we need can be introduced and illustrated with the simplest many-electron atom, helium. As shown in Fig. 4.1, the helium

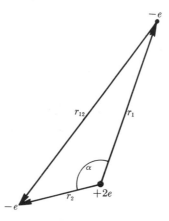

Fig. 4.1. Interparticle vectors in the helium atom.

atom consists of a nucleus with charge $+2e$ and two electrons. Although the electrons are in fact physically indistinguishable, it is convenient in formulating the problem to label them with the indices 1 and 2. The potential energy of the system of three charged particles is given in terms of the three interparticle distances, r_1, r_2, and r_{12}. According to Coulomb's law, the potential energy is

$$V(r_1, r_2, r_{12}) = -\frac{2e^2}{r_1} - \frac{2e^2}{r_2} + \frac{e^2}{r_{12}} \qquad (4.1)$$

where the first and second terms correspond to the attraction between the nucleus and electrons 1 and 2, respectively, and the third term arises from the mutual repulsion of the two electrons. To determine the properties of the helium atom [i.e., of the three-particle system with the potential-energy function given in Eq. 4.1], we must find the form of the Schrödinger equation for this case, analogous to Eq. 3.3 for the hydrogen atom. We assume that the nucleus is stationary because of its large mass relative to that of the electrons (see Section 3.2.1) and write kinetic-energy terms for the electrons only. The resulting Schrödinger equation is

$$-\frac{\hbar^2}{2m}\left(\frac{\partial^2 \psi}{\partial x_1{}^2} + \frac{\partial^2 \psi}{\partial y_1{}^2} + \frac{\partial^2 \psi}{\partial z_1{}^2}\right) - \frac{\hbar^2}{2m}\left(\frac{\partial^2 \psi}{\partial x_2{}^2} + \frac{\partial^2 \psi}{\partial y_2{}^2} + \frac{\partial^2 \psi}{\partial z_2{}^2}\right)$$
$$+ V(r_1, r_2, r_{12})\psi = E\psi \tag{4.2}$$

where x_1, y_1, z_1 and x_2, y_2, z_2 are the Cartesian coordinates, relative to the nucleus as origin, of electrons 1 and 2, respectively, and m is the electron mass. The symbol E represents the total energy of the electrons in the atom and ψ is the electronic wave function,

$$\psi = \psi(x_1, y_1, z_1; x_2, y_2, z_2) \tag{4.3}$$

4.1.1   Operator form of the Schrödinger equation   It is convenient here to introduce some new symbols, which provide a "shorthand" for writing the Schrödinger equation. The symbol ∇^2, read "del-squared" and called the *Laplacian operator*, represents

$$\nabla^2 = \frac{\partial^2}{\partial x^2} + \frac{\partial^2}{\partial y^2} + \frac{\partial^2}{\partial z^2} \tag{4.4}$$

that is, the differential operator ∇^2 operating on the function ψ gives

$$\nabla^2\psi = \frac{\partial^2 \psi}{\partial x^2} + \frac{\partial^2 \psi}{\partial y^2} + \frac{\partial^2 \psi}{\partial z^2}$$

Thus, Eq. 4.2 can be written more compactly by making the substitutions

$$\nabla_1{}^2 = \frac{\partial^2}{\partial x_1{}^2} + \frac{\partial^2}{\partial y_1{}^2} + \frac{\partial^2}{\partial z_1{}^2}$$

$$\nabla_2{}^2 = \frac{\partial^2}{\partial x_2{}^2} + \frac{\partial^2}{\partial y_2{}^2} + \frac{\partial^2}{\partial z_2{}^2}$$

to give

$$-\frac{\hbar^2}{2m}\nabla_1{}^2\psi - \frac{\hbar^2}{2m}\nabla_2{}^2\psi + V(r_1, r_2, r_{12})\psi = E\psi \tag{4.5}$$

The three terms on the left-hand side of Eq. 4.5 can be further condensed by considering them to result from a single operation on the wave function ψ. We use the compact notation

$$H\psi = E\psi \tag{4.6}$$

where the operator H is the sum of the terms

$$H = -\frac{\hbar^2}{2m}\nabla_1{}^2 - \frac{\hbar^2}{2m}\nabla_2{}^2 + V(r_1, r_2, r_{12}) \tag{4.7}$$

H is called the *Hamiltonian operator*, and in view of Eq. 4.5 it is the operator corresponding to the total energy of the system; that is, H operating on the wave function ψ gives the energy E times ψ. The shorthand symbols

∇^2 and H are widely used, particularly for atoms and molecules with a large number of electrons, each of which contributes kinetic- and potential-energy terms to the Schrödinger equation.

Our problem is to find the helium-atom energies E and the associated wave functions ψ which satisfy Eq. 4.5. Because the potential-energy term depends on the radial distances r_1 and r_2 it is again convenient to use spherical polar coordinates rather than Cartesian coordinates. Writing r_1, θ_1, ϕ_1 and r_2, θ_2, ϕ_2 for the spherical polar coordinates of electrons 1 and 2, respectively, we express the wave function Eq. 4.3 as

$$\psi = \psi(r_1, \theta_1, \phi_1; r_2, \theta_2, \phi_2)$$

or, in shorthand notation,

$$\psi = \psi(1, 2)$$

where the 1 and 2 in parentheses represent (r_1, θ_1, ϕ_1) and (r_2, θ_2, ϕ_2), respectively. Introduction of the spherical polar coordinates into the Schrödinger equation yields a kinetic-energy operator for each of the electrons that has the form given in Eq. 3.9 for the hydrogen atom. It should be noted that the symbols ∇^2 and H for the Laplacian and Hamiltonian operators, respectively, do not specify the coordinate system. Thus, for example, ∇_1^2 can be written in spherical polar coordinates; from Eq. 3.9 we know that the result is

$$\nabla_1^2 = \frac{1}{r_1^2} \frac{\partial}{\partial r_1} \left(r_1^2 \frac{\partial}{\partial r_1} \right) + \frac{1}{r_1^2 \sin \theta_1} \frac{\partial}{\partial \theta_1} \left(\sin \theta_1 \frac{\partial}{\partial \theta_1} \right) + \frac{1}{r_1^2 \sin^2 \theta_1} \frac{\partial^2}{\partial \phi_1^2} \quad (4.8)$$

In contrast to the hydrogen-atom Schrödinger equation (Eq. 3.9), the helium-atom equation (Eq. 4.5) has no known exact solution. The source of the difficulty is the presence of the electron-electron repulsion term e^2/r_{12} in the potential energy, in addition to electron-nuclear terms $-2e^2/r_1$ and $-2e^2/r_2$. Although the two-body problem can be solved exactly, a three-body problem of this type, in which the potential energy of each particle depends upon the positions of the other two, cannot be solved in either classical mechanics or quantum mechanics. It is necessary, therefore, to attack the problem by the method of successive approximation.

It has been shown by a series of approximate calculations that both the energy E and the wave function ψ for the helium atom can be obtained to any desired accuracy. The most accurate calculated result for the energy required to remove the two electrons from the atom agrees with the available experimental measurement within experimental error, which is about 1 part in 10 million. Such excellent agreement gives us confidence that the Schrödinger equation is the correct equation for many-electron atoms. However, the mathematical details of these accurate calculations are complicated, and the resulting expressions are so involved that they are not very useful for the general understanding we are trying to achieve here. We look instead at the simplest approximation methods, which are sufficient for obtaining the most important qualitative features of the helium-atom solutions.

4.1.2 Independent-electron approximation We begin by neglecting the potential-energy term e^2/r_{12}, which couples the motion of the two electrons. The resulting equation has the form

$$\left(-\frac{\hbar^2}{2m}\nabla_1^2 - \frac{2e^2}{r_1}\right)\psi + \left(-\frac{\hbar^2}{2m}\nabla_2^2 - \frac{2e^2}{r_2}\right)\psi = E\psi \qquad (4.9)$$

Equation 4.9 may be thought of as the first approximation to Eq. 4.5. Each of the terms in parentheses on the left-hand side of Eq. 4.9 is the Hamiltonian operator for a hydrogen-like atom with nuclear charge $+2e$. Thus, without the e^2/r_{12} term neither electron is affected by the presence of the other and the two of them behave completely independently. This means that, by analogy with the separation of the variables r, θ, and ϕ in the hydrogen atom (Eq. 3.11), the wave function satisfying Eq. 4.9 can be written as the product of two functions

$$\psi(1, 2) = \psi_1(1)\psi_2(2) \qquad (4.10)$$

The factors $\psi_1(1)$ and $\psi_2(2)$ satisfy the equations for a hydrogen-like atom with nuclear charge $+2e$,

$$\left(-\frac{\hbar^2}{2m}\nabla_1^2 - \frac{2e^2}{r_2}\right)\psi_1 = E_1\psi_1 \qquad (4.11a)$$

and

$$\left(-\frac{\hbar^2}{2m}\nabla_2^2 - \frac{2e^2}{r_2}\right)\psi_2 = E_2\psi_2 \qquad (4.11b)$$

Substitution of Eq. 4.10 into Eq. 4.9 yields

$$\psi_2(2)\left[\left(-\frac{\hbar^2}{2m}\nabla_1^2 - \frac{2e^2}{r_1}\right)\psi_1(1)\right] + \psi_1(1)\left[\left(-\frac{\hbar^2}{2m}\nabla_2^2 - \frac{2e^2}{r_2}\right)\psi_2(2)\right]$$
$$= E\psi_1(1)\psi_2(2)$$

where use has been made of the fact that ∇_1^2 operates only on $\psi_1(1)$ and not on $\psi_2(2)$, and similarly for ∇_2^2. Replacing the expression in square brackets by E_1 and E_2 by means of Eqs. 4.11, we have

$$(E_1 + E_2)\psi_1(1)\psi_2(2) = E\psi_1(1)\psi_2(2)$$

In this approximation the total energy of the electrons in the helium atom is thus given by the sum of the two one-electron energies,

$$E = E_1 + E_2 \qquad (4.12)$$

Furthermore, according to Eq. 4.10, the probability density function for the two electrons is the product of the probability densities for each electron in the field of the bare nucleus,

$$\psi^2(1, 2) = \psi_1^2(1)\psi_2^2(2) \tag{4.13}$$

The fact that the one-electron probability densities combine in this way shows clearly that the two electrons move independently of one another in the absence of the e^2/r_{12} term.

The ground state $(E = E_0)$ of the helium atom is seen in this approximation (Eq. 4.12) to have twice the energy of a hydrogen-like atom with nuclear charge $Z = 2$ in its ground $(1s)$ state. Thus $E_0 = 2E_{1s}(Z = 2) = -108.8$ eV. This value is larger than the true helium energy $(E = -79.0$ eV), but still of the right order. The functions ψ_1 and ψ_2 corresponding to this energy are the hydrogen-like $1s$ functions $(Z = 2)$

$$\begin{aligned} \psi_1(1) &= \left(\frac{2^3}{\pi a_0^3}\right)^{1/2} \exp\left(-2r_1/a_0\right) \\ \psi_2(2) &= \left(\frac{2^3}{\pi a_0^3}\right)^{1/2} \exp\left(-2r_2/a_0\right) \end{aligned} \tag{4.14}$$

For the first excited state of helium the same approximation yields the energy $E_1 = E_{1s} + E_{2s}$, and the wave function $\psi(1, 2) = \psi_{1s}(1)\psi_{2s}(2)$, and so on. It is clear that the wave functions and energy levels of helium are very easy to determine in this approximation which ignores the electron-electron repulsion.

4.1.3 Average-shielding approximation

Although the grossest features of the helium atom are given by the above approximation, it is important to understand the nature of the changes produced by the presence of the e^2/r_{12} term. Fortunately, we can include the effects of electron repulsion in a fairly simple fashion so as to obtain the second approximation to the helium-atom solution. To find how to introduce the appropriate correction for the ground-state wave function and energy, let us examine the behavior of one of the electrons, say, electron 2, in the presence of the other—electron 2 "sees" not only the nuclear charge $+2e$ (via the attractive potential-energy term $-2e^2/r_2$), but also the field due to electron 1 (via the repulsive potential-energy term e^2/r_{12}). In the ground state of the atom, electron 1 is in a $1s$ orbital, and it has a spherically symmetric probability distribution with a radial distribution function like that shown at the top of Fig. 3.5c. If electron 2 is far away from the nucleus [i.e., if r_2 is large relative to the distance where the radial distribution function $r_1^2R^2(r_1)$ has its maximum], it will be attracted by the nucleus and repelled by the negative cloud around the nucleus due to electron 1 (see Fig. 4.2a). The result is that electron 2

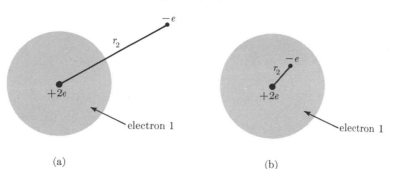

(a)   (b)

·**Fig. 4.2.** Relation of instantaneous position of electron 2 to the time-averaged cloud of electron 1. (**a**) Electron 2 outside region of high charge density of electron 1; (**b**) electron 2 inside region of high charge density of electron 1.

is attracted to the nucleus somewhat less than it would be in the one-electron He^+ ion; that is, the nuclear charge is "shielded" by the part of the electron cloud corresponding to electron 1 that is inside the r_2 distance. If r_2 is very large ($r_2 \rightarrow \infty$), practically all of the electron 1 cloud is closer to the nucleus and the shielding results in an effective nuclear charge Z_e acting on electron 2 that is near unity. On the other hand, if electron 2, which also has a $1s$-type distribution, is close to the nucleus (see Fig. 4.2b), the shielding is much less complete. In the limit as r_2 goes to zero, the effective nuclear charge Z_e approaches the unshielded nuclear charge of 2. Thus, Z_e is a function of r_2 and has the limits

$$1 \leq Z_e \leq 2$$

This suggests that as the second approximation to the helium-atom wave function, we introduce an average Z_e which gives the best approximation; that is, the unshielded wave functions of Eq. 4.14 are replaced by the shielded functions

$$\psi_1(1) = \left(\frac{Z_e^3}{\pi a_0^3}\right)^{1/2} \exp\left(-Z_e r_1/a_0\right)$$

$$\psi_2(2) = \left(\frac{Z_e^3}{\pi a_0^3}\right)^{1/2} \exp\left(-Z_e r_2/a_0\right)$$

$$(4.15)$$

Because the two electrons have equivalent distributions in the helium-atom ground state, the same value of Z_e appears in both functions of Eq. 4.15.

Although the shielding model has shown that the effective charge Z_e must lie somewhere between 1 and 2, we do not know its actual value. There are several ways of making an estimate by means of theoretical calculations (see below) or by use of experimental data. One procedure is based upon the fact that the value of Z_e affects not only the wave function, but also the energy of an electron. For a hydrogen-like atom with nuclear charge $+Z_e e$, Eq. 1.57 shows that the energy for the ground state ($1s$) would be

$$E_{1s} = -\frac{Z_e^2 e^2}{2a_0} \qquad (4.16)$$

If we apply this equation to the helium atom, each electron has an energy given by Eq. 4.16. The first ionization potential, IP_1, which is the energy required to remove one of the electrons in the process (He $\rightarrow$ He$^+$ + e$^-$), should therefore be approximated by

$$IP_1 \simeq -E_{1s} = \frac{Z_e^2 e^2}{2a_0} = (13.6 \; Z_e^2) \text{ eV} \qquad (4.17)$$

The second IP is the energy required to remove an electron from the He$^+$ ion in the process (He$^+$ $\rightarrow$ He^{2+} + e$^-$). Since there is now only one electron, there is no shielding of the nucleus ($Z_e = Z = 2$), and we have

$$IP_2 = (13.6) \; (4) = 54.4 \text{ eV}$$

By experimental measurements using photons of a sufficiently high frequency or electrons with sufficiently high kinetic energy to ionize the atom, we find

$$IP_1 = 24.6 \text{ eV} \qquad IP_2 = 54.4 \text{ eV}$$

The agreement between experiment and theory for IP_2 is to be expected, since He$^+$ is a one-electron ion that can be accurately treated by the methods of Chapter 3. Further, from the much smaller experimental value for IP_1, it is clear that significant shielding must be present. If we choose Z_e such that $IP_1 = Z_e^2 e^2/2a_0$, we find that $Z_e = 1.34$. This value is a reasonable one—it lies between the required limits 1 and 2—and shows that the radial distribution in helium corresponds to partial shielding of the nucleus. In Fig. 4.3, we show $1s$ radial distributions for $Z = 1, 1.34,$ and 2.

4.1.4 Improved approximations to the electron-electron interaction   In the second approximation to the helium atom, in which mutual shielding of the electrons was taken into account, we obtained some qualitative information concerning the effect of the electron-electron interaction on the energies and wave functions of the individual electrons. The total energy was assumed to be the sum of the energies of two independent (but shielded) electrons with $1s$ hydrogen-like wave functions; thus, the energy calculation did not directly involve the electron-electron interaction. For even a qualitative understanding of the helium-atom spectrum, however, it is important to have a more direct theoretical method for estimating the effect of electron-electron interaction upon the total energy. Since this interaction is represented by the term e^2/r_{12} in the potential-energy function (Eq. 4.1) in

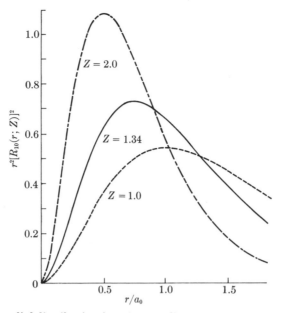

Fig. 4.3. The $1s$ radial distribution functions: Full curve, $Z = 1.34$; dash-dot curve, $Z = 2.0$; dashed curve, $Z = 1.0$.

the Schrödinger equation (Eq. 4.2), we focus on this term. We assume, as in Section 4.1.3, that in the first approximation the helium ground-state wave function is given by Eqs. 4.10 and 4.14. The average contribution of the potential-energy terms to the total energy can then be determined by evaluating the integral (Section 2.7)

$$\langle V \rangle = \int \psi^2(1, 2) \, V(r_1, r_2, r_{12}) dv_1 \, dv_2 \tag{4.18}$$

where dv_1 and dv_2 are the differential volume elements for electrons 1 and 2 in spherical polar coordinates. Substituting from Eqs. 4.10 and 4.1 into 4.18, we have

$$
\begin{aligned}
\langle V \rangle &= \int \psi_1^2(1)\psi_2^2(2) \left(-\frac{2e^2}{r_1} - \frac{2e^2}{r_2} + \frac{e^2}{r_{12}} \right) dv_1 \, dv_2 \\
&= \int \psi_1^2(1) \left(-\frac{2e^2}{r_1} \right) dv_1 \int \psi_2^2(2) dv_2 \\
&\quad + \int \psi_2^2(2) \left(-\frac{2e^2}{r_2} \right) dv_2 \int \psi_1^2(1) dv_1 \\
&\quad + \int \psi_1^2(1)\psi_2^2(2) \left(\frac{e^2}{r_{12}} \right) dv_1 \, dv_2
\end{aligned}
\tag{4.19}
$$

Since $\psi_1^2(1)$ and $\psi_2^2(2)$ are normalized, Eq. 4.19 reduces to

$$\langle V \rangle = \int \psi_1^2(1) \left(-\frac{2e^2}{r_1} \right) dv_1 + \int \psi_2^2(2) \left(-\frac{2e^2}{r_2} \right) dv_2$$
$$+ \int \psi_1^2(1)\psi_2^2(2) \left(\frac{e^2}{r_{12}} \right) dv_1\, dv_2 \tag{4.20}$$

Each of the terms in Eq. 4.20 has a simple interpretation. The first term represents the attractive interaction between the charge distribution $-e\psi_1^2(1)$ and the point nucleus of charge $+2e$; the second term represents the corresponding interaction for electron 2 and the nucleus; and the third term represents the interaction between the charge distribution $-e\psi_1^2(1)$ for electron 1 and $-e\psi_2^2(2)$ for electron 2. The nucleus-electron terms are just the same as for the one-electron hydrogen-like atoms and were already obtained in Section 3.9. Thus

$$\langle V \rangle = 2 \left(-\frac{4e^2}{a_0} \right) + \int \psi_1^2(1)\psi_2^2(2) \left(\frac{e^2}{r_{12}} \right) dv_1\, dv_2$$

To calculate the electron-electron term, we substitute the explicit form for $\psi_1^2(1)$ and $\psi_2^2(2)$ (Eq. 4.14) and obtain

$$\int \psi_1^2(1)\psi_2^2(2) \left(\frac{e^2}{r_{12}} \right) dv_1\, dv_2$$
$$= e^2 \left(\frac{8}{\pi a_0^3} \right)^2 \int \frac{\exp(-4r_1/a_0)\, \exp(-4r_2/a_0)}{r_{12}} dv_1\, dv_2$$
$$= e^2 \left(\frac{8}{\pi a_0^3} \right)^2 \int_0^{2\pi} d\phi_1 \int_0^{\pi} \sin\theta_1\, d\theta_1 \int_0^{\infty} r_1^2 dr_1 \int_0^{2\pi} d\phi_2$$
$$\times \int_0^{\pi} \sin\theta_2\, d\theta_2 \int_0^{\infty} r_2^2 dr_2\, \frac{\exp(-4r_1/a_0)\, \exp(-4r_2/a_0)}{r_{12}}$$
$$= e^2 \left(\frac{32}{a_0^3} \right) \int_0^{\infty} r_1^2 dr_1 \exp(-4r_1/a_0) \left[\frac{1}{r_1} - \exp(-4r_1/a_0) \left(\frac{2}{a_0} + \frac{1}{r_1} \right) \right] \tag{4.21}$$

To prove the last equality of Eq. 4.21 we must express $1/r_{12}$ in terms of the coordinates of the two electrons; the relationship in vector form (Fig. 4.1) is

$$\mathbf{r}_{12} = \mathbf{r}_1 - \mathbf{r}_2$$

or, in terms of the Cartesian components,

$$
\begin{aligned}
x_{12} &= x_1 - x_2 \\
&= r_1 \sin\theta_1 \cos\phi_1 - r_2 \sin\theta_2 \cos\phi_2 \\
y_{12} &= y_1 - y_2 \\
&= r_1 \sin\theta_1 \sin\phi_1 - r_2 \sin\theta_2 \sin\phi_2 \\
z_{12} &= z_1 - z_2 \\
&= r_1 \cos\theta_1 - r_2 \cos\theta_2
\end{aligned}
\tag{4.22}
$$

The distance r_{12} is thus

$$
\begin{aligned}
r_{12} &= (x_{12}{}^2 + y_{12}{}^2 + z_{12}{}^2)^{1/2} \\
&= (r_1{}^2 + r_2{}^2 - 2ar_1r_2)^{1/2}
\end{aligned}
\tag{4.23}
$$

In Eq. 4.23 we have written

$$
\begin{aligned}
a &= \cos \alpha \\
&= \cos \theta_1 \cos \theta_2 + \sin \theta_1 \sin \theta_2 \cos (\phi_1 - \phi_2)
\end{aligned}
\tag{4.24}
$$

where α is the angle between $\mathbf{r}_1$ and $\mathbf{r}_2$ (Fig. 4.1).

If $r_1 > r_2$, the quantity $1/r_{12}$ can be expanded about $(r_2/r_1) = 0$ in a Maclaurin series,

$$
\begin{aligned}
\frac{1}{r_{12}} = \frac{1}{r_1} \Bigg[1 + a \left(\frac{r_2}{r_1}\right) + \frac{1}{2} (3a^2 - 1) \left(\frac{r_2}{r_1}\right)^2 \\
+ \frac{1}{2} (5a^3 - 3a) \left(\frac{r_2}{r_1}\right)^3 + \cdots \Bigg]
\end{aligned}
\tag{4.25}
$$

If $r_2 > r_1$, the expansion takes the same form as Eq. 4.25, but with r_1 and r_2 interchanged; thus, by writing $r_>$ for the greater of (r_1, r_2) and $r_<$ for the smaller, we can express the expansion as

$$
\begin{aligned}
\frac{1}{r_{12}} = \frac{1}{r_>} \Bigg[1 + a \left(\frac{r_<}{r_>}\right) + \frac{1}{2} (3a^2 - 1) \left(\frac{r_<}{r_>}\right)^2 \\
+ \frac{1}{2} (5a^3 - 3a) \left(\frac{r_<}{r_>}\right)^3 + \cdots \Bigg]
\end{aligned}
\tag{4.26}
$$

Substitution of Eq. 4.26 into Eq. 4.21 and integration over the angles θ_1, ϕ_1, θ_2, ϕ_2 gives zero for each term in the expansion except the first. This is true not only for the angle-dependent terms we have actually written down in Eq. 4.26 (Problem 4.7), but it can be shown quite generally that for spherically symmetric wave functions $\psi_1(1)$ and $\psi_2(2)$, contributions to the integral of Eq. 4.21 vanish for all angle-dependent terms in the expansion of $1/r_{12}$ (see Reference 6 in the Additional Reading list, p. 446). Therefore, for the present case in which $\psi_1(1)$ and $\psi_2(2)$ are spherically symmetric 1s hydrogen-like functions, the substitution of $1/r_>$ for $1/r_{12}$ in Eq. 4.21 gives the exact result for the integral; after integration over the angles, the integral becomes

$$
e^2 \left(\frac{32}{a_0{}^3}\right)^2 \int_0^\infty r_1{}^2 dr_1 \int_0^\infty r_2{}^2 dr_2 \, \frac{1}{r_>} \exp\left(-\frac{4r_1}{a_0} - \frac{4r_2}{a_0}\right)
\tag{4.27}
$$

To evaluate this radial integral, we first integrate over r_2 and then over r_1. According to the definition of $r_>$, it is to be replaced by r_1 when r_2 is in the range $0 \leq r_2 \leq r_1$, and by r_2 when r_2 is in the range $r_1 \leq r_2 \leq \infty$. Consequently, the integral of Eq. 4.27 is split into two terms, yielding

$$\int \psi_1^2(1)\psi_2^2(2) \left(\frac{e^2}{r_{12}}\right) dv_1 \, dv_2$$

$$= e^2 \left(\frac{32}{a_0^3}\right)^2 \int_0^\infty r_1^2 dr_1 \exp\left(-4r_1/a_0\right) \left[\frac{1}{r_1} \int_0^{r_1} \exp\left(-4r_2/a_0\right) r_2^2 dr_2\right.$$

$$\left. + \int_{r_1}^\infty \exp\left(-4r_2/a_0\right) r_2 dr_2\right]$$

$$= e^2 \left(\frac{32}{a_0^3}\right) \int_0^\infty r_1^2 dr_1 \exp\left(-4r_1/a_0\right) \left[\frac{1}{r_1} - \exp\left(-4r_1/a_0\right)\left(\frac{2}{a_0} + \frac{1}{r_1}\right)\right]$$

(4.28)

The factor in brackets in the integrand of the last line of Eq. 4.21 is the potential energy of electron 1 due to its electrostatic interaction with electron 2 (apart from the factor e^2). To obtain the total potential energy $V_1(r_1)$ of electron 1, we multiply the bracketed expression in the last line of Eq. 4.21 by e^2 and add the nuclear attraction term $-2e^2/r_1$; that is,

$$V_1(r_1) = \frac{-2e^2}{r_1} + e^2 \left[\frac{1}{r_1} - \exp\left(-4r_1/a_0\right)\left(\frac{2}{a_0} + \frac{1}{r_1}\right)\right] \qquad (4.29)$$

We see that the second term of Eq. 4.29 has the form of the shielding interaction described earlier; for large r_1, the exponential factor [exp $(-4r_1/a_0)$] goes to zero, so that

$$V_1(r_1) \xrightarrow[r_1 \to \infty]{} -\frac{e^2}{r_1}$$

and the nucleus is completely shielded by electron 2; for small $r_1 \ll a_0$, the exponential factor can be expanded to give

$$V_1(r_1) = -\frac{2e^2}{r_1} + e^2 \left[\frac{1}{r_1} - \left(1 - \frac{4r_1}{a_0} + \cdots\right)\right.$$

$$\left. \times \left(\frac{2}{a_0} + \frac{1}{r_1}\right)\right] \xrightarrow[r_1 \to 0]{} -\frac{2e^2}{r_1} + \frac{2e^2}{a_0} \xrightarrow[r_1 \to 0]{} -\frac{2e^2}{r_1}$$

and the shielding effect is negligible. The function $V_1(r_1)$ is plotted in Fig. 4.4 together with the unshielded $-2e^2/r_1$ and the completely shielded $-e^2/r_1$ potential-energy functions. We see that the effective potential goes from one limit to the other as r_1 goes from zero to infinity.

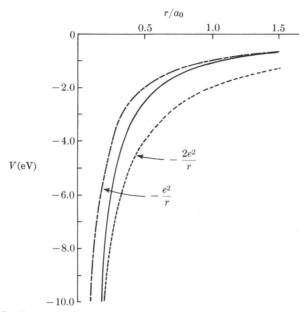

Fig. 4.4. Effective potential energy of electron 1 in the helium-atom ground state according to the hydrogen-like orbital approximation. Full curve, $V_1(r_1)$ of Eq. 4.29; dashed curve, unshielded potential $-2e^2/r_1$; dash-dot curve, completely shielded potential $-e^2/r_1$.

To complete the evaluation of the electron-electron repulsion, we carry out the integration over the coordinates of electron 1 in Eq. 4.21 and obtain

$$\int \psi_1^2(1)\psi_2^2(2)\left(\frac{e^2}{r_{12}}\right) dv_1\, dv_2 = \frac{5}{2}\left(\frac{e^2}{2a_0}\right) \tag{4.30}$$

Adding this term to the energy of the noninteracting electrons, we find for the ground-state energy

$$E_{\mathrm{He}} = E_1 + E_2 + \int \psi_1^2(1)\psi_2^2(2)\left(\frac{e^2}{r_{12}}\right) dv_1\, dv_2$$

$$= -4\left(\frac{e^2}{2a_0}\right) - 4\left(\frac{e^2}{2a_0}\right) + \frac{5}{2}\left(\frac{e^2}{2a_0}\right) \tag{4.31}$$

$$= -\frac{11}{2}\left(\frac{e^2}{2a_0}\right)$$

From Eq. 4.31, the total energy calculated for the helium atom is -74.8 eV. This is to be compared with the energy of the noninteracting electron approximation $(-108.8\ \mathrm{eV})$ and the experimental energy $(-79.0\ \mathrm{eV})$. It is clear that introduction of the e^2/r_{12} term gives a considerably improved result, although there is still an error of 4.2 eV.

The first ionization potential IP_1 is the difference between the atom and ion energy; that is, in the present approximation,

$$IP_1 = E_{He^+} - E_{He} = -54.4 - (-74.8) = 20.4 \text{ eV}$$

compared with the experimental value of 24.6 eV.

4.1.5 Perturbation theory

The procedure we have used for the helium atom in Section 4.1.4 is a particular example of a general approximation method, called *perturbation theory*. It has been widely applied to quantum-mechanical problems that are too difficult to solve exactly. In perturbation theory, the potential-energy function is usually divided into two parts, one of which is chosen so that the Schrödinger equation corresponding to it *can* be solved exactly for the wave function (called the *unperturbed solution*) and the energy (called the *unperturbed energy*). The other part of the potential energy (called the *perturbation*) is introduced as a correction and its average or expectation value is calculated with the unperturbed solutions. The result is added to the unperturbed energy to provide an approximation to the total energy of the system. To formulate this description in mathematical terms, we write the total Hamiltonian H as the sum of an unperturbed Hamiltonian H_0 and the perturbation H',

$$H = H_0 + H' \tag{4.32}$$

The unperturbed Schrödinger equation, namely,

$$H_0\psi_i^{(0)} = E_i^{(0)}\psi_i^{(0)} \tag{4.33}$$

can be solved exactly for the unperturbed wave function $\psi_i^{(0)}$ and the unperturbed energy $E_i^{(0)}$ of the ith state; the superscript zero refers to the unperturbed problem. The total energy E_i for the ith state is then given approximately by the sum

$$E_i = E_i^0 + \langle H' \rangle_i \tag{4.34}$$

where $\langle H' \rangle_i$ is the average of H' for the unperturbed ith state; that is,

$$\langle H' \rangle_i = \int \psi_i^0 H' \psi_i^0 \, dv \tag{4.35}$$

For the helium atom, we separated the total potential energy into a term $-2e^2/r_1 - 2e^2/r_2$ representing electrons interacting with the nucleus but not with each other (the unperturbed or independent-electron problem), and the electron-electron interaction term e^2/r_{12} (the perturbation). In accordance with perturbation theory, the total energy is given to a first approximation by the sum of the unperturbed energy $E_1 + E_2$ and the average of the perturbation, $\langle e^2/r_{12} \rangle$ (Eq. 4.31).

4.1.6 *Variation principle* It is also possible to refine the perturbation calculation by introducing an effective nuclear charge Z_e into the wave function in correspondence with the earlier qualitative treatment (Section 4.1.3). The shielded functions $\psi_1(1)$ and $\psi_2(2)$, given by Eq. 4.15, are hydrogen-like solutions for a potential energy corresponding to a nuclear charge $+ Z_e e$. Since this is not the true nuclear charge, we could treat corrections due to the difference in the potential $[(Z_e - Z)e^2/r_1 + (Z_e - Z)e^2/r_2]$ as a perturbation, as well as the electron-electron interaction term. Alternatively, we can use the functions of Eq. 4.15 to evaluate the average of the entire Hamiltonian operator, which yields an average value for the total energy. We have

$$
\langle H \rangle = \int \psi_1(1)\psi_2(2) \left(- \frac{\hbar^2}{2m}(\nabla_1{}^2 + \nabla_2{}^2) - \frac{2e^2}{r_1} - \frac{2e^2}{r_2} + \frac{e^2}{r_{12}} \right)
$$

$$
\times \psi_1(1)\psi_2(2)\, dv_1\, dv_2
$$

$$
= \int \psi_1(1)\psi_2(2) \left(- \frac{\hbar^2}{2m}\nabla_1{}^2 - \frac{Z_e e^2}{r_1} \right) \psi_1(1)\psi_2(2)\, dv_1\, dv_2
$$

$$
+ \int \psi_1(1)\psi_2(2) \left(- \frac{\hbar^2}{2m}\nabla_2{}^2 - \frac{Z_e e^2}{r_2} \right) \psi_1(1)\psi_2(2)\, dv_1\, dv_2 \qquad (4.36)
$$

$$
+ (Z_e - 2)e^2 \int \psi_1(1)\psi_2(2) \left(\frac{1}{r_1} + \frac{1}{r_2} \right) \psi_1(1)\psi_2(2)\, dv_1\, dv_2
$$

$$
+ e^2 \int \psi_1(1)\psi_2(2) \left(\frac{1}{r_{12}} \right) \psi_1(1)\psi_2(2)\, dv_1\, dv_2
$$

where we have written the wave functions in the integrals in the form required for finding average values of differential operators (see Chapter 2).

The first two integrals on the right-hand side of Eq. 4.36 are easily evaluated if one recognizes that the hydrogen-like wave functions $\psi_1(1)$ and $\psi_2(2)$ with effective nuclear charge Z_e satisfy the one-electron Schrödinger equations

$$
\left(- \frac{\hbar^2}{2m}\nabla_1{}^2 - \frac{Z_e e^2}{r_1} \right) \psi_1(1) = - \left(\frac{Z_e{}^2 e^2}{2a_0} \right) \psi_1(1)
$$

$$
\left(- \frac{\hbar^2}{2m}\nabla_2{}^2 - \frac{Z_e e^2}{r_2} \right) \psi_2(2) = - \left(\frac{Z_e{}^2 e^2}{2a_0} \right) \psi_2(2)
$$

$$(4.37)$$

Use of Eq. 4.37 and the fact that $\psi_1(1)$ and $\psi_2(2)$ are normalized gives $-Z_e{}^2 e^2/2a_0$ for each of these two integrals. The third integral on the right-hand side of Eq. 4.36 is evaluated by substituting from Eq. 4.15, integrating out the dependence on the coordinates of one electron, integrating over angles for the other electron to give a factor of (4π), and use of Eq. 3.26 for the resulting radial integrals; the result is

$$(4\pi)(Z_e - 2)e^2 \left(\frac{Z_e^3}{\pi a_0^3}\right) \left[\int_0^\infty r_1 \exp\left(-2Z_e r_1/a_0\right) dr_1\right.$$
$$\left. + \int_0^\infty r_2 \exp\left(-2Z_e r_2/a_0\right) dr_2\right] \qquad (4.38)$$
$$= 4Z_e(Z_e - 2)\left(\frac{e^2}{2a_0}\right)$$

The remaining integral in Eq. 4.36 involving $1/r_{12}$ is evaluated by a method analogous to that which led to Eq. 4.30; taking account of the screening of the wave functions, we find

$$\int \psi_1(1)\psi_2(2)\left(\frac{e^2}{r_{12}}\right)\psi_1(1)\psi_2(2)\ dv_1\ dv_2 = \frac{5}{4}Z_e\left(\frac{e^2}{2a_0}\right) \qquad (4.39)$$

Substitution of Eqs. 4.37–4.39 into Eq. 4.36 shows that the average value of the total helium energy for the wave functions of Eq. 4.15 is

$$\langle H \rangle = \left(-2Z_e^2 + 4Z_e(Z_e - 2) + \frac{5}{4}Z_e\right)\left(\frac{e^2}{2a_0}\right) \qquad (4.40)$$

where use of the shielded wave function has made each term dependent on the effective charge Z_e. If we plot $\langle H \rangle$ as a function of Z_e (Fig. 4.5), we see that there is a value of Z_e for which the average energy $\langle H \rangle$ exhibits a *minimum*. The values of $\langle H \rangle$ and Z_e corresponding to this minimum are determined by differentiation; that is, from

$$\frac{d\langle H \rangle}{dZ_e} = \left(-4Z_e + 8Z_e - 8 + \frac{5}{4}\right)\left(\frac{e^2}{2a_0}\right) = 0$$

we find

$$Z_e = 2 - \frac{5}{16} = 1.6875$$
$$\langle H \rangle = -2\left(2 - \frac{5}{16}\right)^2\left(\frac{e^2}{2a_0}\right) = -77.5 \text{ eV} \qquad (4.41)$$

Comparing the true helium-atom energy of -79.0 eV with this value of $\langle H \rangle$ and with the unshielded perturbation result of -74.8 eV, we see that the choice of Z_e which makes $\langle H \rangle$ a minimum improves the energy by an amount equal to $\frac{3}{4}$ the error of the unshielded perturbation treatment. Since Eq. 4.41 gives the minimum energy value that can be obtained with a shielded wave function of the form of Eqs. 4.10 and 4.15, the value $Z_e = 1.6875$ is the best possible choice for the effective charge as judged by the calculated energy, the error being greater for any other choice. The

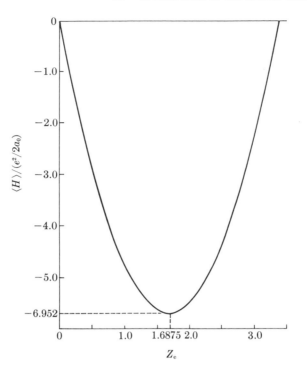

Fig. 4.5. The average energy of the helium atom $\langle H \rangle$ as a function of the effective nuclear charge Z_e appearing in the shielded wave function; see Eq. 4.40.

value 1.6875 is somewhat larger than 1.34, which was the Z_e value determined by fitting the experimental IP (Section 4.1.3). The difference between the two results arises from the fact that the present value corresponds to an actual calculation of the average of H, while the earlier one was obtained from a simplified formula for the energy of the shielded, but otherwise noninteracting, two-electron system.

The approach that we have employed here is an example of a widely used technique called the variation method. It is based on a generalization of the minimization procedure by which the best Z_e was determined. For the ground state of an atom or molecule, it can be shown that the average value of the exact Hamiltonian operator for an approximate wave function is always greater than the exact energy. It follows from this important result, called the *variation principle*, that the best form for an approximate wave function is the one which *minimizes* the average energy. This makes it possible to introduce any number of parameters (such as Z_e in the present example) into a wave function. These parameters are then varied simultaneously so as to find the lowest energy, which is the best value of the system

The square of this function, which gives the probability distribution, is

$$[\psi_{He}(1, 2) + \psi_{He}(2, 1)]^2 = [1s\alpha(1)]^2 [1s\beta(2)]^2$$
$$+ 2[1s\alpha(1)1s\beta(2)1s\alpha(2)1s\beta(1)] \qquad (4.45)$$
$$+ [1s\alpha(2)]^2 [1s\beta(1)]^2$$

From Eq. 4.45 we see that interchanging the labels of electrons 1 and 2 in this probability distribution gives the same distribution; that is,

$$[1s\alpha(2)]^2 [1s\beta(1)]^2 + 2[1s\alpha(2)1s\beta(1)1s\alpha(1)1s\beta(2)] + [1s\alpha(1)]^2 [1s\beta(2)]^2$$

which is identical with Eq. 4.45. Thus, the wave function of Eq. 4.44 corresponds to two indistinguishable electrons—one electron occupies a $1s$ orbital with α spin and one occupies a $1s$ orbital with β spin, but it is impossible to say which electron is which.

Equation 4.44 is not the only way of combining the functions $\psi_{He}(1, 2)$ and $\psi_{He}(2, 1)$ to obtain a probability distribution that does not distinguish between the electrons. The alternative to taking the sum of the functions is to take the difference and write

$$\psi_{He}(1, 2) - \psi_{He}(2, 1) = 1s\alpha(1)1s\beta(2) - 1s\alpha(2)1s\beta(1) \qquad (4.46)$$

which gives the probability distribution

$$[\psi_{He}(1, 2) - \psi_{He}(2, 1)]^2 = [1s\alpha(1)]^2 [1s\beta(2)]^2$$
$$- 2[1s\alpha(1)1s\beta(2)1s\alpha(2)1s\beta(1)] + [1s\alpha(2)]^2 [1s\beta(1)]^2 \qquad (4.47)$$

Here again, interchanging the labels of the two electrons yields a probability distribution that is unaltered,

$$[1s\alpha(2)] [1s\beta(1)]^2 - 2[1s\alpha(2)1s\beta(1)1s\alpha(1)1s\beta(2)] + [1s\alpha(1)]^2 [1s\beta(2)]^2$$

The functions themselves (Eqs. 4.44 and 4.46) do not both have the property that they are unaltered by interchanging electrons. From Eq. 4.44, we obtain upon electron interchange

$$1s\alpha(2)1s\beta(1) + 1s\alpha(1)1s\beta(2)$$

which is the same as Eq. 4.44; therefore, we say that such a function is *symmetric* for electron interchange. By contrast, Eq. 4.46 gives

$$1s\alpha(2)1s\beta(1) - 1s\alpha(1)1s\beta(2) = -[1s\alpha(1)1s\beta(2) - 1s\alpha(2)1s\beta(1)]$$

which is the negative of the original function; the expression in Eq. 4.46 is said to be *antisymmetric* to electron interchange. Thus, there appear to be two different ground-state helium wave functions (Eqs. 4.44 and 4.46) which give different probability distributions (Eqs. 4.45 and 4.47), both of which satisfy the requirement that the electrons be indistinguishable. There is no obvious difficulty with either of the functions and one might expect some ground-state helium atoms to be found in nature with each of these wave functions. However, this is not true—*only helium atoms with probability distributions corresponding to the antisymmetric function are found to exist;* the symmetric function (even as a first approximation) is incorrect. Thus, the properly antisymmetrized helium-atom ground state is represented by

$$
\begin{aligned}
\psi(1, 2) &= (1/2)^{1/2}[1s\alpha(1)1s\beta(2) - 1s\alpha(2)1s\beta(1)] \\
&= (1/2)^{1/2}1s(1)1s(2)[\alpha(1)\beta(2) - \alpha(2)\beta(1)]
\end{aligned}
\tag{4.48}
$$

where we have introduced the factor $(1/2)^{1/2}$ so that the wave function $\psi(1, 2)$ is normalized (see Problem 4.5).

It is important to note that in the second line of Eq. 4.48 we have been able to express the antisymmetrized wave function as a product of the spatial part $[1s(1)1s(2)]$ and the spin part $(1/2)^{1/2}[\alpha(1)\beta(2) - \alpha(2)\beta(1)]$. This is always possible for a two-electron system if spin-orbit coupling is so small that it can be neglected; for systems of three or more electrons such a factorization of the antisymmetric wave function cannot be made.

Another way of writing $\psi(1, 2)$, which is often used in atomic and molecular theory, is in the equivalent determinantal form; that is,

$$
\psi(1, 2) = (1/2)^{1/2} \begin{vmatrix} 1s\alpha(1) & 1s\beta(1) \\ 1s\alpha(2) & 1s\beta(2) \end{vmatrix}
\tag{4.49}
$$

which yields Eq. 4.48 when the determinant is expanded. Equation 4.49 is called a *Slater determinant*, after J. C. Slater, who introduced this convenient form for electronic wave functions. We see that the rows of the determinant are labelled by the electron indices (1, 2) and the columns by the forms of the one-electron wave functions $(1s\alpha, 1s\beta)$.

An important aspect of the antisymmetry of the wave function for electron interchange is that such a function automatically satisfies the Pauli principle. Let us look at the example of two electrons in the same orbital with the same spin, say, $1s\alpha$. Writing the Slater determinant, we have

$$
(1/2)^{1/2} \begin{vmatrix} 1s\alpha(1) & 1s\alpha(1) \\ 1s\alpha(2) & 1s\alpha(2) \end{vmatrix}
$$
$$
= (1/2)^{1/2}[1s\alpha(1)1s\alpha(2) - 1s\alpha(1)1s\alpha(2)] = 0
$$

The wave function is identically zero, corresponding to the Pauli requirement that no such system with two electrons having the same quantum numbers can exist. This result holds for any number of electrons, so that an alternative statement of the Pauli exclusion principle is the following: *A wave function for a system of two or more electrons must be antisymmetric with respect to interchange of the labels of any two electrons;* that is, the function must change sign when any two electron labels are interchanged. A determinant with two or more identical columns (or rows) is zero. Thus, if any two one-electron wave functions are identical (i.e., if any pair of electrons have identical quantum numbers), the antisymmetric wave function for the system vanishes, in accordance with the previous statement of the Pauli exclusion principle.

Returning now to the problem of the ground state of the lithium atom, if we attempted to place all three electrons in a $1s$ orbital (that is, with $n_1 = n_2 = n_3 = 1$, $l_1 = l_2 = l_3 = 0$, and $m_1 = m_2 = m_3 = 0$), three *different* values of m_s would be required to satisfy the Pauli principle. Since m_s may take only the two values $\pm\frac{1}{2}$, the Pauli principle excludes the function $1s(1)1s(2)1s(3)$. Correspondingly, all the antisymmetric functions that can be constructed are zero; that is, we can write the four determinants

$$(1/6)^{1/2} \begin{vmatrix} 1s\alpha(1) & 1s\alpha(1) & 1s\alpha(1) \\ 1s\alpha(2) & 1s\alpha(2) & 1s\alpha(2) \\ 1s\alpha(3) & 1s\alpha(3) & 1s\alpha(3) \end{vmatrix} \qquad (1/6)^{1/2} \begin{vmatrix} 1s\alpha(1) & 1s\alpha(1) & 1s\beta(1) \\ 1s\alpha(2) & 1s\alpha(2) & 1s\beta(2) \\ 1s\alpha(3) & 1s\alpha(3) & 1s\beta(3) \end{vmatrix}$$

$$(1/6)^{1/2} \begin{vmatrix} 1s\alpha(1) & 1s\beta(1) & 1s\beta(1) \\ 1s\alpha(2) & 1s\beta(2) & 1s\beta(2) \\ 1s\alpha(3) & 1s\beta(3) & 1s\beta(3) \end{vmatrix} \qquad (1/6)^{1/2} \begin{vmatrix} 1s\beta(1) & 1s\beta(1) & 1s\beta(1) \\ 1s\beta(2) & 1s\beta(2) & 1s\beta(2) \\ 1s\beta(3) & 1s\beta(3) & 1s\beta(3) \end{vmatrix}$$

all of which yield zero when they are expanded (Problem 4.9). Thus, to obtain a satisfactory wave function, at least one electron must be in an orbital with higher energy. For hydrogen, the next orbitals have $n = 2$ and all of the $n = 2$ orbitals ($2s$, $2p_0$, $2p_{+1}$, $2p_{-1}$) have the same energy; they are said to be degenerate (Section 2.8). However, the shielding effects due to the presence of more than one electron in an atom cause the energy of the $2s$ orbital to be somewhat lower than that of the $2p$ orbitals. (This splitting of the degeneracy with respect to the quantum number l plays an important role in determining the structure of many-electron atoms and is discussed in more detail in the next section.) Thus, in the ground state of lithium, there are two electrons in the $1s$ orbital (with spins "paired"; i.e., $m_{s1} = +\frac{1}{2}$, $m_{s2} = -\frac{1}{2}$) and one electron in the $2s$ orbital. The antisymmetrized approximate wave function can be written either

$$\psi_{\text{Li}}(1, 2, 3) = (1/6)^{1/2} \begin{vmatrix} 1s\alpha(1) & 1s\beta(1) & 2s\alpha(1) \\ 1s\alpha(2) & 1s\beta(2) & 2s\alpha(2) \\ 1s\alpha(3) & 1s\beta(3) & 2s\alpha(3) \end{vmatrix} \qquad (4.50)$$

or

$$\psi_{Li}(1, 2, 3) = (1/6)^{1/2} \begin{vmatrix} 1s\alpha(1) & 1s\beta(1) & 2s\beta(1) \\ 1s\alpha(2) & 1s\beta(2) & 2s\beta(2) \\ 1s\alpha(3) & 1s\beta(3) & 2s\beta(3) \end{vmatrix} \qquad (4.50)$$

Both of these functions have the same energy in the isolated atom, since in the absence of a magnetic field a $2s$ orbital with α spin is degenerate with one of β spin.

The assignment of the electrons in an atom to the available orbitals is called the *electron configuration*. As a shorthand for the determinant, we often specify the configuration simply in terms of the occupied orbitals with or without explicit designation of the spins. Thus, for the ground state of helium, we can write

$$1s\alpha1s\beta \quad \text{or} \quad 1s^2$$

and for the ground state of lithium

$$1s\alpha1s\beta2s\alpha \quad \text{or} \quad 1s^22s$$

where no superscript on the orbital implies that it is occupied by one electron (with either spin), while a superscript 2 indicates that it is occupied by two electrons with paired spins. By a corresponding procedure, we can make use of the exclusion principle to work out ground-state configurations for the other atomic elements, provided we know the relative energies of the various atomic orbitals (Sections 4.4 and 4.5).

4.3   EXCITED STATES OF HELIUM

The excited states of an atom can be treated in the same way as the ground state. For example, in the case of helium, if the (e^2/r_{12}) term in the Hamiltonian operator is ignored so that the Schrödinger equation takes the form of Eq. 4.9 and its solutions take the form of Eq. 4.10, we would obtain the first excited state $1s2s$ by setting $\psi_1(1) = \psi_{1s}(1)$ and $\psi_2(2) = \psi_{2s}(2)$. The two-electron wave function would then be

$$\psi(1, 2) = \psi_{1s}(1)\psi_{2s}(2) \qquad (4.51)$$

and the energy would be

$$E = E_{1s} + E_{2s} \qquad (4.52)$$

in this first approximation. The two-electron probability density corresponding to the wave function of Eq. 4.51 is

$$P(1, 2) = \psi^2(1, 2) = \psi_{1s}^2(1)\psi_{2s}^2(2) \qquad (4.53)$$

When multiplied by the volume element $dv_1\, dv_2$ for the two electrons this function gives the probability of simultaneously finding electron 1 in the volume dv_1 at the point r_1, θ_1, ϕ_1, and electron 2 in the volume dv_2 at r_2, θ_2, ϕ_2. If we wish to know the probability of finding electron 1 at r_1, θ_1, ϕ_1 *irrespective* of the position of electron 2, we must sum [or rather integrate, since $P(1, 2)$ is a continuous function of r_2, θ_2, ϕ_2] over all the possible positions of electron 2. The probability density $P(1)$ for electron 1 alone is thus the integral

$$P(1) = \int \psi^2(1, 2)\, dv_2 = \int \psi_{1s}^2(1)\, \psi_{2s}^2(2)\, dv_2$$
$$= \psi_{1s}^2(1)\int\psi_{2s}^2(2)\, dv_2 = \psi_{1s}^2(1)$$

where the last equality follows from the fact that ψ_{2s} is normalized. Similarly, the probability density for electron 2 is

$$P(2) = \psi_{2s}^2(2)$$

Since the wave function of Eq. 4.51 gives different distribution functions for the two electrons, it violates the principle of indistinguishability and is unacceptable.

By analogy with our discussion of the helium ground state (Eqs. 4.43–4.49), it is clear that we can form linear combinations of the function $\psi(1, 2)$ given in Eq. 4.51 and the function with the electrons interchanged

$$\psi(2, 1) = 1s(2)2s(1) \tag{4.51'}$$

such that the electrons are indistinguishable. Writing the symmetric and antisymmetric linear combinations

$$\psi(1, 2) + \psi(2, 1) = 1s(1)2s(2) + 1s(2)2s(1) \tag{4.54a}$$

$$\psi(1, 2) - \psi(2, 1) = 1s(1)2s(2) - 1s(2)2s(1) \tag{4.54b}$$

we can form probability distributions by squaring these functions and demonstrate that neither electron is associated with a specific orbital (Problem 4.15).

From our knowledge of the Pauli principle, which requires that electronic wave functions be antisymmetric with respect to electron interchange, we might conclude that of the two functions given in Eqs. 4.54 only the second is a satisfactory description of the He (1s 2s) excited state. However, such a conclusion would be premature because Eqs. 4.54 include only the spatial part of the wave function and the spin must be introduced to obtain a complete description of the state. Considering the spin functions separately, we have the four possibilities

$$\alpha(1)\alpha(2) \quad \beta(1)\beta(2) \quad \alpha(1)\beta(2) \quad \beta(1)\alpha(2)$$

or, requiring indistinguishability of the electrons with respect to spin,

$$\alpha(1)\alpha(2)$$
$$\beta(1)\beta(2)$$
$$\alpha(1)\beta(2) + \beta(1)\alpha(2)$$
$$\alpha(1)\beta(2) - \beta(1)\alpha(2)$$

All four spin functions are possible for this case since the orbital quantum numbers of the two electrons are different, so that the spin quantum numbers can be the same. Of the four functions, the first three are symmetric with respect to electron interchange and the fourth is antisymmetric. To combine these with the spatial functions in such a way that the complete wave function is antisymmetric, we must multiply the symmetric spatial function by antisymmetric spin functions and vice versa; that is,

$$\psi_1(1, 2) = (1/2)^{1/2}[1s(1)2s(2) + 2s(1)1s(2)](1/2)^{1/2}[\alpha(1)\beta(2) - \beta(1)\alpha(2)]$$

$$\psi_2(1, 2) = (1/2)^{1/2}[1s(1)2s(2) - 2s(1)1s(2)](1/2)^{1/2}[\alpha(1)\beta(2) + \beta(1)\alpha(2)]$$

$$\psi_3(1, 2) = (1/2)^{1/2}[1s(1)2s(2) - 2s(1)1s(2)]\alpha(1)\alpha(2)$$

$$\psi_4(1, 2) = (1/2)^{1/2}[1s(1)2s(2) - 2s(1)1s(2)]\beta(1)\beta(2) \tag{4.55}$$

where we have introduced factors of $(1/2)^{1/2}$ so that the space and spin parts are separately normalized. Equation 4.55 shows that for He($1s\,2s$) there are three functions whose space parts are antisymmetric and one function whose space part is symmetric.

We can also write the He($1s\,2s$) functions in the determinantal form introduced previously. This can be done directly for $\psi_3(1, 2)$ and $\psi_4(1, 2)$,

$$\psi_3(1, 2) = (1/2)^{1/2} \begin{vmatrix} 1s\alpha(1) & 2s\alpha(1) \\ 1s\alpha(2) & 2s\alpha(2) \end{vmatrix} \tag{4.56a}$$

$$\psi_4(1, 2) = (1/2)^{1/2} \begin{vmatrix} 1s\beta(1) & 2s\beta(1) \\ 1s\beta(2) & 2s\beta(2) \end{vmatrix} \tag{4.56b}$$

but not for $\psi_1(1, 2)$ and $\psi_2(1, 2)$. However, it is possible to show that the latter two can be expressed as the sum or difference of two determinants; that is,

$$\psi_1(1, 2) = (1/2)^{1/2} \left\{ (1/2)^{1/2} \begin{vmatrix} 1s\alpha(1) & 2s\beta(1) \\ 1s\alpha(2) & 2s\beta(2) \end{vmatrix} \right.$$
$$\left. - (1/2)^{1/2} \begin{vmatrix} 1s\beta(1) & 2s\alpha(1) \\ 1s\beta(2) & 2s\alpha(2) \end{vmatrix} \right\} \tag{4.56c}$$

gives higher weight to the regions of small r_{12}, the positive values of the integrand predominate (i.e., those with r_1 and r_2 simultaneously less than or greater than the $2s$ nodal surface value), yielding a positive value of K. The degeneracy of $^1\psi$ and $^3\psi$ is therefore split by the *exchange interaction* in the manner shown in Fig. 4.6, with the triplet energy lower than the singlet energy by the amount $2K$.

no electron interaction   Coulomb interaction   exchange interaction

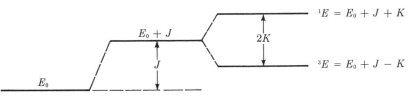

Fig. 4.6. Splitting of singlet and triplet states arising from the first excited helium configuration $1s2s$.

This ordering of the singlet and triplet levels arising from the assignment of the electrons to two different orbitals holds for all of the excited states of helium, as can be seen from the level diagram of Fig. 4.7. The $1s2p$ levels shown in Fig. 4.7 are higher than the corresponding $1s2s$ levels. The origin of this difference in energy is discussed in Section 4.4; it arises primarily from the difference in $\langle e^2/r_{12}\rangle$ for $1s$, $2s$ and $1s$, $2p$.

4.3.1 Parahelium and orthohelium A transition between states of different multiplicity requires a change of spin. Since the electromagnetic fields that induce electronic transitions act only very weakly upon electron spins, except when spin and orbital motions are strongly coupled by relativistic effects [for atoms with large Z ($\gtrsim 40$)], singlet to triplet and triplet to singlet transitions are very improbable and are said to be *forbidden*. In helium, the singlet states (called parahelium) and the triplet states (called orthohelium) may for many purposes be treated as separate forms of the element. Thus, the lowest triplet level (the ground state of orthohelium) is metastable, with a lifetime governed chiefly by the rate of collisions with other atoms, molecules, and the walls of the container.

Multiplicities greater than unity (i.e., spin degeneracies) indicate that an atom or molecule has a permanent magnetic dipole moment due to the unpaired electron spin. Substances containing such atoms or molecules are *attracted* by a magnetic field and are said to be *paramagnetic*. If the multiplicity is unity (i.e., spin zero), no permanent magnetic moment exists unless there is orbital degeneracy. Substances without a permanent magnetic moment are *repelled* by magnetic fields (by forces arising from the

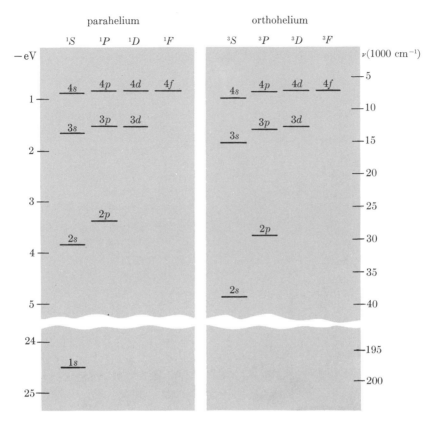

Fig. 4.7. The energy levels of helium.

interaction of the field with induced magnetic moments) and are said to be *diamagnetic*. From this discussion, we see that parahelium is diamagnetic and orthohelium is paramagnetic.

4.3.2 Doubly excited helium states Each of the excited helium states in Fig. 4.7 is a singly excited state; that is, one electron occupies the $1s$ ground-state orbital and one occupies an excited orbital. Doubly excited states in which both electrons occupy excited orbitals [e.g., He $(2s^2)$] have energies greater than the ionization limit of 79 eV above the ground state. Thus, these doubly excited states of helium can undergo so called auto-ionization transitions forming a He^+ ion and a free electron, whose kinetic energy is specified by the requirement of energy conservation (see Problem 4.11). The presence of such auto-ionizing states in the (He^+, e^-) continuum can have an observable effect on the absorption spectrum of the system; for example, at the proper energy for the He $(1s^2) \rightarrow$ He $(2s\,2p)$ excitation, there is found an increased absorption superimposed on the smooth $He(1s^2) \rightarrow He^+(1s) + e^-$ continuum absorption (Fig. 4.8). Such spectral *resonances*, which can be both positive and negative, have become the subject of intensive investigation in recent years. As might be expected, resonances are found also when an electron of the right energy collides with a He atom.

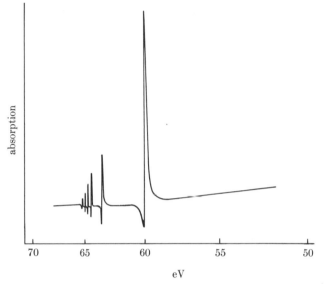

Fig. 4.8. Resonances in the helium continuum absorption spectrum (175–245 Å region); the large peak arises from the $2s2p$ state. [Redrawn from R. B. Madden and K. Codling, *Astrophys. J.* **141,** 364 (1965). Courtesy of The University of Chicago Press.]

4.4 SCREENING AND THE ORBITAL ENERGIES IN MANY-ELECTRON ATOMS

The principles used for determining the ground-state configurations of hydrogen, helium, and lithium apply to all of the atoms in the periodic table. An essential step in the application of these principles to many-electron atoms is a knowledge of the ordering of the orbital energies.

In the previous section we noted that the $2s$ and $2p$ levels, which are degenerate in a one-electron atom, are split when more than one electron is present. To understand the origin of the energy difference, we compare the two lithium configurations $1s^2 2s$ and $1s^2 2p$, where $2p$ can be any of the three p orbitals. In both configurations, the electron in the $n = 2$ orbital is partially shielded from the nuclear attraction by the two $1s$ electrons. However, the extent of shielding is different because it depends upon the radial distribution of the electron. Although a quantitative determination of the wave functions would require calculations similar to those done for helium (Section 4.1), we can make use of the qualitative features of the radial distribution functions for the $2s$ and $2p$ electrons of hydrogen (i.e., $4\pi r^2 R_{20}{}^2$ and $4\pi r^2 R_{21}{}^2$, respectively) shown in Fig. 3.5. In addition to the major maximum at $r = (3 + 5^{1/2})(a_0/Z)$, the $2s$ distribution has a "hump" close to the nucleus at $r = (3 - 5^{1/2})(a_0/Z)$, while the $2p$ distribution function is very small in this region. A $2s$ electron thus penetrates inside the $1s$ distribution to a greater extent than does a $2p$ electron. This results in less shielding of the $2s$ than the $2p$ electron by the inner-shell $1s$ electrons, so that the effective nuclear charge Z_e for a $2s$ electron is greater than that for a $2p$ electron. Employing the hydrogen-like energy equation for an electron in the $n = 2$ level,

$$E_2 = -\frac{Z_e{}^2 e^2}{8a_0}$$

we see that the $2s$ electron with a larger Z_e is lower in energy (more stable) than the $2p$ electron.

A crude estimate of the $2s$, $2p$ splitting can be made by perturbation theory. For the Li atom in the $1s^2 2s$ state, we have

$$E_{\mathrm{Li}}(1s^2 2s) = 2E_{1s} + E_{2s} + \langle V_{ee}\rangle_{1s^2 2s}$$

where the electron-electron repulsion term is

$$\langle V_{ee}\rangle_{1s^2 2s} = \int \psi_{\mathrm{Li}}(1, 2, 3)\left(\frac{e^2}{r_{12}} + \frac{e^2}{r_{23}} + \frac{e^2}{r_{13}}\right)$$
$$\times \psi_{\mathrm{Li}}(1, 2, 3)\, dv_1\, dv_2\, dv_3$$

and $\psi_{\mathrm{Li}}(1, 2, 3)$ represents the $1s^2 2s$ wave function given in Eq. 4.50. The corresponding treatment with the $1s^2 2p$ wave function for Li yields $E_{\mathrm{Li}}(1s^2 2p) = 2E_{1s} + E_{2p} + \langle V_{ee}\rangle_{1s^2 2p}$. Since $E_{2s} = E_{2p}$, the energy difference between the Li atom in the $1s^2 2s$ and the $1s^2 2p$ states is given by the difference in the electron-repulsion terms $\langle V_{ee}\rangle_{1s^2 2s}$ and $\langle V_{ee}\rangle_{1s^2 2p}$. The perturbation-theory result is 0.41 eV, while the experimental value is 1.85 eV. Quantitative calculations for different atoms show that the magnitude of the $2s$, $2p$ splitting increases rapidly with Z; for example in Na the splitting is about 30 eV and in Cu^+ it is about 160 eV.

A corresponding argument applied to the l levels for higher values of n demonstrates that they also have different energies in a many-electron atom. For example, for $n = 3$, the $3s$ level is lowest, $3p$ is next, and $3d$ is highest. The energy levels for orbitals in many-electron atoms are shown schematically in Fig. 4.9. Notice that in some cases the intrashell splitting (same n, different l) is larger than the intershell splitting (different n), so that an "inversion" of level order occurs. Thus, the $4s$ level is lower than the

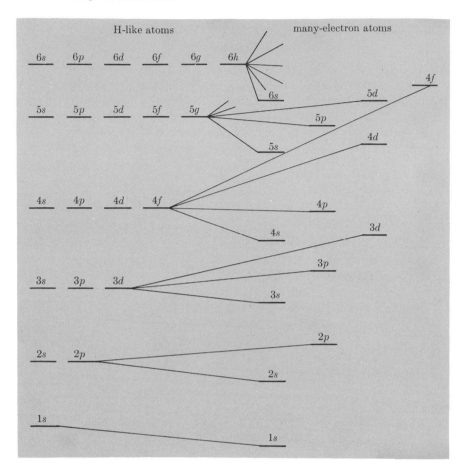

Fig. 4.9. Splitting of orbital degeneracy in many-electron atoms.

3d level in the diagram. Similarly, the 5s level is lower than the 4d, and the 6s is lower than both the 5d and 4f levels. The level ordering shown in Fig. 4.9 is common for neutral atoms with enough electrons to fill the orbitals that have inverted order. However, a detailed examination of spectra, careful calculations, or both, are required to establish the exact order in each particular atom or ion. Qualitatively, we expect that as electrons are removed from a neutral atom, the resulting ion should become more hydrogen-like; the limiting case that obtains when all but one electron have been removed is, of course, the hydrogen-like level ordering with all *l* values associated with a given *n* having the same energy. Consequently, when a certain number of electrons have been removed, the number depending on the atom, the intrashell splittings becomes smaller than the intershell splittings and the energy-level inversions disappear.

In illustrating the splitting of levels within the shells of many-electron atoms by means of Fig. 4.9, no indication is given of the dependence of the orbital energies upon atomic number. As Z increases, the electron-nuclear attraction becomes larger since the shielding by the added electrons is incomplete. Thus the inner-orbital levels decrease sharply in energy with increasing Z, while the energy levels of the more shielded outer orbitals decrease less sharply with Z (see Fig. 4.10).

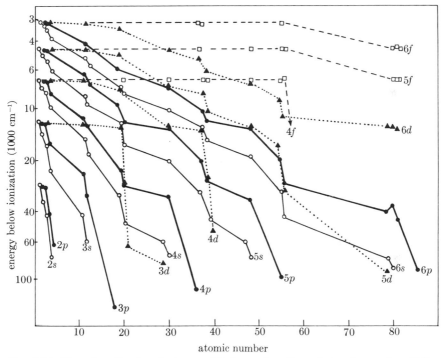

Fig. 4.10. Energies of outer electron orbitals as a function of atomic number. [After W. Kauzmann, *Quantum Chemistry* (Academic Press, New York, 1957), p. 326.]

4.5   THE AUFBAU PRINCIPLE AND THE PERIODIC TABLE

We are now equipped with the information necessary to determine the electronic structure of atoms and to discuss the periodic table. What is required is a knowledge of the order of the atomic energy levels, which we considered in the previous section, and of the limitation placed by the Pauli principle on the number of electrons associated with each level.

From the treatment of the hydrogen atom, it is evident that the meaning

fully occupied). For many applications a knowledge of such details of the electronic structure is important; it is discussed in Section 4.6.

For $n = 3$, we see from Fig. 4.9 that the level ordering is $3s < 3p < 3d$, so that the $3s$ and $3p$ orbitals are expected to be filled with eight electrons before the $3d$ subshell is occupied. This makes possible a description of the electronic configurations of the atoms Na, Mg, . . . , Ar, which have 11, 12, . . . , 18 electrons. All of these atoms have the orbitals with principal quantum numbers $n = 1$ and $n = 2$ filled; this is often described by saying they have a Ne-like core of 10 electrons corresponding to the configuration $1s^2 2s^2 2p^6$. Suppressing the core description, we can write the $n = 3$ configurations:

Na	Mg	Al	Si
$3s$	$3s^2$	$3s^2 3p$	$3s^2 3p^2$

P	S	Cl	Ar
$3s^2 3p^3$	$3s^2 3p^4$	$3s^2 3p^5$	$3s^2 3p^6$

Combining these results with those obtained above for H, He, . . . , Ne, we have the first three rows (periods) of the periodic table:

H							He
$1s$							$1s^2$

Li	Be	B	C	N	O	F	Ne
$2s$	$2s^2$	$2s^2 2p$	$2p^2$	$2p^3$	$2p^4$	$2p^5$	$2p^6$

Na	Mg	Al	Si	P	S	Cl	Ar
$3s$	$3s^2$	$3s^2 3p$	$3p^2$	$3p^3$	$3p^4$	$3p^5$	$3p^6$

Elements in the same row are seen to fill orbitals with a given value of n. Since the average distance from the nucleus of an electron in an orbital is determined largely by n, a set of orbitals with the same value of n is called a *shell*. Standard notation for the shells corresponding to the various values of n is as follows:

$$n = 1 \quad 2 \quad 3 \quad 4 \quad 5 \quad 6$$
$$\updownarrow \quad \updownarrow \quad \updownarrow \quad \updownarrow \quad \updownarrow \quad \updownarrow$$
$$K \quad L \quad M \quad N \quad O \quad P$$

The lengths of the first two periods correspond to the number of electrons, $2n^2$, which can be accommodated in the K and L shells. The similarity of properties of elements in the same column (e.g., Li and Na, Be and Mg) follows from the fact that they have the same number and type of outer-shell electrons.

The lengths of the periods beginning with the third can be explained on the basis of the level inversions due to the relative magnitudes of the intershell and intrashell splitting, as illustrated in Fig. 4.9. Since the $4s$ orbital is more stable than $3d$, the next elements after Ar add $4s$ electrons, rather than $3d$. Thus, K with 19 electrons has the ground-state configuration $1s^2 2s^2 2p^6 3s^2 3p^6 4s$, or ([Ar]$4s$) in terms of the Ar core [Ar], and is an alkali metal like Na([Ne]$3s$); correspondingly, Ca with 20 electrons has the configuration [Ar]$4s^2$ and has properties similar to those of the other alkaline earth elements like Mg ([Ne]$3s^2$). Since the $4s$ orbital is now doubly occupied, the filling of the $3d$ orbitals begins with the next element, Sc. The five $3d$ orbitals can hold 10 electrons, so that the fourth period, which corresponds to the filling of the $4s$, $3d$, and $4p$ orbitals, contains 18 elements. The fourth period ends at Kr when the sixth $4p$ electron has been added. The next element has one $5s$ electron and therefore is an alkali metal (Rb). The fifth period is identical to the fourth, except that it begins by filling $5s$, then $4d$, and ends when the sixth $5p$ electron has been added (Xe). Thus period 5, like period 4, consists of 18 elements. Period 6 is similar to periods 4 and 5, except that the $4f$ orbitals are also filled, adding $2(2l + 1) = 2 \times (2 \times 3 + 1) = 14$ more elements to give a total of 32. Period 7, in which the $7s$, $7p$, $6d$, and $5f$ subshells are successively filled, presumably also consists of 32 elements, although only 17 of them have been discovered by the late 1960's.

Although many chemical and physical properties of the elements depend upon the number of electrons in the outermost shell (i.e., the shell with largest n), the elements Sc, Ti, . . . , Zn whose $3d$ subshell is being filled, behave similarly even though they lie in the same period rather than in the same group (see Chapter 5). These elements with related properties are called the *first transition series*. Correspondingly, the elements Y, Zr, . . . , Cd (whose $4d$ subshell is filling) have similar properties and are called the *second transition series*. The *third transition series* involves the $5d$ subshell and consists of Hf, Ta, . . . , Hg. It is preceded by 14 elements whose $4f$ subshell is being filled. These elements have outer-shell configurations similar to that of La, and are called the lanthanides, or rare earths; the 14 elements following Ac (including the man-made transuranic elements), whose $5f$ shells are being occupied, are called the actinides.

The ground-state configurations of all the elements are given in Table 4.2. The configurations listed have been determined experimentally. In most cases they correspond to the configurations that are obtained from the energy-level ordering in Fig. 4.9. There are, however, a number of exceptions. The first two of these occur in the lowest transitions series; thus, Cr has the configuration [Ar]$4s3d^5$ instead of [Ar]$4s^2 3d^4$, and Cu is [Ar]$4s3d^{10}$ instead of [Ar]$4s^2 3d^9$, implying that in each case the former configuration is more stable than the latter. Corresponding exceptions occur for some of the elements with higher atomic number.

In both Cr and Cu, we see that a $4s$ electron has been shifted to a $3d$

Table 4.2 *Atomic properties*

Z	Atom	Mass (amu)[a]	Configuration[b,c]	Term[b,c]	IP(eV)[b]	EA(eV)[d]	Crystal Radius (Å)[e]	Orbital Radius (Å)[e]
1	H	1.00797	$1s$	$^2S_{1/2}$	13.595	[0.75415]	0.25	0.529
2	He	4.0026	$1s^2$	$^1S_0$	24.580		1.53	0.291
3	Li	6.939	[He]$2s$	$^2S_{1/2}$	5.390	~0.6, (0.62)[f]	1.45	1.586
4	Be	9.0122	[He]$2s^2$	$^1S_0$	9.320		1.05	1.040
5	B	10.811	[He]$2s^22p$	$^2P_{1/2}$	8.296	(0.3)	0.85	0.776
6	C	12.01115	[He]$2s^22p^2$	$^3P_0$	11.264	1.25	0.70	0.620
7	N	14.0067	[He]$2s^22p^3$	$^4S_{3/2}$	14.54		0.65	0.521
8	O	15.9994	[He]$2s^22p^4$	$^3P_2$	13.614	1.465 or 1.478	0.60	0.450
9	F	18.9984	[He]$2s^22p^5$	$^2P_{3/2}$	17.42	3.400 or 3.448	0.50	0.396
10	Ne	20.183	[He]$2s^22p^6$	$^1S_0$	21.559		1.60	0.354
11	Na	22.9898	[Ne]$3s$	$^2S_{1/2}$	5.138	(0.54), (0.47)	1.80	1.713
12	Mg	24.312	[Ne]$3s^2$	$^1S_0$	7.644		1.50	1.279
13	Al	26.9815	[Ne]$3s^23p$	$^2P_{1/2}$	5.984	(0.52)	1.25	1.312
14	Si	28.086	[Ne]$3s^23p^2$	$^3P_0$	8.149	(1.39)	1.10	1.068
15	P	30.9738	[Ne]$3s^23p^3$	$^4S_{3/2}$	11.0	(0.78)	1.00	0.919
16	S	32.064	[Ne]$3s^23p^4$	$^3P_2$	10.357	2.07	1.00	0.810
17	Cl	35.453	[Ne]$3s^23p^5$	$^2P_{3/2}$	13.01	3.613	1.00	0.725
18	Ar	39.948	[Ne]$3s^23p^6$	$^1S_0$	15.755		1.92	0.659
19	K	39.102	[Ar]$4s$	$^2S_{1/2}$	4.339	0.49 or 0.75	2.20	2.162
20	Ca	40.08	[Ar]$4s^2$	$^1S_0$	6.111		1.80	1.690

Table 4.2 (Continued)

Z	Atom	Mass (amu)[a]	Configuration[b,c]	Term[b,c]	IP(eV)[b]	EA(eV)[d]	Crystal Radius (Å)[e]	Orbital Radius (Å)[e]
21	Sc	44.956	$[\text{Ar}]4s^23d$	$^2D_{3/2}$	6.56		1.60	1.570
22	Ti	47.90	$[\text{Ar}]4s^23d^2$	$^3F_2$	6.83		1.40	1.477
23	V	50.942	$[\text{Ar}]4s^23d^3$	$^4F_{3/2}$	6.74		1.35	1.401
24	Cr	51.996	$[\text{Ar}]4s3d^5$	$^7S_3$	6.76	0.980	1.40	1.453
25	Mn	54.9380	$[\text{Ar}]4s^23d^5$	$^6S_{5/2}$	7.432		1.40	1.278
26	Fe	55.847	$[\text{Ar}]4s^23d^6$	$^5D_4$	7.896	0.582	1.40	1.227
27	Co	58.9332	$[\text{Ar}]4s^23d^7$	$^4F_{9/2}$	7.86	0.936	1.35	1.181
28	Ni	58.71	$[\text{Ar}]4s^23d^8$	$^3F_4$	7.633	1.276	1.35	1.139
29	Cu	63.546	$[\text{Ar}]4s3d^{10}$	$^2S_{1/2}$	7.723	1.801	1.35	1.191
30	Zn	65.37	$[\text{Ar}]4s^23d^{10}$	$^1S_0$	9.391		1.35	1.065
31	Ga	69.72	$[\text{Ar}]4s^23d^{10}4p$	$^2P_{1/2}$	6.00		1.30	1.254
32	Ge	72.59	$[\text{Ar}]4s^23d^{10}4p^2$	$^3P_0$	8.13		1.25	1.090
33	As	74.9216	$[\text{Ar}]4s^23d^{10}4p^3$	$^4S_{3/2}$	10		1.15	1.001
34	Se	78.96	$[\text{Ar}]4s^23d^{10}4p^4$	$^3P_2$	9.750		1.15	0.918
35	Br	79.904	$[\text{Ar}]4s^23d^{10}4p^5$	$^2P_{3/2}$	11.84	3.363	1.15	0.851
36	Kr	83.80	$[\text{Ar}]4s^23d^{10}4p^6$	$^1S_0$	13.996		1.97	0.795
37	Rb	85.47	$[\text{Kr}]5s$	$^2S_{1/2}$	4.176	(0.42)	2.35	2.287
38	Sr	87.62	$[\text{Kr}]5s^2$	$^1S_0$	5.692		2.00	1.836
39	Y	88.905	$[\text{Kr}]5s^24d$	$^2D_{3/2}$	6.6		1.80	1.693
40	Zr	91.22	$[\text{Kr}]5s^24d^2$	$^3F_2$	6.95		1.55	1.593

Table 4.2 (Continued)

Z	Atom	Mass (amu)[a]	Configuration[b,c]	Term[b,c]	IP(eV)[b]	EA(eV)[d]	Crystal Radius (Å)[e]	Orbital Radius (Å)[e]
41	Nb	92.906	$[\text{Kr}]5s^14d^4$	$^6D_{1/2}$	6.77		1.45	1.589
42	Mo	95.94	$[\text{Kr}]5s^14d^5$	$^7S_3$	7.18	1.0	1.45	1.520
43	Tc	(97)	$[\text{Kr}]5s^24d^5$	$^6S_{5/2}$			1.35	1.391
44	Ru	101.07	$[\text{Kr}]5s^14d^7$	$^5F_5$	7.5		1.30	1.410
45	Rh	102.905	$[\text{Kr}]5s^14d^8$	$^4F_{9/2}$	7.7		1.35	1.364
46	Pd	106.4	$[\text{Kr}]4d^{10}$	$^1S_0$	8.33		1.40	0.567
47	Ag	107.868	$[\text{Kr}]5s^14d^{10}$	$^2S_{1/2}$	7.574		1.60	1.286
48	Cd	112.40	$[\text{Kr}]5s^24d^{10}$	$^1S_0$	8.991		1.55	1.184
49	In	114.82	$[\text{Kr}]5s^24d^{10}5p$	$^2P_{1/2}$	5.785		1.55	1.382
50	Sn	118.69	$[\text{Kr}]5s^24d^{10}5p^2$	$^3P_0$	7.332		1.45	1.240
51	Sb	121.75	$[\text{Kr}]5s^24d^{10}5p^3$	$^4S_{3/2}$	8.64		1.45	1.193
52	Te	127.60	$[\text{Kr}]5s^24d^{10}5p^4$	$^3P_2$	9.01		1.40	1.111
53	I	126.9044	$[\text{Kr}]5s^24d^{10}5p^5$	$^2P_{3/2}$	10.44	3.063 or 3.076	1.40	1.044
54	Xe	131.30	$[\text{Kr}]5s^24d^{10}5p^6$	$^1S_0$	12.127		2.60	0.986
55	Cs	132.905	$[\text{Xe}]6s$	$^2S_{1/2}$	3.893	(0.39)	2.66	2.518
56	Ba	137.34	$[\text{Xe}]6s^2$	$^1S_0$	5.210		2.15	2.060
57	La	138.91	$[\text{Xe}]6s^25d$	$^2D_{3/2}$	5.61		1.95	1.915
58	Ce	140.12	$[\text{Xe}](6s^24f5d)$	$(^3H_5)$	6.91		1.85	1.978
59	Pr	140.907	$[\text{Xe}](6s^24f^3)$	$(^4I_{9/2})$	5.76		1.85	1.942
60	Nd	144.24	$[\text{Xe}]6s^24f^4$	$^5I_4$	6.31		1.85	1.912

Table 4.2 (Continued)

Z	Atom	Mass (amu)[a]	Configuration[b,c]	Term[b,c]	IP(eV)[b]	EA(eV)[d]	Crystal Radius (Å)[e]	Orbital Radius (Å)[e]
61	Pm	(145)	$[\text{Xe}](6s^24f^5)$	$(^6H_{5/2})$			1.85	1.882
62	Sm	150.35	$[\text{Xe}]6s^24f^6$	$^7F_0$	5.6		1.85	1.854
63	Eu	151.96	$[\text{Xe}]6s^24f^7$	$^8S_{7/2}$	5.67		1.85	1.826
64	Gd	157.25	$[\text{Xe}]6s^24f^75d$	$^9D_2$	6.16		1.80	1.713
65	Tb	158.924	$[\text{Xe}]6s^24f^9$	$(^6H_{15/2})$	6.74		1.75	1.775
66	Dy	162.50	$[\text{Xe}](6s^24f^{10})$	$(^5I_8)$	6.82		1.75	1.750
67	Ho	164.930	$[\text{Xe}](6s^24f^{11})$	$(^4I_{15/2})$			1.75	1.727
68	Er	167.26	$[\text{Xe}](6s^24f^{12})$	$(^3H_6)$			1.75	1.703
69	Tm	168.934	$[\text{Xe}]6s^24f^{13}$	$^2F_{7/2}$			1.75	1.681
70	Yb	173.04	$[\text{Xe}]6s^24f^{14}$	$^1S_0$	6.2		1.75	1.658
71	Lu	174.97	$[\text{Xe}]6s^24f^{14}5d$	$^2D_{3/2}$	5.0		1.75	1.553
72	Hf	178.49	$[\text{Xe}]6s^24f^{14}5d^2$	$^3F_2$	5.5		1.55	1.476
73	Ta	180.948	$[\text{Xe}]6s^24f^{14}5d^3$	$^4F_{3/2}$	7.88		1.45	1.413
74	W	183.85	$[\text{Xe}]6s^24f^{14}5d^4$	$^5D_0$	7.98	0.5	1.35	1.360
75	Re	186.2	$[\text{Xe}]6s^24f^{14}5d^5$	$^6S_{5/2}$	7.87	0.15	1.35	1.310
76	Os	190.2	$[\text{Xe}]6s^24f^{14}5d^6$	$^5D_4$	8.7		1.30	1.266
77	Ir	192.2	$[\text{Xe}]6s^24f^{14}5d^7$	$^4F_{9/2}$	9.2		1.35	1.227
78	Pt	195.09	$[\text{Xe}]6s4f^{14}5d^9$	$^3D_3$	8.96		1.35	1.221
79	Au	196.967	$[\text{Xe}]6s4f^{14}5d^{10}$	$^2S_{1/2}$	9.223		1.35	1.187
80	Hg	200.59	$[\text{Xe}]6s^24f^{14}5d^{10}$	$^1S_0$	10.434		1.50	1.126

orbital to produce a half-filled ($3d^5$) or completely filled ($3d^{10}$) subshell. This is a manifestation of the fact that for the outermost electrons, completely filled or half-filled shells or subshells lead to special stability for the atom. Quantitative justification of this result can be given by perturbation theory in exactly the same way that we made an approximate calculation of the energy of the helium atom, that is, by obtaining the average electron-nuclear and electron-electron interaction energies for the wave functions corresponding to each of the configurations. Qualitatively, it is evident that because the electrons in the same subshell have equivalent spatial distributions, their shielding of one another is relatively small (as we have seen in the case of He, the $1s$ electrons provide a shielding effect of about $0.30e$ upon one another). Thus, the effective charge Z_e increases as Z increases during the filling of the subshell, with a concomitant rise in stability, until the subshell is completed. The next electron then goes into a subshell with a larger value of l or n, and usually has a radial distribution that lies mostly outside the completed subshells. Consequently, the outermost electron is substantially shielded by those in the closed subshells and has an IP that is relatively low.

When a subshell is being filled, the lowest energy can be achieved if the electrons go into different orbitals insofar as is possible. This reduces the electron-electron repulsion, since two electrons are less likely to be found close together when they are in separate orbitals than when they occupy the same orbital. Thus, when a subshell is half-filled, one electron will usually occupy each of the degenerate orbitals. In this way, the electron-nuclear attraction is maximized while the electron-electron repulsion is minimized.

The electron configurations of atoms with very large values of Z are hard to explain on the basis of the simple rules we have used above. One reason for this is that many of the energy levels become very close in energy, so that it becomes more difficult to determine which orbitals are occupied. Also, as Z increases the nuclei attract the electrons more strongly, and a balance between the electrostatic and centrifugal forces requires an increase in the electronic velocities. At very large Z, the velocities of the electrons are so great that relativistic effects become important and alter the description that we have given here. One result is that the interaction of the spin and orbital magnetic moments prevents us from factoring the wave function into separate space and spin parts, so that the occupation of orbitals within a subshell is not so well defined.

4.6 RUSSELL–SAUNDERS COUPLING AND TERM VALUES

In previous paragraphs we have discussed how the configurations of atoms are determined from a knowledge of the energy-level ordering and the limitations set by the Pauli principle. The ground-state configurations are given in terms of n and l of the occupied orbitals. For closed shells and subshells, all orbitals with the same n and l values are, by definition, doubly occupied. However, for partly filled shells, it is not clear from a configurational specification alone how the electrons with given n and l are distributed among the different possible m and m_s values. It is necessary to have this additional information to completely specify the state of an atom.

The m and m_s quantum numbers are indicative of the orientation of the orbital and spin angular momentum of the electrons. In Chapter 3, we discussed the determination of the orientation of the angular momenta in a one-electron atom relative to an external magnetic field. Here we are considering an isolated atom containing several open-shell electrons, so that the important question concerns the relative orientation of the individual electronic angular momenta with respect to each other. Specification of this information defines the atomic *term values* corresponding to a given electronic configuration. Because of electron-electron repulsion, the several terms that arise from a configuration have different energies. Furthermore, the spin-orbit coupling splits them into components with different total angular momentum. For the one-electron case, the relative orientation of the spin angular momentum **s** and the orbital angular momentum **l** of the electron determines the total electronic angular momentum **j** of the atom. For a many-electron atom, we have to consider the orbital and spin angular momenta of all the electrons and show how to couple them together to obtain the total angular momentum.

If we consider an open shell with k electrons, we can associate an orbital angular momentum $\mathbf{l}_i$ and spin angular momentum $\mathbf{s}_i$ with the ith electron; as before, the label i is introduced only for convenience and we have to be careful to remember that the electrons are really indistinguishable (Section 4.2). It is now possible to couple together all of the orbital angular momenta $\mathbf{l}_1, \mathbf{l}_2, \ldots, \mathbf{l}_k$, to form a total orbital angular momentum **L**, and similarly all of the spin angular momenta $\mathbf{s}_1, \mathbf{s}_2, \ldots, \mathbf{s}_k$ to form a total spin angular momentum **S**, and then to couple **L** and **S** together to form a total angular momentum **J** for the atom. Alternatively, we could couple together the orbital and spin angular momenta of each electron to obtain a total angular momentum for each electron $(\mathbf{l}_1, \mathbf{s}_1 \rightarrow \mathbf{j}_1; \mathbf{l}_2, \mathbf{s}_2 \rightarrow \mathbf{j}_2; \ldots)$, and then couple the $\mathbf{j}_1, \mathbf{j}_2, \ldots, \mathbf{j}_k$ together to give the total **J** for the atom. The first coupling scheme is called *Russell–Saunders (or L-S) coupling* and the second scheme is called *j-j coupling*. Which is the more appropriate scheme for a given atom can be determined by perturbation theory from the relative magnitudes of the coupling terms; that is, if the interaction energies coupling the $\mathbf{l}_i$ to each other and the $\mathbf{s}_i$ to each other are larger than the interaction energy between individual members of each pair $(\mathbf{l}_i, \mathbf{s}_i)$, Russell–Saunders coupling is the better approximation; conversely, if the $\mathbf{l}_i, \mathbf{s}_i$ interaction energy is larger than that among the different l_i and the different s_i, then j-j coupling is best. It turns out that for atoms with small atomic number ($Z \lesssim 40$) Russell–Saunders coupling is used; for ($Z \gtrsim 40$) j-j coupling is the more appropriate scheme for the ground and low-lying excited states.

To see how this difference in treating light and heavy atoms arises, we have to determine the origin of the coupling energy. From our discussion of the one-electron case, we know that the orbital and spin angular momenta are coupled together by the interaction between the magnetic moments associated with the angular momenta, the so-called *spin-orbital coupling.* By contrast, the dominant term in the coupling of individual l_i or individual s_i is *electrostatic* in origin. It arises from the dependence of the electron-electron repulsion on the orbitals occupied by the electrons. If, for example, we consider two electrons in p orbitals of the same subshell, each has the orbital angular momentum quantum number $l = 1$ and the magnetic quantum numbers m can be 0, ± 1. The two electrons have a large average mutual repulsion if they are in the same orbital; in this case, they have the same m value and their two orbital angular momenta will add. Correspondingly, if they have different m quantum numbers (which means that the l vectors point in different directions), the electron-electron repulsion is lowered. If m_1 and m_2 are the magnetic quantum numbers of the occupied p orbitals, the z component of the total orbital angular momentum M_L for the two electrons is $(m_1 + m_2)\hbar$. Thus M_L has its maximum value ($2\hbar$) when m_1 and m_2 are both equal to unity; that is, the two electrons are paired in the same p_{+1} orbital and the average electron-electron repulsion is large. On the other hand, if one electron occupies the p_{+1} orbital and the other occupies the p_0 orbital, M_L is equal to $\hbar$ and the average electron-electron repulsion is smaller. If the two electrons are paired in the p_0 orbital, they both have $m = 0$, so that M_L is zero. In this case the electron-electron repulsion is large; in fact it is larger than when the electrons are paired in p_{+1}, since the "doughnut" shape of the probability density of p_{+1} allows the average distance between the two electrons to be somewhat greater than for the p_0 orbitals (see Figs. 3.10c and 3.12). The above examples illustrate how the coupling of orbital angular momenta is related to the electrostatic repulsion, although we have not given a quantitative treatment of this effect. The spin angular momentum is similarly related to the electron-electron repulsion, because electrons that have parallel spins must be in different orbitals and electrons in different orbitals have a smaller average electron-electron repulsion than do those in the same orbital.

A calculation of the electrostatic interactions between two p electrons in the same subshell shows that the smallest electron-electron repulsion occurs when p_{+1} and p_0 are occupied and the spins are parallel (maximum spin, maximum orbital angular momentum for parallel spin). The next lowest energy occurs when the electrons are paired in the p_{+1} orbital (zero spin, maximum orbital angular momentum), and the pairing of the electrons in the p_0 orbital (zero spin, zero orbital angular momentum) corresponds to the highest energy. A generalization of these results, known as Hund's rules, is given in Section 4.6.2.

Russell–Saunders coupling is appropriate when the electrostatic electron-electron energy is larger than the magnetic spin-orbit energy, while j-j coupling is appropriate when the converse is true. Since the magnetic interaction depends on the electronic velocities (the current associated with the orbital motion), which increase with atomic number, it is reasonable that the spin-orbit interaction becomes more important for the heavier atoms.

4.6.1   Terms for the lighter atoms   Having described some of the general features of the angular-momentum interactions, we concentrate on Russell–Saunders coupling since we are primarily concerned with the structure of the lighter atoms. For a group of k electrons, the possible values of the total spin angular momentum are obtained by writing

$$
\begin{aligned}
S &= s_1 + s_2 + \cdots + s_k,\, s_1 + s_2 + \cdots + s_k - 1, \\
&\quad s_1 + s_2 + \cdots + s_k - 2, \cdots \\
&= \frac{k}{2}, \frac{k}{2} - 1, \frac{k}{2} - 2, \cdots, 0 \qquad (k \text{ even}) \\
&= \frac{k}{2}, \frac{k}{2} - 1, \frac{k}{2} - 2, \cdots, \frac{1}{2} \qquad (k \text{ odd})
\end{aligned}
\tag{4.63}
$$

That is, S goes from a maximum, corresponding to all the spins being parallel, in integer steps down to a minimum which is zero for an even number of electrons and $\frac{1}{2}$ for an odd number of electrons. The orbital angular momentum possibilities are slightly more complicated because the individual orbital angular momentum quantum numbers l_1 can have different values. If the quantum numbers for the individual electrons are l_1, $l_2, \ldots, l_k$, the possible values of the total orbital angular momentum quantum number L are obtained by writing

$$
\begin{aligned}
L &= l_1 + l_2 + \cdots + l_k,\, l_1 + l_2 + \cdots + l_k - 1, \\
&\quad l_1 + l_2 + \cdots + l_k - 2, \ldots
\end{aligned}
\tag{4.64}
$$

That is, L goes from a maximum, corresponding to all of the $\mathbf{l}_i$ lined up parallel, in integer steps down to a minimum value, which is greater than or equal to zero and depends on the individual l_i values. If all l_i are equal, the minimum is zero; if one of the l_i is larger than the others, the minimum value is that given by orienting the other angular momenta to oppose it, subject to the condition that $L \geq 0$. Thus, for three p electrons ($l_1 = 1$, $l_2 = 1$, $l_3 = 1$), we have the possible L and S values

$$
L = 3, 2, 1, 0 \qquad S = \frac{3}{2}, \frac{1}{2}
$$

while for one f electron ($l_1 = 3$) and two p electrons ($l_2 = 1$, $l_3 = 1$) we have

$$L = 5, 4, 3, 2, 1 \qquad S = \frac{3}{2}, \frac{1}{2}$$

Once the possible L and S values are known, the possible total angular momentum quantum numbers are obtained by writing

$$J = L + S, L + S - 1, L + S - 2, \ldots, |L - S| \qquad (4.65)$$

That is, J goes from a maximum, corresponding to addition of L and S, in integer steps down to a minimum corresponding to the absolute value of the difference between L and S (i.e., to $L - S$ if $L \geq S$ and $S - L$ if $S \geq L$). Thus, for $L = 3$, $S = \frac{3}{2}$, we have

$$J = \frac{9}{2}, \frac{7}{2}, \frac{5}{2}, \frac{3}{2}$$

while for $L = 1$, $S = \frac{3}{2}$

$$J = \frac{5}{2}, \frac{3}{2}, \frac{1}{2}$$

For each set of values, L, S, and J, the term symbol is written as in the one-electron case discussed in Section 3.14; that is,

$$^{2S+1}L_J$$

where for the integer values of L we use the capital letters

$$
\begin{array}{cccccc}
L = 0 & 1 & 2 & 3 & 4 & 5 \ldots \\
\updownarrow & \updownarrow & \updownarrow & \updownarrow & \updownarrow & \updownarrow \\
S & P & D & F & G & H \ldots
\end{array}
$$

Thus, the term symbol for an atom with $L = 3$, $S = \frac{3}{2}$, $J = \frac{5}{2}$ is $^4F_{5/2}$, while an atom with $L = 1$, $S = \frac{1}{2}$, $J = \frac{3}{2}$ has the term symbol $^2P_{3/2}$. These symbols are read "quartet F five halves" and "doublet P three halves," respectively; that is, the electron spin superscript is read as follows:

$$
\begin{array}{ccccc}
2S + 1 = & 1 & 2 & 3 & 4 \quad \ldots \\
& \text{singlet} & \text{doublet} & \text{triplet} & \text{quartet}
\end{array}
$$

corresponding to the fact that for $L > S$, the number of possible J levels (the multiplicity of the level) is equal to $2S + 1$.

In the discussion of term symbols and possible terms associated with a given configuration, we have ignored the Pauli principle up to this point. As long as the n and/or l quantum numbers of all the electrons in the open-shell set are different, all of the possible terms can occur; that is, each combination of L and S is possible, as is each J derivable from a given L and S.

However, if some of the electrons have the same n and l quantum numbers, the Pauli principle restricts the m and m_s quantum numbers; a limit is thereby placed on the different terms that can exist in these cases. To make these limitations concrete, let us look at carbon in the ground-state configuration $(1s^2 2s^2 2p^2)$. Since in this configuration the $1s$ and $2s$ electrons make up closed shells or subshells, the only electrons that can contribute to nonzero values of L, S, and J are the two $2p$ electrons. According to our previously discussed rules, without considering the Pauli principle, we have for the p^2 case $(l_1 = 1, l_2 = 1, s_1 = \frac{1}{2}, s_2 = \frac{1}{2})$ the possible terms given in Table 4.3.

Table 4.3 *Terms arising from the configuration $np\ n'p$*

L	S	
	0	1
2	$^1D_2$	$^3D_{3,2,1}$
1	$^1P_1$	$^3P_{2,1,0}$
0	$^1S_0$	$^3S_1$

All of these would occur for excited configurations with $n_1 \neq n_2$ (e.g., $1s^2 2s^2 2p3p$). For $n_1 = n_2$ and $l_1 = l_2$, however, the Pauli principle limits the possible m, m_s values. To relate these to the different terms, we make a chart giving all allowed m_1, m_2, m_{s1}, m_{s2}, that is, all combinations that satisfy the Pauli principle. There are 15 possible combinations, which we represent here by arrows on the appropriate line to indicate occupancy of an orbital with quantum number m by an electron. The arrow points up if the electron has $m_s = +\frac{1}{2}$ and down if it has $m_s = -\frac{1}{2}$ (see Table 4.4). Nine of the combinations have $m_{s1} = \pm\frac{1}{2}$, $m_{s2} = \mp\frac{1}{2}$, so that all m_1, m_2 are allowed $(3 \times 3 = 9)$. The remaining six have $m_{s1} = m_{s2} = \pm\frac{1}{2}$, corresponding to the requirement that the m_1, m_2 values must be different $(3 \times 2 = 6)$, since the spin quantum numbers are the same. In making this count it must be remembered that, because electrons are indistinguishable, it does not matter which electron has which quantum number; for example, $m_1 = 1$, $m_{s1} = +\frac{1}{2}$, $m_2 = 0$, $m_{s2} = -\frac{1}{2}$ is the same as $m_1 = 0$, $m_{s1} = -\frac{1}{2}$, $m_2 = 1$, $m_{s1} = -\frac{1}{2}$.

Having written down all of the sets of z-component spin and orbital angular momentum quantum numbers consistent with the Pauli principle, we can associate with each of them a value of M_L and M_S, the quantum

Table 4.4 Allowed m, m_s for p^2 configuration

M_S	0	0	0	0	0	0	0	0	0	1	-1	1	-1	1	-1
M_L	-2	0	2	-1	-1	0	0	1	1	-1	-1	0	0	1	1

numbers for the z component of total orbital and spin angular momentum; that is, as shown in Table 4.4

$$M_L = m_1 + m_2 \qquad M_S = m_{s_1} + m_{s_2}$$

As for the one-electron angular momenta, for each L and S there are $2L + 1$ values of M_L and $2S + 1$ values of M_S, respectively; that is,

$$M_L = L, L - 1, L - 2, \ldots, -L$$
$$M_S = S, S - 1, S - 2, \ldots, -S \qquad (4.66)$$

Thus, a term corresponding to a given value of L and S has $(2L + 1) \times (2S + 1)$ combinations of M_L, M_S associated with it. To find the L and S values that can be constructed from the M_L, M_S combinations in Table 4.4, we begin with the maximum M_S and find the maximum M_L associated with it. From Table 4.4, this is $(M_S)_{max} = 1$, $(M_L)_{max} = 1$, since $M_L > 1$ occurs only for $M_S = 0$ in the present case. According to Eq. 4.66, these correspond to a term with

$$L = 1 \quad M_L = 1, 0, -1 \qquad S = 1 \quad M_S = 1, 0, -1$$

These values of M_L and M_S account for nine of the M_L, M_S combinations in the table and leave

M_S	0	0	0	0	0	0
M_L	-2	0	2	-1	0	1

From these combinations, we again pick out the maximum M_S and M_L (here $M_S = 0$, $M_L = 2$), and associate them with

$$L = 2 \qquad M_L = 2, 1, 0, -1, -2 \qquad S = 0 \quad M_S = 0$$

Five more of the combinations have been used up and there remain only $M_L = 0$, $M_S = 0$, which must correspond to $L = 0$, $S = 0$. Thus, we find that, consistent with the Pauli principle, the p^2 configuration gives

$$L = 1, S = 1 \qquad L = 2, S = 0 \qquad L = 0, S = 0$$

After finding the values of J for each allowed L, S pair, we obtain the term symbols for p^2 with $n_1 = n_2$,

$$^3P_{2,1,0} \quad ^1D_2 \quad ^1S_0$$

Note that once the appropriate restrictions have been applied to limit the L, S values, all of the J values that can be formed from them are allowed. Comparing these terms with those given in Table 4.3, we see that in the present example $^1P_1$, $^3S_1$, and $^3D_{3,2,1}$ are eliminated by the Pauli principle; for example, ^3D would require $M_S = 1$, $M_L = 2$, which corresponds to $m_{s_1} = m_{s_2} = \frac{1}{2}$, $m_1 = m_2 = 1$, and clearly violates the Pauli principle.

4.6.2   Ground-state terms; Hund's rules

The method for identifying the possible p^2 terms for two electrons with the same n and l can be used for other open-shell systems; some examples are given in the problems. Once the terms are known, it is necessary to find their relative energy in order to choose which term characterizes the ground state of the atom. This can be done by a set of simple rules, called *Hund's rules:*

(a) The terms are ordered according to their S values, the term with maximum S being most stable and the stability decreasing with decreasing S. Thus, the ground state has *maximum spin multiplicity.*

(b) For a given value of S (given spin multiplicity), the state with *maximum L* is most stable.

(c) For given S and L, the *minimum J* value is most stable if there is a single open subshell that is *less than half-full* and the *maximum J* is most stable if the subshell is *more than half-full.*

Rules (a) and (b) arise from the electron-electron (electrostatic) interaction between the electrons, while rule (c) is a consequence of the spin-orbit (magnetic) interaction. Rule (a) is easy to understand, in that maximum S implies parallel spin (i.e., the same m_s quantum numbers for $M_S = S$). This means that the m quantum numbers must be different, corresponding to the fact that the electrons occupy different orbitals, which minimizes the repulsive interaction (see Section 4.6). In addition, parallel spins *per se* lead to a stabilizing effect due to the exchange interaction (see Section 4.3). The maximum L rule has a corresponding origin, as described in Section 4.6. The J rule derives from the relationship between the sign of the total orbital magnetic moment and the number of electrons present in a subshell relative to the number that can be accommodated.

When a subshell is more than half-full, it is usually more convenient to work out terms with respect to *electron holes* (i.e., vacancies in the various spin and orbital states) rather than electrons. The terms obtained in this manner are the correct ones, but one must remember to invert the multiplets so that the energies of electron-hole terms increase with decreasing J, in accordance with rule (c). For example, the configurations np^2 and np^4 both give rise to the terms $^3P_{2,1,0}$, $^1D_2$, $^1S_0$; for np^2 the ground-state term is $^3P_0$, while for np^4 it is $^3P_2$.

An idea of the relative energies involved in the term splittings can be obtained from Fig. 4.11, which gives the experimental results for carbon in

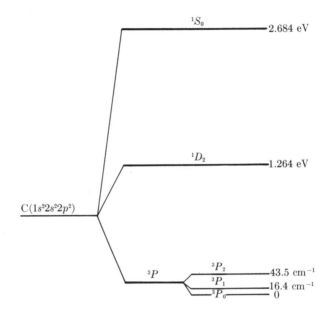

Fig. 4.11. Term splitting in the ground-state $(1s^22s^22p^2)$ configuration of carbon.

the $1s^22s^22p^2$ configuration. It is clear that for a light atom like carbon, the spin-orbit splitting is much smaller than the electrostatic interaction. The next lowest configuration for carbon is $1s^22s2p^3$, for which the most stable term $(^5S_2)$ has an energy about 4.2 eV above the ground state; thus, the difference between configurations is of the same order as, though often somewhat larger than, the splittings between terms of the same configuration.

Knowing how to find the possible terms for a configuration and their relative energies, we can work out the ground-state terms for the atoms. As an example, we consider the first two rows of the periodic table. In Table 4.5, we have written down the lowest configuration, already discussed in Section 4.5; only the open-subshell electrons are included since the closed subshells do not affect the term.

For H, He, Li, and Be, the terms are self-evident; that is, the orbital angular momentum must be zero and the spin is $S = 0$ for the closed sub-

Table 4.5 *Ground-state configuration and terms for first- and second-row elements*

H	He	Li	Be	B	C	N	O	F	Ne
$1s$	$1s^2$	$2s$	$2s^2$	$2p$	$2p^2$	$2p^3$	$2p^4$	$2p^5$	$2p^6$
$^2S_{1/2}$	$^1S_0$	$^2S_{1/2}$	$^1S_0$	$^2P_{1/2}$	$^3P_0$	$^4S_{3/2}$	$^3P_2$	$^2P_{3/2}$	$^1S_0$

shells (He, Be) or $\frac{1}{2}$ for the open subshells (H, Li). The remaining atoms are best treated by again making a chart (Table 4.6) for the m, m_s values satisfying Hund's first two rules subject to the limitations of the Pauli principle; that is, we choose as many as we can of the spin quantum numbers m_s equal $[(M_S)_{max}]$ and then, subject to that condition, we maximize M_L. Each of the resulting M_S, M_L combinations are allowed and yield

Table 4.6 *Assignment of electrons in the ground-state configuration according to Hund's rules*

m	B	C	N	O	F	Ne
$+1$	↑	↑	↑	↑↓	↑↓	↑↓
0		↑	↑	↑	↑↓	↑↓
-1			↑	↑	↑	↑↓
M_S	$\frac{1}{2}$	1	$\frac{3}{2}$	1	$\frac{1}{2}$	0
M_L	1	1	0	1	1	0

a term for which $S = M_S$ and $L = M_L$. Given L and S, the J value is obtained from Eq. 4.65 and the third Hund's rule; for $2p$ and $2p^2$, we require the minimum J and for $2p^4$ and $2p^5$ the maximum J. The resulting term symbols are given in the last row of Table 4.5. The ground-state terms for all of the atoms, which can be obtained by corresponding procedures, are listed in Table 4.2.

4.7 IONIZATION POTENTIALS AND ELECTRON AFFINITIES

We briefly discussed the ionization potentials of simple atoms when we first introduced the concept of effective charge. Because the IP is one of the properties of an atom which is important in determining the nature of its chemical bonds, we wish to see how the IP varies across and down the periodic table. Another property, which is of corresponding importance but has not been considered so far, is the *electron affinity* (EA). Just as the IP (actually the *first* IP) is defined as the energy required for the process

$$A \rightarrow A^+ + e^- : IP \qquad (4.67)$$

The EA (actually the *first* EA) is defined as the energy released in the process

$$A + e^- \rightarrow A^- : EA \qquad (4.68)$$

That is, in the capture by a neutral atom of a free electron that initially has zero translational energy, the final energy of the electron is negative. The values of IP and EA that have been measured are listed in Table 4.2. Far fewer electron affinities than ionization potentials are known because of experimental difficulties in measuring the energy of A^-. One of the most accurate methods of determining EA is by deducing, from the absorption spectrum of the negative ion A^-, the energy that a photon must have to remove the extra electron in a transition to an unbound state with zero translational energy (photodetachment). A limitation of the method is that other spectral lines can obscure the photodetachment spectrum. For some atoms (e.g., hydrogen and the alkali metal atoms) the best-known values of EA are those that have been obtained from calculations using the variation principle and empirical corrections.

In Fig. 4.12, the ionization potentials are plotted as a function of atomic number. There are a number of evident trends. As one goes down a column of the periodic table and compares atoms with corresponding outer-shell

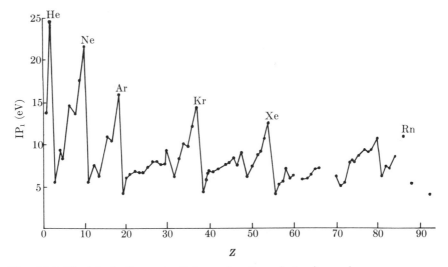

Fig. 4.12. First ionization potential as a function of atomic number.

configurations (e.g., Ne and Ar, Li and Na) the ionization potentials are similar but tend to decrease; for example, the IP for Ne is 21.6 eV and that for Ar is 15.8 eV, for Li the IP is 5.4 eV and that for Na is 5.1 eV. This decrease of IP is a relatively small effect but is of some importance in bonding. By analogy with Eq. 3.38 we can write

$$\text{IP} \simeq \frac{Z_e^2}{n^2}\left(\frac{e^2}{2a_0}\right) \tag{4.69}$$

In terms of this approximate formula, the decrease within a group results from the fact that Z_e for the outermost electrons does not increase as rapidly as the principle quantum number n. Much more important is the increase in IP across the periodic table for a given n. Comparing Li with Ne, we see that the IP rises from 5.4 eV for Li to 21.6 eV for Ne. As one proceeds across a period, the successive increases in Z result in corresponding increases in Z_e, the effective nuclear charge seen by the outer-shell electrons. The accompanying addition of electrons as a shell is being filled does relatively little, since electrons in the same shell have approximately the same average radius and therefore do not shield one another very effectively. From Eq. 4.69 one sees that as Z_e increases across a period, while n remains constant, the IP will increase. Furthermore, the initiation of a new shell (with a corresponding increase in n) will cause an abrupt decrease in IP, in agreement with the experimental results shown in Fig. 4.12.

The general increase of the IP in a given row (fixed n) is not completely monotonic. For $n = 2$, we see that IP of Be is slightly larger than that of B and that of N is greater than that of O. These inversions occur with regularity in other rows as well, and are another manifestation of the extra stability of filled or half-filled shells and subshells; for example, in Be($1s^2 2s^2$) the $2s$ subshell is filled and in N($1s^2 2s^2 2p^3$) the $2p$ subshell is half-filled.

An atom containing fewer electrons than its atomic number has a net positive charge. It is often useful to compare the electron configurations of these ions with those of *isoelectronic* neutral atoms, that is, those having the same number of electrons. Because of the relative stability of closed-shell configurations, positive ions that are isoelectronic with the rare gases are particularly stable (for example, Li^+, Na^+, Mg^{2+}).

The electron affinities follow trends similar to those for the IP. This is understandable, since if we write Eq. 4.68 in reverse

$$X^- \rightarrow X + e^-$$

the EA of X is the IP of X^-. In a given row (fixed n) the EA increases with increasing atomic number; for example, for Li, EA is estimated to be 0.6 eV, while for F it is $\sim$3.4 eV. The EA is particularly large for neutral atoms with a single hole in the outermost shell (i.e., those in which the added electron yields a closed-shell configuration). Correspondingly, an atom like O $(1s^2 2s^2 2p^4)$, which lacks two electrons to form a closed shell, can also form a stable negative ion, with EA = 1.48 eV. Although the O$^=$ ion has a closed shell, it appears to be unstable in the free state, so that the second electron affinity of O is "zero." The rare-gas negative ions would have the extra electron in the next higher n shell and so also have zero electron affinity; for example, for Ne$^-$, the configuration is $(1s^2 2s^2 2p^6 3s)$, which is unstable relative to Ne and a free electron. The EA, like IP, tends to decrease with increasing n in a given column of the periodic table; for example, EA for F is $\sim$3.4 eV, while that of I is 3.1 eV. However, EA for Cl (3.6 eV) is slightly larger than that for F.

4.8 ATOMIC RADII

The "radius" of an atom is a quantity that is not exactly defined, its magnitude depending upon the method of measurement. As discussed in the one-electron case, it is nevertheless useful for many problems to have estimates of an average radius that are based upon the interatomic distances in various compounds. It can be seen from the average values of the crystal radii in Table 4.2 that they follow a periodicity roughly similar to that of the first IP. This periodicity can be explained by writing an approximate equation for the effective radius of the electron cloud by analogy with Eq. 1.55:

$$r_{\text{eff}} \simeq \frac{n^2}{Z_e} a_0 \tag{4.70}$$

On the basis of Eq. 4.70, one expects the radii to decrease across a period, because Z_e increases and n remains constant. The increase in n when one goes from one period to another similarly causes an abrupt increase in radius. Since the reported radii depend upon the types of bonds formed by the various elements, large deviations from the trends predicted by Eq. 4.70 are to be expected. Thus the van der Waals radii (see Section 5.4) reported for the rare gases are considerably larger than the average of the covalent radii and ionic radii (see Section 5.6 and 5.3.5, respectively) reported for other atoms in the same period. Also, the radius for Na participating in an ionic bond as in NaCl is expected to be different from that of Na in a molecule like Na$_2$. This means that a variety of definitions for atomic radii can be useful, but that interpretation of the corresponding values must be based on some understanding of chemical bonding, as discussed in Chapter 5.

The radial distribution function (Section 3.10) for a many-electron atom is given by the sum of the distribution functions of the occupied orbitals. As an example, we consider the ground state of He, which can be represented by a wave function of the form (Eq. 4.49)

$$\psi(1,2) = (1/2)^{1/2} \begin{vmatrix} 1s\alpha(1) & 1s\beta(1) \\ 1s\alpha(2) & 1s\beta(2) \end{vmatrix}$$

The probability distribution is

$$[\psi(1,2)]^2 = \tfrac{1}{2}\{[1s\alpha(1)]^2[1s\beta(2)]^2$$

$$-2[1s\alpha(1)\ 1s\beta(2)\ 1s\alpha(2)\ 1s\beta(1)]$$

$$+[1s\alpha(2)]^2[1s\beta(1)]^2\}$$

To find the probability distribution for electron 1 independent of its spin and of the position and spin of electron 2, we integrate over the coordinates of electron 2 and the spins of both electrons. This yields the probability density $P_1(\mathbf{r})$ of electron 1 at the point $\mathbf{r}$,

$$P_1(\mathbf{r}) = [1s(\mathbf{r})]^2$$

since the $1s$ functions are normalized and the spin functions are normalized and orthogonal. A corresponding procedure for electron 2 gives

$$P_2(\mathbf{r}) = [1s(\mathbf{r})]^2$$

which is identical in form with $P_1(\mathbf{r})$, as it must be if the two electrons are to be indistinguishable. Adding together $P_1(\mathbf{r})$ and $P_2(\mathbf{r})$ yields a probability distribution function corresponding to two electrons in the $1s$ orbital (one with spin α and the other with spin β). This is exactly what would have been obtained simply by summing the distribution functions of the occupied orbitals. To find the radial distribution function, we integrate the resulting sum over the angular coordinates and multiply by r^2.

The form of the distribution functions for He and for other atoms is expected to be considerably changed from that for hydrogen. The orbitals are significantly contracted because of the attraction of the larger nuclear charge. However, the outermost filled orbitals of atoms in the same group are fairly similar in extent, there being an approximate balance between the increased nuclear charge and the shielding by the inner electrons. For the sequence He, Ne, . . . , Xe, the outermost maximum increases only from 0.291 to 0.986 Å (see Table 4.2).

A good approximation to the radial distribution can be obtained from the variational principle. Considering Ar as an example, we write down the determinant corresponding to the lowest energy configuration of Ar,

$$\psi(1, 2, \ldots, 18) = \frac{1}{(18!)^{\frac{1}{2}}} \begin{vmatrix} 1s\alpha(1)\ 1s\beta(2) \cdots 3p_z\beta(18) \end{vmatrix}$$

where we have used the shorthand expression for the determinant in terms of its principal diagonal.[1] We now minimize the energy of this wave function

$$E = \int\int \cdots \int \psi(1, 2, \ldots, 18)\, H\psi(1, 2, \ldots, 18)\, d\tau_1\, d\tau_2 \cdots d\tau_{18}$$

where $\int d\tau_i$ indicates integration over both coordinates and spin of electron i, to determine the functional form of the individual orbitals.[2] This procedure for finding one-electron orbitals is called the *Hartree–Fock* or *self-consistent-field* (SCF) method (see Reference 11 of the Additional Reading

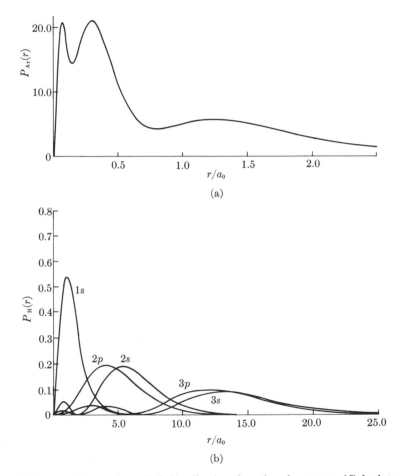

Fig. 4.13. (a) Electronic radial distribution function for argon. [Calculated by D. R. Hartree and W. Hartree, *Proc. Roy. Soc.* (*London*) **A166,** 450 (1938).] (b) Hydrogen-orbital ($Z = 1$) radial distributions for comparison.

[1] The expression $\left|\, 1s\alpha(1)\ 1s\beta(2)\ 2s\alpha(3)\,\right|$ represents the determinant

$$\begin{vmatrix} 1s\alpha(1) & 1s\beta(1) & 2s\alpha(1) \\ 1s\alpha(2) & 1s\beta(2) & 2s\alpha(2) \\ 1s\alpha(3) & 1s\beta(3) & 2s\alpha(3) \end{vmatrix}$$

[2] Integrals over the spin variable γ are given in Eq. 4.42.

list). For a simple example of an SCF-type calculation for He, see Problem 4.14. The resulting radial distribution curve for Ar is given in Fig. 4.13a and the hydrogen orbitals ($Z = 1$) are shown in Fig. 4.13b. It is evident in Fig. 4.13a that there is a sharp separation of the individual electron shells. Most important, the last filled shell (the M shell, $3s^2 3p^6$) completely dominates the outer part of the radial distribution curve. This suggests that its form should determine the "size" of the atom.

4.9 SPECTRA OF MANY-ELECTRON ATOMS

Although the discussion of electron configurations of atoms has emphasized the ground states, absorption of radiation or a collision with an electron, an atom, or molecule, can provide enough energy to raise an atom to an excited state. Transitions between the various states of an atom give rise to spectra. In contrast to the hydrogen-atom spectrum, which is due to transitions of its one electron, the spectra of many-electron atoms can result from transitions of either inner-shell or outer-shell electrons.

4.9.1 X-Ray spectra The inner-shell electrons have relatively widely spaced energy levels, so that the radiation corresponding to transitions between them has wavelengths on the order of 10^{-1} to 10^2 Å (x rays) and energies of 10^2 to 10^5 eV. A common method of producing x rays is to bombard a piece of metal with electrons of sufficient energy to knock an inner-shell electron out of a metal atom. The remaining electrons in the atom rearrange themselves by falling into the "holes" created and emit x radiation in accordance with the Bohr frequency rule. Since the inner-shell energies change from metal to metal, the wavelengths of the x rays vary with the target substance. Variations of x ray spectra with Z have no simple periodicity, since the inner shells of the various metals are occupied to the same extent. This contrasts with outer-shell spectra, which are similar for elements of the same group of the periodic table. Moseley (1913) found that a plot of the square root of the frequencies of the most energetic lines versus atomic number is nearly linear. A convenient empirical formula is *Moseley's*

$$\left(\frac{\tilde{\nu}}{R}\right)^{1/2} = \left(\frac{1}{n_2{}^2} - \frac{1}{n_1{}^2}\right)^{1/2}(Z - \sigma) \tag{4.71}$$

law, where R is the Rydberg constant and Z is the atomic number. The empirical constant σ can be interpreted as a *screening constant;* that is, its value is the amount of nuclear charge screened or shielded from the electron involved in the transition by the other electrons in the atom. For the transitions ($n_1 = 2$, $n_2 = 1$) of each element, Eq. 4.71 gives an adequate fit to experimental data if σ is assigned the value 1.13.

A more detailed treatment applicable to a wide range of lines requires recognition of the fact that the effective nuclear charge is different for the two shells involved in the transition; moreover, relativistic effects (i.e., spin-orbit coupling) that become more important for large Z must be included. The more accurate formula is

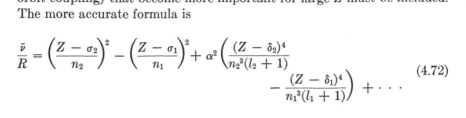

$$\frac{\tilde{\nu}}{R} = \left(\frac{Z - \sigma_2}{n_2}\right)^2 - \left(\frac{Z - \sigma_1}{n_1}\right)^2 + \alpha^2 \left(\frac{(Z - \delta_2)^4}{n_2{}^3(l_2 + 1)} - \frac{(Z - \delta_1)^4}{n_1{}^3(l_1 + 1)}\right) + \cdots \quad (4.72)$$

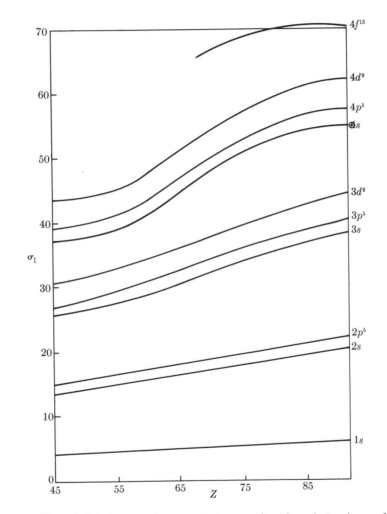

Fig. 4.14. Nonrelativistic screening constant σ as a function of atomic number Z. [From A. H. Compton and S. K. Allison, *X-Rays in Theory and Experiment* (D. Van Nostrand, New York, 1935), 2nd ed., p. 626.]

where α is the dimensionless *fine-structure constant*

$$\alpha = \frac{e^2}{\hbar c} = 7.29720 \times 10^{-3} \qquad (4.73)$$

When Eq. 4.72 is fitted to observed x-ray spectra, it is found that the non-relativistic screening constant σ depends upon the configuration of the relevant shell and increases with increasing Z (see Fig. 4.14). The value of the relativistic screening constant δ also depends upon the shell configuration, but is independent of Z (see Table 4.7). The designation of x-ray lines

Table 4.7   *Relativistic shielding constants*[a]

Configuration	δ
$2p^5$	3.50
$3p^5$	8.5
$3d^9$	13.0
$4p^5$	17.0
$4d^9$	24
$4f^{13}$	34

[a] A. H. Compton and S. K. Allison, *X-Rays in Theory and Experiment* (Van Nostrand, Princeton, N.J., 1935), 2nd ed., p. 612.

is shown in Table 4.8.

Although x-ray spectra involve inner-shell electrons of high energies, it is interesting to note that the emitted wavelengths are altered by the distribution of the outer-shell electrons; for example, the K_α lines of sulfur occur at different wavelengths depending on whether the S atom is bonded to other S atoms, as in crystalline sulfur, or to O atoms, as in Na_2SO_4. The shift is small (0.0028 Å) but it is large enough to be measured and has recently begun to be used to obtain information about the electron distribution in chemical bonds [see R. Manne, *J. Chem. Phys.* **46**, 4645 (1967)].

4.9.2   Outer-shell transitions   The spectra arising from transitions of outer-shell electrons can be analyzed in terms of the electron configurations corresponding to the initial and final states. For polyelectronic atoms, the simplest spectra are those of the alkali metals, in which all the electrons but one occupy closed inner shells. In ordinary atomic spectra, the configura-

Table 4.8 *Designation of x-ray lines*

n_2	n_1	Designation
1	2	K_α
1	3	K_β
1	4	K_γ
2	3	L_α
2	4	L_β
3	4	M_α
.	.	.
.	.	.
.	.	.

tion of these inner-shell electrons does not change, so that the transitions of the single outer-shell electron give rise to a spectrum which is similar to that of hydrogen.

In the case of sodium, for example, a $3s$ electron lies outside the closed $1s^2 2s^2 2p^6$ configuration. The excited-state configurations of interest are thus $1s^2 2s^2 2p^6 3p$, $1s^2 2s^2 2p^6 3d$, $1s^2 2s^2 2p^6 4s$, etc. These and other levels are shown in Fig. 4.15. Just as in hydrogen, the allowed transitions between the various levels are determined by selection rules. They are the same as those given in Eq. 3.50, namely,

$$\Delta n \quad \text{arbitrary} \qquad \Delta l = \pm 1$$

Although the allowed transitions shown in Fig. 4.15 are similar to those for hydrogen shown in Fig. 3.13, there are two important differences. First, the splitting of the Na subshell levels causes lines due to transitions terminating at different levels to overlap; for example, the $3s\ ^2S - 3p\ ^2P$ transition is at $\sim$5890 Å, the $3p\ ^2P - 5s\ ^2S$ transition is at $\sim$6160 Å, and the $3p\ ^2P - 4d\ ^2D$ transition is at $\sim$5690 Å. Second, there are no highly energetic transitions. All the lines involving the valence electron have wavelengths above the ultraviolet region of the spectrum. This makes the sorting of lines into the different series a slightly more difficult task than in the case of hydrogen. The names given to the series are

$np \leftrightarrow 3s$	principal	$(n \geq 3)$
$ns \leftrightarrow 3p$	sharp	$(n \geq 4)$
$nd \leftrightarrow 3p$	diffuse	$(n \geq 3)$
$nf \leftrightarrow 3d$	fundamental	$(n \geq 4)$

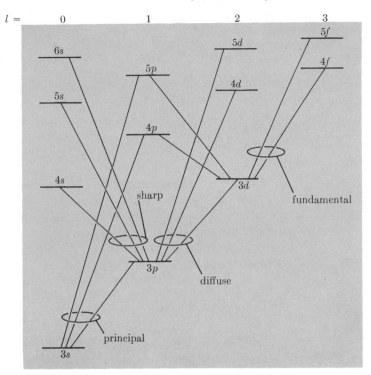

Fig. 4.15. Energy levels and allowed one-electron transitions of the sodium atom.

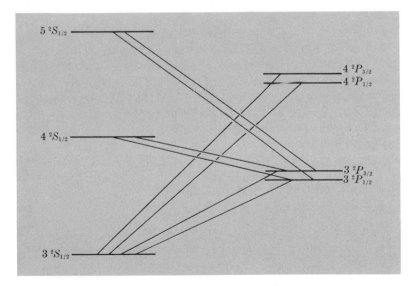

Fig. 4.16. Transitions among lowest terms of the sodium atom.

where n is the principal quantum number of the upper state. The letters we have been using to designate the different values of the quantum number l are in fact derived from these names for the spectral series of alkali atoms.

The actual transitions in an alkali atom are not simply between different configurations but rather between the terms associated with the configurations. For the Na atom, we show some of the lowest terms in Fig. 4.16 and indicate the allowed transitions between the components. The $3p^2P_{3/2} \rightarrow 3s^2S_{1/2}$ and $3p^2P_{1/2} \rightarrow 3s^2S_{1/2}$ transitions, which differ in wavelength by 17.2 cm^{-1}, form the intense sodium D-line doublet that appears in the visible (yellow) range.

For other many-electron atoms, which have several electrons in the outer shell, the spectra become more complicated. They can be understood, however, by simple extension of the techniques we have described here (see References 4, 7, and 9 of the Additional Reading list).

4.10 SCREENING CONSTANT RULES

Many of the concepts concerning atomic structure, such as the spacing of the energy levels, the ionization potentials, and so on, were discussed in this chapter in terms of the effective nuclear charge resulting from the shielding of the nucleus by the other electrons. These effective charges Z_e clearly depend on the (nl) quantum numbers of the level, the total nuclear charge Z, and the number of electrons N present in the atom or ion. One procedure for determining Z_e values is to do a variational calculation of the type described for the helium atom; that is, the configuration of interest for the atom is chosen, a determinantal wave function is formed from the occupied orbitals, a Z_e value is introduced for each occupied (nl) level, [i.e., $Z_e(nl)$] and all of the $Z_e(nl)$ are varied systematically by the Hartree–Fock procedure so as to minimize the total energy of the atom. This approach has been applied to neutral atoms and to ions. The functions used for the individual orbitals, which were introduced by Slater and have been widely applied to many atomic and molecular problems, were chosen to have a separable form similar to that obtained for hydrogen-like systems, (Eq. 3.44); that is,

$$\psi_{nlm}(r, \theta, \phi, \zeta) = R_{nl}'(r, \zeta)\Theta_l^m(\theta)\Phi_m(\phi) \qquad (4.74)$$

where $\zeta = Z_e/n$, and the normalized radial function is

$$R_{nl}'(r, \zeta) = N_{nl}\, r^{n-1}\, e^{-\zeta r} \qquad (4.75)$$

with r expressed in units of the Bohr radius, a_0. From Eq. 4.75, we see that instead of the entire polynomial of the hydrogen-like result given for $n = 1, 2, 3$ in Table 3.1, only the maximum power of r is included. This is often a satisfactory approximation, particularly in molecular applications for which the outer parts of the orbitals are most important. Orbitals of the type given in Eqs. 4.74 and 4.75 are called Slater or Slater-type orbitals (abbreviated STO's). The values of the exponent ζ resulting from such variational calculations for the atoms He through Kr are shown in Table 4.9. It is evident from Table 4.9 that the Z_e values vary rather systematically for a given atom, in that they generally decrease with increasing n and, for a fixed n, with increasing l; above Ne, the Z_e values for $2s$ and $2p$ are inverted, however. Moreover, for specified n and l, the Z_e value increases with increasing atomic number.

Table 4.10 indicates the changes that occur in the effective charges in an isoelectronic sequence as the nuclear charge increases from C^0 to Mg^{6+}. This table demonstrates the important result that the screening of the nucleus by the electrons in a given configuration is nearly independent of the total charge Z; that is, $Z - \zeta$ is approximately constant for the entire isoelectronic series.

Since the calculations required for determining the effective charges for the orbitals of each individual atom are very time consuming, it is convenient to have a set of rules which provide a reasonable approximation. Such a set of rules for the Z_e values was particularly useful before high-speed digital computers came into common usage. In those days, variational calculations with the orbitals of Eq. 4.74 had been performed only for the first row atoms. In examining the results for these systems, Slater noticed regularities of the type outlined above. Furthermore, he realized that the regularities could serve as a basis for expressing the screening constant s, defined by $Z_e = Z - s$, as a simple function of quantities that are known once the atom or ion and its configuration have been specified. As we stated above, the value of Z_e (and therefore of s) depends upon the particular shell or subshell under consideration. For a given (nl) the screening is determined primarily by the number of electrons present in inner shells or in the same shell or subshell (i.e., with n and l values less than or equal to the ones of interest). Of these, the shielding by the inner electrons ($n' < n$, or $n' = n$ and $l' < l$) is more effective than that of electrons in the same shell or subshell. We have seen this already in our discussion of the variation of electronic properties (e.g., IP and atomic radius) across the periodic table. On the other hand, the electrons outside the subshell under consideration ($n' > n$, or $n' = n$ and $l' > l$) have only a very small shielding effect.

With these ideas and certain assumptions about the functional dependence of s upon the number of occupied orbitals of each type, Slater in 1930 introduced a widely used set of rules for estimating the screening constants of atoms and ions. According to these rules, for a $1s$ electron in a filled $1s$ subshell the screening constant, s_{1s} is

$$s_{1s} = 0.3$$

Table 4.9 *Best values of ζ for the ground state of neutral atoms*[a]

	Z	1s	2s	2p	3s	3p	4s	3d	4p
He	2	1.6875							
Li	3	2.6906	0.6396						
Be	4	3.6848	0.9560						
B	5	4.6795	1.2881	1.2107					
C	6	5.6727	1.6083	1.5679					
N	7	6.6651	1.9237	1.9170					
O	8	7.6579	2.2458	2.2266					
F	9	8.6501	2.5638	2.5500					
Ne	10	9.6421	2.8792	2.8792					
Na	11	10.6259	3.2857	3.4009	0.8358				
Mg	12	11.6089	3.6960	3.9129	1.1025				
Al	13	12.5910	4.1068	4.4817	1.3724	1.3552			
Si	14	13.5745	4.5100	4.9725	1.6344	1.4284			
P	15	14.5578	4.9125	5.4806	1.8806	1.6288			
S	16	15.5409	5.3144	5.9885	2.1223	1.8273			
Cl	17	16.5239	5.7152	6.4966	2.3561	2.0387			
Ar	18	17.5075	6.1152	7.0041	2.5856	2.2547			
K	19	18.4895	6.5031	7.5136	2.8933	2.5752	0.8738		
Ca	20	19.4730	6.8882	8.0207	3.2005	2.8861	1.0995		
Sc	21	20.4566	7.2868	8.5273	3.4466	3.1354	1.1581	2.3733	
Ti	22	21.4409	7.6883	9.0324	3.6777	3.3679	1.2042	2.7138	
V	23	22.4256	8.0907	9.5364	3.9031	3.5950	1.2453	2.9943	
Cr	24	23.4138	8.4919	10.0376	4.1226	3.8220	1.2833	3.2522	
Mn	25	24.3957	8.8969	10.5420	4.3393	4.0364	1.3208	3.5094	
Fe	26	25.3810	9.2995	11.0444	4.5587	4.2593	1.3585	3.7266	
Co	27	26.3668	9.7025	11.5462	4.7741	4.4782	1.3941	3.9518	
Ni	28	27.3526	10.1063	12.0476	4.9870	4.6950	1.4277	4.1765	
Cu	29	28.3386	10.5099	12.5485	5.1981	4.9102	1.4606	4.4002	
Zn	30	29.3245	10.9140	13.0490	5.4064	5.1231	1.4913	4.6261	
Ga	31	30.3094	11.2995	13.5454	5.6654	5.4012	1.7667	5.0311	1.5554
Ge	32	31.2937	11.6824	14.0411	5.9299	5.6712	2.0109	5.4171	1.6951
As	33	32.2783	12.0635	14.5368	6.1985	5.9499	2.2360	5.7928	1.8623
Se	34	33.2622	12.4442	15.0326	6.4678	6.2350	2.4394	6.1590	2.0718
Br	35	34.2471	12.8217	15.5282	6.7395	6.5236	2.6382	6.5197	2.2570
Kr	36	35.2316	13.1990	16.0235	7.0109	6.8114	2.8289	6.8753	2.4423

[a] E. Clementi and D. L. Raimondi, *J. Chem. Phys.* **38**, 2686 (1963); computed for ground state configuration, except for Cr([Ar]$4s^2 3d^4$) and Cu([Ar]$4s^2 3d^9$).

Table 4.10 *Effective charges for inner-most electron (1s) in iso-electronic series C^0 to Mg^{6+}*[a]

Ion	Z	$\zeta[1s(1)]$	$Z - \zeta$
C^0	6	5.377	0.623
N^+	7	6.382	0.618
O^{2+}	8	7.387	0.613
F^{3+}	9	8.393	0.607
Ne^{4+}	10	9.398	0.602
Na^{5+}	11	10.404	0.596
Mg^{6+}	12	11.409	0.591

[a] Calculated from polynomial expansions of ζ given by E. Clementi, *J. Chem. Phys.* **38**, 996 (1963).

For electrons with $n > 1$ and $l = 0,1$ s_{nl} is given by

$$s_{nl} = 0.35x + 0.85y + 1.00z$$

where x is the number of other electrons in the same shell as the screened electron of interest, y is the number of electrons in the shell with principal quantum number $(n - 1)$, and z is the number of electrons with principal quantum number $\leq (n - 2)$.

For $3d$ electrons,

$$s_{3d} = 0.35x + 1.00y$$

where x is the number of $3d$ electrons and y is the number of electrons with $n \leq 3$ and $l < 2$.

Now that variational values for Z_e and s are available for many more atoms and ions than were used in Slater's original formulation of the screening rules (values of Z_e given in Table 4.9 for neutral ground-state atoms with $2 \leq Z \leq 36$ were reported in 1963 and values for $37 \leq Z \leq 86$ in 1967), it is possible to obtain a somewhat better fit for screening constants of valence electrons (i.e., those in the outermost shell) from a revised version of the rules (Problem C4.5). Furthermore, the dependence of the screening constants upon all of the electrons (including those external to the subshell under consideration) can be systematized as an extension of Slater's rules. For the screening constant s_{nl} of each l orbital, Clementi and Raimondi used an expression of the form

$$\begin{aligned} s_{nl} = a_0 &+ a_1[N(nl) - 1] + a_2 N(n, l + 1) \cdots a_i N(n, n - 1) \\ &+ a_{i+1} N(n + 1, 0) + a_{i+2} N(n + 1, 1) + \cdots + a_8 N(4p) \end{aligned}$$

$$(4.76)$$

where $N(n'l')$ is the number of electrons occupying the $(n'l')$ subshell and $a_0, a_1, \ldots, a_8$ are constants to be determined for each nl. The $4p$ subshell is the highest one included because the rules were devised for ground-state atoms from He through Kr. The equations obtained by Clementi and Raimondi are

$$s_{1s} = 0.3[N(1s) - 1] + 0.0072[N(2s) + N(2p)]$$
$$+ 0.0158[N(3s) + N(3p) + N(4s) + N(3d) + N(4p)]$$
$$s_{2s} = 1.7208 + 0.3601[N(2s) + N(2p) - 1]$$
$$+ 0.2062[N(3s) + N(3p) + N(4s) + N(3d) + N(4p)]$$
$$s_{2p} = 2.5787 + 0.3326[N(2p) - 1] - 0.0773N(3s)$$
$$- 0.0161[N(3p) + N(4s)] - 0.0048N(3d) + 0.0085N(4p)$$
$$s_{3s} = 8.4927 + 0.2501[N(3s) + N(3p) - 1] + 0.0778N(4s)$$
$$+ 0.3382N(3d) + 0.1978N(4p).$$
$$s_{3p} = 9.3345 + 0.3803[N(3p) - 1] + 0.0526N(4s)$$
$$+ 0.3289N(3d) + 0.1558N(4p)$$
$$s_{4s} = 15.505 + 0.0971[N(4s) - 1] + 0.8433N(3d)$$
$$+ 0.0687N(4p)$$
$$s_{3d} = 13.5894 + 0.2693[N(3d) - 1] - 0.1065N(4p)$$
$$s_{4p} = 24.7782 + 0.2905[N(4p) - 1]$$

The accuracy of the Clementi–Raimondi rules can be checked by calculating ζ for several orbitals and comparing the results with the values given in Table 4.9.

Although the Slater orbitals differ from the hydrogen-like functions by retaining only the maximum power of r in the polynomial which multiplies the exponential factor, a reasonable estimate of orbital energies can be obtained by substituting Z_e into the hydrogen-like energy formula, Eq. 3.38. To indicate the utility of this approximation, we employ it to calculate the ionization potentials of lithium and its positive ions in their ground states. For the configurations of the atoms and ions we have

$$\underline{\text{Li}(1s^2 2s)}$$

$$s(1s) = 0.3(2 - 1) + 0.0072(1) = 0.3072$$
$$Z_e(1s) = 2.6928$$

$$s(2s) = 1.7208 + 0.360(0) = 1.7208$$
$$Z_e(2s) = 1.2792$$

$$E_{\text{Li}} = 2E_{1s} + E_{2s}$$
$$= -[2(2.6928/1)^2 + (1.2792/2)^2](13.6) \text{ eV}$$
$$= -203 \text{ eV}$$

$$\underline{\text{Li}^+(1s^2)}$$

$$s(1s) = 0.3(2 - 1) = 0.3$$
$$Z_e(1s) = 2.7$$
$$E_{\text{Li}^+} = -2(2.7/1)^2(13.6)\text{eV} = -198\text{eV}$$

$\underline{Li^{2+}(1s)}$

$$s(1s) = (0.3)(1 - 1) = 0$$
$$Z_e(1s) = 3 - 0 = 3$$
$$E_{Li^{2+}} = -(3/1)^2 \ (1.36) \text{ eV} = -122.4 \text{ eV}$$

The ionization potentials corresponding to these energies are thus

$$IP_1 = E_{Li}^+ - E_{Li} = 5 \text{ eV} \quad (\text{expt } 5.39 \text{ eV})$$
$$IP_2 = E_{Li}^{2+} - E_{Li}^+ = 75.6 \text{ eV} \quad (\text{expt } 75.7 \text{ eV})$$
$$IP_3 = -E_{Li}^{2+} = 122.4 \text{ eV} \quad (\text{expt } 122.4 \text{ eV})$$

Thus, the screening constant rules give results in satisfactory agreement with experiment for this case.

Another application of the screening constant rules is to the estimation of orbital radii. For the $2p$ orbital of carbon ($1s^2 2s^2 2p^2$), we have

$$Z_e = 2(1.5679) = 3.1358$$

Using for r_{eff} twice the maximum value of the radial distribution function for the $2p$ Slater orbital, we find

$$r_{eff} = \frac{n^2}{Z_e} a_0 = \frac{4a_0}{3.1358} = 1.28a_0 = 0.68 \text{ Å}$$

in approximate agreement with the estimated carbon atom $2p$ orbital radius of 0.62 A given in Table 4.2.

SUMMARY

In this chapter we have learned how to use one-electron orbitals to construct the wave functions for and to describe the structures of many-electron atoms. The electron-electron repulsion, which is the essential complicating element in such systems, was treated in terms of average shielding and perturbation theory. An improvement in the calculated energy of many-electron atoms by use of the variation principle was demonstrated. For ground-state Li, excited states of He, and for more complex atoms, the Pauli principle was seen to be a necessary guide in writing electron configurations and approximate wave functions; in the case of excited He it was shown to give rise to an exchange energy which splits singlet and triplet terms. Periodicity of the ionization potentials and electron affinities of the elements were discussed in terms of the *Aufbau* principle. Rules for determining atomic term values and identifying ground-state terms were given. Principles of x-ray emission and outer-electron spectra of the atoms were discussed, and rules for calculating screening constants were presented.

In the next chapter we introduce the concepts underlying the ways in which atoms combine to form molecules and crystals.

ADDITIONAL READING

[1] R. M. HOCHSTRASSER, *Behavior of Electrons in Atoms* (Benjamin, New York, 1964).

[2] M. W. HANNA, *Quantum Mechanics in Chemistry* (Benjamin, New York, 1965), Chap. 6.

[3] F. L. PILAR, *Elementary Quantum Chemistry* (McGraw-Hill, New York, 1968), Chap. 8.

[4] G. HERZBERG, *Atomic Spectra and Atomic Structure* (Dover, New York, 1944).

[5] J. C. DAVIS, *Advanced Physical Chemistry* (The Ronald Press, New York, 1965), Chap. 7.

[6] L. PAULING AND E. B. WILSON, JR., *Introduction to Quantum Mechanics* (McGraw-Hill, New York, 1935).

[7] E. U. CONDON AND G. H. SHORTLEY, *The Theory of Atomic Spectra* (Cambridge University Press, Cambridge, 1959).

[8] J. C. SLATER, *Quantum Theory of Atomic Structure* (McGraw-Hill, New York, 1960), Vol. I. There is a second volume which contains additional material of a more specialized nature.

[9] B. W. SHORE AND D. H. MENZEL, *Principles of Atomic Spectra* (Wiley, New York, 1968).

[10] J. W. LINNETT, *Wave Mechanics and Valency* (Methuen, London, 1960), Chaps. V and VI.

[11] D. R. HARTREE, *The Calculation of Atomic Structures* (Wiley, New York, 1957).

[12] R. CASTANG, J. DESCAMPS, AND J. PHILIBERT, Eds., *X-Ray, Optics and Microanalysis* (Hermann, Paris, 1966), pp. 328–338.

PROBLEMS

4.1 Write down the ground-state electron configurations for the unipositive ions Li^+ through Na^+ and the dipositive ions Na^{2+} through K^{2+}; use the *Aufbau* principle, but do not refer to the ground-state configurations of the neutral atoms.

4.2 For each of the following, use the value of IP or EA in Table 4.2 to estimate Z_e from the Bohr formula; then use Z_e to estimate the orbital radius r: Li, Be, N, O^-, F, F^-, Ne, Na, Mg, S, S^-, Cl, Cl^-.

4.3 Nitrogen atoms have the ground-state electron configuration $[He]2s^2 2p^3$. Write down the wave function for the $2p$ subshell as a Slater determinant of hydrogen-like orbitals and spin functions. Derive the probability density $P(1)$ for a single electron irrespective of position of the other two electrons. Comment on the angular dependence of $P(1)$. Repeat the same calculation for the highest energy subshell of the ground state of Pd.

4.4 Calculate in cm sec^{-1} the root-mean-square velocity of a $1s$ electron in an atom with nuclear charge Z, neglecting shielding effects. What value of Z would make the $1s$ electron travel with the velocity of light? For what range of Z do you expect relativistic effects to be important for the innermost electrons? For example, for what range of Z is $(1 - v^2/c^2)^{-1/2} > 1.1$?

4.5 Show that the He-atom ground-state wave function of Eq. 4.48 is normalized if $1s$, α, β are normalized and α and β are orthogonal; that is, show that

$$\int\int [\psi(1, 2)]^2 \, d\tau_1 \, d\tau_2 = 1$$

4.6 Consider a one-dimensional oscillator that is essentially harmonic (see Section 2.10), but whose motion is described by a potential-energy expression containing a small quartic term in addition to the much larger quadratic term; that is,

$$V(x) = \tfrac{1}{2} \, kx^2 + bx^4$$

with $bx^2 \ll k$ over the effective range of x. Show that the presence of the quartic term raises the ground-state energy above the harmonic value approximately by the amount $\tfrac{3}{4}b\hbar^2/(mk)$, where m is the mass of the oscillator. (*Hint:* Find the average value of bx^4 for a ground-state harmonic oscillator with force constant k and mass m.)

4.7 Derive Eq. 4.27 by performing the integrations over θ_1, ϕ_1, θ_2, ϕ_2, indicated in Eq. 4.21 after substituting for $1/r_{12}$ from Eqs. 4.26 and 4.24.

4.8 Show that each of the wave functions of Eq. 4.56 is a solution to Eq. 4.9, with energy $E = E_{1s} + E_{2s}$.

4.9 Show that the possible determinantal wave functions with three electrons in $1s$ orbitals are zero.

4.10 Evaluate J and K in Eqs. 4.60a and 4.60b, respectively, by substituting for $1/r_{12}$ from Eqs. 4.26 and 4.24 and performing the integration over angles to give equations analogous to Eq. 4.27; the radial integrals can then be evaluated by analogy with Eq. 4.28. (This is a long problem).

4.11 Calculate the velocity in cm sec^{-1} of the electron produced by the autoionization of a helium atom in the state $2s \, 4d$. For present purposes, the electron-electron interaction can be assumed to take the form of complete shielding of the $4d$ electron by the $2s$ electron.

4.12 Verify that the following configurations give rise to the indicated terms.
  (a) $ns \, np$: ^3P, ^1P
  (b) $np \, n'd$ or $np \, nd$: ^3P, ^3D, ^3F, ^1P, ^1D, ^1F

(c) nd^2: ^3P, ^3F, ^1S, ^1D, ^1G

(d) np^3: ^4S, ^2P, ^2D

Determine the J values associated with each term.

4.13 Two terms are missing from each of the following lists of terms deriving from the indicated configuration. Find the missing terms.

(a) np^5 $n'p$: ^3P, ^3D, ^1P, ^1D

(b) ns np^5: ^1P

(c) np $n'g$: ^3F, ^3H, ^1F, ^1G

(d) nd^9 $n's^2$ $n'p^5$: ^3P, ^3D, ^1P, ^1D

4.14 (a) Derive the effective potential-energy function for electron 1 of a two-electron atom with nuclear charge Z if the probability density of electron 2 is given by the square of the normalized wave function

$$\phi_2(r_2) = \left(\frac{\varsigma_2^3}{\pi}\right)^{1/2} \exp\left(-\varsigma_2 r_2\right)$$

where $\varsigma_2 a_0$ is the effective charge seen by electron 2.

(*Hint:* Calculate the analog of Eq. 4.29 for this more general case.)

(b) Use the result from part (a) to construct the effective Hamiltonian for electron 1.

$$Ans.\ H_1 = -\frac{\hbar^2}{2m}\nabla_1^2 - \frac{Ze^2}{r_1} + \frac{e^2}{r_1}[1 - (1 + \varsigma_2 r_1)\exp\left(-2\varsigma_2 r_1\right)]$$

(c) Solve the one-electron Schrödinger equation $H_1\phi_1(r_1) = \epsilon_1\phi_1(r_1)$ by the variation principle, using the normalized trial function

$$\phi_1(r_1) = \left(\frac{\varsigma_1^3}{\pi}\right)^{1/2} \exp\left(-\varsigma_1 r_1\right)$$

with ς_1 as the variational parameter. [*Hint:* Calculate the average of H_1 for the given trial function,

$$\langle H_1 \rangle = \int \phi_1(r_1) H_1 \phi_1(r_1) dv_1 = \langle \epsilon_1 \rangle$$

then, keeping ς_2 fixed, set the derivative of $\langle H_1 \rangle$ with respect to ς_1 equal to zero; you may delay the solution of the resulting equation and proceed to part (d).]

(d) Find the common value of ς_1 and ς_2 for the ground state of the two-electron atom by setting $\varsigma_2 = \varsigma_1 = \varsigma$ in the equation for ς_1, and solving it; calculate $\langle \epsilon_1 \rangle$ in terms of Z.

$$Ans.\ \langle \epsilon_1 \rangle = -(e^2/2a_0)(Z^2 - 5Z/4 + 75/256)$$

(e) Show that the average total energy of the atom in the present approximation to the wave function ($\varsigma_2 = \varsigma_1 = \varsigma$) is given by

$$\langle E \rangle = \langle \epsilon_1 \rangle + \langle \epsilon_2 \rangle - \left\langle \frac{e^2}{r_{12}} \right\rangle$$

where

$$\left\langle \frac{e^2}{r_{12}} \right\rangle = \int \phi_1(r_1)\, \phi_2(r_2)\, \frac{e^2}{r_{12}}\, \phi_1(r_1)\, \phi_2(r_2)\, dv_1\, dv_2$$

$$= \int \phi_1(r_1) \left(\frac{e^2}{r_1} [1 - (1 + \zeta r_1) \exp(-2\zeta r_1)] \right) \phi_1(r_1)\, dv_1$$

(*Hint:* Show that $\langle \epsilon_1 \rangle$ and $\langle \epsilon_2 \rangle$ each contain the term $2\langle e^2/r_{12} \rangle$.)

(f) Use the results of parts (e) and (d) and the fact that $\langle \epsilon_2 \rangle = \langle \epsilon_1 \rangle$ to calculate $\langle E \rangle$ in terms of Z. Compare your answer with Eq. 4.41 for $Z = 2$.

(g) This problem is a simple example of a SCF calculation for a many-electron atom. Koopmans' theorem asserts that the one-electron energy of the most energetic occupied orbital, as obtained from an SCF calculation, is a reasonably good approximation to $-\text{IP}_1$ for an atom or molecule. Compare $-\langle \epsilon_1 \rangle$ with IP_1 for He ($Z = 2$) and with EA for H ($Z = 1$).

Ans. $-\langle \epsilon_1 \rangle = 24.39$ eV and 0.5846 eV for $Z = 2$ and 1, respectively

(h) Compare the results of part (g) with IP_1 for He calculated from $\langle E \rangle$ of part (f) with $Z = 2$ and E_{He^+}, and with EA for H calculated from $\langle E \rangle$ of part (f) with $Z = 1$ and E_{H}. Which result is closer to the experimental value for He and H?

Ans. $E_{\text{He}^+} - \langle E \rangle_{Z=2} = 23.07$ eV
$E_{\text{H}} - \langle E \rangle_{Z=1} = -0.744$ eV

4.15 Derive an expression for the normalized probability densities for electron 1 irrespective of the position of electron 2 for a two-electron system described by the wave function of Eq. 4.54a and for one described by the wave function of Eq. 4.54b. Repeat the derivation to obtain for each case the normalized probability density for electron 2 irrespective of the position of electron 1. Are the two electrons distinguishable by having different spatial distributions?

4.16 The values of ζ listed in Table 4.9, together with energy data, can be used to estimate values of ζ for excited configurations of atoms and ions. Consider, for example, the Be excited configuration $\text{Be}(1s^2 2s 2p)$. According to J. C. Slater [*Quantum Theory of Atomic Structure* (McGraw-Hill, New York, 1960), Vol. 1, p. 339], the excitation energy (i.e., energy above the ground state) of this configuration obtained by taking the average of the terms arising from it is $0.24721\, R_{\text{Be}}$, where R_{Be} is the Rydberg constant for Be. Using the Rydberg formula $\epsilon_{nl} = -\zeta_{nl}^2 R$ for the one-electron energies and ignoring electron-electron interactions, we can approximate the excitation energy (in units of R_{Be}) by

$$E[\text{Be}(1s^2 2s 2p)] - E[\text{Be}(1s^2 2s^2)] \simeq\; - 2\zeta_{1s}^2[\text{Be}(1s^2 2s 2p)] - \zeta_{2s}^2[\text{Be}(1s^2 2s 2p)]$$
$$- \zeta_{2p}^2[\text{Be}(1s^2 2s 2p)] + 2\zeta_{1s}^2[\text{Be}(1s^2 2s^2)] + 2\zeta_{2s}^2[\text{Be}(1s^2 2s^2)]$$

where, for example, $\zeta_{2s}^2[\text{Be}(1s^22s2p)]$ represents the value of ζ_{2s}^2 calculated for the Be configuration $1s^22s2p$.

One of the three unknown excited configuration ζ's can be calculated from this formula if the other two can be estimated independently. Since the screening constant s_{1s} does not change appreciably as $2s$ and $2p$ electrons are added (Problems C4.3 and C4.4), no great error results if we assume that ζ_{1s} is the same in both the excited and ground configurations of Be. We can estimate the value of $\zeta_{2s}[\text{Be}(1s^22s2p)]$ by assuming that the screening effects of the other electrons upon the $2s$ electron are additive and independent of Z; thus $s_{2s}(1s^22s2p)$ is lower than $s_{2s}(1s^22s^22p)$ by the screening contribution of the second $2s$ electron. The latter can be approximated by the difference between $s_{2s}(1s^22s^2)$ and $s_{2s}(1s^22s)$; that is,

$$s_{2s}(1s^22s2p) \simeq s_{2s}(1s^22s^22p) - [s_{2s}(1s^22s^2) - s_{2s}(1s^22s)]$$

(a) Use the formula $n\zeta_{nl} = Z - s_{nl}$ and the values of ζ_{nl} given in Table 4.9 for Li$(1s^22s)$, Be$(1s^22s^2)$, and B$(1s^22s^22p)$ to calculate $s_{2s}(1s^22s)$, $s_{2s}(1s^22s^2)$, and $s_{2s}(1s^22s^22p)$, respectively. From your results, estimate the value of $s_{2s}(1s^22s2p)$ and $\zeta_{2s}[\text{Be}(1s^22s2p)]$.

Ans. $\zeta_{2s}[\text{Be}(1s^22s2p)] \simeq 0.972$

(b) From the values of ζ_{1s} and ζ_{2s} for Be$(1s^22s^2)$ given in Table 4.9, the excitation energy of Be$(1s^22s2p)$, and your result for part (a), estimate the value of $\zeta_{2p}[\text{Be}(1s^22s2p)]$ and $s_{2p}(1s^22s2p)$.

Ans. $s_{2p}(1s^22s2p) \simeq 2.405$

4.17 In principle, Table 4.9 provides a means of rapidly estimating energies of excited configurations of atoms and ions from the Rydberg formula $\epsilon_{nl} = -\zeta_{nl}^2 R$ for the energies of the individual electrons. To estimate the configuration energy of C$(1s^22s2p^3)$ relative to that of C$(1s^22s^22p^2)$ in units of R_C, we use the Rydberg formula to write

$$\begin{aligned}\Delta E &= E[\text{C}(1s^22s2p^3)] - E[\text{C}(1s^22s^22p^2)] \\ &\simeq -\zeta_{2s}^2[\text{C}(1s^22s2p^3)] - 3\zeta_{2p}^2[\text{C}(1s^22s2p^3)] \\ &\quad + 2\zeta_{2s}^2[\text{C}(1s^22s^22p^2)] + 2\zeta_{2p}^2[\text{C}(1s^22s^22p^2)]\end{aligned}$$

(We have assumed that ζ_{1s} is the same in both configurations.)

(a) Write approximations for the screening constants that can be obtained from Table 4.9 and the result of Problem 4.16.

Ans. $s_{2s}(1s^22s2p^3) \simeq s_{2s}(1s^22s^22p^3) - s_{2s}(1s^22s^2) + s_{2s}(1s^22s)$; $s_{2p}(1s^22s2p^3) \simeq s_{2p}(1s^22s^22p^3) - s_{2p}(1s^22s^22p) + s_{2p}(1s^22s2p)$

(b) From appropriate values of ζ_{nl} given in Table 4.9 and the results of Problems 4.16 and 4.17(a), estimate values of $\zeta_{2s}[\text{C}(1s^22s2p^3]$ and $\zeta_2[\text{C}(1s^22s2p^3)]$. Calculate the excitation energy of C$(1s^22s2p^3)$ in units of R_C, in eV, and in kcal/mole.

Compare your result with the value of 190 kcal/mole estimated from spectroscopic data.

Ans. $\Delta E \simeq 226$ kcal/mole

4.18   A possible method for estimating $\zeta_{2p}[\text{Be}(1s^2 2s 2p)]$, in addition to that used in Problem 4.16, is to obtain an approximate value of $s_{2p}(1s^2 2s 2p)$ from the excitation energy of $\text{Li}(1s^2 2p)$ by writing

$$s_{2p}(1s^2 2s 2p) \simeq \tfrac{1}{2}[s_{2p}(1s^2 2p) + s_{2p}(1s^2 2s^2 2p)]$$

Is this a reasonable approximation to make? If the excitation energy of $\text{Li}(1s^2 2p)$ is 0.13583 in units of R_{Li} (reference, Problem 4.16), estimate $s_{2g}(1s^2 2s 2p)$ from the above formula by the methods of Problems 4.16 and 4.17. Compare this value with the result of Problem 4.16. How much does the estimated value of the excitation energy of $\text{C}(1s^2 2s 2p^3)$ (Problem 4.17) change if this alternate result for $s_{2p}(1s^2 2s 2p)$ is used? What is the value of the excitation energy of $\text{C}(1s^2 2s 2p^3)$ estimated from the average of the two results for $s_{2p}(1s^2 2s 2p)$? Comment on the self-consistency of the methods of Problems 4.16–4.18.

Ans. The average for $s_{2p}(1s^2 2s 2p)$ leads to an excitation energy of
$\simeq 120$ kcal/mole for $\text{C}(1s^2 2s 2p^3)$

For the following problems a programmable desk calculator or digital computer will be useful.

C4.1   From Eqs. 4.62 and the experimental He spectrum [C. E. Moore, *Atomic Energy Levels*, Nat. Bur. Std. (U.S.) Circ. No. 467 (U.S. Government Printing Office, Washington, D.C., 1949)] compute "experimental" values of $K(1snl)$ and $J(1snl)$ in eV for $2 \le n \le 6$ and $0 \le l \le n - 1$. Write a brief statement summarizing the apparent trends.

C4.2   Show that for the $1s2s$ state of He, the orbital probability densities for the singlet and triplet in the hydrogen-like approximation are

$$[^1\psi(1, 2)]^2 = \frac{8}{\pi}[(1 - r_2)\exp(-r_1) + (1 - r_1)\exp(-r_2)]^2 \exp[-2(r_1 + r_2)]$$

$$[^3\psi(1, 2)]^2 = \frac{8}{\pi}[(1 - r_2)\exp(-r_1) - (1 - r_1)\exp(-r_2)]^2 \exp[-2(r_1 + r_2)]$$

where r_1 and r_2 are in units of a_0. Calculate a table of values of $(^1\psi)^2$ and $(^3\psi)^2$ for values of r_1, r_2 between 0 and 4 at intervals of 0.10. Construct a plot of lines of constant probability density (contour map) in the $r_1 r_2$ plane, either by using a program for calculating contour maps or by setting down the calculated values at the appropriate grid points and sketching in the contour lines. Use your results to comment qualitatively on the nature of the correlation of electronic positions displayed by the singlet and triplet functions.

C4.3 For each of the values of ζ_{nl} listed in Table 4.9 calculate the corresponding screening constant s_{nl} from the definition $Z_e = n\zeta_{nl} = Z - s_{nl}$. Plot s_{nl} as a function of Z for each value of n and l of the screened electron.

C4.4 Find values of a_{nl} that minimize the sum of the squares of the differences between the values of s_{1s} calculated in Problem C4.3 and the corresponding values given by the formula

$$s_{1s} = a_{1s}(N_{1s} - 1) + a_{2s}N_{2s} + a_{2p}N_{2p} + a_{3s}N_{3s}$$
$$+ a_{3p}N_{3p} + a_{4s}N_{4s} + a_{3d}N_{3d} + a_{4p}N_{4p}$$

Compare your result with the Clementi–Raimondi rules for s_{1s}.

C4.5 The original Slater rules for obtaining atomic screening constants express s_{nl} as the sum

$$s_{nl} = a_{nl}x + b_{nl}y + c_{nl}z$$

where x is the number of other electrons occupying orbitals in the same shell as the screened electron of interest, y is the number of electrons in the next inner shell, and z is the number of electrons in shells farther in than the first; that is, if N_n is the number of electrons in the nth shell, then $x = N_n - 1$, $y = N_{n-1}$, and $Z = N_{n-2} + N_{n-3} + \cdots$. Improved results are obtained if the $3d$ subshell is treated as a separate shell, with $3s$, $3p$ considered as $n - 1$ relative to $3d$, and $1s$, $2s$, $2p$ treated as $n - 2$ relative to $3d$.

(a) From values of ζ_{nl} given in Table 4.9 and the definition of the screening constant s_{nl}, construct a table listing x, y, z, and s_{nl} for electrons in the ground-state atoms with $2 \leq Z \leq 36$.

(b) According to the Slater rules, the screening constant is independent of the number of electrons in shells exterior to the shielded electron under consideration. Calculate the standard deviation from the mean of s_{nl} for configurations with the same values of x, y, and z, but different numbers of external electrons. Would you expect the rules to apply better to valence (i.e., outermost shell) electrons or inner electrons?

(c) Considering the values of s_{nl} for only valence electrons, and assuming that $a_{nl} = a$, $b_{nl} = b$, and $c_{nl} = c$ are independent of n and l for $n \leq 3, l \leq 1$, find values of a, b, and c that give the best simultaneous fit for s_{1s}, s_{2s}, s_{2p}, s_{3s}, and s_{3p}. Is your value for c relative to that of b physically reasonable?

Ans. $a \simeq 0.366$, $b \simeq 0.854$, $c \simeq 0.711$. (Slater: $a = 0.35$, $b = 0.85$, $c = 1.00$)

(d) Is the value for s_{3d} in Sc$([\text{Ar}]4s^23d)$ as predicted by the values of b and c found in part (c) in satisfactory agreement with the value obtained from Table 4.9? Find a value of a_{3d} that gives a satisfactory fit to s_{3d} for various numbers of $3d$ electrons.

Ans. $a_{3d} \simeq 0.171$ (Slater: $a_{3d} = 0.35$, $b_{3d} = c_{3d} = 1.00$)

(e) Does the value of $a_{4s} = a_{4p} = a$, where a is the value found in part (c) satisfactorily represent the dependence of s_{4s} and s_{4p} upon the number of electrons with $n = 4$? What value of $b_{4s} = b_{4p} = c_{4s} = c_{4p} = b_4$ gives the best fit to s_{4s} for K([Ar]4s) and s_{4p} for Ga([Ar]$4s^2 3d^{10} 4p$)?

Ans. $b_4 \simeq 0.861$

(f) Summarize your results for the revised Slater rules. Tabulate values of s_{nl} for valence electrons obtained from Table 4.9, those obtained by your revised rules, and those obtained from Slater's original rules. To what extent are the Slater rules reliable?

5

Molecules and chemical bonds

When two atoms approach, they exert forces upon one another. At large distances the forces tend to be attractive, while at short distances they are repulsive. For some pairs of atoms the attractive forces are very weak, while for other pairs the attraction is strong enough to lead to the formation of stable molecules. In this chapter we examine the origin of the interatomic forces and the nature of chemical bonds in simple molecules.

5.1 CLASSIFICATION OF INTERACTIONS

Since atoms are composed of charged particles, we expect the interatomic forces to be largely electrostatic in nature. Accordingly, we anticipate that the interaction of two atoms will depend upon (1) the charge state of the atoms (i.e., whether they are neutral atoms or ions), and (2) the electronic structure of the atoms (or ions). It will be useful for our development of a theory of chemical bonding to classify atoms and ions according to whether their outermost electron shell is complete (closed-shell atoms) or incomplete (open-shell atoms). Examples of closed-shell atoms are neutral rare-gas atoms such as He, Ne, Kr, and ions with rare-gas structures such as Li^+, Na^+, Mg^{++}, F^-, Cl^-; most atoms, such as H, Na, Mg, C, F, are of the open-shell type.

It is easy to show that the dominant interactions of closed-shell atoms must differ from those of open-shell atoms by considering a simple example of each type. First, let us see what happens when two He atoms are brought together. Each atom has the closed-shell configuration $1s^2$ with a spherically symmetric electron cloud. As the atoms approach, the two clouds begin to overlap as indicated in Fig. 5.1. If the electron clouds were to remain as they are in the separate atoms, this would mean that the electrons of one atom get near the nucleus of the other atom, and become stabilized by the additional attraction. However, according to the Pauli principle, the K shell ($n = 1$) of each He atom can hold only two electrons; if the electrons of one atom are to be accommodated on the other atom, they must occupy orbitals belonging to a shell of higher energy (for He, the $n = 2$ shell). There results a distortion of the electron clouds which reduces the overlap that would violate the Pauli principle. The electrons in part occupy less stable orbitals and there is an over-all increase in energy; that is, the two He atoms repel one another.

The closed-shell repulsive interaction of He atoms can be contrasted with the interaction of two H atoms, both of which have an unfilled $1s$ shell. As the two H atoms are brought together the electron clouds overlap, and there is a possibility of one or the other or both electrons being near either nucleus. Since the H atom has an open-shell structure, there is no Pauli exclusion effect if the two electrons have opposite spin, and the electrons of both atoms can be accommodated in the $1s$ orbital of either.[1] Thus, in contrast to the He-He case, there is no destablizing "promotion" of electrons. As we shall see, the resulting interaction between two open-shell atoms is not necessarily repulsive and can be strongly attractive.

---

[1] If the spins are parallel, there would be a "closed-shell" repulsion similar to that for He-He.

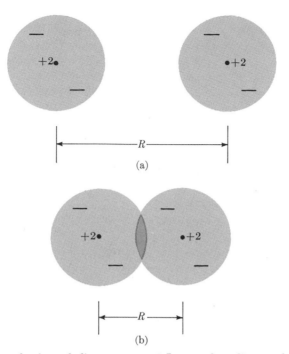

Fig. 5.1. Approach of two helium atoms. **(a)** Internuclear distance large, no overlap of electron clouds; **(b)** internuclear distance small, overlap of electron clouds.

From the discussion so far, it is evident that open- and closed-shell systems can be very different in their behavior. For an analysis of the chemical bond, therefore, it is convenient to make the following classification of interactions.

(1) Interaction between a closed-shell positive ion and a closed-shell negative ion; for example, Na^+ and Cl^-. This gives rise to an "ionic bond."

(2) Interaction between two closed-shell neutral atoms; for example, He-He. This is the "van der Waals interaction."

(3) Interaction between an open-shell ion (commonly positive) and one or more closed-shell ions (commonly negative). This occurs, for example, in transition-metal complexes of the type $M^{2+}(X^-)_6$.

(4) Interactions between two open-shell neutral atoms; for example, H-H. This gives rise to the "covalent bond."

Although the above classification is useful, it must be remembered that any such separation into different categories is somewhat arbitrary. Thus, a bond between H and F involves two open-shell atoms (type 4) but there are significant ionic contributions to the interaction energy (type 1).

5.2 THE IONIC BOND

In Chapter 4 we saw that the ionization potentials of the alkali metals are small, while the electron affinities of the halogens are large. Thus, the positive ions Li^+, Na^+, K^+, Rb^+, Cs^+, and the negative ions F^-, Cl^-, Br^-, I^- are relatively easy to make. When brought together, such positive and negative ions interact strongly and form a bond that is essentially ionic. As an example, we consider the sodium chloride molecule, Na^+Cl^-. The ionization process to form Na^+ requires that an energy equal to the ionization potential IP be added to the system, while the electron capture process to form Cl^- releases an amount equal to the electron affinity EA. Thus, we can write

$$Na(1s^22s^22p^63s) \rightarrow Na^+(1s^22s^22p^6) + e^- \quad \Delta E = 5.14\,eV$$
$$e^- + Cl(1s^22s^22p^63s^23p^5) \rightarrow Cl^-(1s^22s^22p^63s^23p^6) \quad \Delta E = -3.65\,eV$$
$$Na + Cl \rightarrow Na^+ + Cl^- \quad \Delta E = 1.49\,eV$$

(The EA value given here differs slightly from that for the ground-state in Table 4.2, since we require a weighted average for the terms contributing to the lowest electron configuration of Cl.) Even for this case involving an easily ionizable atom like Na and an atom with a high electron affinity like Cl, energy is required to form $Na^+ + Cl^-$; that is, 1.49 eV is needed for each ion pair formed, or 34.4 kcal mole^{-1} of ion pairs.

In calculating this energy, we have not considered the electrostatic attraction between the ions; that is, 1.49 eV is the energy required for the formation of the pair when there is an infinite separation between the two ions. As the ions are brought together, the ion-pair system is stabilized by the electrostatic interaction energy. Since Na^+ and Cl^- have spherically symmetric closed shells, we can for the moment ignore their electronic structure and calculate this energy by treating them as point charges. At an interionic separation R, sufficiently large to neglect overlap effects, we have

$$\Delta E(R) = -\frac{e^2}{R} + \Delta E(\infty) \tag{5.1}$$

where $\Delta E(\infty) = 1.49$ eV. As can be seen from Fig. 5.2, when R is decreased below 9.66 Å ($= 18.26\ a_0$), the energy difference ΔE becomes negative. This means that when the ion pair Na^+ and Cl^- is formed from the neutral atoms Na and Cl at a distance $R < 9.66$Å, energy is released.

At small interionic distances, the electron clouds of the two ions overlap as they do in the He-He case. Because the ions have closed-shell configurations, the Pauli principle requires that the additional electrons near each nucleus behave as though they occupied empty atomic orbitals of the ions; that is, the valence orbital $3s$, and the excited orbitals, $3p$, $3d$, . . . of

Na$^+$ and the excited orbitals $3d$, $4s$, $4p$, . . . of Cl$^-$. The energy of the ion pair thus increases with increasing overlap or "interpenetration" of the electron clouds. A repulsive-energy term must therefore be added to Eq. 5.1. Since atomic orbitals have an exponential dependence upon distance from the nucleus (see Table 3.1), the overlap, and hence the repulsive energy, is expected to depend exponentially upon R. Although the exact ex-

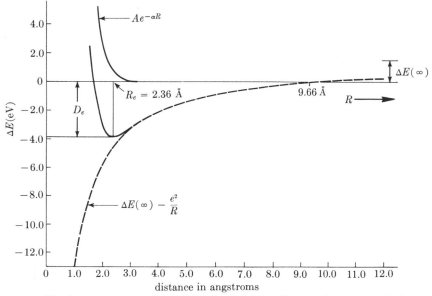

Fig. 5.2. The interaction energy of the ion pair Na$^+$—Cl$^-$ as a function of the inter-ionic distance.

pression is complicated, we can approximate it by a simple exponential of the form Ae^{-aR} and write the energy of the ion pair relative to that of the neutral atoms as

$$\Delta E(R) = Ae^{-aR} - \frac{e^2}{R} + \Delta E(\infty) \tag{5.2}$$

Here the constants A and a can be determined theoretically from the wave functions for the atoms or estimated experimentally by fitting known properties of the NaCl molecule or of the NaCl crystal. The potential-energy function in Eq. 5.2 has the form shown in Fig. 5.2. It follows Coulomb's law at large distances, is repulsive at small distances, and has a minimum at some intermediate point, the equilibrium internuclear distance R_e. At the point $R = R_e$ the attractive and repulsive forces are just balanced, so that the potential-energy curve has zero slope. If we take the value of ΔE at R_e to be $-D_e$, then D_e is the *bond energy*, that is, the amount of energy that would have to be supplied to separate the molecule initially fixed at R_e into neutral atoms at infinity:

$$NaCl \rightarrow Na + Cl \qquad \Delta E = E(Na) + E(Cl) - E(NaCl) = D_e$$

From Fig. 5.2, it is seen that to dissociate the NaCl molecule into the ions Na^+ and Cl^- requires the additional energy $(IP - EA) = 1.49$ eV:

$$NaCl \rightarrow Na^+ + Cl^- \quad \Delta E = E(Na^+) + E(Cl^-) - E(NaCl)$$
$$= D_e + IP - EA = D_e + 1.49 \text{ eV}$$

The distance R_e has been accurately measured by microwave spectroscopy (see Chapter 7); its value is 2.3609 Å. The bond energy D_e is rather difficult to determine directly for the NaCl molecule, although D_e has been measured spectroscopically to a few parts in ten thousand for other molecules (such as H_2 and I_2) with well developed vibrational bands (see Chapter 7). A value of $D_e = 4.22$ eV $= 97.3$ kcal mole^{-1} for NaCl can be derived from thermochemical data, but this result may be in error by a few hundredths of an eV ($\sim$1 kcal mole^{-1}). Another property of the potential-energy function that can be determined experimentally is the value of the second derivative of the interaction energy $\Delta E(R)$ at $R = R_e$, which is equal to the force constant k_e in the harmonic-oscillator approximation to the NaCl molecule (Eq. 2.110). For a harmonic oscillator the vibration frequency ν_e is related to k_e by the equation (see Eq. 2.114)

$$\nu_e = \frac{1}{2\pi} \left(\frac{k_e}{\mu} \right)^{1/2} \tag{5.3}$$

where μ is the reduced mass associated with the relative motion of the two atoms. The vibrational frequency is usually expressed in wave numbers (cm^{-1}) and given the symbol ω_e. Although Eq. 5.2 deviates from a harmonic-oscillator curve at large and small R, Eq. 5.3 is a good approximation for the spectroscopically observed frequency. For NaCl^{35}, $\nu_e = 1.093 \times 10^{13}$ sec^{-1} ($\omega_e = 364.6$ cm^{-1}) and $\mu = 2.30314 \times 10^{-23}$ g; thus, $k_e = 1.086 \times 10^5$ erg cm^{-2} (or 6.780 eV Å^{-2}). We now have three experimental numbers, R_e, D_e, and k_e, from which we can find the constants a and A in Eq. 5.2. Since we have more knowns than unknowns, the system is overdetermined for the simple two-parameter formula given in Eq. 5.2. We use the two more accurately known quantities R_e and ν_e to determine a and A, and then calculate D_e from Eq. 5.2 to check our treatment by comparing the result with the thermochemical value. Setting $d(\Delta E)/dR = 0$ at $R = R_e$ in Eq. 5.2, we obtain

$$\frac{d(\Delta E)}{dR}\bigg|_{R=R_e} = -aA\exp(-aR_e) + \frac{e^2}{R_e^2} = 0 \tag{5.4}$$

Solving Eq. 5.4 for A, we have

$$A = \left(\frac{e^2}{aR_e{}^2}\right) \exp(aR_e) \tag{5.5}$$

Making use of Eq. 5.5, we can write Eq. 5.2 in the form

$$\Delta E(R) = \left(\frac{e^2}{aR_e{}^2}\right) \exp\left[-a(R-R_e)\right] - \frac{e^2}{R} + \Delta E(\infty) \tag{5.6}$$

The second derivative of Eq. 5.6 at $R = R_e$ is equal to the force constant

$$\frac{d^2(\Delta E)}{dR^2}\bigg|_{R=R_e} = \left(\frac{e^2}{R_e{}^3}\right)(aR_e - 2) = k_e \tag{5.7}$$

From Eq. 5.7, we calculate aR_e to be

$$aR_e = 2 + \frac{k_e R_e{}^3}{e^2} = 2 + \frac{(1.086 \times 10^5)(2.3609 \times 10^{-8})^3}{(4.80298 \times 10^{-10})^2}$$
$$= 8.195$$

Thus

$$a = \frac{8.195}{2.3609} \text{ Å}^{-1} = 3.47 \text{ Å}^{-1}$$

The constant multiplying the exponential function in Eq. 5.6 can now be evaluated:

$$\frac{e^2}{aR_e{}^2} = \frac{(4.80298 \times 10^{-10})^2}{(8.195)(2.3609 \times 10^{-8})} \text{ erg} = 1.192 \times 10^{-12} \text{ erg}$$

Or, using the fact that

$$\frac{e^2}{R_e} = \left(\frac{e^2}{a_0}\right)\left(\frac{a_0}{R_e}\right) = \frac{(27.2107 \text{ eV})(0.529167 \text{ Å})}{(2.3609 \text{ Å})} = 6.099 \text{ eV}$$

we can write

$$\frac{e^2}{aR_e{}^2} = \frac{6.099}{8.195} \text{ eV} = 0.744 \text{ eV}$$

so that from Eq. 5.5 $A = 2700$ eV. Introducing these values for A and a into Eq. 5.2, we find for $R = R_e$

$$D_e = -\Delta E(R_e) = -A \exp(-aR_e) + \frac{e^2}{R_e} - \Delta E(\infty)$$
$$= -0.744 + 6.099 - 1.49 \text{ eV} = 3.87 \text{ eV}$$

Thus, the very simple ionic model of the NaCl molecule gives a binding energy that is only about 8% less than the thermochemical value. For other alkali halide molecules, correspondingly good results are obtained (see Problem C5.1). This shows that the most important interaction terms are included in Eq. 5.2, although the discrepancy between theory and experiment suggests that not all contributions to the binding have been taken into account. To obtain an idea of what we have neglected, we look in the next section at another property of the NaCl molecule.

5.3   DIPOLE MOMENTS AND POLARIZATION

In the simple ionic model, we have assumed that, except for the overlap repulsion, the charge clouds of Na^+ and Cl^- can be treated as point charges. To determine whether this is a valid picture, we need to look at a property of the NaCl molecule that is sensitive to the charge distribution. Since the energy is well approximated by Eq. 5.2, it clearly is not a good choice for such a test. However, another measurable physical property of the NaCl molecule, the electric dipole moment $\mathbf{\mu}_{el}$, is well suited for this purpose.

5.3.1   *Electric dipole moment*   The electric dipole moment of a system of charges $q_1, q_2, \ldots, q_n$ located at position vectors $\mathbf{r}_1, \mathbf{r}_2, \ldots, \mathbf{r}_n$ measured from any suitable origin is defined as

$$\mathbf{\mu}_{el} = q_1\mathbf{r}_1 + q_2\mathbf{r}_2 + \cdots + q_n\mathbf{r}_n \tag{5.8}$$

According to the simple ionic picture of the NaCl molecule, it consists of a sodium ion of charge $+q$ separated by a distance R_e from a chloride ion of charge $-q$. The electric dipole moment of such a system is a vector pointing collinearly with the charges and in the direction of the positive charge; its magnitude is

$$\mu_{el} = qR_e \tag{5.9}$$

From Eq. 5.9 the electric dipole moment has units of charge times distance (e.g., esu cm.). An alternative unit used for molecular dipole moments is the *Debye* (abbreviated D), after the eminent physical chemist Peter Debye (1884–1966), who made important contributions to the theory of molecular dipole moments as well as to many other areas of physical chemistry. The Debye is taken to be 10^{-18} esu cm, which is the order of magnitude of most molecular electric dipole moments, since the electronic charge is 4.8×10^{-10} esu and interionic distances are of the order 1 Å $= 10^{-8}$ cm.

An important quantity related to the dipole moment is the electric field generated by it, the so-called *dipole field*. For a point charge $\pm q$, we know that the electrostatic potential at a distance r is $\pm q/r$ and that the electric field is directed radially with strength $\mp q/r^2$. To derive an expression for the dipole field, we consider a point P located at a large distance r from the center O of a dipole $\mu = qR$, and let $\overline{OP}$ make an angle θ with the dipole

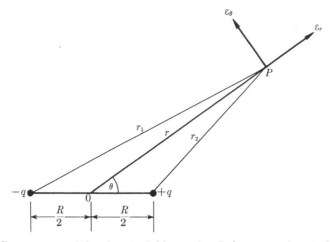

Fig. 5.3. Components of the electric field at point P due to an electric dipole.

axis (see Fig. 5.3). Since the dipole is composed of two point charges, the electrostatic potential at P is

$$\mathcal{V} = q\left(\frac{1}{r_2} - \frac{1}{r_1}\right) = q\left(\frac{r_1 - r_2}{r_1 r_2}\right) \tag{5.10}$$

where r_2 and r_1 are the distances to P, from the charges q and $-q$, respectively. Expressing this potential in terms of R, r, and θ, by the law of cosines, we have

$$r_1^2 = r^2(1 + \rho \cos\theta + \tfrac{1}{4}\rho^2) \tag{5.11a}$$
$$r_2^2 = r^2(1 - \rho \cos\theta + \tfrac{1}{4}\rho^2) \tag{5.11b}$$

where $\rho = R/r$ and $r \gg R$. If we take square roots of both sides of Eqs. 5.11 and expand the radicals $(1 \pm \rho \cos\theta + \tfrac{1}{4}\rho^2)^{1/2}$ about $\rho = 0$, we obtain

$$r_1 = r(1 + \tfrac{1}{2}\rho \cos\theta + \cdots) \tag{5.12a}$$
$$r_2 = r(1 - \tfrac{1}{2}\rho \cos\theta + \cdots) \tag{5.12b}$$

Keeping only the first two terms and introducing these expressions for r_1 and r_2 into Eq. 5.10, we find

$$\mathcal{V} = q\frac{(r_1 - r_2)}{r_1 r_2} \simeq \frac{qR \cos\theta}{r^2} = \frac{\mu \cos\theta}{r^2} \tag{5.13}$$

since for $r \gg R$, the product $r_1 r_2$ in the denominator can be approximated by r^2. The electric field can be resolved into a component $\mathcal{E}_r$ in the direction of r and a component $\mathcal{E}_\theta$ perpendicular to the r direction; the components are given by the appropriate derivatives of the electric potential:

to be sufficiently small so that it contributes negligibly to the dipole moment of the molecule. Considering only the Cl^- induced moment, we have from Eqs. 5.16 and 5.17

$$\mu_{ind} \simeq \mu_{Cl^-} = \alpha_{Cl^-}\mathcal{E}_{Cl^-} \simeq \frac{q\alpha_{Cl^-}}{R^2} \tag{5.18}$$

Using the value $\alpha_{Cl^-} = 2.974$ Å³ obtained from measurements of the index of refraction of alkali halide crystals, we find that at $R = R_e$

$$\mu_{ind} \simeq \frac{(4.80)(2.97)}{(2.36)^2} D = 2.56D$$

The numerical value is close to the difference of 2.34D between the simple ionic and the measured dipole moment of NaCl.

In addition to modifying the dipole moment, inclusion of polarization also lowers the energy of the molecule. This energy decrease can be obtained by calculating the work E_1 required to induce the dipole moment μ_{ind} in the chloride ion by the field (q/R^2) of the sodium ion (treated as a point charge at distance R), and the energy of interaction E_2 between the sodium ion and the induced dipole of the chloride ion. The work required to induce the dipole μ_{ind} is given by the integral

$$E_1 = \int_0^\mathcal{E} \mu_{ind}\, d\mathcal{E}' = \alpha_{Cl^-} \int_0^\mathcal{E} \mathcal{E}'\, d\mathcal{E}' = \frac{1}{2}\alpha_{Cl^-}\mathcal{E}^2$$

$$= \frac{1}{2}\alpha_{Cl^-}\left(\frac{q}{R^2}\right)^2 = \frac{1}{2}\frac{q^2\alpha_{Cl^-}}{R^4} \tag{5.19}$$

According to Eqs. 5.13 and 5.18 the potential at the Na^+ ion ($r = R, \theta = \pi$) due to the induced dipole is[1]

$$\mathcal{v} = -\frac{\mu_{ind}}{R^2} = -\frac{q\alpha_{Cl^-}}{R^4} \tag{5.20}$$

Thus, the interaction energy is

$$E_2 = q\mathcal{v} = -\frac{q^2\alpha_{Cl^-}}{R^4} \tag{5.21}$$

From Eqs. 5.19 and 5.21, the total polarization energy is

$$\Delta E_{pol} = E_1 + E_2 = -\frac{1}{2}\frac{q^2\alpha_{Cl^-}}{R^4} \tag{5.22}$$

If this polarization term is included in the interaction energy $\Delta E(R)$, the expression becomes (with q^2 replaced by its value e^2)

---

[1] If both ions are assumed to be polarizable, the dipole-dipole term $-\mu_{Na^+}\mu_{Cl^-}/qR^3$ must be added to Eq. 5.20 to give

$$\mathcal{v} = -\frac{\mu_{Na^+} + \mu_{Cl^-}}{R^2} - \frac{\mu_{Na^+}\mu_{Cl^-}}{qR^3}$$

$$\Delta E(R) = A e^{-aR} - \frac{e^2}{R} - \frac{1}{2}\frac{e^2 \alpha_{Cl^-}}{R^4} + \Delta E(\infty) \qquad (5.23)$$

Calculation of D_e from Eq. 5.23 after fitting a and A to R_e and ω_e gives the result 4.07 eV, a value somewhat closer to the thermochemical estimate of 4.22 eV. If polarization of both Na^+ and Cl^-, the van der Waals attraction (Section 5.4), and the additional induced dipole-induced dipole term in the expansion of ΔE_{pol} are all included (see Eq. 5.36), the calculated value is 4.09 eV. The remaining discrepancy of 0.1 eV can be attributed to the indistinguishability of the electrons and the overlap of their charge clouds, which we discuss in more detail later in connection with covalent bonding (Section 5.6).

5.3.3 Simple model for polarizability

In calculating the induced dipole moment, we introduced the polarizability as an empirical parameter. To obtain an idea of what determines its magnitude, we discuss a crude model of a polarizable atom. We simplify the structure of the atom by replacing the Coulomb attraction of the electrons to the nucleus with a one-dimensional Hooke's-law potential energy of the form $V = \frac{1}{2}kx^2$. The substitution of the correct potential by another is clearly arbitrary, except for the fact that both are attractive and that the force constant k can be adjusted so that the two potentials yield corresponding energies (see below).

When an external electric field $\mathcal{E}$ is applied in the x direction, the total force on an electron (with charge $-e$) is the sum of the forces due to the Hooke's-law potential and the electric field; that is,

$$F = -kx - e\mathcal{E}$$

The equilibrium position of the electron (i.e., center of the electron cloud relative to the position of the nucleus) is the value of x at which the force vanishes,

$$x_e = -\frac{e\mathcal{E}}{k}$$

The induced dipole moment is thus

$$\mu_{ind} = -ex_e = \frac{e^2}{k}\mathcal{E} \qquad (5.24)$$

Since $\mu_{ind} = \alpha\mathcal{E}$ by Eq. 5.17, Eq. 5.24 shows that the polarizability of the one-electron atom is

$$\alpha = \frac{e^2}{k} \qquad (5.25)$$

To estimate the order of magnitude of k for an atom, we assume arbitrarily that if the electron is "pulled out" to twice the value of r for which the where ε is the applied field and $\mathbf{P}$ is the *polarization* (dipole moment per unit volume) of the medium (see Chapter 17 of Reference 8 in the Additional Reading list). If there are N atoms (or nonpolar molecules) per unit volume, each with an induced dipole moment $\mathbf{\mu}_{ind}$, then

$$\mathbf{P} = N\mathbf{\mu}_{ind} = N\alpha\varepsilon_{eff} \tag{5.33}$$

where ε_{eff} is the effective electric field (i.e., the applied field plus the field due to the polarization of the medium). Lorentz showed that for cubic

Table 5.1 *Experimental alkali atom polarizabilities in units of* $Å^3$

Atom	ε deflection[a]	ε-$\mathscr{B}$ balance[b]
Li	22 ± 2	20 ± 3
Na	21.5 ± 2	20 ± 2.5
K	38 ± 4	36 ± 4.5
Rb	38 ± 4	40 ± 5
Ca	48 ± 6	52 ± 6.5

[a] G. E. Chamberlain and J. C. Zorn, *Phys. Rev.* **129**, 677 (1963).
[b] A. Salop, E. Pollack, and B. Bedersen, *Phys. Rev.* **124**, 1431 (1961).

lattices and for media such as dilute gases for which the distribution of the induced dipoles is random and their interaction can be neglected,

$$\varepsilon_{eff} = \varepsilon + \frac{4\pi}{3}\mathbf{P} \tag{5.34}$$

Upon combining Eqs. 5.32–5.34 and solving for α, the Lorentz-Lorenz equation is obtained (Problem 5.12):

$$\alpha = \frac{3}{4\pi N}\frac{n^2 - 1}{n^2 + 2} \tag{5.35}$$

Values of α for the rare gases determined from refraction measurements are given in Table 5.2.

Because ions or atoms of the same species have different electrical environments in different molecules or crystals, the polarizability of an ion

Table 5.2 *Polarizabilities of rare-gas atoms*[a]

Atom	α (Å^3)
He	0.2051
Ne	0.395
Ar	1.64
Kr	2.48
Xe	4.04

[a] From index of refraction; A. Dalgarno and A. E. Kingston, *Proc. Roy. Soc.* (London) **A259**, 424 (1961).

or atom varies somewhat with the substance. The dependence of α upon crystal structure is illustrated by the fact that $\alpha(\text{CsCl}) = 6.235$ Å^3 in a simple cubic lattice and 5.829 Å^3 in a body-centered cubic lattice. Nonetheless, a reasonably consistent set of values of polarizabilities of the ions can be obtained by a least-squares fit to the crystal data for all the alkali halides. Values of α_+ for the alkali metal ions and of α_- for the halide ions determined in this way are given in Table 5.3. The results listed are "static polarizabilities" obtained by extrapolating the refraction measurements to zero frequency; they are therefore applicable to the case of polarization by static electric fields.

Theoretical values of α can be calculated by treating the electric field as a perturbation and applying perturbation theory (which must include corrections to the wave function in this case) or variational techniques (see Problem 5.29). An approach that gives good accuracy is one in which the applied electric field is included in the Hamiltonian used for a Hartree-Fock calculation (see Section 4.8). Some polarizabilities obtained in this way are given in Table 5.4. For systems containing only two electrons such as H^-, Li^+, and He, C. Schwartz [*Phys. Rev.* **123**, 1700 (1961)] has calculated α from very accurate variational wave functions. His value for He of 0.2050 Å^3 agrees with the gas-phase refraction measurements. For Li^+, Schwartz obtains 0.0283 Å^3, compared with the value 0.0286 Å^3 derived from refraction measurements on alkali halides. Similarly, Schwartz' value of 30.4 Å^3 for H^- agrees quite well with an independent estimate of 30.2 Å^3 by S. Geltman; the Hartree-Fock value for H^- is only 11.8 Å^3, which indicates that the simple one-electron theory is inadequate for this case (see Problem 5.35).

Table 5.3 *Data for electrostatic model
of alkali halide molecules*

M	IP_1[a] (eV)	IP_2[a] (eV)	α_+[b] (Å³)
Li	5.390	75.619	0.0286
Na	5.138	47.29	0.255
K	4.339	31.81	1.201
Rb	4.176	27.5	1.797
Cs	3.893	25.1	3.137

X	EA[c] (eV)	$\overline{EA}$[d] (eV)	α_-[b] (Å³)
F	3.400 or 3.448	3.417 or 3.465	0.759
Cl	3.613	3.649	2.974
Br	3.363	3.515	4.130
I	3.063 or 3.076	3.377 or 3.390	6.199

[a] C. E. Moore, *Atomic Energy Levels,*
Natl. Bur. Std. (U.S.) Circ. No. 467 (U.S.
Government Printing Office, Washington, D.C., 1949).
[b] J. R. Tessman, A. H. Kahn, and W.
Shockley, *Phys. Rev.* **92,** 890 (1953);
from index of refraction of alkali halide
crystals.
[c] R. S. Berry and C. W. Reimann, *J. Chem.
Phys.* **38,** 1540 (1963).
[d] Weighted average for two terms ($^2P_{3/2}$,
$^2P_{1/2}$) contributing to lowest ns^2np^5 configuration of X.

5.3.5 The Rittner model for alkali halide molecules

An equation of the form
developed for the energy of the NaCl molecule is applicable to all of the
alkali halides. Generalizing Eq. 5.23 to include the van der Waals attraction (Section 5.4) and the polarizability contribution from both ions and
retaining terms only through $1/R^7$, one obtains (see Problem 5.2)

$$\Delta E(R) = A e^{-aR} - C_6 R^{-6} - \frac{e^2}{R}$$
$$- \frac{1}{2} e^2 \frac{(\alpha_+ + \alpha_-)}{R^4} - 2e^2 \frac{\alpha_+ \alpha_-}{R^7} + \Delta E(\infty) \tag{5.36}$$

Here α_+ and α_- refer to the polarizabilities of the positive and negative ion, respectively; a consistent set of experimental values for some ions are listed in Table 5.3. The constant A is evaluated for each ion pair from some molecular property, as described above (Section 5.2). By contrast the constant a, which gives the range of the repulsive interaction, can be chosen the same for all of the ions; the value $a = 2.96$ Å^{-1} is often used.

Instead of an exponential repulsion, a simple inverse power of R is sometimes introduced. Equation 5.36 then becomes

$$\Delta E(R) = \frac{A'}{R^n} - C_6 R^{-6} - \frac{e^2}{R} - \frac{1}{2} e^2 \frac{\alpha_+ + \alpha_-}{R^4} - 2e^2 \frac{\alpha_+ \alpha_-}{R^7} \tag{5.37}$$

where A' is a constant and n, the *Born repulsion parameter*, is usually set equal to an integer between 9 and 12. In addition to the charge-induced dipole interaction terms ($\sim \alpha/R^4$) for each ion, Eqs. 5.36 and 5.37 also contain an induced dipole-induced dipole term ($\alpha_+ \alpha_- / R^7$) which is usually small but not negligible. Correspondingly, the induced molecular dipole moment becomes

$$\mu_{\text{ind}} = \frac{e(\alpha_+ + \alpha_-)}{R^2} + \frac{4e\alpha_+ \alpha_-}{R^5} \tag{5.38}$$

E. S. Rittner [*J. Chem. Phys.* **19**, 1030 (1951)] has applied an electrostatic model based on Eq. 5.36 to all the alkali halide molecules. The values

Table 5.4 *Polarizabilities of atoms and ions calculated by the coupled self-consistent-field method*[a]

Atom	α (Å^3)	Atom	α (Å^3)
H$^-$	13.8	F$^-$	1.56
He	0.196	Ne	0.351
Li$^+$	0.0280	Na$^+$	0.140
Be^{2+}	0.00765	Mg^{2+}	0.0695
B^{3+}	0.00289	Al^{3+}	0.0393
C^{4+}	0.00132	Be	6.75

[a] H. D. Cohen and C. C. J. Roothaan, *J. Chem. Phys.* **43**, S34 (1965); H. D. Cohen, *ibid.* **43**, 3558 (1965); **45**, 10 (1966).

of D_e predicted by the model agree with the thermochemical values surprisingly well, considering the simplicity of the assumptions made (see Problems 5.15 and C5.1). Even the deviations from harmonicity, as measured by infrared absorption spectra (see Chapter 7), are reproduced quite accurately by the Rittner model. The dipole moments predicted by the model are somewhat lower than the measured values. This may indicate that the ionic polarizabilities are reduced slightly in the diatomic fields, which are higher than those in the crystals from which the values of α were obtained.

Modifications of the Rittner model give reasonable predictions for the geometry (planar) and binding energy ($\sim$45 kcal mole^{-1}) of the alkali halide dimers that are found in the gas phase. (For a simplified treatment of dimers which neglects polarization, see Problem 5.4.)

Numerical data required for the application of Eqs. 5.36–5.38 to the alkali halide molecules are given in Tables 5.3 and 5.5.

5.4   THE VAN DER WAALS ATTRACTION

In the previous section we considered a positive and a negative closed-shell ion and saw that the attractive Coulomb term provided sufficient energy to produce a stable molecule. Now we discuss the other type of closed-shell, closed-shell interaction which is of importance, namely, that of two *neutral* closed-shell atoms such as He.

Both the Coulomb term and the charge-induced-dipole interaction appearing in Eq. 5.23 for the ionic case are missing for neutral atoms. However, the repulsive term Ae^{-aR} in the expression for $\Delta E(R)$ is still present, because the overlap of the electron clouds when the two atoms approach has the same effect whether the atoms are charged (ions) or neutral. Since the exponential-like repulsive potential is the only term remaining in Eq. 5.23 [$\Delta E(\infty) = 0$ because we are considering the neutral atoms themselves], it is expected to determine the form of the interaction energy, particularly at small internuclear separations. As before, the exact values of the constants A and a in the exponential terms depend on the wave function and therefore on the species involved.

One can ask whether the ions formed by transferring an electron from one He atom to the other, yielding He$^+$, He$^-$ and having a Coulomb attraction corresponding to that in Na$^+$Cl$^-$, make a significant contribution here. We can easily see that they do not, since the energy required to form the ion pair is approximately the IP of He (the EA being essentially zero). Thus $\Delta E(\infty)$ equals $\sim$24.6 eV, which would be counterbalanced by the Coulomb attraction only at a distance of $\sim a_0$, at which the overlap repulsion is already very large (see Problems 5.6 and 5.9).

Careful measurements of pressure-volume-temperature relations for

Table 5.5 Properties of alkali halide molecules

M	X	C_6^a (eV Å^6)	C_8^a (eV Å^8)	R_e^b (Å)	ω_e^b (cm^{-1})	μ_{el} (D)	D_e^c (eV)	D_e^c (kcal mole^{-1})
Li	F	0.5	0.4	1.5639[b,d]	910.34[e]	6.3248[f]	5.99	138
	Cl	1.2	1.5	2.0207[g]	641[h]	7.1289[f,g]	4.85	112
	Br	1.6	2.1	2.1704	563[h]	7.268[i]	4.36	101
	I	2.1	3.3	2.3919	498[h]	6.25[i]	3.66	84
Na	F	2.8	2.4	1.9260	536.1	8.1558[k]	4.94	114
	Cl[35]	7.0	8.7	2.3609	364.6	9.0020[f]	4.22	97
	Br[79]	8.7	12	2.5020	298.5	9.1183[f]	3.74	86
	I	11.9	19	2.7114	259.2	9.2357[f]	3.43	79
K	F	12.2	13	2.1716	426.0	8.5926[l]	5.08	117
	Cl[35]	39[m]	46	2.6668	279.8	10.269[n,f]	4.37	101
	Br[79]	52[m]	62	2.8208	219.17	10.628[l]	3.92	90
	I	76[m]	97	3.0478	186.53	11.05[i]	3.45	80
Rb[85]	F	19	25	2.2704	373.3	8.5465[f]	5.02	116
	Cl[35]	49	84	2.7869	223.3	10.515[f]	4.31	99
	Br[79]	62	112	2.9447	169.46		3.89	90
	I	84	175	3.1768	138.51		3.31	76
Cs	F	32	49	2.3455	352.6	7.8839[f,o]	5.17	119
	Cl[35]	80.5	156	2.9064	214.2	10.387[f]	4.59	106
	Br[79]	102	212	3.0722	149.50		4.19	97
	I	140	324	3.3152	119.20	12.1[i]	3.57	82

[a] Van der Waals coefficients calculated from ultraviolet spectra of alkali halide crystals; J. E. Mayer, *J. Chem. Phys.* **1**, 270 (1933). For the form of the C_8 term, see problem C5.4.

[b] From microwave spectra. The ω_e values given here are those for the most abundant isotope of M and X; J. R. Rusk and W. Gordy, *Phys. Rev.* **127**, 817 (1962); P. L. Clouser and W. Gordy, *ibid.* **134**, A863 (1964); S. E. Veazey and W. Gordy, *ibid.* **138**, A1303 (1965), except where otherwise noted.

[c] Values of D_e for dissociation into neutral atoms are derived from values of ΔH_0° for dissociation into atomic ions given by L. Brewer and E. Brackett, *Chem. Rev.* **61**, 425 (1961), by correcting for zero-point energy, IP, and EA. Error limits are approximately ± 0.04 eV $= \pm 1$ kcal.

[d] L. Wharton et al., *J. Chem. Phys.* **38**, 1203 (1963); measured for Li^{6}F^{19}.

[e] G. L. Vidale, *J. Phys. Chem.* **64**, 314 (1960).

[f] A. J. Hebert, F. J. Lovas, C. A. Melendres, C. D. Hollowell, T. L. Story, Jr., and K. Street, Jr., *J. Chem. Phys.* **48**, 2824 (1968).

[g] D. D. Lide, Jr., P. Cahill, and L. P. Gold, *J. Chem. Phys.* **40**, 156 (1964); A. J. Hebert et al., as quoted by R. L. Matcha, *J. Chem. Phys.* **47**, 4595 (1967).

[h] W. Klemperer et al., *J. Chem. Phys.* **33**, 1534 (1960). Natural isotopic mixtures were used.

[i] A. J. Hebert, F. W. Breivogel, Jr., and K. Street, Jr., *J. Chem. Phys.* **41**, 2368 (1964); measured for Li^{6}Br^{79} and Li^{6}Br^{81}.

[j] A. Honig et al., *Phys. Rev.* **96**, 629 (1954).

[k] C. D. Hollowell, A. J. Hebert, and K. Street, Jr., *J. Chem. Phys.* **41**, 3540 (1964).

[l] R. van Wachem, F. H. deLeeuw, and A. Dymanus, *J. Chem. Phys.* **47**, 2256 (1967).

[m] D. W. Lynch, *J. Phys. Chem. Solids* **28**, 1941 (1967).

[n] R. van Wachem and A. Dymanus, *J. Chem. Phys.* **46**, 3749 (1967).

[o] Measured for Cs^{133}F^{17}.

helium gas show that there does exist a weak attractive component to the interaction. The fact that helium liquefies at 4.3°K further attests to the existence of an attractive force and to the fact that it is very weak. This force is often called the *van der Waals attraction* after J. D. van der Waals, who introduced the gas law which takes both attractive and repulsive forces into account. The attractive force is also sometimes referred to as the London *dispersion force* after Fritz London, who first gave a simple explanation for it in 1930. Since such forces exist between all atoms, we can dis-

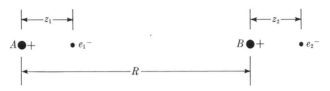

Fig. 5.5. Definition of distances for interaction of two one-dimensional, one-electron atoms.

cuss the origin of the dispersion force in terms of the interaction between a pair of one-electron atoms. Moreover, because the dispersion force involves the polarization of the atoms, we can adopt the simple, one-dimensional, harmonic-oscillator model of Section 5.3.3 for its description.

Let the two "one-dimensional" atoms approach to an internuclear distance R which is close enough for the electrons to interact with each other and the opposite nuclei, but not so close as to cause any appreciable repulsion due to the overlapping of the electron clouds. The interaction potential V' between the two atoms is the sum of the electrostatic potentials for the pairs of particles considered; that is, from Fig. 5.5

$$V'(z_1, z_2) = e^2 \left(\frac{1}{R} + \frac{1}{R + z_2 - z_1} - \frac{1}{R - z_1} - \frac{1}{R + z_2} \right) \quad (5.39)$$

where the internuclear axis is taken to be the z axis. In writing Eq. 5.39, we have included only the interaction terms between electron e_1 and nucleus B and between electron e_2 and nucleus A; the harmonic terms that we use to describe the binding of e_1 to A and e_2 to B have to be added to obtain the complete potential for the system. Thus, the total quantum-mechanical Hamiltonian for the one-dimensional motion along the internuclear axis is

$$H = -\frac{\hbar^2}{2m} \frac{\partial^2}{\partial z_1^2} + \frac{1}{2} k z_1^2 - \frac{\hbar^2}{2m} \frac{\partial^2}{\partial z_2^2} + \frac{1}{2} k z_2^2 + V'(z_1, z_2) \quad (5.40)$$

Since R is assumed to be large and each electron remains near its own nucleus due to the binding force, z_1 and z_2 are small on the average compared with R. We can therefore simplify the Hamiltonian by expanding $V'(z_1, z_2)$ about $z_1 = z_2 = 0$; that is,

$$V'(z_1, z_2) = e^2 \left\{ \frac{1}{R} + \left[R\left(1 + \frac{z_2 - z_1}{R}\right) \right]^{-1} - \left[R\left(1 - \frac{z_1}{R}\right) \right]^{-1} \right.$$
$$\left. - \left[R\left(1 + \frac{z_2}{R}\right) \right]^{-1} \right\}$$
$$= e^2 \left\{ \frac{1}{R} + \frac{1}{R}\left[1 - \left(\frac{z_2 - z_1}{R}\right) + \left(\frac{z_2 - z_1}{R}\right)^2 + \cdots \right] \right.$$
$$\left. - \frac{1}{R}\left[1 + \frac{z_1}{R} + \left(\frac{z_1}{R}\right)^2 + \cdots \right] - \frac{1}{R}\left[1 - \frac{z_2}{R} - \left(\frac{z_2}{R}\right)^2 + \cdots \right] \right\}$$
$$= - \frac{2e^2 z_1 z_2}{R^3} + \cdots$$

The terms in R^{-1} and R^{-2} have cancelled and we approximate V' by keeping only the first nonzero term. Thus, the Hamiltonian becomes

$$H = - \frac{\hbar^2}{2m}\left(\frac{\partial^2}{\partial z_1^2} + \frac{\partial^2}{\partial z_2^2}\right) + \frac{1}{2}k(z_1^2 + z_2^2) - \frac{2e^2}{R^3}z_1 z_2 \qquad (5.41)$$

In this simplified form, H is seen to be the Hamiltonian of a two-dimensional harmonic oscillator (the variables being z_1 and z_2) with a "cross term" involving the product of the two variables ($z_1 z_2$) added to the potential energy. Such a cross term can be eliminated by a change of variables. We introduce the new variables η and ξ by the definitions

$$\eta = z_2 - z_1 \qquad \xi = z_1 + z_2 \qquad (5.42)$$

Determining the expression for H in terms of η and ξ (Problem 5.32), we find

$$H = - \frac{\hbar^2}{2\mu}\frac{\partial^2}{\partial \eta^2} + \frac{1}{4}\left(k + \frac{2e^2}{R^3}\right)\eta^2 - \frac{\hbar^2}{2\mu}\frac{\partial^2}{\partial \xi^2} + \frac{1}{4}\left(k - \frac{2e^2}{R^3}\right)\xi^2 \qquad (5.43)$$

With the coordinates η and ξ, H is again a two-dimensional harmonic-oscillator Hamiltonian but now there is no cross term present; that is, each of the "dimensions" (corresponding to the variables ξ and η) can be treated as an independent oscillator of the type already studied in Section 2.10. We note from Eq. 5.43 that the effective mass μ for each oscillator is one-half as large as the electron mass m and that the effective force constant for one oscillator is $k_1 = \frac{1}{2}k(1 + 2e^2/kR^3)$, while that of the other oscillator is $k_2 = \frac{1}{2}k(1 - 2e^2/kR^3)$. Since the classical frequency of an oscillator with force constant k and reduced mass μ is (see Eq. 2.114),

$$\nu = \frac{1}{2\pi}\left(\frac{k}{\mu}\right)^{1/2} \qquad (5.44)$$

we have

$$\nu_1 = \frac{1}{2\pi}\left(\frac{k_1}{\mu}\right)^{1/2} = \frac{1}{2\pi}\left(\frac{k}{m}\right)^{1/2}\left(1 + \frac{2e^2}{kR^3}\right)^{1/2} \tag{5.45}$$

and

$$\nu_2 = \frac{1}{2\pi}\left(\frac{k_2}{\mu}\right)^{1/2} = \frac{1}{2\pi}\left(\frac{k}{m}\right)^{1/2}\left(1 - \frac{2e^2}{kR^3}\right)^{1/2} \tag{5.46}$$

For large R, we can expand the radicals in Eqs. 5.45 and 5.46 about $(1/R) = 0$; thus,

$$\left(1 \pm \frac{2e^2}{kR^3}\right)^{1/2} = \left[1 \pm \frac{e^2}{kR^3} - \frac{1}{8}\left(\frac{2e^2}{kR^3}\right)^2 + \cdots\right]$$

$$\nu_1 = \nu\left[1 + \frac{e^2}{kR^3} - \frac{1}{8}\left(\frac{2e^2}{kR^3}\right)^2 + \cdots\right] \tag{5.47}$$

and

$$\nu_2 = \nu\left[1 - \frac{e^2}{kR^3} - \frac{1}{8}\left(\frac{2e^2}{kR^3}\right)^2 + \cdots\right] \tag{5.48}$$

where ν is the frequency of the noninteracting oscillators (i.e., the oscillators of Eq. 5.41 with the $z_1 z_2$ term neglected).

As we saw in Section 2.10, the ground state of an oscillator has a zero-point energy of $\frac{1}{2}h\nu$. From Eqs. 5.47 and 5.48, this means that in the present case the zero-point energy is

$$E_0 = \frac{1}{2}h(\nu_1 + \nu_2) = h\nu - \frac{1}{2}h\nu\frac{e^4}{k^2R^6} + \cdots \tag{5.49}$$

The first term on the right-hand side of Eq. 5.49 corresponds to the ground-state of the harmonic-oscillator atoms when there is no interaction between them. The interaction energy, called the dispersion energy, is in the lowest-order approximation

$$\Delta E(R)_{\text{dis}} = -\frac{1}{2}h\nu\frac{e^4}{k^2R^6} \tag{5.50}$$

and is seen to be attractive. Since the polarizability of a Hooke's-law atom is given by Eq. 5.25, we can write Eq. 5.50 in the form

$$\Delta E(R)_{\text{dis}} = -\frac{1}{2}h\nu\frac{\alpha^2}{R^6} \tag{5.51}$$

Equation 5.51 demonstrates the important result that the dispersion energy varies as the square of the polarizability and as the inverse sixth power of the distance. The presence of the square of the polarizability suggests an interpretation in terms of the interaction of two so-called *fluctuating dipoles:* If one considers a He atom, which has, on the average, a zero dipole moment corresponding to its spherical charge distribution, one can say that at an instant of time the electron is located at a certain point relative to the nucleus and yields an instantaneous dipole moment. This moment induces a moment in the second atom; by analogy with the induced dipole-induced dipole interaction for the alkali halide molecules, the instantaneous correlated moments stabilize the system. Averaging over all possible instantaneous moments yields the total interaction energy. This is not expected to be large because it involves correlated fluctuations that are very small. In the simple treatment outlined above, we have restricted the oscillators to one-dimensional motion. If this restriction is removed (see Problem 5.7), the numerical factor is slightly larger because of the contribution of transverse motion and there results the expression

$$\Delta E(R)_{\text{dis}} = -\frac{3}{4} h\nu \frac{\alpha^2}{R^6} \tag{5.52}$$

For purely classical oscillators, the ground state would have corresponded to harmonic motion with zero amplitude; that is, the electrons would have the respective positions $z_1 = 0$ and $z_2 = 0$, and there would have been no interaction energy. Thus, the attractive interaction energy is a quantum-mechanical effect that arises as a direct result of the zero-point motion, which is a manifestation of the uncertainty principle.

A more realistic approximate treatment of the dispersion forces between two atoms gives the same result as Eq. 5.52, except that the oscillator energy $h\nu$ is replaced by the first IP of the atom; that is,

$$\Delta E(R)_{\text{dis}} = -\frac{3}{4} \frac{(\text{IP})\alpha^2}{R^6} \tag{5.53a}$$

For a pair of unlike atoms A and B, the formula becomes

$$\Delta E(R)_{\text{dis}} = -\frac{3}{2} \frac{(\text{IP}_A)(\text{IP}_B)}{\text{IP}_A + \text{IP}_B} \frac{\alpha_A \alpha_B}{R^6} \tag{5.53b}$$

This approximate equation was first derived by London; it is often written in the form $\Delta E(R)_{\text{dis}} = -C_6/R^6$, where C_6 is called the van der Waals coefficient.

The complete curve for the He-He interaction energy is obtained by combining the exponential repulsion Ae^{-aR} with the dispersion-force attraction; that is,

$$\Delta E(R) = Ae^{-aR} - \frac{C_6}{R^6} \tag{5.54}$$

To see the order of magnitude expected for the van der Waals attraction, we use Eq. 5.53a to calculate the constant factor C_6;

Table 5.6 *Repulsive potential parameters and van der Waals interaction constants for rare-gas atom pairs*

$$\text{Best } \Delta E(R) = Ae^{-aR} - C_6 R^{-6} \qquad \text{Best } C_6 \text{ for long-range interaction}$$

System	From transport data			Calculated C_6 (eV Å⁶)	Low-temperature transport data C_6 (eV Å⁶)	Low-velocity molecular beams	
	A (eV)	a (Å⁻¹)	C_6 (eV Å⁶)			C_5 (eV Å⁶)	$C_6/C_{6\mathrm{Ar}}$ [a]
He-He	1657[b]	5.05[b]	0.879[b]	0.879			
He-Ne	909	4.28	2.8	1.8			
He-Ar	1307	3.79	9.5	5.7			0.14
He-Kr	717	3.49	13.1	8			
He-Xe	1665	3.44	20.4	13			
Ne-Ne	4583	4.61	5.4	3.9	4.1		
Ne-Ar	6648	4.12	18.3	12			0.34
Ne-Kr	3604	3.82	25.4	16			
Ne-Xe	5865	3.76	39.5	28			
Ar-Ar	9575	3.62	62.0	38	40	36	1.00
Ar-Kr	5224	3.33	85.8	53		51	1.13
Ar-Xe	8521	3.27	134.	90			2.13
Kr-Kr	2854	3.03	119.	74	78	70	
Kr-Xe	4630	2.98	185.	126			
Xe-Xe	7555	2.92	287.	213	165		

[a] Directly measured ratios of the C_6 value for the listed atom pair to that for a pair of Ar atoms.

[b] A and a were chosen to give a good fit to the van der Waals minimum with C_6 equal to the accurate theoretical value.

series, with two exceptions, have the electron configuration $[KL]3s^2 3p^6$-$3d^n 4s^2$, where $[KL]$ represents the closed-shell Ne core, and the number of d electrons (n) ranges from 0 to 10. In this series the $4s$ subshell fills before the $3d$ subshell because shielding effects make the intrashell splitting of the $3p$ and $3d$ orbitals greater than the intershell splitting between $3p$ and $4s$. This appears to be true, at least, for K, Ca, and the first few elements in the transition series. However, as one continues across the periodic table, the stability of the $3d$ orbitals relative to $4s$ increases because the additional $3d$ electrons shield each other less well than they do the $4s$ electrons. Moreover, the increase in the effective nuclear charge leads to a contraction of the $3d$ orbitals which further increases their stability. Thus, in the later elements of the transition series the $3d$ and $4s$ orbital energies are very close to each other and an analysis of the ground-state configuration requires consideration of the electron-electron interaction terms as well.

When two electrons are removed to form the dipositive ions of the transition metals, the $4s$ level definitely lies above the $3d$ level. For a qualitative explanation of this reordering of the $3d$ and $4s$ levels, we consider the isoelectronic species Sc and V^{2+}. In the Sc atom, a $3d$ electron is shielded from the nucleus so much more than is a $4s$ electron, which penetrates the Ar core, that the effective nuclear charge "seen" by the $3d$ electron is less than that seen by a $4s$ electron; that is, the effect of intershell splitting between the $3d$ and $4s$ levels is overridden, the $4s$ level lies lower, and the ground configuration of Sc is thus $[Ar]3d4s^2$. In the isoelectronic V^{2+} ion, the nuclear charge is greater by two. The effective charges seen by a $4s$ and a $3d$ electron are thus greater than in Sc. Since the orbital energy is roughly proportional to $-Z_e^2/n^2$, the difference in shielding of a $3d$ and a $4s$ electron would now have to be larger to make the $4s$ orbital more stable than $3d$ (see Problem 5.37). Since the shielding for a given orbital is expected to be about the same in Sc and V^{2+} (compare Table 4.9), the $3d$ level lies below the $4s$ level, and the ground configuration for V^{2+} is $[Ar]3d^3$. Experiments have shown that the $3d$ level lies below the $4s$ level for all of the dipositive ions from Sc^{2+} through Zn^{2+}. As a result, the two $4s$ electrons are removed and the M^{2+} electron configuration is $[KL]\,3s^2 3p^6 3d^n$. In the special cases of Cr ($[KL]\,3s^2 3p^6 3d^5 4s$) and Cu ($[KL]\,3s^2 3p^6 3d^{10} 4s$), which have only one $4s$ electron in their ground configurations, the ions Cr^{2+} ($[KL]\,3s^2 3p^6 3d^4$) and Cu^{2+} ($[KL]\,3s^2 3p^6 3d^9$) are formed by removal of the $4s$ electron and one of the $3d$ electrons. Table 5.7 gives the values of n for the first transition series.

When a transition-metal ion is surrounded by several closed-shell negative ions (or dipolar molecules with negative poles toward the central metal ion) as in Fig. 5.7, we expect the dominant interaction between the positive and negative ions to correspond to that in the $Na^+ Cl^-$ molecule; that is, there is a strong electrostatic attraction between the central ion and each of

Table 5.7 *Number of 3d electrons for dipositive ions of the first transition series.*

Ion	Ca^{2+}	Sc^{2+}	Ti^{2+}	V^{2+}	Cr^{2+}	Mn^{2+}	Fe^{2+}	Co^{2+}	Ni^{2+}	Cu^{2+}	Zn^{2+}
n	0	1	2	3	4	5	6	7	8	9	10

the ligands and an overlap repulsion of the form Ae^{-aR}, which becomes important at small R. In addition, there is both electrostatic and overlap repulsion between ligands, though the ligand-ligand distances are usually sufficiently large that the overlap term is relatively unimportant.

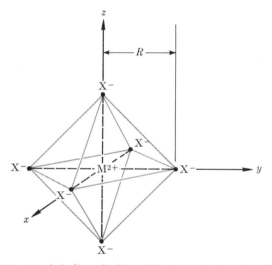

Fig. 5.7. Arrangement of six ligands (X^-) about a metal ion (M^{2+}).

If we assume that the variation in the central-ion, ligand distance R (see Fig. 5.7) as a function of the central ion for a given type of ligand (say, F^-) is dependent primarily upon the extension of the $3d$ orbitals of the central ion (which determines the onset of the overlap repulsion), the simple electrostatic model predicts a fairly regular decrease in R as we go across the series Sc^{2+} to Zn^{2+}. The $3d$ electrons are shielded from the nucleus mainly by the electrons in the closed inner shells and subshells, while the shielding from the other $3d$ electrons is incomplete; that is, according to the rules of Section 4.10, the effective nuclear charge Z_e for the $3d$ electrons increases monotonically with the atomic number Z. Since the most probable radius of the $3d$ electrons is $9a_0/Z_e$, the metal ions "shrink" as Z increases (Problem 5.36). This allows the ligands to approach closer to the central ion before the overlap repulsion becomes important. The observed dependence of R upon the central ion atomic number is shown in Fig. 5.8.

Corresponding to the decrease in internuclear distance in going from Sc^{2+} to Zn^{2+}, we would expect an increase in the electrostatic stabilization energy. On this basis, the total binding energy of the complexes should increase with Z as shown by the dashed line in Fig. 5.9. Although the general trend is in the expected direction, the actual binding energies tend to lie

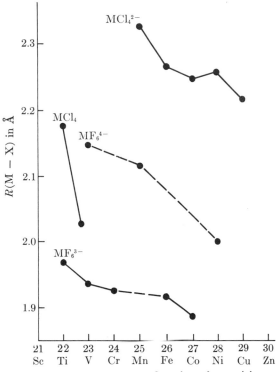

Fig. 5.8. Ion-ligand distance R for several series of transition-metal complexes [H. Basch, A. Viste, and H. B. Gray, *J. Chem. Phys.*, 44, 10 (1966).]

above this line, with the exception of the complexes of Ca^{2+}, Mn^{2+}, and Zn^{2+}. For example, the experimental hydration energies for $M(H_2O)_6^{2+}$ complexes are plotted in Fig. 5.9. It is clear that the binding energy as a function of Z is more complicated than the prediction of the simple ionic model. To explain this apparent anomaly, we must take account of the fact that the central ion does not have a closed-shell structure and look at the behavior of the open-shell $3d$ electrons in the electrostatic field due to the ligands.

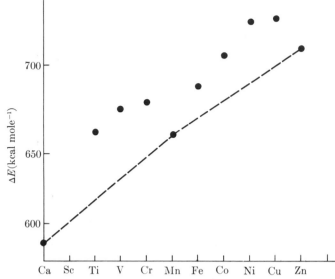

Fig. 5.9. Total energy released by the reaction $M^{2+} + 6H_2O \rightarrow [M(H_2O)_6]^{2+}$ for metal ions of the first transition series. (No octahedrally coordinated hydrate of Sc^{2+} has been prepared.)

We begin by considering a single $3d$ electron outside the closed-shell core $[KL]3s^2 3p^6$, as in Sc^{2+} and its octahedral complexes. The $3d$ electron could occupy any one of the orbitals $3d_{3z^2-r^2}$, $3d_{x^2-y^2}$, $3d_{xy}$, $3d_{yz}$, or $3d_{zx}$. In the free ion, the five orbitals are degenerate and the wave function of the electron could be any linear combination of these orbitals. When the ligands are introduced, their field partly lifts the degeneracy. In the ground state of the octahedral complex, the $3d$ electron will occupy whichever orbital has minimum energy. Comparison of Figs. 5.7 and 5.10 shows that the $d_{3z^2-r^2}$ and $d_{x^2-y^2}$ orbitals point along metal-ligand axes, while the d_{xy}, d_{yz}, and d_{zx} electron clouds point between metal-ligand axes. Since there is an electrostatic repulsion between an electron and the negative ligands (or the negative end of the dipole, in the case of neutral ligands), the energy of a $3d$ electron in any of the five orbitals will be greater (less negative) in the ligand field than in the free ion. However, because a $3d_{3z^2-r^2}$ or $3d_{x^2-y^2}$ is on the average closer to the ligands than a $3d_{xy}$, $3d_{yz}$, or $3d_{zx}$ orbital, an electron in the former is more destabilized than one in the latter. Figure 5.11 shows the splitting of the d electrons in the octahedral ligand field. The notation t_{2g} for the d_{xy}, d_{yz}, and d_{zx} degenerate lower-lying orbitals and e_g for the $d_{3z^2-r^2}$ and $d_{x^2-y^2}$ degenerate upper orbitals follows that of Mulliken; the letters t and e refer to the fact that the former are still triply degenerate, while the latter are doubly degenerate.

If we calculate the electrostatic interaction between six octahedrally

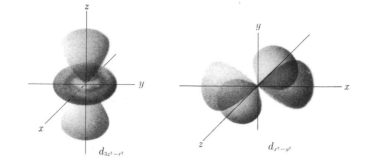

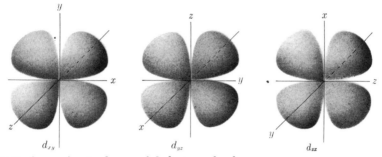

Fig. 5.10. Approximate shapes of d-electron clouds.

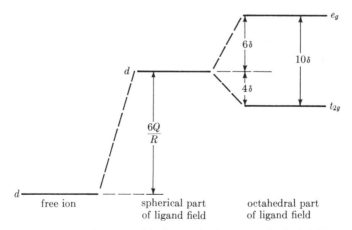

Fig. 5.11. Splitting of d-electron orbital energies in an octahedral field.

placed ligands each with point charge Q and an electron in one of the d orbitals, we find for the t_{2g} orbitals

$$V_t = \frac{6Q}{R}\left(1 - \frac{1}{9}\frac{\langle r^4 \rangle}{R^4}\right) \tag{5.56a}$$

and for the e_{2g} orbitals

$$V_e = \frac{6Q}{R}\left(1 + \frac{1}{6}\frac{\langle r^4 \rangle}{R^4}\right) \tag{5.56b}$$

Here $\langle r^4 \rangle$ is the average of r^4 over the normalized radial part of a $3d$ orbital; that is, for a hydrogen-like $3d$ orbital with effective charge Z_e, the result is

$$\langle r^4 \rangle = \frac{1}{6!}\left(\frac{2Z_e}{3a_0}\right)^7 \int_0^\infty \left[r^2 \exp\left(-\frac{Z_e r}{3a_0}\right)\right] r^4 \left[r^2 \exp\left(-\frac{Z_e r}{3a_0}\right)\right] r^2 \, dr$$
$$= 25515 \left(\frac{a_0}{Z_e}\right)^4 \tag{5.57}$$

Equations 5.56 and 5.57 show that for typical values of $R(\sim 4a_0)$ and $Z_e(\sim 10)$, the spherically symmetric Coulomb repulsion between the electron and the six point charges dominates their interaction. Added to this is the smaller nonspherical term which splits the degeneracy of the t- and e-type orbitals, namely,

$$\Delta E(t_{2g}) = V_t - \frac{6Q}{R} = -\frac{2Q}{3R^5}\langle r^4 \rangle$$
$$\Delta E(e_g) = V_e - \frac{6Q}{R} = \frac{Q}{R^5}\langle r^4 \rangle \tag{5.58}$$

That the d_{xy}, d_{yz}, and d_{xz} orbitals all have the same energy is evident from their equivalent distribution with respect to the octahedral ligands; the degeneracy of $d_{3z^2-r^2}$ and $d_{x^2-y^2}$ is not obvious from the form of the orbitals but follows from a calculation (see Problem 5.40). If all five $3d$ orbitals are doubly occupied, the total charge cloud is spherically symmetric (if distortion due to polarization by the ligand ions is neglected) and the energy per electron must reduce to the Coulomb term $6Q/R$. This implies that the total stabilization energy of the six t_{2g} electrons, namely, $6 \times \Delta E(t_{2g})$, must equal the total "destabilization" energy of the four e_g electrons, namely, $4 \times \Delta E(e_g)$. Thus the ratio of the two energies is expected to be $4:6$, in agreement with Eq. 5.58. A parameter δ, such that the $\Delta E(e_g) - \Delta E(t_{2g}) = 10\delta$, is often introduced to describe the magnitude of the splitting. From Eq. 5.58, $\delta = \frac{1}{6}(Q/R^5)\langle r^4 \rangle$ and clearly depends both on the form of the $3d$ function and on the distance R.

Having determined the relative energies of the orbitals in the octahedral field, we proceed as in the atomic problem and find the ground state of the complex formed from the ions Ca^{2+} through Zn^{2+} by placing the $3d$ electrons in the most stable orbitals, taking into account the Pauli principle. From the diagram in Fig. 5.11, we can then calculate the net ligand-field stabilization energy (LFSE) for each ion. The result for the first four ions is given in Table 5.8. Note that the electrons have been placed in separate orbitals and that the electrons have been assigned parallel spins (as indi-

Table 5.8 Octahedral complexes

Ion	Ca²⁺	Sc²⁺	Ti²⁺	V²⁺
n	0	1	2	3
e_g	— —	— —	— —	— —
t_{2g}	— — —	↑ — —	↑ ↑ —	↑ ↑ ↑
LFSE	0	4δ	8δ	12δ

cated by the direction of the arrows which represent occupancy of an orbital by an electron). This is in accordance with Hund's rule for the atoms, namely, that the state of maximum spin for a given configuration corresponds to the ground state (Section 4.6.2).

When we reach Cr²⁺, we must decide whether to pair the fourth electron with one of the electrons in the singly occupied t_{2g} orbitals, or to place it in an empty e_g orbital with a spin parallel to that of the other three electrons. If there were no ligand field to split the $3d$ levels, the latter choice would be appropriate, in accordance with Hund's rule. On the other hand, if the splitting between the two levels (10δ) is very large, placing the fourth electron in one of the t_{2g} orbitals would give a lower energy. Thus one must ascertain whether there is a *strong* ligand-field (whether 10δ is more important than the electron-electron repulsion energy) or a *weak* ligand field (10δ is less important than the electron-electron repulsion energy). Both possibilities are found, depending upon the nature of the ligands and the charge state of the central ion. As we shall see by completing the present analysis and comparing with Fig. 5.9, the ions $[M(H_2O)_6]^{2+}$ are weak-field complexes.

Assigning the remaining $3d$ electrons on the assumption that $[M(H_2O)_6]^{2+}$ is a weak-field case, we have the results given in Table 5.9.

Table 5.9 *Weak-field* octahedral complexes

Ion	Cr²⁺	Mn²⁺	Fe²⁺	Co²⁺	Ni²⁺	Cu²⁺	Zn²⁺
n	4	5	6	7	8	9	10
e_g	↑ —	↑ ↑	↑ ↑	↑ ↑	↑ ↑	↑↓ ↑	↑↓ ↑↓
t_{2g}	↑ ↑ ↑	↑ ↑ ↑	↑↓ ↑ ↑	↑↓ ↑↓ ↑	↑↓ ↑↓ ↑↓	↑↓ ↑↓ ↑↓	↑↓ ↑↓ ↑↓
LFSE	6δ	0	4δ	8δ	12δ	6δ	0

A plot of LFSE (in units of δ) versus n, the number of $3d$ electrons, is shown for the octahedral weak-field case in Fig. 5.12. When this stabilization is superimposed upon the dashed line of Fig. 5.9 the experimental points are qualitatively accounted for. In particular, we note that the three spherically symmetric ions [$Ca^{2+}(d^0)$, $Mn^{2+}(d^5)$, $Zn^{2+}(d^{10})$] have no additional crystal field stabilization and so fall on the line corresponding to the simple Z dependence of the binding energy.

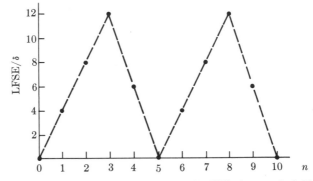

Fig. 5.12. Ligand field stabilization energies (LFSE) for weak-field octahedral complexes.

It is clear from Tables 5.8 and 5.9 that only for d^4 through d^7 configurations is there a difference between the strong-field and weak-field cases; for all other configurations (d^1 through d^3, d^8 through d^{10}), there is no competition between the crystal field and the electron repulsion. If the field is weak, the electrons go into different orbitals and keep their spins parallel subject to the limits set by the Pauli principle. This means that the total spin of the complex has its maximum value (e.g., for d^4, we have $S = 2$) and the system is often described as a high-spin complex. Correspondingly, the strong-field case leads to pairing of the electrons and results in so-called low-spin complexes.

In attempting a quantitative description of crystal field splittings, it must be borne in mind that δ is a function of the strength of the ligand field, which in turn depends upon the nature of both the central ion and the ligands. The best experimental determinations of δ come from analyses of the absorption spectra of the complexes (see Chapter 7) rather than from the heats of formation. In many cases, the absorption band with the longest wavelength corresponds approximately to the transition of a single electron from the t_{2g} level to the e_g level, so that 10δ can be determined directly; in others, the situation is more complicated but a detailed interpretation of the spectrum still yields a value for 10δ. When the spectroscopically determined LFSE is subtracted from the hydration energies of Fig. 5.9, the resulting points lie very close to the dashed line.

From studies of different metal ions with a variety of ligands, it is found that, although the specific value of δ depends on both the metal ion and the ligand, it is possible to arrange the ligands in a series corresponding to increasing δ values independent of the metal ion. Some members of the series, called the *spectrochemical series*, are listed in order of increasing δ:

$$I^- < Br^- < Cl^- < F^- < OH^- < H_2O < NH_3 < NO_2^- < CN^-$$

Although qualitative arguments for the relative position of an ion or molecule in this series can be given, a detailed quantitative interpretation does not exist. For any given ligand, the tripositive metal ions (M^{3+}) have larger crystal-field splittings than the dipositive ions (M^{2+}) since the internuclear distance is shorter for the former than for the latter. Some typical complexes with their δ values and an indication of whether they are high spin or low spin are listed in Table 5.10.

An interesting point is raised by the Cu^{2+} complexes, which have the electronic configuration $(t_{2g})^6(e_g)^3$ according to Table 5.9. Since there are three electrons in the e_g orbitals, their configuration could be $(d_{x^2-y^2})^2$

Table 5.10 *Values of δ for some octahedral transition-metal complexes.*[a]

	6F⁻	6Cl⁻	6Br⁻	6O²⁻	6CN⁻	6H₂O	6NH₃
Ti³⁺	0.217	0.171	0.161			0.252	
V³⁺	0.197	0.172				0.228	
V²⁺	0.149					0.154	
Cr³⁺	0.188	0.171	0.164	0.201	0.331	0.216	0.268
Mn³⁺	0.270						
Mn²⁺	0.104	0.093					
Fe³⁺	0.174						
Fe²⁺					0.402*		
Co³⁺	0.162				0.415*	0.226*	0.285*
Co²⁺				0.119		0.115	0.125
Ni²⁺	0.091	0.091	0.087			0.105	0.134
Cu²⁺						0.156	0.187
Mo³⁺		0.238					
Rh³⁺		0.252*	0.234*			0.335*	0.420*
Ir³⁺		0.309*	0.286*				
Pt⁴⁺		0.360*	0.298*				

[a] Values of δ are in eV; low-spin (strong-field) complexes are indicated by an asterisk. Reported by H. Basch, A. Viste, and H. B. Gray, *J. Chem. Phys.* **44**, 10 (1966); and L. E. Orgel, *An Introduction to Transition-Metal Chemistry: Ligand-Field Theory* (Wiley, New York, 1966), 2nd ed.

$d_{3z^2-r^2}$ or $(d_{3z^2-r^2})^2 d_{x^2-y^2}$, or any linear combination of the two; that is, for an octahedral complex the two configurations are degenerate. There is a general theorem due to H. Jahn and E. Teller (the so-called Jahn–Teller theorem) that demonstrates that any nonlinear polyatomic molecule in which there is orbital degeneracy will distort so as to remove that degeneracy. Such distortions are shown in Fig. 5.13. In the present case, the

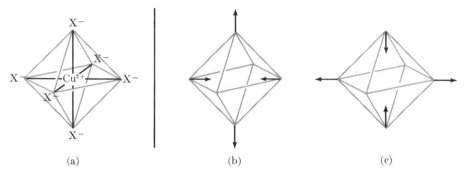

(a)　　　　　　　　　　　(b)　　　　　　　　　　　(c)

Fig. 5.13. Distortions of Cu^{2+} complexes. **(a)** Undistorted octahedron; **(b)** extension; **(c)** compression.

degeneracy can be removed most simply by elongating the complex (Fig. 5.13b), which stabilizes the $d_{3z^2-r^2}$ orbital relative to $d_{x^2-y^2}$, or by compressing the complex (Fig. 5.13c), which stabilizes the $d_{x^2-y^2}$ orbital relative to $d_{3z^2-r^2}$. Although the Jahn–Teller theorem does not say which of the two possibilities will occur, structural studies suggest that case (b) is applicable to some six-coordinated Cu complexes. In the presence of other additional cations, however, the distortion can correspond to case (c); for example, it is observed for K_2CuF_4, where the four long-bonded F^- ions are shared between two Cu^{2+} ions, while the two short-bonded F^- ions are bonded to only one Cu^{2+} ion (the K^+ ions are interstitial). An extreme case of the type-(b) distortion is that in which the two ligands along the z axis are removed completely, yielding so-called "square planar" complexes; examples are $(PtCl_4)^{2-}$ and $[Ni(CN)_4]^{2-}$.

5.5.1 Tetrahedral complexes For complexes with the ligands in tetrahedral positions (Fig. 5.14), the $3d$ orbitals are again split into two sets, corresponding to $d_{3z^2-r^2}$, $d_{x^2-y^2}$ (designated e) and d_{xy}, d_{yz}, d_{zz} (designated t_2). However, in contrast to the octahedral case, the $d_{3z^2-r^2}$, $d_{x^2-y^2}$ pair now have the lower energy (see Fig. 5.15). This result is seen to be qualitatively reasonable from a comparison of ligand positions for octahedral and tetrahedral complexes shown in Fig. 5.14; in the tetrahedral case, it is the $d_{3z^2-r^2}$ and $d_{x^2-y^2}$ clouds which "avoid" the ligands. The strong- and weak-field configuration and LFSE are obtained in the same manner as for octa-

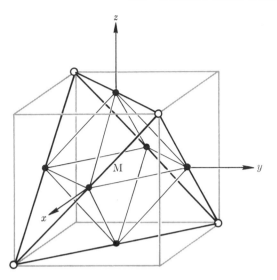

Fig. 5.14. Comparison of octahedral and tetrahedral ligand positions; ($\bullet$) octahedral positions; ($\circ$) tetrahedral positions.

hedral complexes (Problem 5.21). Some known examples of tetrahedral complexes are listed in Table 5.11.

The simple electrostatic picture of transition-metal complexes is a very useful one that provides a qualitative interpretation of data for a wide

Table 5.11 *Values of δ for some tetrahedral transition-metal complexes*[a]

	$4Cl^-$	$4Br^-$	$4O^{2-}$	$4S^{2-}$
Ti^{4+}	0.108	0.094		
V^{4+}	0.112			
V^{3+}	0.069			
Cr^{4+}			0.322	
Mn^{7+}			0.322	
Mn^{6+}			0.236	
Nn^{5+}			0.183	
Mn^{2+}	0.045			
Fe^{3+}	0.062			
Fe^{2+}	0.050			
Co^{2+}	0.046	0.038	0.047	0.040
Ni^{2+}	0.043			

[a] Values of δ are given in eV; reported by H. Basch, A. Viste, and H. B. Gray, *J. Chem. Phys.* **44**, 10 (1966).

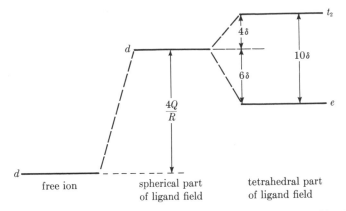

Fig. 5.15. Splitting of d-electron orbital energies in a tetrahedral field. The tetrahedral δ is approximately half that of the corresponding octahedral complex with the same M—X distances.

variety of systems. However, detailed studies of the electron distribution and certain other properties (e.g., spin-orbit coupling) show that a more refined treatment, which includes the structure of the ligands, is frequently required for an adequate description. Since this introduces complications comparable to those arising in the interaction between two open-shell systems we defer our discussion until we have had a chance to study simple examples of the latter type. It is important to note, however, that the concept of stabilization energies is left largely intact by the more sophisticated treatment (see Section 6.4.8).

5.6 OPEN-SHELL INTERACTIONS

The interaction of two atoms, both of which have unfilled valence shells, leads to the most common type of chemical bond. As the simplest example, we examine the formation of a hydrogen molecule from two ground-state hydrogen atoms, each of which has only one electron in its valence-shell $1s$ orbital. As the two nuclei approach one another, their electron clouds overlap. If the electron spins are paired, the electron of each of the atoms can occupy the $1s$ orbital of the other, without violating the Pauli principle, and thereby be near both nuclei. This results in an electrostatic interaction between each of the electrons and the two nuclei that is attractive relative to the interaction between an electron and the single nucleus of an isolated atom. Thus, when two open-shell atoms come together, there is a significant stabilization in contrast to the dominantly repulsive interaction between closed-shell He atoms.

The discussion of open-shell atoms leads naturally to the idea of "sharing" or "exchanging" of electrons; that is, each electron for part of the time is associated with each of the nuclei. This concept, which is the quantum-mechanical analog of the Lewis "shared-pair" or "covalent" bond (H : H), serves as the basis for our understanding of the binding energy of hydrogen and of many other molecules.

To develop a more quantitative description, we write down the potential-energy function for two hydrogen atoms, a system consisting of two protons a distance R apart and two electrons (see Fig. 5.16). The total electrostatic

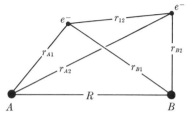

Fig. 5.16. Interparticle distances in the H_2 molecule.

potential-energy function is the sum of the six pair-wise attractions and repulsions,

$$V = - \frac{e^2}{r_{A1}} - \frac{e^2}{r_{B2}} - \frac{e^2}{r_{A2}} - \frac{e^2}{r_{B1}} + \frac{e^2}{r_{12}} + \frac{e^2}{R} \qquad (5.59)$$

The complete Hamiltonian for the system includes, in addition to V of Eq. 5.59, the kinetic-energy terms for each of the four particles. Since we are interested here primarily in the electronic properties of the system, we neglect the nuclear motion by assuming the nuclei to be fixed at the internuclear distance R; as we shall point out later (see Section 7.3), this is a valid approximation and the motion of the nuclei can be treated after the electronic problem for fixed nuclei has been solved. The fixed-nucleus Hamiltonian is

$$H = - \frac{\hbar^2}{2m} (\nabla_1^2 + \nabla_2^2) + V \qquad (5.60)$$

where V is given by Eq. 5.59. Since the Schrödinger equation corresponding to the Hamiltonian in Eq. 5.60 cannot be solved exactly, we proceed—as we did for the He atom—by examining a number of different approximations.

At very large values of R, one electron is found almost exclusively near nucleus A, while the other electron is found almost exclusively near nucleus B. Labelling these electrons 1 and 2, respectively, we may neglect all but the first two terms of V (Eq. 5.59) and write the Hamiltonian for the two (noninteracting) atoms as

$$H_0 = -\frac{\hbar^2}{2m} \nabla_1^2 - \frac{e^2}{r_{A1}} - \frac{\hbar^2}{2m} \nabla_2^2 - \frac{e^2}{r_{B2}} \qquad (5.61)$$

The corresponding Schrödinger equation

$$H_0 \psi(1, 2) = E\psi(1, 2) \qquad (5.62)$$

is easily solved since H_0 is the sum of two one-electron atomic Hamiltonians; the first two terms in Eq. 5.61 correspond to a hydrogen atom composed of electron 1 on nucleus A and the second two terms to a hydrogen atom composed of electron 2 on nucleus B. The ground-state solution is the product of the ground-state wave functions for the two hydrogen atoms, namely,

$$\psi(1, 2) = 1s_A(1)\, 1s_B(2) \qquad (5.63)$$

where

$$
\begin{aligned}
1s_A(1) &= (\pi a_0^3)^{-1/2} \exp\left(-\frac{r_{A1}}{a_0}\right) \\
1s_B(2) &= (\pi a_0^3)^{-1/2} \exp\left(-\frac{r_{B2}}{a_0}\right)
\end{aligned}
\qquad (5.64)
$$

Since $1s_A(1)$ and $1s_B(2)$ are each ground-state eigenfunctions for a hydrogen atom, we have

$$
\begin{aligned}
H_0\, 1s_A(1)\, 1s_B(2) &= \left(-\frac{\hbar^2}{2m}\nabla_1^2 - \frac{e^2}{r_{A1}}\right) 1s_A(1)\, 1s_B(2) \\
&+ \left(-\frac{\hbar^2}{2m}\nabla_2^2 - \frac{e^2}{r_{B2}}\right) 1s_A(1)\, 1s_B(2) \\
&= 1s_B(2)\left(-\frac{\hbar^2}{2m}\nabla_1^2 - \frac{e^2}{r_{A1}}\right) 1s_A(1) \\
&+ 1s_A(1)\left(-\frac{\hbar^2}{2m}\nabla_2^2 - \frac{e^2}{r_{B2}}\right) 1s_B(2) \\
&= 2E_H\, 1s_A(1)\, 1s_B(2)
\end{aligned}
\qquad (5.65)
$$

that is, the total energy is equal to that of two hydrogen atoms.

As the atoms move close enough to each other to interact, one contribution to the energy comes from the van der Waals attraction. Since it is a small term compared with the measured bond energy of H_2, we ignore it. The next approximation to the interaction energy can be obtained by perturbation theory. We write the Hamiltonian in the form

$$H = H_0 + V_I \qquad (5.66)$$

where H_0 is the Hamiltonian appropriate for infinite R given by Eq. 5.61, and V_I is the *interaction operator*

$$V_I = e^2 \left(-\frac{1}{r_{B1}} - \frac{1}{r_{A2}} + \frac{1}{r_{12}} + \frac{1}{R} \right) \qquad (5.67)$$

whose contribution goes to zero as R becomes large. Approximating the true wave function for the interacting atoms by Eq. 5.63, the solution corresponding to H_0, we can use Eq. 5.65 to write

$$H \, 1s_A(1) \, 1s_B(2) = 2E_H \, 1s_A(1) \, 1s_B(2) + V_I \, 1s_A(1) \, 1s_B(2) \qquad (5.68)$$

Multiplying both sides of the equation by $1s_A(1) \, 1s_B(2)$ and integrating over the coordinates of electrons 1 and 2, we find

$$\begin{aligned} \int &1s_A(1) \, 1s_B(2) \, H \, 1s_A(1) \, 1s_B(2) \, dv_1 \, dv_2 \\ &= 2E_H \int [1s_A(1)]^2 \, [1s_B(2)]^2 \, dv_1 \, dv_2 + \int [1s_A(1)]^2 \, V_I \, [1s_B(2)]^2 \, dv_1 \, dv_2 \end{aligned} \qquad (5.69)$$

The first integral on the right-hand side of Eq. 5.69 can be factored to give

$$\int [1s_A(1)]^2 \, [1s_B(2)]^2 \, dv_1 \, dv_2 = \int [1s_A(1)]^2 \, dv_1 \int [1s_B(2)]^2 \, dv_2$$

since the integrand is a product of two one-electron functions. It is clear that the integral is merely a product of the normalization integrals for the $1s_A$ and $1s_B$ orbitals. Since these orbitals are already normalized, the integrals are both unity, and Eq. 5.69 reduces to

$$\begin{aligned} \int &1s_A(1) \, 1s_B(2) \, H \, 1s_A(1) \, 1s_B(2) \, dv_1 \, dv_2 \\ &= 2E_H + \int [1s_A(1)]^2 \, V_I \, [1s_B(2)]^2 \, dv_1 \, dv_2 \end{aligned} \qquad (5.70)$$

that is, the average value of the complete Hamiltonian for the wave function $1s_A(1) \, 1s_B(2)$ is equal to the energy of the two separated hydrogen atoms plus the average of the interaction operator V_I for the assumed form of the wave function. Equation 5.70 has the usual perturbation-theory form with H_0 as the unperturbed Hamiltonian and V_I as the perturbation (Section 4.1.5). The perturbation integral does not factor into one-electron integrals since V_I includes the electron-electron repulsion term e^2/r_{12}, which couples the motion of the two electrons. However, we can write the integral as a sum of terms:

$$\begin{aligned} \int [1s_A(1)]^2 \, V_I \, [1s_B(2)]^2 \, dv_1 \, dv_2 &= \frac{e^2}{R} \int [1s_A(1)]^2 \, [1s_B(2)]^2 \, dv_1 \, dv_2 \\ &\quad - \int \frac{e^2}{r_{B1}} [1s_A(1)]^2 \, [1s_B(2)]^2 \, dv_1 \, dv_2 \end{aligned} \qquad (5.71)$$

$$- \int \frac{e^2}{r_{A2}} [1s_A(1)]^2 \, [1s_B(2)]^2 \, dv_1 \, dv_2 + \int \frac{e^2}{r_{12}} [1s_A(1)]^2 \, [1s_B(2)]^2 \, dv_1 \, dv_2$$

In the first three terms on the right-hand side of Eq. 5.71, we can evaluate the integrals over the electron coordinates not involved in the operator and obtain

$$\int 1s_A(1)^2 \, V_I \, 1s_B(2)^2 \, dv_1 \, dv_2 = \frac{e^2}{R} - \int \frac{e^2}{r_{B1}} 1s_A(1)^2 \, dv_1$$

$$- \int \frac{e^2}{r_{A2}} 1s_B(2)^2 \, dv_2 + \int 1s_A(1)^2 \frac{e^2}{r_{12}} 1s_B(2)^2 \, dv_1 \, dv_2 \tag{5.72}$$

The first and last terms on the right-hand side of Eq. 5.72 correspond to the nucleus-nucleus and the electron-electron interaction, respectively; they are both positive and therefore make a repulsive contribution to the interaction energy. It is the second and third integrals, arising from the interaction of the electron of one nucleus with the other nucleus, that are attractive. To find the total interaction energy, the integrals must be evaluated; this can be done by standard methods described in texts on quantum mechanics (see References 6–10 of the Additional Reading list).

Equation 5.72 provides an expression for the interaction energy ΔE, the difference in energy between the molecule described by the wave function $1s_A(1)1s_B(2)$ and the separated atoms. If we evaluate the interaction energy for different values of the internuclear distance R, we obtain the result shown as the solid curve of Fig. 5.17. As we see from the figure, ΔE goes to zero for large R, has a minimum, and then rises steeply for smaller values of R. The minimum, which occurs at $R = 0.93$ Å, is equal to -12.0 kcal mole^{-1} relative to the separated atoms. Also shown in the figure is a dashed curve representing the experimental ΔE for the hydrogen molecule. It has a minimum at a distance $R = 0.742$ Å, which is not very different from the calculated value. However, the depth of the minimum is equal to 109.5 kcal mole^{-1}, much larger than the value obtained from Eq. 5.72. The fact that the wave function $1s_A(1)1s_B(2)$ yields only about 10% of the dissociation energy shows that it does not provide an adequate description of the hydrogen molecule.

Equation 5.72, which gives ΔE, is equivalent to the classical electrostatic energy of two point nuclei A and B and two charge distributions of the form $[1s_A(1)]^2$ and $1s_B(2)]^2$. What this expression suggests, as does the wave function $1s_A(1)1s_B(2)$, is that we know that electron 1 is on atom A and electron 2 is on atom B. However, as we saw in Section 4.2, there is no way of labelling electrons so as to distinguish one from the other. Consequently, a wave function of the form $1s_A(1)1s_B(2)$ implies that we know more than we can measure. The way out of this difficulty, which is also the cause of the large error in the calculated dissociation energy, is the same as that described previously for the He atom: We must use a wave function that takes account of the indistinguishability of the electrons.

5.6.1 *The Heitler–London or valence-bond treatment of* H_2 Since the two electrons of the interacting hydrogen atoms are indistinguishable, an acceptable wave function for the system must correspond to probability

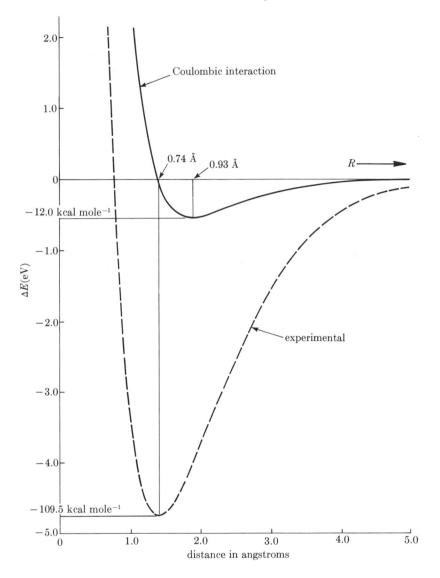

Fig. 5.17. Interaction energy for the ground state of H_2. Solid line, $\triangle E$ calculated from the wave function of Eq. 5.63; dashed line, experimental $\triangle E$.

densities which are identical for the two electrons. The function $1s_A(1)1s_B(2)$ does not meet this criterion because electron 1 has a density centered on nucleus A, while electron 2 has a density centered on nucleus B. The assignment of electron 2 to nucleus A and electron 1 to nucleus B is equally likely. We can build the required indistinguishability into the approximate wave function by writing

$$\psi_\pm(1, 2) = (\tfrac{1}{2})^{1/2}[1s_A(1)\, 1s_B(2) \pm 1s_B(1)\, 1s_A(2)] \qquad (5.73)$$

If we now exchange the electron labels 1 and 2, the function ψ_+ does not change, while the function ψ_- changes sign:

$$\psi_+(2, 1) = \psi_+(1, 2) \qquad \psi_-(2, 1) = -\psi_-(1, 2) \qquad (5.74)$$

In either case, the two-electron density $\psi_\pm^2$ is independent of the labels assigned to the electrons, since

$$[\psi_\pm(2, 1)]^2 = [\psi_\pm(1, 2)]^2 \qquad (5.75)$$

Thus, the wave functions of Eq. 5.73, unlike that of Eq. 5.63, do not imply that it is possible to distinguish between the two electrons. The factor $2^{-1/2}$ in Eq. 5.73 provides for the approximate normalization of the wave function; that is,

$$\int [\psi_\pm(1, 2)]^2\, dv_1\, dv_2 \simeq 1 \qquad (5.76)$$

One can calculate the interaction energy ΔE from $\psi_\pm$ by the method used in the previous section. The result is (Problem 5.24)

$$\begin{aligned} \Delta E_\pm &= \int [1s_A(1)]^2\, V_I\, [1s_B(2)]^2\, dv_1\, dv_2 \\ &\pm \int 1s_A(1)\, 1s_B(1)\, V_I\, 1s_A(2)\, 1s_B(2)\, dv_2\, dv_2 \end{aligned} \qquad (5.77)$$

where the $+$ and $-$ signs correspond to the wave functions ψ_+ and ψ_-, respectively. The first integral is usually called a *Coulomb integral* and is designated by the symbol Q. The Coulomb integral represents the classical Coulombic interaction of the charge cloud $[1s_A(1)]^2$ with $[1s_B(2)]^2$, of the charge cloud $[1s_A(1)]^2$ with nucleus B, of the charge cloud $[1s_B(1)]^2$ with nucleus A, and of the nuclei with one another. Thus, Q is the interaction energy obtained previously for the wave function $1s_A(1)1s_B(2)$. (Compare the atomic Coulomb integral Q for excited He, Section 4.3.) The second integral cannot be interpreted as a classical electrostatic interaction of two charge clouds, since $1s_A(1)1s_B(1)$ is not an electron density in the ordinary sense. Because it arises as a result of exchanging electrons between the two atoms (i.e., one function has electron 1 on nucleus A and electron 2 on nucleus B and the other function has electron 1 on nucleus B and electron 2 on nucleus A), the integral is called an *exchange integral;* it is usually given the symbol J. (Compare the atomic exchange integral K, Section 4.3.) The interaction energy for the H_2 molecule becomes

$$\Delta E_\pm = Q \pm J \qquad (5.78)$$

where $\Delta E_\pm$, Q, and J are all functions of R.

We have already plotted Q in Fig. 5.17 and seen that it approaches zero for large R and is negative except at very small values of R. Calculation of J shows that it approaches zero for large R, is negative in the region near R_e, and that the absolute value of J in this region is several times larger than that of Q (Fig. 5.18). This means that ψ_+, which yields ΔE with the

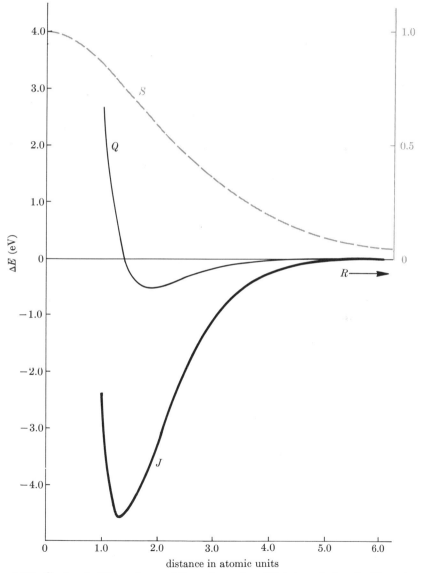

Fig. 5.18. Coulomb (Q), exchange (J), and overlap (S) integrals for the H₂ molecule.

+ sign in Eq. 5.78, is the state with lower energy. When the − sign is used in Eq. 5.78, the slight binding due to the Q term is counteracted by the large positive $-J$ term, resulting in a repulsive potential. The value of D_e as calculated from ψ_+ is smaller than the experimental value, but is sufficiently close to demonstrate that the present treatment includes the dominant

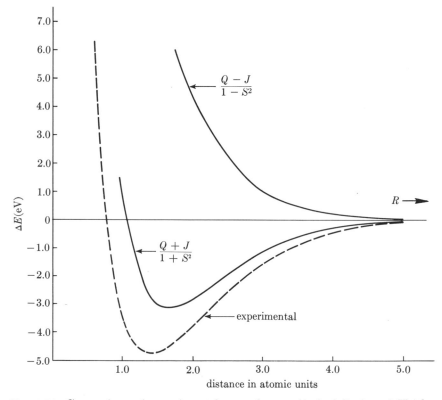

Fig. 5.19. Comparison of experimental ground-state (dashed line) and Heitler–London ground-state and excited-state (solid lines) $\triangle E$ curves for the H_2 molecule.

terms contributing to the covalent bond. The major portion of the binding energy is due to the exchange integral, which arises from the required indistinguishability of the electrons involved in the bond.

In the foregoing discussion, we have concentrated on the role of the integrals Q and J, and tacitly assumed that the atomic orbitals $1s_A$ and $1s_B$ are orthogonal. Actually the integral of the product of these two functions, usually designated as S,

$$S = \int 1s_A(1)\, 1s_B(1)\, dv_1 \tag{5.79}$$

is, like Q and J, a function of R (Fig. 5.18). The value of S is unity at $R = 0$, since at that point the nuclei A and B coincide and the normalized functions $1s_A$ and $1s_B$ become identical. As $R \to \infty$, the value of $S \to 0$. Thus S is a measure of the degree of "overlap" of the functions $1s_A$ and $1s_B$, and is called the *overlap integral*.

When S is included in the formulation (see Problem 5.30), it can be shown that the normalized wave function is

$$\psi_\pm(1, 2) = [2(1 \pm S^2)]^{-1/2}[1s_A(1)\,1s_B(2) \pm 1s_B(1)\,1s_A(2)] \quad (5.80)$$

The expression for $\Delta E_\pm$ becomes

$$\Delta E_\pm = \frac{Q \pm J}{1 \pm S^2} \quad (5.81)$$

Again the $+$ sign corresponds to the bonding ground state, and the $-$ sign corresponds to the excited repulsive state. The forms of ΔE_+ and ΔE_- are shown as the lower and upper curves, respectively, in Fig. 5.19. The experimental potential-energy curve for the ground state is the dashed line in the figure. For the ground state, the Heitler-London result including overlap is $D_e = 72.8$ kcal/mole and $R_e = 0.87$ Å, compared with the experimental values of $D_e = 109.5$ kcal/mole and $R_e = 0.742$ Å.

This treatment of the covalent bond, first worked out by Heitler and London for H_2, and further developed by Slater and Pauling for extension to more complicated molecules, is based upon a bonding wave function for the two electrons which are shared between the two atoms. It is often called the *valence-bond approach*. The bonding function ψ_+ corresponds to a state with paired electron spins, while the repulsive function ψ_- has parallel spins (see Section 5.6.5 and Problems 5.31 and 5.34); that is, the bonding function is a singlet and the antibonding function is a triplet. Because the bonding function has paired spins, one often refers to it as describing an *electron-pair bond*.

By examining the total electron density corresponding to the Heitler-London wave functions, we can gain additional insight into the nature of the chemical bond. The square of the wave functions $[\psi_\pm(1, 2)]^2$ gives the probability per unit volume that electron 1 is at x_1, y_1, z_1 when electron 2 is at x_2, y_2, z_2. To obtain the distribution function for electron 1 *regardless* of the position of electron 2, we must integrate over all possible positions of electron 2; that is, if we call $P_\pm(1)$ the probability density for electron 1, we have

$$P_\pm(1) = \int[\psi_\pm(1,2)]^2\,dv_2 \quad (5.82)$$

Substituting Eq. 5.80 into Eq. 5.82 and making use of Eq. 5.79 and the fact that $1s_A$ and $1s_B$ are normalized, we obtain

$$P_\pm(1) = [2(1 \pm S^2)]^{-1}[1s_A(1)^2 + 1s_B(1)^2 \\ \pm 2S\,1s_A(1)\,1s_B(1)] \quad (5.83)$$

Similarly, if we integrate $[\psi_\pm(1, 2)]^2$ over the coordinates of electron 1, the result is

$$P_\pm(2) = [2(1 \pm S^2)]^{-1}[1s_A(2)^2 + 1s_B(2)^2 \\ \pm 2S \, 1s_A(2) \, 1s_B(2)] \tag{5.84}$$

Comparison of Eqs. 5.83 and 5.84 shows that the probability of finding electron 1 at any point in space is the same as the probability of finding electron 2 at that point; in other words, the wave function $\psi_\pm$ properly treats the two electrons equivalently.

The total charge density at the point x, y, z is the sum of the two individual electron densities, multiplied by the electronic charge e:

$$\rho_\pm(x, y, z) = 2e \, P_\pm(x, y, z) \tag{5.85}$$

When Eqs. 5.83 and 5.84 are substituted into Eq. 5.85 with explicit expressions for the $1s$ orbitals, the density becomes

$$\rho_\pm(x, y, z) = \frac{e}{\pi a_0^3(1 \pm S^2)} \left[\exp\left(-\frac{2r_A}{a_0}\right) + \exp\left(-\frac{2r_B}{a_0}\right) \right. \\ \left. \pm 2S \exp\left(-\frac{r_A + r_B}{a_0}\right) \right] \tag{5.86}$$

where r_A and r_B are the distances from the point (x, y, z) to nucleus A and nucleus B, respectively. In Fig. 5.20a the first two terms in Eq. 5.86 are plotted as a function of position along the internuclear axis. The sum of these terms is just the sum of two exponential functions centered on the respective nuclei. The overlap term in Eq. 5.86 is plotted in Fig. 5.20b. This function is constant between the nuclei since there $r_A + r_B = R$; outside the nuclei, the function drops exponentially to zero. The total density, shown in Fig. 5.20c, is the sum or the difference of the plots a and b, depending on whether ψ_+ or ψ_- is the wave function. For ψ_+, electronic charge is shifted from the region outside the nuclei into the region between the two nuclei. The closer the atoms approach, the larger S becomes, and therefore, according to Eq. 5.86, the larger is the build-up of charge.

To see what effect the shift of electronic charge into the overlap region has upon the energy of the pair of atoms, we look at the average kinetic and potential energies as a function of the internuclear distance R. If the total energy E and its variation with R is known, we can use the virial theorem (Section 3.9) to calculate the average kinetic energy $\langle T \rangle$ and the average potential energy $\langle V \rangle$. The expression for the virial theorem given in Section 3.9 for the hydrogen atom must be modified for molecules to include the forces on the nuclei. For a diatomic molecule, the virial theorem is

$$\langle T \rangle = -\tfrac{1}{2}\langle V \rangle - \tfrac{1}{2}R\frac{dE}{dR} \tag{5.87}$$

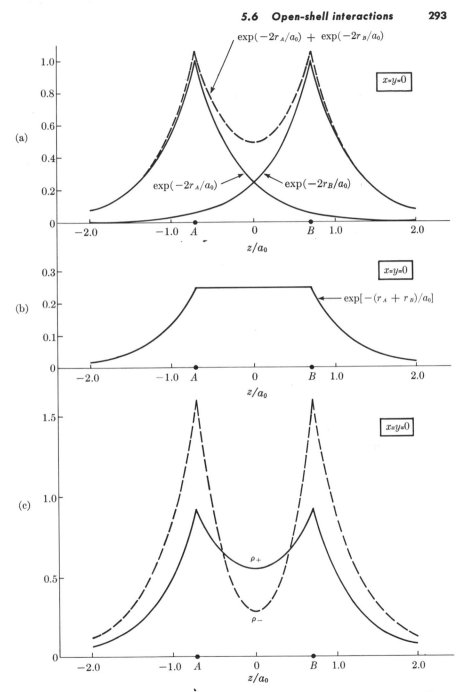

Fig. 5.20. Electron density along $\dot{H}_2$ axis in the simple VB approximation. **(a)** First two terms of Eq. 5.86; **(b)** overlap term of Eq. 5.86; **(c)** solid line, electron density for ψ_+; dashed line, electron density for ψ_-.

At equilibrium $(R = R_e)$, the force on the nuclei, $-dE/dR$, is zero (see Section 5.2) and the equation reduces to the simpler form $\langle T \rangle = -\frac{1}{2}\langle V \rangle$ given in Section 3.9. Since the total energy is the sum of the expectation values of the kinetic and potential energies $(E = \langle T \rangle + \langle V \rangle)$, we can rewrite Eq. 5.87 in the form

$$\langle T \rangle = - \left(E + R\frac{dE}{dR} \right) \tag{5.88a}$$

$$\langle V \rangle = 2E + R\frac{dE}{dR} \tag{5.88b}$$

The values of $\langle T \rangle$, $\langle V \rangle$, and E as functions of R calculated for the ground state of H_2 are given in Fig. 5.21 [W. Kolos and L. Wolniewicz, *J. Chem. Phys.* **43**, 2429 (1965)]. The behavior of the curves can be understood qualitatively on the basis of our previous description of the build up of electronic charge in the overlap region. At large internuclear distances $(R \gtrsim 3.5a_0)$, the build up of charge in the overlap region has the effect of extending the atomic charge clouds along the internuclear axis. Since the volume in which the electrons are confined is expanded, the average kinetic energy of the electrons is *lowered* in accord with the uncertainty principle. Because parts of the charge clouds are displaced from regions near one of the nuclei (a region of low potential energy) to the overlap region between the nuclei (a region of relatively larger potential energy), the shift of charge at large internuclear distances *raises* the average electronic potential energy.

At small internuclear distances $(1.0a_0 \leq R \leq 2.5a_0)$, the build up of charge between the nuclei and the consequent reduction of charge outside the nuclei is such as to effectively reduce the size of the region available to the electrons below that of the separated atoms. This results in an increase in the average kinetic energy. However, because the nuclei are close together, the overlap electron cloud is near both nuclei and its average electronic potential energy is decreased. At very small internuclear distances $(R < 0.6a_0)$, the mutual repulsion of the nuclei becomes important and the potential energy increases rapidly.

The total energy of a molecule is given by the sum of its average kinetic and potential energies. At large internuclear distances, the kinetic energy falls with decreasing R slightly more rapidly than the potential energy rises. The total energy thus decreases as R is reduced at large internuclear distances. At very small internuclear distances, both the kinetic and potential energies increase with decreasing R, so that the total energy increases rapidly in this region. At an intermediate value of $R(1.401a_0$ for $H_2)$, the attractive and repulsive forces are balanced, and the total energy is a minimum. At this point $(R = R_e)$, the energy relative to that of the sepa-

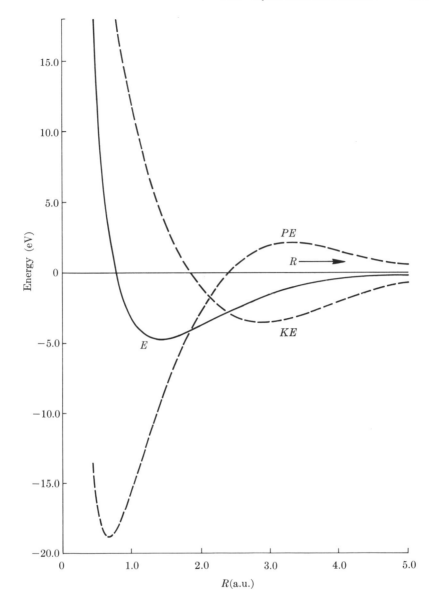

Fig. 5.21. Total, kinetic, and potential energy of H_2 as a function of internuclear distance.

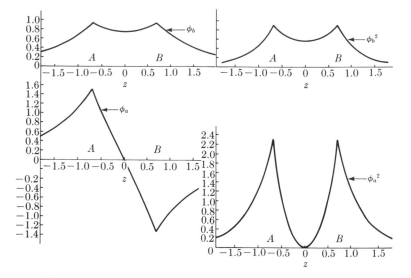

Fig. 5.22. Plots of the molecular orbitals ϕ_b and ϕ_a and the corresponding probability densities ϕ_b^2 and ϕ_a^2 along the internuclear axis.

$$E_b = \int \phi_b(1)\ h_e(1)\ \phi_b(1)\ dv_1 = \frac{1}{2}\ (\int 1s_A(1)\ h_e(1)\ 1s_A(1)\ dv_1$$
$$+ \int 1s_B(1)\ h_e(1)\ 1s_B(1)\ dv_1 + \int 1s_A(1)\ h_e(1)\ 1s_B(1)\ dv_1 \qquad (5.92)$$
$$+ \int 1s_B(1)\ h_e(1)\ 1s_A(1)\ dv_1$$

Since $1s_A$ and $1s_B$ are equivalent and h_e is symmetric in the two nuclei, the first two integrals on the right-hand side of Eq. 5.92 are identical, as are the second pair of integrals. Thus, we can write

$$E_b = \int 1s_A(1)\ h_e(1)\ 1s_A(1)\ dv_1 + \int 1s_A(1)\ h_e(1)\ 1s_B(1)\ dv_1$$

The first integral is called an *atomic integral*,[2] since it is the average of the Hamiltonian over the orbitals of a single atom. The second integral, which includes atomic orbitals on atoms A and B in the integrand, is called a *resonance integral*. Using the standard symbols α and β for the atomic and resonance integral, respectively, we write

$$E_b = \alpha + \beta \qquad (5.93)$$

In a similar way, the energy corresponding to the excited orbital ϕ_a of Eq. 5.91 can be written as

$$E_a = \alpha - \beta \qquad (5.94)$$

When calculations of the integral α are carried out, the major contribution to α is found to be the energy of the electron on one of the separated atoms (see Fig. 5.23). Thus, the interaction energy for the orbital ϕ_b is

[2] The integral $a = \int 1s_A(1)h_e(1)1s_A(1)\ dv_1$ is called a *Coulomb integral* by some authors.

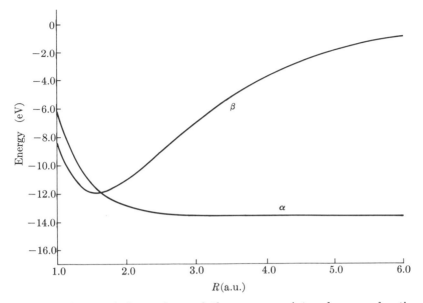

Fig. 5.23. The atomic integral α and the resonance integral β as a function of R for H_2^+.

$$\Delta E_b = E_b - E_H \simeq \beta \qquad (5.95)$$

and for the orbital ϕ_a is

$$\Delta E_a = E_a - E_H \simeq -\beta \qquad (5.96)$$

Since calculated values of β are found to be negative (see Fig. 5.23), ϕ_b leads to a negative interaction energy for the molecule (i.e., bonding), while ϕ_a leads to a positive interaction energy (i.e., repulsion or antibonding). Accordingly, ϕ_b and ϕ_a are designated *bonding* and *antibonding* orbitals, respectively. These results are in agreement with the notion that ϕ_b corresponds to a slight build up of charge density between the nuclei, while ϕ_a has a nodal plane that bisects the nuclear axis and therefore corresponds to a decrease in charge density between the nuclei.

Other bonding and antibonding molecular orbitals can be formed in a corresponding way by using excited atomic orbitals (e.g., $2s_A$, $2p_A$, . . . , $2s_B$, $2p_B$, . . .) in expressions like Eqs. 5.90 and 5.91; see Section 6.1.7.

The atomic and resonance integrals α and β play a fundamental role in MO theory, analogous to that of the Coulomb and exchange integrals Q and J in the VB approach. As $R \to \infty$, the integral α approaches the isolated atom value and the integral β approaches zero. Consequently, $\Delta E \to 0$ for both the bonding and antibonding orbitals. For intermediate and small values of R, the behavior of ΔE is qualitatively similar to the VB results for H_2 (see Fig. 5.24).

As in the first VB approximation, we have neglected the overlap of $1s_A$ and $1s_B$ in writing Eqs. 5.90 and 5.91. Introducing the correctly normalized functions

$$\phi_b = (2 + 2S)^{-1/2}(1s_A + 1s_B)$$
$$\phi_a = (2 - 2S)^{-1/2}(1s_A - 1s_B)$$

(5.97)

we obtain for the energies

$$E_b = \frac{\alpha + \beta}{1 + S} \qquad E_a = \frac{\alpha - \beta}{1 - S}$$

(5.98)

For short distances, S is not small so that the corrections embodied in Eq. 5.98, as compared with Eqs. 5.93 and 5.94, are not negligible (Problem 5.26 and Fig. 5.18). However, to simplify the writing we continue to use

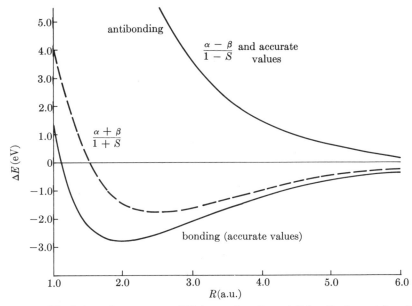

Fig. 5.24. The interaction energy of $H_2{}^+$ as a function of R for the lowest bonding and antibonding states.

the more approximate functions. (Eqs. 5.90 and 5.91) and the energies (Eqs. 5.95 and 5.96).

To employ the simple molecular orbitals for a treatment of H_2, we make use of the *Aufbau* principle introduced in Chapter 4 for atoms. By analogy with the He atom, the ground state of H_2 has the two electrons paired in the molecular orbital of lowest energy, ϕ_b; that is,

$$\psi_1(1, 2) = \phi_b(1)\phi_b(2)$$
$$= \frac{1}{2}[1s_A(1) + 1s_B(1)][1s_A(2) + 1s_B(2)]$$

(5.99)

Since the two electrons are in the same orbital, the Pauli principle requires that they have paired spins; that is, the H_2 ground state in MO theory is a singlet, as it was in the VB treatment.

In addition to the ground-state configuration, there are several other possible assignments of the two electrons to the two lowest-lying molecular orbitals of H_2. If we assign the spin function α to an electron with $m_s = \frac{1}{2}$ and the function β to an electron with $m_s = -\frac{1}{2}$, as in Chapter 4, the possible one-electron wave functions including both the space and spin parts are

$$\phi_b\alpha \quad \phi_b\beta \quad \phi_a\alpha \quad \phi_a\beta$$

To form electronic states of the H_2 molecule, we consider that the two electrons can occupy any two of these four functions. Since we can select two functions for occupancy from the four available functions in

$$\frac{4!}{2!2!} = \frac{4 \times 3}{1 \times 2} = 6$$

different ways, there are six possible choices. The function

$$\phi_b\alpha(1) \; \phi_b\beta(2)$$

corresponds to the ground state, as we have just seen. In accord with the discussion of the Pauli principle in Chapter 4, the wave function must be antisymmetric with respect to electron interchange; that is,

$$\psi_1 = -(\tfrac{1}{2})^{1/2} \begin{vmatrix} \phi_b\alpha(1) & \phi_b\beta(1) \\ \phi_b\alpha(2) & \phi_b\beta(2) \end{vmatrix} = \phi_b(1)\phi_b(2) \, (\tfrac{1}{2})^{1/2} \, [\alpha(1)\beta(2) - \beta(1)\alpha(2)]$$

The form of the spin part of the function reflects the fact that no measurement can be made to determine which of the two electrons has spin α and which has spin β. This is similar to the result for the VB H_2 wave function (Eq. 5.73) when the impossibility of determining which electron is associated with nucleus A and which with nucleus B is recognized.

The state having the highest energy is formed by choosing the antibonding spin orbitals $\phi_a\alpha$ and $\phi_a\beta$ for occupancy. The antisymmetric wave function for this state is

$$\psi_2 = (\tfrac{1}{2})^{1/2} \begin{vmatrix} \phi_a\alpha(1) & \phi_a\beta(1) \\ \phi_a\alpha(2) & \phi_a\beta(2) \end{vmatrix} = \phi_a(1)\phi_a(2) \, (\tfrac{1}{2})^{1/2} \, [\alpha(1)\beta(2) - \beta(1)\alpha(2)]$$

$$(5.100)$$

States of intermediate energy result when the orbitals ϕ_a and ϕ_b are selected for single occupancy. Since in this case the orbitals are singly occupied, there is no restriction on the spin functions associated with the orbitals; that is, we assume that simple approximations to the wave functions can be obtained by taking appropriate linear combinations of the products

$$\phi_a\alpha(1)\ \phi_b\alpha(2) \qquad \phi_a\beta(1)\ \phi_b\alpha(2)$$
$$\phi_a\alpha(1)\ \phi_b\beta(2) \qquad \phi_a\beta(1)\ \phi_b\beta(2)$$

The linear combinations must satisfy the Pauli principle (the over-all wave functions must be antisymmetric with respect to exchange of one electron for another), and the electrons must be indistinguishable with respect to both spatial coordinates and spin. The procedure for doing this is exactly analogous to the method used in Section 4.3 to determine wave functions for the first excited state of He (Eqs. 4.55), except that we replace the functions $1s$ and $2s$ by ϕ_a and ϕ_b, respectively. As in Eq. 4.55, we obtain a single function with an antisymmetric spin part, and three functions with a common spatial part and different symmetric spin parts, namely,

$$\psi_3 = \left(\frac{1}{2}\right)^{1/2}[\phi_a(1)\phi_b(2) + \phi_b(1)\phi_a(2)]\left(\frac{1}{2}\right)^{1/2}[\alpha(1)\beta(2) - \beta(1)\alpha(2)] \quad (5.101)$$

$$\psi_4 = (\tfrac{1}{2})^{1/2}\,[\phi_a(1)\phi_b(2) - \phi_b(1)\phi_a(2)]\,(\tfrac{1}{2})^{1/2}\begin{Bmatrix} \alpha(1)\alpha(2) \\ [\alpha(1)\beta(2) + \beta(1)\alpha(2)] \\ \beta(1)\beta(2) \end{Bmatrix}$$

Thus, there are three possible spin states that can be associated with the orbital part of ψ_4, while only one spin state is associated with the orbital parts of ψ_1, ψ_2, and ψ_3. Since the direct influence of the electron spins on the energy can be neglected to a first approximation (Section 4.3), ψ_4 is a triply degenerate state, while the other three are nondegenerate; that is, ψ_4 is a triplet state (multiplicity of three), while ψ_1, ψ_2, and ψ_3 are singlet states (multiplicity of one). From our discussion of the He excited states (Section 4.3), we remember that multiplicites greater than one indicate that the atom or molecule has a permanent magnetic dipole moment due to the unpaired electron spin and is paramagnetic. If the multiplicity is one, no permanent electronic magnetic moment exists unless there is orbital degeneracy. Substances without a permanent magnetic moment are diamagnetic. From this discussion, we see that H_2 is diamagnetic in the singlet ground state ψ_1.

A calculation of the ground-state energy corresponding to ψ_1, including overlap, gives the results $D_e = 62.1$ kcal mole^{-1} and $R_e = 0.85$ Å. Thus, although the R_e value from MO theory is very near the VB result, the D_e value is in somewhat greater disagreement with experiment. The simple MO method is nevertheless very important in the discussion of chemical bonding. It is particularly suited for the interpretation of molecular spectra in terms of excitations of a single electron from one orbital to another. In Fig. 5.25 a schematic diagram of the four levels derived from the atomic $1s$ orbitals for H_2 is given, and the one-electron transition from the ground state to an excited state of the same multiplicity is indicated. It is of interest to note that three of the states are bound (ψ_1, ψ_2, ψ_3), while only one (ψ_4) is repulsive.

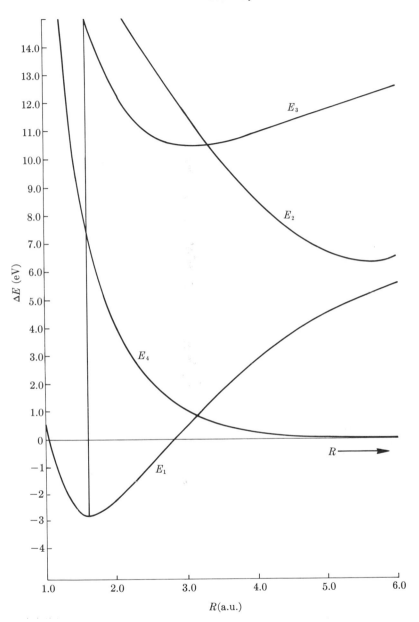

Fig. 5.25. Interaction energies as functions of R for the electronic states of H_2 obtained by the simple MO treatment with $1s$ atomic orbitals.

5.6.3 Comparison of the VB and MO approaches to H₂ Although different
arguments were used to introduce the VB and MO wave functions for H_2
and the resulting expressions do not look very similar, the two kinds of
wave functions are related to one another. A study of this relationship
shows that the best ground-state H_2 wave function which can be constructed from hydrogen atom $1s$ functions lies somewhere between the
simple VB and MO limits.

Multiplying out the spatial part of the MO function ψ_1 given in Eq. 5.99,
we have

$$\psi_1(1,\,2) = \tfrac{1}{2}[1s_A(1)\,1s_B(2) + 1s_B(1)\,1s_A(2)]$$
$$+ \tfrac{1}{2}[1s_A(1)\,1s_A(2) + 1s_B(1)\,1s_B(2)] \qquad (5.102)$$

Comparison of the first term in brackets on the right-hand side of Eq. 5.102
with Eq. 5.73 shows that it is equivalent to the VB ground-state function
ψ_+. The second term in Eq. 5.102 can be regarded as a linear combination of
two states, one with both electrons on atom A and the other with both
electrons on atom B. These ionic states correspond to the chemical formulas
$H_A^-\,H_B^+$ and $H_A^+\,H_B^-$. If the notation ψ_i is adopted for the ionic wave
function or *ionic structure*

$$\psi_i = (\tfrac{1}{2})^{1/2}\,[1s_A(1)\,1s_A(2) + 1s_B(1)\,1s_B(2)] \qquad (5.103)$$

then Eq. 5.102 takes the form

$$\psi_1(1,\,2) = (\tfrac{1}{2})^{1/2}\,[\psi_+(1,\,2) + \psi_i(1,\,2)] \qquad (5.104)$$

The VB function ψ_+ represents the so-called *covalent* structure (H—H) because the two electrons are shared equally between the atoms A and B. By
contrast the MO ground state is an equal mixture of covalent and ionic
structures for the H_2 molecule.

If we expand the excited MO function ψ_2 in a similar way, we obtain

$$\psi_2(1,\,2) = -\tfrac{1}{2}[1s_A(1)\,1s_B(2) + 1s_B(1)\,1s_A(2)]$$
$$+ \tfrac{1}{2}[1s_A(1)\,1s_A(2) + 1s_B(1)\,1s_B(2)]$$
$$= (\tfrac{1}{2})^{1/2}\,[-\psi_+(1,\,2) + \psi_i(1,\,2)] \qquad (5.105)$$

This state is also a linear combination of the covalent and ionic VB functions, but with opposite signs. Equations 5.104 and 5.105 can be solved for
the VB covalent wave function ψ_+ to give

$$\psi_+(1,\,2) = (\tfrac{1}{2})^{1/2}\,[\psi_1(1,\,2) - \psi_2(1,\,2)] \qquad (5.106)$$

The function ψ_+ is thus an equal mixture (with opposite signs) of the
ground-state MO electron configuration $\phi_b{}^2$ and the doubly excited-state

MO electron configuration $\phi_a{}^2$. Such a mixture of configurations used to describe the state of an atom or molecule is called a *configuration-interaction wave function*.

5.6.4   Linear variation principle   There is no reason to believe that a linear combination of equal parts covalent and ionic structures, as in the MO function, or equal parts ground–state and doubly excited configurations, as in the VB function, is the best approximation to the H_2 ground state. From our previous description of the variational method, it is clear that we can make up linear combinations with arbitrary ratios of ψ_+, ψ_i or of $\phi_b{}^2$, $\phi_a{}^2$ and minimize the energy as a function of the ratios to find the best value. To see what this involves we consider a variation function of the form

$$\psi = c_1\phi_1 + c_2\phi_2$$

where ϕ_1 and ϕ_2 are arbitrary functions (e.g., ψ_+ and ψ_i) and minimize the corresponding expectation value of the energy, which is given by

$$E = \frac{\int \psi H \psi \, dv}{\int \psi \psi \, dv} = \frac{c_1{}^2\int \phi_1 H \phi_1 \, dv + 2c_1c_2\int \phi_1 H \phi_2 \, dv + c_2{}^2\int \phi_2 H \phi_2 \, dv}{c_1{}^2\int \phi_1{}^2 \, dv + 2c_1c_2\int \phi_1\phi_2 \, dv + c_2{}^2\int \phi_2{}^2 \, dv} \quad (5.107)$$

The term $\int \psi^2 \, dv$ in the denominator of Eq. 5.107 accounts for the normalization of ψ; that is, for arbitrary real ψ, the function $\psi/[\int \psi^2 \, dv]^{1/2}$ is normalized. To simplify the writing, we introduce symbols for the integrals,

$$H_{ij} = \int \phi_i H \phi_j \, dv \qquad S_{ij} = \int \phi_i \phi_j \, dv \qquad i, j = 1, 2$$

so that Eq. 5.107 becomes

$$E = \frac{c_1{}^2 H_{11} + 2c_1c_2 H_{12} + c_2{}^2 H_{22}}{c_1{}^2 S_{11} + 2c_1c_2 S_{12} + c_2{}^2 S_{22}} \quad (5.108)$$

Equation 5.108 shows that E is a function of c_1 and c_2 for fixed values of H_{ij} and S_{ij}, which are determined by the choice of *basis functions* ϕ_1 and ϕ_2. To minimize E with respect to the coefficients c_1 and c_2, we differentiate E with respect to c_1 and c_2 and set the derivatives equal to zero. This is done most simply by first multiplying both sides of Eq. 5.108 by the denominator on the right-hand side to obtain $E(c_1{}^2 S_{11} + 2c_1c_2 S_{12} + c_2{}^2 S_{22}) = c_1{}^2 H_{11} + 2c_1c_2 H_{12} + c_2{}^2 H_{22}$. Differentiation of this expression with respect to c_1 gives

$$(2c_1 S_{11} + 2c_2 S_{12})E + (c_1{}^2 S_{11} + 2c_1c_2 S_{12} + c_2{}^2 S_{22})\frac{\partial E}{\partial c_1} \quad (5.109a)$$
$$= 2c_1 H_{11} + 2c_2 H_{12}$$

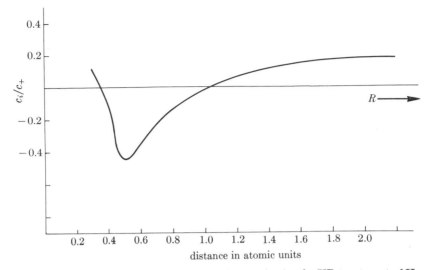

Fig. 5.27. Ratio of ionic and covalent coefficients c_i/c_+ for the VB treatment of H_2 as a function of internuclear distance R in the region $0.3 \leq R/a_0 \leq 2.0$. As $R \to \infty$, the ratio $c_i/c_+ \to 0$.

value of c_i/c_+ is 0.16; that is, the best function is primarily covalent. Moreover, as $R \to \infty$, $c_i/c_+ \to 0$ and the function becomes entirely covalent, corresponding to the fact that for no interaction between the ions the energy of the ionic state H^+H^- is 12.84 eV above that of two H atoms.

A similar calculation can be carried out with the configurations ϕ_b^2 and ϕ_a^2 by writing

$$\psi_{MO} = c_b\phi_b^2 + c_a\phi_a^2 \tag{5.119}$$

and minimizing the energy as a function of c_b and c_a. Plots of $E_\pm$ are shown in Fig. 5.26. The best value of the ratio c_a/c_b is bound to be -0.73 at $R = R_e = 0.88$ Å. When the best value of c_i/c_+ is used in Eq. 5.118 and the best value of c_a/c_b is used in Eq. 5.119, the functions ψ_{VB} and ψ_{MO} are found to be identical (Problem 5.33). Thus, the VB approach including ionic structures is equivalent to the MO approach with configuration interaction.

Since the MO and VB techniques are both approximate, it is important that in the next order of approximation they lead to identical formulas. In many applications to complex molecules, only the simplest treatment of the VB or MO type is possible. It is then useful to do both—if they give the same or similar results, it is likely that the answer is meaningful; if they give very different results, one should be very cautious about believing either of them.

Since the simple VB ground state has been shown to be a linear combination of two MO singlet states, the VB ground state must also be a singlet, as was pointed out previously. This can also be shown by direct application of the general form of the Pauli principle to the VB approach (Problem 5.34). It is left as an exercise (see Problem 5.27) for the student to show that the VB repulsive state is identical to the MO excited triplet state ψ_4.

5.6.6 Improved hydrogen molecule wave functions The simple VB and MO treatments of H_2 described in Sections 5.6.1 and 5.6.2 are analogous to the perturbation treatment of He discussed in Section 4.1.5; that is, the simplest form for the wave function that is a reasonable approximation is constructed from solutions for the "unperturbed" problem (noninteracting atoms in VB, H_2^+ in MO) and the energy is estimated by calculating the average of the true Hamiltonian, $\langle H \rangle$, for the assumed wave function. The results obtained are satisfactory in that they give a reasonable qualitative picture of the interaction between two hydrogen atoms. It is important to show also that quantum mechanics can give accurate quantitative results for H_2, in order to demonstrate that the Schrödinger equation is sufficient for an essentially complete understanding of this, the simplest of chemical bonds. Moreover, it is useful to have an understanding of the principles used to find improved wave functions.

In Section 4.1.6, we saw that a description of the He atom, better than the perturbation result, is obtained if the effective nuclear charge is chosen to make $\langle H \rangle$ a minimum, in accordance with the variation principle. The same is true for a VB or MO treatment of H_2; that is, for each internuclear distance R, there is a value of the effective nuclear charge that makes $\langle H \rangle$ a minimum. If the orbitals $1s_A$ and $1s_B$ are written in the more general form that includes an effective nuclear charge

$$1s_A(1) = \left(\frac{\zeta^3}{\pi a_0^3}\right)^{1/2} \exp\left(-\zeta r_{1A}/a_0\right) \quad 1s_B(1) = \left(\frac{\zeta^3}{\pi a_0^3}\right)^{1/2} \exp\left(-\zeta r_{1B}/a_0\right)$$

the value of $\langle H \rangle$ calculated by either the VB or MO method is a function of ζ. Minimizing $\langle H \rangle$ with respect to ζ, one obtains the "best" possible energy for wave functions of the assumed functional form. For the covalent VB treatment of H_2 (without ionic terms), the binding energy is found to be 3.782 eV at $R_e = 1.406a_0$ and $\zeta = 1.166$; thus, varying the effective charge lowers the energy by 0.62 eV in this case. In the single-configuration MO treatment of H_2, varying ζ lowers the H_2 energy by 0.79 eV to -3.488 eV at $R_e = 1.38a_0$ and $\zeta = 1.197$. If the VB wave function with ionic terms (or, equivalently, the MO wave function including configuration interaction) is used, the best value of ζ ($\zeta = 1.194$) is the same for both the covalent and ionic contributions at $R_e = 1.43a_0$, giving an energy of -4.024 eV, an improvement of 0.242 eV over the VB, variable-ζ treatment with covalent terms alone.

that a correspondingly accurate treatment is not possible at the present time. Nevertheless, most chemists believe that quantum mechanics in its present form is, in principle, able to provide a complete description of chemical phenomena.

SUMMARY

The interactions of atoms are classified according to their charge state and whether their electron shells are closed or open. Many of the properties of diatomic molecules formed from closed-shell ions are adequately described by an electrostatic model in which the polarizability of the ions and their mutual overlap repulsion are taken into account. Closed-shell neutral atoms attract one another weakly by dispersion interactions (van der Waals or London forces) that correspond to in-phase fluctuations of instantaneous dipoles. Open-shell ions of the transition series form complexes with ligand ions or polar molecules; stability, magnetic behavior, and other properties of these complexes can be understood in terms of the splitting of the d-orbital degeneracy by the electrostatic ligand field, and the application of Hund's rules for orbital occupancy. The covalent bonds formed by open-shell neutral atoms can be described to a first approximation by valence-bond (VB) and molecular-orbital (MO) models in which the indistinguishability of the pair of electrons forming the bond plays an important role. When the VB and MO treatments are carried to higher-order approximations, they are equivalent.

Having developed the principles of covalent bonding in this chapter by treating the simple case of the H_2 molecule, we employ them in Chapter 6 to describe bonding in many-electron diatomic and polyatomic molecules.

ADDITIONAL READING

[1] E. A. MOELWYN-HUGHES, *Physical Chemistry* (Macmillan, New York, 1964), 2nd ed., Chaps. 7 and 9.

[2] L. E. ORGEL, *An Introduction to Transition-Element Chemistry: Ligand-Field Theory* (Wiley, New York, 1966) 2nd ed.

[3] C. A. COULSON, *Valence* (Oxford University Press, Oxford, 1961), 2nd ed., Chaps. 4–6.

[4] C. J. BALLHAUSEN, *Introduction to Ligand-Field Theory* (McGraw-Hill, New York, 1962).

[5] J. C. DAVIS, *Advanced Physical Chemistry* (The Ronald Press, New York, 1965), Chap. 10.

[6] F. L. PILAR, *Elementary Quantum Chemistry* (McGraw-Hill, New York, 1968), Chap. 16.

[7] L. PAULING and E. B. WILSON, JR., *Introduction to Quantum Mechanics* (McGraw-Hill, New York, 1935), Chaps. 12 and 14.

[8] H. EYRING, J. WALTER, and G. E. KIMBALL, *Quantum Chemistry* (Wiley, New York, 1944), Chaps. 11 and 12.

[9] J. C. SLATER, *Quantum Theory of Molecules and Solids* (McGraw-Hill, New York, 1963), Vol. I.

[10] W. KAUZMANN, *Quantum Chemistry* (Academic, New York, 1957), Chaps. 11 and 13.

PROBLEMS

5.1 Derive an expression for the potential energy of a charge $+q$ at a distance r from the center of an electric dipole μ, such that the line between q and μ makes an angle θ with the dipole axis.

$$Ans.\ V = (\mu q/r^2)\cos\theta.$$

5.2 From Eqs. 5.16 and 5.17, show that the induced dipole moments μ_+ and μ_- of the M^+ and X^- ions, respectively, in an ionic molecule MX with internuclear distance R, are given by

$$\mu_+ = \frac{e(\alpha_+ R^4 + 2\alpha_+\alpha_- R)}{R^6 - 4\alpha_+\alpha_-}$$

$$\mu_- = \frac{e(\alpha_- R^4 + 2\alpha_+\alpha_- R)}{R^6 - 4\alpha_+\alpha_-}$$

where α_+ and α_- are the polarizabilities of M^+ and X^-, respectively, and e is the magnitude of the electronic charge. Expand these expressions in powers of R^{-1} and derive Eq. 5.36 for the potential energy of an ionic MX molecule in which both atoms are polarizable. (*Hint:* The derivation given for Eq. 5.23 can be used as a guide, but the dipole-dipole term $-\mu_+\mu_-/qR^3$ must be included in the expression for the electrostatic potential analogous to Eq. 5.20.)

5.3 Although the exponential repulsive term in Eq. 5.2 is easier to justify theoretically, some calculations are simpler if the term Br^{-n} is used instead. Find numerical values of B and n which are consistent with the observed values of R_e and ν_e for the NaCl molecule, neglecting α_+.

5.4 Dimers of alkali halides are known to form in the gas phase with energies of a few eV for dissociation into two monomers. Assuming that the structure of $(NaCl)_2$ is that of Fig. 5.29, that the overlap repulsion (Born–Mayer) occurs only for nearest-neighbor pairs, and neglecting polarization, write down the expression for the potential energy $V(x, y)$ for $(NaCl)_2$. At the equilibrium distances x_0, y_0, use the fact that

$$\left.\frac{\partial V}{\partial x}\right|_{x_0,y_0} = \left.\frac{\partial V}{\partial y}\right|_{x_0,y_0} = 0$$

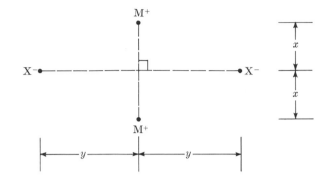

Fig. 5.29. Geometry of an alkali-halide dimer.

to show that $x_0 = y_0$; then show that x_0 satisfies the equation

$$x_0^2 \exp\left(-\sqrt{2}ax_0\right) = \frac{4 - \sqrt{2}}{8Aa}e^2$$

Using the values of A and a obtained from the point-charge model for NaCl, find x_0 by graphical or numerical means, and calculate the dissociation energy of $(NaCl)_2$ by subtracting the point-charge value of $2D_e$ from $-V(x_0, x_0)$.

Ans. 5 eV; experimental value $\sim$2 eV

5.5 Show that a displacement of the origin by the vector **a** does not change the value of the dipole moment of a charge distribution if the sum of all the charges $(q_1 + q_2 + \cdots + q_n)$ is zero.

5.6 In principle, a He_2 molecule could be formed in the same way that NaCl is formed, namely, by the transfer of an electron to form $He^+ + He^-$, which mutually attract one another. From high-energy scattering measurements it is found that the repulsive term for two He atoms at moderate separations is approximately B/R^5, where B $= 3.5$ eV $Å^5$. Find the value of R for which the He^+-He^- interaction energy is a minimum, according to this simple picture, and determine the interaction energy at the minimum, taking the zero of energy to be that of two infinitely separated neutral He atoms. Does the ionic mechanism lead to a He_2 molecule that is stable with respect to the separated neutral atoms? (You may assume that the EA of He is much smaller than the first IP.)

5.7 Two hydrogen atoms are separated by a distance R along the z direction of Cartesian coordinate systems centered on each atom.
(a) Show that the potential energy of interaction between the atoms is approximately given by

$$V' = \frac{e^2}{R^3}\left(x_1 x_2 + y_1 y_2 - 2z_1 z_2\right)$$

for the three-dimensional case, where x_1, y_1, z_1, and x_2, y_2, z_2 are the Cartesian coordinates of the electron of atoms 1 and 2, respectively, each measured with its nucleus as origin.

(b) Using a three-dimensional isotropic harmonic-oscillator model for the binding of each electron to its own atom, show that the van der Waals attractive energy for this system is

$$-\frac{3}{4} h\nu \frac{e^4}{k^2 R^6}$$

where k is the force constant and ν is the classical frequency of the oscillator.

5.8 High-energy molecular-beam scattering experiments show that the repulsive potential for the He-He interaction can be approximated in the form BR^{-n}. The values of B and n for the indicated ranges of R are as follows.

R range (Å)	B (eV Å^n)	n
0.52–1.02	2.88	1.79
0.97–1.48	3.47	5.03
1.27–1.59	4.71	5.94

Plot the experimental He-He repulsive potential over the range 0.5–1.6Å and choose values of A and a which give a good fit for the experimental form of the repulsion Ae^{-aR}. (*Hint:* On a plot of ln ΔE versus R, the intercept of the best straight line is ln A and the slope is $-a$.)

5.9 Plot the interaction energy $\Delta E(R)$ for an "ionic" He^+He^- molecule using Eq. 5.2, assuming the EA to be negligible and that A and a are the same as for two neutral He atoms (see Table 5.6). Does the ionic model for He_2 predict stability with respect to dissociation into 2He?

5.10 The index of refraction of a pure NaBr crystal of density 3.227 g cm^{-3} was measured at three wavelengths:

λ(Å)	n
6563	1.6355
5893	1.6412
4861	1.6555

(a) Find the value of α(NaBr) at 5893 Å.

(b) Plot $n(\lambda)$ versus λ^{-2} and extrapolate linearly to $\lambda = \infty$. Determine the static polarizability of NaBr and compare with $\alpha_+ + \alpha_-$ from Table 5.3.

5.11 In measurements of α by the $\mathcal{E}$-$\mathcal{B}$ gradient balance method described in Section 5.3.4, beams of K atoms with negative $(\mu_m)_z$ are observed to suffer null deflection at the following values of $\mathcal{E}$ and $\mathcal{B}$.

5.25 Integrals of functions centered on different atoms are most easily calculated in confocal elliptic coordinates (see Fig. 5.29). In these coordinates, a point P is assigned the coordinates (ξ,η,ϕ) such that

$$\xi = (r_A + r_B)R^{-1}$$
$$\eta = (r_A - r_B)R^{-1}$$

where r_A and r_B are the distances from P to the nucleus A and B, respectively, and R is the AB distance. The azimuthal angle ϕ is the angle between the plane ABP and an arbitrary plane through AB. The differential volume element in elliptic coordinates is

$$dv = dx\, dy\, dz = \frac{1}{8} R^3(\xi^2 - \eta^2)\, d\xi\, d\eta\, d\phi$$

Show that the ranges of ξ,η, and ϕ for integration over all space are

$$1 \le \xi < \infty \qquad -1 \le \eta \le 1 \qquad 0 \le \phi \le 2\pi$$

Use confocal elliptic coordinates to evaluate the overlap integral

$$S = \int 1s_A 1s_B\, dv$$

where

$$1s_A = \left(\frac{Z^3}{\pi a_0^3}\right)^{1/2} \exp\left(-\frac{Zr_A}{a_0}\right)$$

$$1s_B = \left(\frac{Z^3}{\pi a_0^3}\right)^{1/2} \exp\left(-\frac{Zr_B}{a_0}\right)$$

$$Ans. \left(1 + \rho + \frac{1}{3}\rho^2\right) e^{-\rho}, \text{ where } \rho = ZR/a_0$$

5.26 Use the result of problem 5.25 to find the percent error in using Eqs. 5.93 and 5.94 rather than Eqs. 5.98 for E_a and E_b in MO theory for the H_2 molecule at the experimental value of R_e ($= 1.4006$ a.u.).

5.27 Show that the simple VB wave function ψ_- and the simple MO wave function ψ_4 for H_2 are identical.

5.28 If the N-H bonds of NH_3 are kept fixed at their equilibrium lengths l and the three bond angles are kept equal but allowed to vary between $0°$ and $120°$, the effect is to allow the distance x from the N atom to the H_3 plane to vary between $\pm l$ and 0. Since a NH_3 molecule with N-H_3 distance x is indistinguishable from one with distance $-x$ (i.e., an inverted molecule), the potential energy function $V(x)$ is symmetric about the origin and exhibits minima at values of x equal to $\pm a$, the equilibrium height of the NH_3 pyramid. In the harmonic approximation, $V(x)$ has the form

$$V(x) = \frac{1}{2} k(x + a)^2 \quad x \le 0$$

$$= \frac{1}{2} k(x - a)^2 \quad x \ge 0$$

where k is a constant.

Taking μ to be the appropriate reduced mass for the problem, F. T. Wall and G. Glockler [*J. Chem. Phys.* **5**, 314 (1937)] found approximate solutions to the Schrödinger equation

$$\left(-\frac{\hbar^2}{2\mu} \frac{d^2}{dx^2} + V(x) \right) \psi(x) = E\psi(x)$$

by using the variational function

$$\psi(x) = c_1 \phi_1(x) + c_2 \phi_2(x)$$

where

$$\phi_1(x) = \left(\frac{\alpha}{\pi} \right)^{1/4} \exp\left[-\tfrac{1}{2} \alpha(x + a)^2 \right]$$

$$\phi_2(x) = \left(\frac{\alpha}{\pi} \right)^{1/4} \exp\left[-\tfrac{1}{2} \alpha(x - a)^2 \right]$$

$$\alpha = \frac{(\mu k)^{1/2}}{\hbar}$$

Show that in the variational secular equation

$$H_{11} = H_{22} = h\nu_0 \left[\frac{1}{2} - a \left(\frac{\alpha}{\pi} \right)^{1/2} \exp(-\alpha a^2) + a^2\alpha[1 - \mathrm{erf}(\sqrt{\alpha a})^{1/2}] \right]$$

$$H_{12} = H_{21} = h\nu_0 \left[\frac{1}{2} - a \left(\frac{\alpha}{\pi} \right)^{1/2} \right] \exp(-\alpha a^2)$$

$$S_{11} = S_{22} = 1$$

$$S_{12} = S_{21} = \exp(-\alpha a^2)$$

where

$$\nu_0 = \frac{1}{2\pi} \left(\frac{k}{\mu} \right)^{1/2} \qquad \mathrm{erf}(t) \equiv \frac{2}{\pi^{1/2}} \int_0^t \exp(-y^2)\, dy$$

Find values of c_2/c_1 corresponding to the two roots of the secular equation, and show that the splitting between these two lowest-lying levels corresponds to the Bohr frequency

$$\nu = 2\nu_0 \left[a \left(\frac{\alpha}{\pi} \right)^{1/2} [1 - \exp(-\alpha a^2)] + \alpha a^2[1 - \mathrm{erf}(\sqrt{\alpha a})^{1/2}] \right] \frac{\exp(-\alpha a^2)}{1 - \exp(-2\alpha a^2)}$$

find an expression for the effective one-electron Hamiltonian $h_e(1)$ for electron 1 in the field of the nuclei and the average field of the other electron.

$$Ans. \ h_e(1) \ = \ -\frac{\hbar^2}{2m}\nabla_1^2 - \frac{e^2}{r_{A1}} - \frac{e^2}{r_{B1}} - \int\phi(2)\frac{1}{r_{12}}\phi(2) \ dv_2$$

(b) If ϕ is approximated by

$$\phi = (2 + 2S)^{-1/2}[1s_A + 1s_B]$$

where

$$S = \int 1s_A 1s_B \ dv$$

and

$$\int 1s_A^2 \ dv = \int 1s_B^2 \ dv = 1$$

use your result for part (a) to find an expression for h_e in terms of the functions $1s_A$ and $1s_B$.

5.40 To show that for an octahedral complex, the energy of an electron in a $d_{3z^2-r^2}$ orbital is the same as one in a $d_{x^2-y^2}$ orbital.
(a) Following Section 3.10, determine how the normalized $d_{3z^2-r^2}$ orbital can be expressed as a sum of the normalized orbitals $d_{z^2-y^2}$ and $d_{z^2-x^2}$.
(b) Compare the expected energy of a $d_{z^2-y^2}$ and a $d_{z^2-x^2}$ electron with that of a $d_{x^2-y^2}$ electron.
(c) Use parts (a) and (b) to find the energy of a $d_{3z^2-r^2}$ electron.

For the following problems a programmable desk calculator or digital computer will be useful.

C5.1 Using the data in Tables 5.3 and 5.5, find values of a and A that fit Eq. 5.36 to R_e and ω_e for each of the diatomic alkali halides. Calculate the electric dipole moment μ_{el} and the dissociation energy D_e and compare with experimental values. Find the average value of a and the mean square deviation.

C5.2 Using the values of α_+ given in Table 5.3, find values of α_- for which Eq. 5.38 gives the experimental μ_{el} for each diatomic alkali halide at the experimental value of R_e. Use the average value of a found in Problem C5.1 and values of A found by fitting the experimental R_e together with the fitted values of α_- and other constants from Tables 5.3 and 5.5 in Eq. 5.36 to calculate ω_e and D_e; compare with experimental values.

C5.3 Repeat problem C5.2, except determine a common value of a by requiring the quantity $[(\omega_e)_{calc} - (\omega_e)_{expt}]^2$ summed over all the diatomic alkali halides to be a minimum by using a standard least-squares routine. How do values of A determined in this way compare with those calculated in Problems C5.1 and C5.2? How well does Eq. 5.36 describe diatomic alkali halide bonding?

C5.4   In addition to the term $-C_6 R^{-6}$, dispersion forces also give rise to a term $-C_8 R^{-8}$. From the estimated values of C_8 in Table 5.5, determine the contribution of this term to the binding energy D_e of the diatomic alkali halide molecules. (Note that the presence of the C_8 term changes the value of A which gives an energy minimum at R_e.)

C5.5   If terms through R^{-10} are kept in the derivation of Eq. 5.36, the additional term $-2e^2 \alpha_+ \alpha_- (\alpha_+ + \alpha_+) R^{-10}$ appears. Calculate the coefficient of R^{-10} for each diatomic alkali halide and determine the effect of including this term upon A and D_e.

C5.6   Repeat Problem C5.1 using BR^{-n} in place of Ae^{-aR} in Eq. 5.36. Are B or n approximately the same for all the diatomic alkali halides?

6

Diatomic and polyatomic molecules

In Chapter 5, we employed the quantum theory for an analysis of the interactions that can occur when two atoms come together. We saw how the requirements of the Pauli principle led to a number of different limiting cases depending on the closed-shell or open-shell character of the combining atoms. For the interactions of open-shell systems, which result in the most common types of chemical bonds, we developed two approaches by examining the H_2 molecule. One of these is called the molecular-orbital (MO) method and the other, the valence-bond (VB) method. Both methods can be extended to more complex systems. In this chapter we show how to employ them more generally for the interpretation of bonding in diatomic and polyatomic molecules.

6.1 HOMONUCLEAR DIATOMIC MOLECULES

The essential aspects of the electronic structure of homonuclear diatomic molecules can be understood by application of the MO and VB methods. We outline the results of both approaches in this section; emphasis is placed on the MO method because it leads to a somewhat more systematic treatment.

6.1.1 Simple MO treatment   Although an exact calculation requires solution of the many-electron Schrödinger equation for each different molecule, a qualitative description can be obtained by use of the one-electron orbital energy levels and the *Aufbau* principle, by analogy with the discussion of many-electron atoms given in Chapter 4. As in the case of H_2, the treatment begins with the atomic orbitals of the two identical atoms, which we label A and B for convenience. Let χ_A be an atomic orbital of atom A and χ_B be an atomic orbital of atom B at a distance R from atom A. For the first-row diatomics in which only the $n = 1$ and 2 shells are occupied, the atomic orbitals are

$$\chi_A = 1s_A, 2s_A, 2p_{xA}, 2p_{yA}, \text{ and } 2p_{zA}$$
$$\chi_B = 1s_B, 2s_B, 2p_{xB}, 2p_{yB}, \text{ and } 2p_{zB}$$

Although the direction of the axes is arbitrary, it is conventional to choose the internuclear axis along the z axis (see Fig. 6.1). Because we are consider-

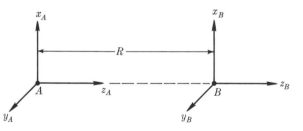

Fig. 6.1. Coordinate system for a diatomic molecule.

ing a homonuclear molecule, the corresponding orbitals on the two separated atoms have the same energy; that is,

$$E(1s_A) = E(1s_B)$$
$$E(2s_A) = E(2s_B)$$
$$E(2p_A) = E(2p_B)$$

We can now combine each of the pairs into molecular orbitals,

$$\frac{1}{\sqrt{2}}\,(1s_A + 1s_B) \qquad \frac{1}{\sqrt{2}}\,(1s_A - 1s_B)$$

$$\frac{1}{\sqrt{2}}\,(2s_A + 2s_B) \qquad \frac{1}{\sqrt{2}}\,(2s_A - 2s_B)$$

$$\cdots \qquad\qquad\qquad \cdots$$

or, in general,

$$\frac{1}{\sqrt{2}}\,(\chi_A + \chi_B) \qquad \frac{1}{\sqrt{2}}\,(\chi_A - \chi_B)$$

where χ_A and χ_B represent the same type of orbital on the two different atoms; as in Section 5.6.2, we have simplified the expression by neglecting the overlap integral in the normalization factor. As $R \to \infty$, the $(+)$ and $(-)$ combinations of each pair of atomic orbitals have the same energy; that is, they are degenerate. However, as R decreases, the orbitals on the two atoms interact and the $(+)$ and $(-)$ molecular orbitals are split in energy, one combination being stabilized (bonding) and the other combination being destabilized (antibonding). To determine the splittings, we can make the same calculation as we did for the $1s$ orbitals of H_2. Atomic integrals α and resonance integrals β are defined for each type of orbital; for example,

$$\alpha_{1s} = \int 1s_A(1)\, h_e(1)\, 1s_A(1)\, dv_1 \quad \beta_{1s} = \int 1s_A(1)\, h_e(1)\, 1s_B(1)\, dv_1$$
$$\alpha_{2s} = \int 2s_A(1)\, h_e(1)\, 2s_A(1)\, dv_1 \quad \beta_{2s} = \int 2s_A(1)\, h_e(1)\, 2s_B(1)\, dv_1 \qquad (6.1)$$

where h_e is an effective one-electron Hamiltonian operator. Knowledge of the exact form of h_e, which depends upon the nature of the occupied molecular orbitals, is not necessary here, since we do not attempt to calculate the α's and β's but regard them as known parameters. However, it can be written down by a generalization of the H_2 discussion given in Problem 5.39 (see also Section 6.1.7).

The energy of each of the bonding orbitals ϕ_b lies below the corresponding level for the two separated atoms by the amount $|\beta|$, while the antibonding orbital ϕ_a has an energy above that of the separated atoms by the amount $|\beta|$, as shown in Fig. 6.2 (see also Eqs. 5.93 and 5.94). Thus the energy difference between ϕ_b and ϕ_a is equal to $2|\beta|$. The $|\beta|$ values, which depend on the internuclear distance R, are expected from their definition (Eq. 6.1) to be roughly proportional to the overlap integral S,

$$S = \int \chi_A(1)\, \chi_B(1)\, dv_1 \qquad (6.2)$$

(Because of the form of h_e, the integrals β and S usually have opposite signs.) This implies that a contribution to binding can result only if the overlap integral is nonzero; if S is large, the splitting $2|\beta|$ is large and if S is small, then $2|\beta|$ is small. For the second-row diatomics, the inner-shell $1s$ orbitals

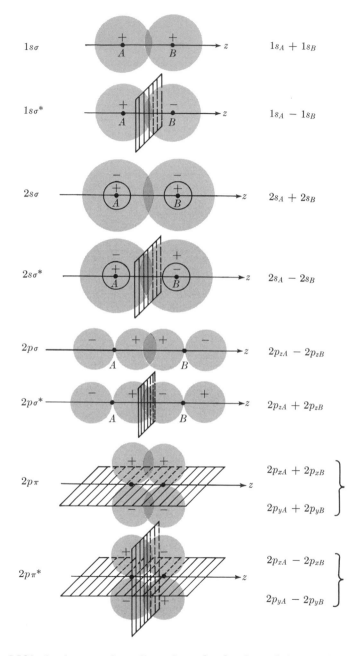

Fig. 6.3. MO's for homonuclear diatomic molecules formed from pairs of AO's. Nodal planes are shaded.

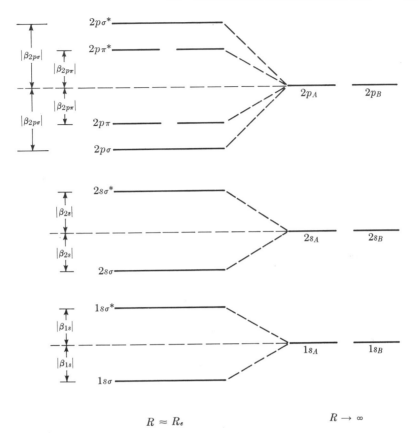

Fig. 6.4. Splitting of MO's for first- and second-row homonuclear diatomic molecules.

plane located at the midpoint of the internuclear axis and perpendicular to it; the corresponding plane in the united atom is the perpendicular one through the atom nucleus. Orbitals are either symmetric (do not change sign) on reflection or antisymmetric (do change sign) on reflection in these planes. From Fig. 6.3, it is clear that $(1s_A + 1s_B)$, $(2s_A + 2s_B)$, $(2p_{zA} - 2p_{zB})$, $(2p_{xA} + 2p_{xB})$, and $(2p_{yA} + 2p_{yB})$ are all symmetric, and $(1s_A - 1s_B)$, $(2s_A - 2s_B)$, $(2p_{zA} + 2p_{zB})$, $(2p_{xA} - 2p_{xB})$, and $(2p_{yA} - 2p_{yB})$ are all antisymmetric. In going from the $R \to \infty$ limit to the $R = 0$ limit, orbitals retain their symmetry. This is why the bonding orbital $(1s_A + 1s_B)$ goes to $1s$, but the antibonding orbital $(1s_A - 1s_B)$ goes to $2p_z$. As is evident from the diagram, in each case the correlation is between the lowest energy orbitals of each type that are not already correlated with other orbitals. In Fig. 6.5, the various scales are to be regarded as diagrammatic (i.e., the energy differences of the separated atom and united atom levels vary from

Table 6.2 Homonuclear diatomic molecules and ions

Molecule	No. of electrons	Simple MO ground-state configuration[a,b]	Bond order	Simple MO approximation to D_e[a]	R_e(Å)	D_e (eV)
H_2^+	1	$(1s\sigma)$	$\frac{1}{2}$	$-\beta_{1s}$	1.060	2.793
H_2	2	$(1s\sigma)^2$	1	$-2\beta_{1s}$	0.7412	4.7476
He_2^+	3	$(1s\sigma)^2(1s\sigma^*)$	$\frac{1}{2}$	$-\beta_{1s}$	1.080	2.5
He_2	4	$(1s\sigma)^2(1s\sigma^*)^2$	0	0	$\cdots$	$\cdots$
Li_2	6	$[He_2](2s\sigma)^2$	1	$-2\beta_{2s}$	2.673	1.14
Be_2	8	$[He_2](2s\sigma)^2(2s\sigma^*)^2$	0	0	$\cdots$	$\cdots$
B_2	10	~~$[Be_2](2p\sigma)^2$~~ $[Be_2](2p\pi)^2$	1	~~$+2\beta_{2p\sigma}$~~ $-2\beta_{2p\pi}$	1.589	$\sim$3.0
C_2	12	~~$[Be_2](2p\sigma)^2(2p\pi)^2$~~ $[Be_2](2p\pi)^4$	2	~~$-2\beta_{2p\sigma}-2\beta_{2p\pi}$~~ $-4\beta_{2p\pi}$	1.242	6.36
N_2^+	13	~~$[Be_2](2p\sigma)^2(2p\pi)^3$~~ $[Be_2](2p\pi)^4(2p\sigma)$	$2\frac{1}{2}$	~~$+\beta_{2p\sigma}-4\beta_{2p\pi}$~~ $+2\beta_{2p\sigma}-4\beta_{2p\pi}$	1.116	8.86
N_2	14	$[Be_2](2p\pi)^4(2p\sigma)^2$	3	$+2\beta_{2p\sigma}-4\beta_{2p\pi}$	1.094	9.902
O_2^+	15	$[Be_2](2p\sigma)^2(2p\pi)^4(2p\pi^*)$	$2\frac{1}{2}$	$+2\beta_{2p\sigma}-3\beta_{2p\pi}$	1.1227	6.77
O_2	16	$[Be_2](2p\sigma)^2(2p\pi)^4(2p\pi^*)^2$	2	$+2\beta_{2p\sigma}-2\beta_{2p\pi}$	1.2074	5.213
F_2	18	$[Be_2](2p\sigma)^2(2p\pi)^4(2p\pi^*)^4$	1	$+2\beta_{2p\sigma}$	1.435	1.34
Ne_2	20	$[Be_2](2p\sigma)^2(2p\pi)^4(2p\pi^*)^4(2p\sigma^*)^2$	0	0	$\cdots$	$\cdots$

[a] In several cases the configuration predicted by the simple MO theory appears to be incorrect; spectroscopic results suggest that $2p\pi_u$ lies below the $2p\sigma_g$ orbital for all second-row homonuclear diatomics except O_2 and O_2^+ (see Fig. 6.11). In the table we list both the simple MO result with a cross through it when it is wrong, and underneath what is thought to be the correct configuration (see Section 6.1.7).

[b] Only the outer shell or subshell is listed; the symbol in square brackets corresponds to a filled core.

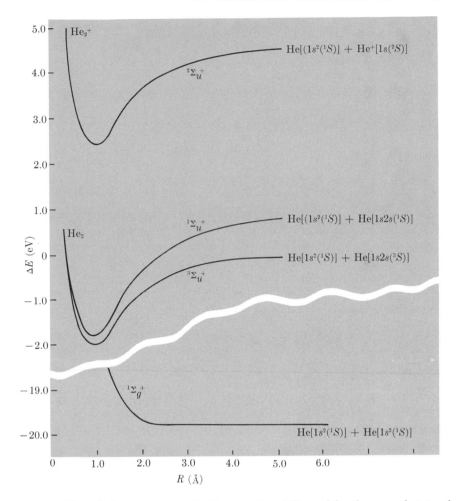

Fig. 6.6. Potential-energy curves for three states of He_2 and for the ground state of He_2^+.

more general self-consistent-field method of combining AO's to form MO's (Section 6.1.7). Moreover, certain exceptions to the simple relation between number of bonds and dissociation energy do occur. The most striking is that for Li_2 and Li_2^+. These two species differ in that Li_2^+ has one bonding electron and Li_2 has two bonding electrons in the $2s\sigma$ orbital. Thus, $D_e(Li_2)$ is expected to be larger than $D_e(Li_2^+)$. However, detailed calculations for these species and experimental data for the corresponding pairs Na_2, Na_2^+ and K_2, K_2^+ suggest that Li_2^+ has the larger dissociation energy, although the equilibrium bond length is greater. One explanation that has been suggested is that because the internuclear distance is very large and the effective β value is very small in these systems [compare $\beta(\text{Li}_2)$ with $\beta(\text{H}_2)$ in Table 6.2] the destabilizing effect of the repulsion between the two bonding electrons in Li_2 overbalances the difference between β and 2β in going from Li_2^+ to Li_2. An examination of H_2 and H_2^+ at a distance $R \sim 2.5a_0$, where the β values are considerably smaller in magnitude than at R_e, shows that the binding energy of H_2^+ at that distance is greater than that of H_2.

6.1.5 *VB treatment* Application of the VB method to homonuclear diatomics is based on direct extension of the formulation given for H_2. Each bond is treated individually and an electron-pair function is written for it. The AO's are categorized either as doubly occupied orbitals which do not contribute to bonding or singly occupied orbitals which do. Taking N_2 as an example, we have to consider the three singly occupied $2p$ orbitals on each atom, the $1s$ and $2s$ orbitals being doubly occupied. Using the same overlap arguments as in MO theory, we expect valence bonds to be formed between $(2p_{xA}, 2p_{xB})$, $(2p_{yA}, 2p_{yB})$, and $(2p_{zA}, 2p_{zB})$. The resulting wave function for the N_2 bonds is

$$[2p_{xA}(1)\ 2p_{xB}(2) + 2p_{xA}(2)\ 2p_{xB}(1)]$$
$$\times\ [2p_{yA}(3)\ 2p_{yB}(4) + 2p_{yA}(4)\ 2p_{yB}(3)]$$
$$\times\ [2p_{zA}(5)\ 2p_{zB}(6) + 2p_{zA}(6)\ 2p_{zB}(5)]$$

where antisymmetrization of the over-all function (including doubly occupied orbitals and spin) would be required to give a complete description satisfying the Pauli principle. The VB result agrees with the MO conclusion that there is a triple bond in N_2. Corresponding VB functions can be written down for other homonuclear diatomics (Problem 6.2.).

6.1.6 *Spin multiplicity and term symbols* There is no ambiguity in the spin states predicted by the simple MO ground-state configurations for the molecules and ions in Table 6.2, with the exception of C_2 and O_2. For example, the molecules H_2, He_2, Li_2, Be_2, B_2, N_2, F_2, and Ne_2 all have two electrons in each occupied orbital, and since the Pauli principle requires that their spins be paired, these molecules are all predicted by the simple MO theory to be singlets. The molecule ions H_2^+, He_2^+, and O_2^+ have one singly occupied orbital and are doublets. In the ground-state configurations of C_2 and O_2, however, a twofold degenerate level ($2p\pi$ in the case of C_2 and $2p\pi^*$ in the case of O_2) is occupied by two electrons in the simple MO theory. If both electrons were to occupy the same orbital, the spins would have to be paired and the state would be a singlet; if each of the two orbitals were singly occupied, the state could be a singlet *or* a triplet, depending on whether the spins are opposed or parallel. Extension of Hund's first rule (Section 4.6.2), which was introduced for the atomic case, to MO's suggests that both C_2 and O_2 are expected to be triplets. From our discussion of $\text{He}(1s2s)$ in

Chapter 4, we know that both states with two singly occupied orbitals should be more stable than that with a doubly occupied orbital, since the electron repulsion is less if two electrons are in different orbitals. Of the two states with singly occupied orbitals, the triplet is more stable because the antisymmetric triplet wave function is small when the two electrons are near one another; that is, the average electron-electron repulsion energy $\langle e^2/r_{12}\rangle$ is expected to be smaller for a pair in a triplet state than in a singlet state. Since all of these arguments are equally valid for atoms and molecules, Hund's rule of maximum spin multiplicity should hold for both.

Term symbols for homonuclear diatomic molecules are obtained by a procedure corresponding to that used for many-electron atoms. The term symbol is written $^{2S+1}\Lambda$, where $2S + 1$ is the multiplicity (1 for singlet, 2 for doublet, etc.) and Λ is the magnitude of M, the total electronic angular momentum about the molecular axis; that is,

$$M = m_1 + m_2 + \cdots$$
$$\Lambda = |M|$$

where $m_1, m_2, \ldots$ are the z-component quantum numbers of the occupied MO's. As in the orbital designations, Greek letters are used for Λ; that is,

$$\Lambda = \begin{array}{cccc} 0 & 1 & 2 & 3 \\ \updownarrow & \updownarrow & \updownarrow & \updownarrow \\ \Sigma & \Pi & \Delta & \Phi \end{array} \cdots$$

Also, a subscript g or u is often included to indicate the symmetry of the total wave function with respect to inversion through the origin, chosen at the center of symmetry (i.e., midpoint) of the molecule. By *inversion* we mean that a point P (see Fig. 6.7) is replaced by its inverse $\bar{P}$; that is, if the

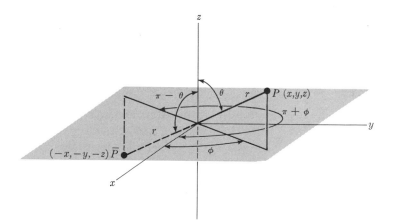

Fig. 6.7. The inversion operation in Cartesian and polar coordinates. The point $\bar{P}$ is obtained by inversion of the point P.

coordinates of P are x,y,z, those of $\bar{P}$ are $-x, -y, -z$. From Fig. 6.7 we see that in polar coordinates inversion replaces the wave function $\chi(r, \theta, \phi)$ by $\chi(r, \pi - \theta, \pi + \phi)$. Each MO either does or does not change sign under inversion; if it does change sign, it is given the symbol u (ungerade) and if it does not, it is given the symbol g (gerade). The g, u symmetries of the MO's are shown in Figs. 6.8 and 6.9. All σ bonding orbitals are g and σ antibond-

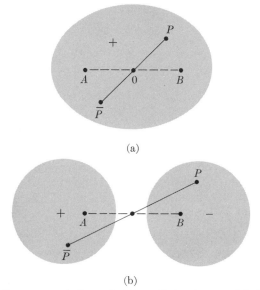

(a)

(b)

Fig. 6.8. Schematic representation of **(a)** a diatomic σ_g orbital (bonding) and **(b)** a diatomic σ_u orbital (antibonding), showing the surfaces of constant value of the functions and the relative signs of the functions in the different regions of space. Radial nodes are not represented.

ing orbitals are u. However, the π orbitals behave oppositely; that is, the π bonding orbitals are u and the π antibonding orbitals are g, as is evident from Fig. 6.9. If there are any number of electrons in orbitals of g symmetry and an even number of electrons with u symmetry, the over-all wave function is g (i.e., does not change sign if all x, y, z, are replaced by $-x$, $-y$, $-z$), while if there are an odd number of electrons in orbitals of u symmetry the over-all wave function is u (does change sign if all x, y, z, are replaced by $-x$, $-y$, $-z$).

As examples of the determination of term symbols for homonuclear diatomics, we consider the ground state of O_2^+ and several states of O_2. The ground-state term symbol for O_2^+ requires consideration only of $(2p\pi)^4$ $(2p\pi^*)$ since all lower-energy electrons are in closed shells or subshells. The bonding $2p\pi$ orbitals are doubly occupied so that the electron spins are paired, and the unpaired electron in the antibonding orbital gives rise to a

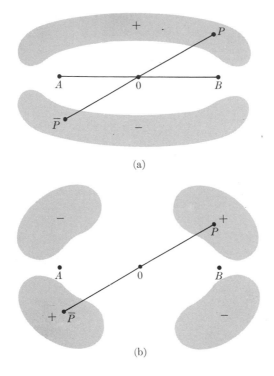

(a)

(b)

Fig. 6.9. Schematic representation of **(a)** a diatomic π_u orbital (bonding) and **(b)** a diatomic π_g orbital (antibonding), showing the surfaces of constant value of the functions and the relative signs of the functions in the different regions of space. Radial nodes are not represented.

doublet because $2S + 1 = 2$ for $S = \frac{1}{2}$. According to Fig. 6.9 the $2p\pi$ orbital is ungerade (u) and the $2p\pi^*$ orbital is gerade (g); since there are an even number of electrons in the ungerade orbitals, the over-all symmetry is gerade. Only the π electrons can contribute to the angular momentum about the z axis. In the filled $2p\pi$ level, two electrons have $m = -1$ and two have $m = +1$; thus the contribution from the $2p\pi$ electrons is zero. The unpaired $2p\pi^*$ electron therefore determines the orbital electronic angular momentum of the molecule; that is, $\Lambda = |m| = 1$. The resulting term symbol for O_2^+ is $^2\Pi_g$.

The lowest O_2 electron configuration is $(2p\pi)^4 (2p\pi^*)^2$. According to Hund's rules, in the ground state the additional electron (relative to O_2^+) goes in the other of the two degenerate $2p\pi^*$ orbitals with its spin parallel to that of the first electron. Consequently, $S = 1$ and the state is a triplet. Since there are still an even number of electrons in u orbitals, the over-all symmetry is g. The quantum number $\Lambda = 0$ because one $2p\pi^*$ electrons

has $m = 1$ and the other $m = -1$. The term symbol is $^3\Sigma_g$. There is, however, an additional symmetry property that must be considered to completely specify Σ states of linear molecules, namely, whether the wave
function changes sign $(-)$ or does not change sign $(+)$ upon reflection in a
plane through the nuclei (e.g., a vertical plane such as that shown in Fig.
6.10). To find out whether the over-all wave function is $+$ or $-$, consider
the effect of such a reflection upon each orbital. If there is a nonzero angular
momentum about the z axis associated with an orbital (e.g., π, δ, . . .),
reflection in a plane through the z axis converts the orbital into its degenerate partner (see Fig. 6.10). In the case of a σ orbital, the reflection leaves

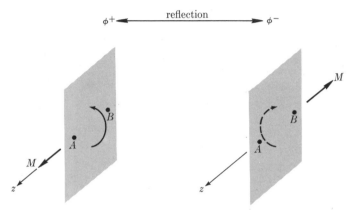

Fig. 6.10. Reflection in a plane through the nuclei of a linear molecule converts
an orbital with angular momentum quantum number $m > 0$ (ϕ^+) into its degenerate
partner with $m < 0$ (ϕ^-) and vice versa.

the orbital unchanged. The ground-state O_2 triplet wave function must, of
course, be antisymmetric with respect to exchange of the coordinates of any
pair of electrons; moreover, the *spatial* part of the wave function must be
antisymmetric, since for a triplet state the *spin* part is symmetric (see
Section 4.3). Since reflection in a vertical plane interchanges the orbitals $(2p\pi^*)_{+1}$ and $(2p\pi^*)_{-1}$, where the subscript indicates the m quantum
number, the effect is the same as an exchange of the labels of the two electrons occupying these orbitals, and the over-all antisymmetrized wave
function changes sign. Thus, the complete ground-state term symbol for
O_2 is $^3\Sigma_g^-$. Note that the $\pm$ symmetry need be considered only for Σ states,
since for states with $\Lambda \neq 0$, reflection of the wave function in a plane
through the nuclei changes the sign of M; that is, one member of the degenerate pair of states is transformed into the other.

   In addition to the ground-state term $^3\Sigma_g^-$, the configuration $(2p\pi^*)^2$
gives rise to two others. If both electrons occupy the same degenerate or-

bital, the value of Λ is 2 and the state is $^1\Delta_g$. If the electrons occupy different degenerate orbitals and have the antisymmetric singlet spin function $[\alpha(1)\beta(2) - \beta(1)\alpha(2)]$, the coordinate wave function is symmetric with respect to exchange of electrons between the two orbitals (see Section 4.3) and the term symbol is $^1\Sigma_g^+$.

Application of the rules given above to the homonuclear diatomics yields the term symbols listed in Table 6.3. From a comparison between the predicted and observed term symbols, we see that the simple MO's appear to be appropriate for most of the diatomic species (i.e., H_2^+, H_2, He_2^+, Li_2, N_2, O_2, O_2^+, F_2) but not all (i.e., the predicted term symbols are wrong for the ground states of B_2, C_2, and N_2^+). To find the source of these discrepancies, we recall that in the homonuclear diatomic molecule correlation diagram (Fig. 6.5) there is an uncertainty about the relative energies of the bonding MO's formed from the $2p\sigma$ and $2p\pi$ AO's; that is, although in the large distance limit $(R \rightarrow \infty)$, $2p\sigma$ is lower than $2p\pi$, the energies are expected to cross at a smaller distance. In determining the molecular configurations on which the term symbols are based, we have assumed that $2p\sigma$ is more stable than $2p\pi$ in all cases (Table 6.2). If we reverse the order, we obtain the configurations listed in Table 6.3, where the orbitals are identified by their g and u symmetry rather than by their bonding and antibonding character. The term symbols resulting from these configurations agree with the observed values given in Table 6.3. For ground-state N_2, O_2^+, and O_2, both $2p\sigma_g$ and $2p\pi_u$ are completely filled, so that the configuration and term symbol are independent of the level ordering. Experimental data for excited states suggest that in N_2 the $2p\pi_u$ orbital is more stable than $2p\sigma_g$, while in O_2^+ and O_2 the reverse order holds. This behavior appears reasonable in terms of the correlation diagram since the O_2^+ and O_2 equilibrium internuclear distances are larger than that of N_2.

A quantitative understanding of the factors determining the energy-level ordering requires a more detailed consideration of MO theory. We present an introductory discussion in the following subsection.

6.1.7 Self-consistent-field orbitals

In formulating the MO's for homonuclear diatomic molecules, we have so far assumed that only the same orbitals on the two atoms interact (only $2s_A$ with $2s_B$, $2p_{xA}$ with $2p_{xB}$, etc.) To obtain a more accurate description, we have to combine these orbitals with other AO's whose overlap with them is nonzero and which can thus contribute to the MO. It turns out that for homonuclear diatomic molecules, all AO's with a given m quantum number should be included. The resulting MO then is identified by the quantum number $\lambda = |m|$ (σ, π, δ, etc.). For example, the σ-type MO's ($\lambda = 0$) are constructed from the "occupied" σ orbitals $1s_A$, $1s_B$, $2s_A$, $2s_B$, and $2p\sigma_A$, and $2p\sigma_B$, and even the "unoccupied" orbitals such as $3s_A$, $3s_B$, $3p\sigma_A$, $3p\sigma_B$, $3d\sigma_A$, $3d\sigma_B$, ... Since in

Table 6.3 Ground-state configurations and term symbols for first-row homonuclear diatomics

Molecule	No. of electrons	Simple MO prediction	Observed	Configuration[a],[b]
H_2^+	1	$^2\Sigma_g^+$	$^2\Sigma_g^+$	$(1s\sigma_g)$
H_2	2	$^1\Sigma_g^+$	$^1\Sigma_g^+$	$(1s\sigma_g)^2$
He_2^+	3	$^2\Sigma_u^+$	$^2\Sigma_u^+$	$(1s\sigma_g)^2(1s\sigma_u)$
Li_2	6	$^1\Sigma_g^+$	$^1\Sigma_g^+$	$(1s\sigma_g)^2(1s\sigma_u)^2(2s\sigma_g)^2$
B_2	10	~~$^3\Sigma_g$~~	$(^3\Sigma_g^-)$[b]	$(1s\sigma_g)^2(1s\sigma_u)^2(2s\sigma_g)^2(2s\sigma_u)^2(2p\pi_u)^2$
C_2	12	~~$^3\Sigma$~~	$^1\Sigma_g^+$	$(1s\sigma_g)^2(1s\sigma_u)^2(2s\sigma_g)^2(2s\sigma_u)^2(2p\pi_u)^4$
N_2^+	13	~~$^2\Pi$~~	$^2\Sigma_g^+$	$(1s\sigma_g)^2(1s\sigma_u)^2(2s\sigma_g)^2(2s\sigma_u)^2(2p\pi_u)^4(2p\sigma_g)$
N_2	14	$^1\Sigma_g^+$	$^1\Sigma_g^+$	$(1s\sigma_g)^2(1s\sigma_u)^2(2s\sigma_g)^2(2s\sigma_u)^2(2p\pi_u)^4(2p\sigma_g)^2$
O_2^+	15	$^2\Pi_g$	$^2\Pi_g$	$(1s\sigma_g)^2(1s\sigma_u)^2(2s\sigma_g)^2(2s\sigma_u)^2(2p\sigma_g)^2(2p\pi_u)^4(2p\pi_g)$
O_2	16	$^3\Sigma_g^-$	$^3\Sigma_g^-$	$(1s\sigma_g)^2(1s\sigma_u)^2(2s\sigma_g)^2(2s\sigma_u)^2(2p\sigma_g)^2(2p\pi_u)^4(2p\pi_g)^2$
F_2	18	$^1\Sigma_g^+$	$^1\Sigma_g^+$	$(1s\sigma_g)^2(1s\sigma_u)^2(2s\sigma_g)^2(2s\sigma_u)^2(2p\sigma_g)^2(2p\pi_u)^4(2p\pi_g)^4$

[a] The configurations given here are based on experimental results and differ in some cases from the simple MO prediction (see text and Table 6.2).
[b] The B_2 ground state is not certain since there is also a very low-lying quintet state.

such a more general description there is no simple way of determining the coefficients of the AO's in a given MO, we have to resort once again to the variational method. By analogy with the Hartree–Fock procedure for atoms (Section 4.8 and Problem 4.14) we minimize the energy of a determinant constructed from the occupied MO's corresponding to the ground-state configuration of the molecule. Each MO is expressed as a linear combination of AO's, and the mixing coefficients are the unknowns to be varied to minimize the energy. This procedure is termed the *self-consistent-field linear-combination-of atomic-orbitals molecular-orbital* method, identified by the letters SCF-LCAO-MO.

As in the atomic case the molecular wave function for a closed-shell system of $2n$ electrons is assumed to have the determinantal form

$$\Psi(1, 2, \ldots, 2n) = \frac{1}{(2n!)^{1/2}} \left| \phi_1\alpha(1) \; \phi_1\beta(2) \; \phi_2\alpha(3) \cdots \phi_n\beta(2n) \right|$$

which corresponds to n doubly occupied molecular orbitals $\phi_i(i = 1, 2, \ldots, n)$; each orbital has one electron with spin α and one with spin β. If the total energy is minimized by varying the form of all the MO's, there results the one-electron Schrödinger-like equation of the form

$$h_e(1) \; \phi_i(1) = \epsilon_i \; \phi_i(1)$$

It is called the Hartree–Fock or SCF equation and has as its solutions the orbitals ϕ_i with orbital energies ϵ_i. The equation is analogous to the Schrödinger equation for a one-electron molecule (e.g., H_2^+) but differs in that the effective Hamiltonian $h_e(1)$ replaces the one-electron Hamiltonian operator $H_0(1)$

$$H_0(1) = -\frac{\hbar^2}{2m} \nabla_1^2 - \frac{Z_A e^2}{r_{1A}} - \frac{Z_B e^2}{r_{1B}}$$

where Z_A and Z_B are the nuclear charges. A generalization of the result for H_2 obtained in Problem 5.39 shows that $h_e(1)$ has the form

$$h_e(1) = H_0(1) + J(1) - K(1)$$

The first term in $h_e(1)$ is just $H_0(1)$, the "bare-nucleus" Hamiltonian given above, namely, the sum of the kinetic-energy operator for electron 1 and its interactions with the two nuclei. The second term $J(1)$ is the sum of Coulomb interactions between electron 1 and the charge clouds of all the other electrons; for the closed-shell system of $2n$ electrons occupying the orbitals $\phi_1, \phi_2, \ldots, \phi_n$, we have

$$J(1) = 2 \sum_{j=1}^{n} \int \phi_j{}^*(2) \frac{e^2}{r_{12}} \phi_j(2) \; dv_2$$

where the factor 2 appears because the orbitals are doubly occupied. The third term $K(1)$ in the expression for $h_e(1)$ arises from the requirement that the wave function be antisymmetric with respect to exchange of any pair of electrons; it is the sum of the exchange interactions of electron 1 in orbital ϕ_i with the electrons with the same spin in other orbitals. The form of $K(1)$ acting on $\phi_i(1)$ is

$$K(1)\ \phi_i(1) = \sum_{j=1}^{n} \int \phi_j{}^*(2)\ \frac{e^2}{r_{12}}\ \phi_i(2)\ dv_2\ \phi_j(1)$$

that is, the operator exchanges electron 1 in orbital ϕ_i with electron 2 in orbital ϕ_j. The reader can verify that the average of $K(1)$ over the orbital ϕ_i gives the sum of the exchange integrals involving ϕ_i;

$$\int \phi_i{}^*(1)\ K(1)\ \phi_i(1)\ dv_1 = \sum_{j=1}^{n} K_{ij}$$

where

$$K_{ij} = \int \phi_i{}^*(1)\ \phi_j{}^*(2)\ \frac{e^2}{r_{12}}\ \phi_i(2)\ \phi_j(1)\ dv_1\ dv_2$$

Since the operator $h_e(1)$ depends on the form of all the orbitals through $J(1)$ and $K(1)$, the Hartree–Fock equation must be solved for the ϕ_i and ϵ_i by a self-consistent process. An initial set of orbitals $\phi_1{}^0, \phi_2{}^0, \ldots, \phi_n{}^0$ is chosen and the operator $h_e(1)$ corresponding to them is calculated. The Hartree–Fock equations corresponding to this $h_e(1)$ are solved and a new set of $\phi_i(\phi_1{}^1, \phi_2{}^1, \ldots, \phi_n{}^1)$ is determined. From the $\phi_i{}^1$ an improved operator $h_e(1)$ is obtained. The iteration process is continued until the $\phi_i(1)$ obtained by solving the Hartree–Fock equations are the same as those used in constructing $h_e(1)$.

In the SCF-LCAO-MO method, the MO's are expressed as a linear combination of AO's; for example, if ϕ_1 and ϕ_2 were σ MO's, they could have the form

$$\phi_1 = a_1(1s_A) + a_2(1s_B) + a_3(2s_A) + a_4(2s_B) + a_5(2p_{zA}) + a_6(2p_{zB}) + \cdots$$
$$\phi_2 = b_1(1s_A) + b_2(1s_B) + b_3(2s_A) + b_4(2s_B) + b_5(2p_{zA}) + b_6(2p_{zB}) + \cdots$$

where the coefficients $a_1, a_2, \ldots, b_1, b_2, \ldots$ are the variational parameters with respect to which the total energy is to be minimized. One obtains a set of equations for the coefficients analogous to the Hartree–Fock equation for the orbitals themselves. The simple MO scheme would be obtained from this procedure if the coefficients were restricted to be nonzero for only one type of orbital; if for ϕ_1 and ϕ_2, say, a_1, a_2 and b_1, b_2 are allowed to vary and $a_3, a_4, \ldots, b_3, b_4, \ldots$ are all set equal to zero, these orbitals reduce to $1s_A + 1s_B$ and $1s_A - 1s_B$. The more general solutions are considered in what follows.

The basic results obtained from an SCF-LCAO-MO calculation for a molecule are the MO's ϕ_1, ϕ_2, . . . and their energies ϵ_1, ϵ_2, ... Because of the mixing of different AO's, the order of the orbital energies can be changed from the simple scheme shown in Fig. 6.4, though it must correspond to the possibilities included in the correlation diagram (Fig. 6.5). Since the individual MO's are no longer composed only of s orbitals or of p

Table 6.4 *Minimal-basis SCF-LCAO-MO's for N_2*[a]

		-15.72176	-1.45241	-0.54451
	ζ	$1\sigma_g$	$2\sigma_g$	$3\sigma_g$
$1s\sigma_g$	6.70	0.70447	-0.16890	-0.06210
$2s\sigma_g$	1.95	0.00842	0.48828	0.40579
$2p\sigma_g$	1.95	0.00182	0.23970	-0.60324

		-15.71965	-0.73066
	ζ	$1\sigma_u$	$2\sigma_u$
$1s\sigma_u$	6.70	0.70437	-0.16148
$2s\sigma_u$	1.95	0.01972	0.74124
$2p\sigma_u$	1.95	0.00857	-0.76578

		-0.57951
	ζ	$1\pi_u$
$2p\pi_u$	1.95	-0.62450

[a] B. J. Ransil, *Rev. Mod. Phys.* **32**, 245 (1960), with $R_e = 1.0456$ Å. The orbital energies ϵ_i in a.u. are given above the orbital designations.

orbitals, they are identified simply by their quantum numbers (σ,π,δ), their g or u symmetries, and their relative energies (e.g., for those of σ_g type, the lowest energy orbital is written $1\sigma_g$, the next $2\sigma_g$, and so on). The designation gives no information about the atomic origin nor of the bonding or antibonding character. Moreover, the simple interpretation of binding energies in terms of resonance integrals β is forfeited when the SCF-LCAO-MO method is used. However, although significant mixing of different AO's does occur, the dominant contributions are often those given by the simple treatment (see Table 6.4).

We illustrate the SCF-LCAO-MO calculation by listing the results of a relatively simple and a very extensive treatment for the N_2 molecule. The first was done some years ago by B. J. Ransil [*Rev. Mod. Phys.* **32**, 245 (1960)] and the second more recently by P. E. Cade, K. D. Sales, and A. C. Wahl [*J. Chem. Phys.* **44**, 1973 (1966)]. The work of Ransil used what is called a *minimal basis set;* that is, only the occupied inner shell and valence orbitals of the N atoms were used for constructing the MO's. The exponents for the Slater-type orbitals (Eqs. 4.74 and 4.75) were determined by the Slater rules (Problem C4.5). Thus, the σ orbitals are mixtures of $1s_A$, $1s_B$, $2s_A$, $2s_B$, $2p_{zA}$, $2p_{zB}$ and the π orbitals of $2p_{xA}$, $2p_{xB}$ and $2p_{yA}$, $2p_{yB}$. In Table 6.4 are given the resulting MO's with their energies. The orbital designations down the first column indicate the simple MO's (unnormalized symmetry orbitals) formed from the sum or difference of pairs of Slater AO's; e.g., $1s\sigma_g$ represents $(1s_A + 1s_B)$ and $1s\sigma_u$ is $(1s_A - 1s_B)$. The next column, headed by ζ, gives the orbital exponents for the AO's comprising these simple MO's. The remaining columns list the contribution made by each of the symmetry orbitals to the SCF-MO designated at the top of the column $(1\sigma_g, 2\sigma_g, \ldots, 1\pi_u)$; e.g., the MO $1\sigma_g$ has the form

$$1\sigma_g = 0.70447\ 1s\sigma_g + 0.00842\ 2s\sigma_g + 0.00182\ 2p\sigma_g$$

Above each SCF-MO symbol is the orbital energy in a.u. (e.g., $1\sigma_g$ has the energy -15.72176 a.u.). From the results in Table 6.4 the ground-state configuration and term symbol are expected to be

$$N_2 \quad (1\sigma_g)^2\ (1\sigma_u)^2\ (2\sigma_g)^2\ (2\sigma_u)^2\ (1\pi_u)^4\ (3\sigma_g)^2 \quad {}^1\Sigma_g{}^+$$

where we have listed the occupied orbitals in order of increasing energy (see also Fig. 6.11). It is important to note that the energy of $1\pi_u$ and $3\sigma_g$ are reversed from the order given previously, with $1\pi_u$ more stable here than $3\sigma_g$ (but see below); this inversion is caused by the fact that the $2\sigma_g$ orbital, which is $2s_A + 2s_B$ in the simple scheme, is stabilized by mixing with $2p_{zA} + 2p_{zB}$ and the $3\sigma_g$ orbital is consequently destabilized (Problem 6.5). Each of the occupied MO's in the present calculation is still dominated by the AO's expected from the simple treatment. Thus, $1\sigma_g$ and $1\sigma_u$ are mainly $1s_A$ and $1s_B$, $2\sigma_g$ and $2\sigma_u$ are mainly $2s_A$ and $2s_B$, and $3\sigma_g$ is mainly $2p_{zA}$, $2p_{zB}$; however, the unoccupied orbital $3\sigma_u$ is composed almost equally of $2s_A$, $2s_B$ and $2p_{zA}$, $2p_{zB}$.

In the more complete calculation of Cade, Sales, and Wahl, Slater-type orbitals $1s \ldots 4f$ were combined first into unnormalized symmetry orbitals, as in the Ransil calculation, and then linear combinations of these simple MO's of each symmetry type were taken (see Sections 6.1.2 and 6.1.6). The orbital exponent ζ and the coefficient of each simple symmetry MO was varied so as to minimize the total energy. To allow additional degrees of freedom, several Slater orbitals with different exponents were included for the same *nlm* values; for example $1s$ and $1s'$ differ only in exponents ζ and

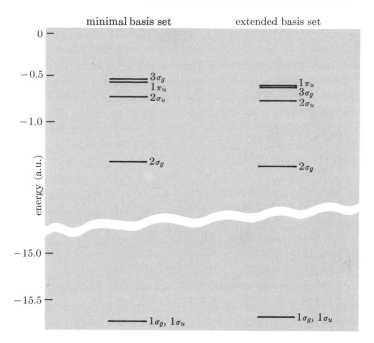

Fig. 6.11. Energies of occupied MO's for N_2 from minimal basis set (Table 6.4) and extended basis set (Table 6.5) calculations.

ζ', and similarly for $2p$, $2p'$, and $2p''$. The final results for the N_2 orbitals are given in Table 6.5 and the relative energies are shown in Fig. 6.11. The total energy obtained by this procedure for N_2 is close to the experimental result (see Problem 6.6) although the calculated dissociation energy D_e is only 5.27 eV, as compared with the experimental value of 9.902 eV. Note that even here the $1\sigma_g$ orbital is mainly $1s\sigma_g$ and that $1\sigma_u$ is mainly $1s\sigma_u$; the $3\sigma_g$ orbital is composed almost equally of $2s\sigma_g$, $2p\sigma_g$, and $2p\sigma_g'$; while the $1\pi_u$ orbital is almost entirely $2p\pi_u$ and $2p\pi_u'$. These two orbitals again have comparable energies, with the $3\sigma_g$ lying slightly lower. Thus, in this more extensive calculation, the $1\pi_u$, $3\sigma_g$ ordering is reversed from that given in Table 6.4. Since the energy difference is small, it is not clear whether $3\sigma_g$ or $1\pi_u$ actually lies lower, nor can one be sure that it is meaningful to speak of a relative ordering; that is, interactions between electrons may be sufficiently important to break down the orbital picture for this case. Since the term symbol for N_2 is independent of the ordering, an experimental determination is not possible.

The ion N_2^+ is observed to have a $^2\Sigma_g^+$ ground state, so that it has the ground-state configuration

$$N_2^+ \quad (1\sigma_g)^2 (1\sigma_u)^2 (2\sigma_g)^2 (2\sigma_u)^2 (1\pi_u)^4 (3\sigma_g) \quad ^2\Sigma_g^+$$

bonding electrons ($1s_H$, $2p_{zF}$) can be written as a linear combination of covalent and ionic structures:

$$\psi_{VB}(1, 2) = N[A\psi_{cov}\,(\text{H-F}) + B\psi_{ion}\,(\text{H}^+\text{F}^-)] \tag{6.3}$$

where

$$\psi_{cov} = \frac{1}{\sqrt{2}}\,[1s_H(1)\,2p_{zF}(2) + 2p_{zF}(1)\,1s_H(2)] \tag{6.4}$$

and

$$\psi_{ion} = 2p_{zF}(1)\,2p_{zF}(2) \tag{6.5}$$

The additional ionic term $1s_H(1)\,1s_H(2)$ corresponding to H$^-$F$^+$ is not included because the relative values of IP and EA for H and F show that it should be unimportant (see Table 4.2). If overlap between the $1s_H$ and $2p_{zF}$ orbitals is neglected in Eq. 6.3, the normalization factor is

$$N = (A^2 + B^2)^{-1/2} \tag{6.6}$$

since ψ_{cov} and ψ_{ion} are individually normalized. The constants A and B can be determined by a variational calculation. However, it is useful to estimate their ratio from the measured dipole moment. If we assume that the non-

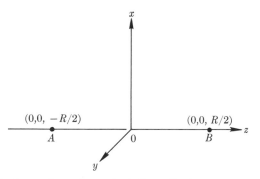

Fig. 6.12. Coordinate system for calculating dipole moments of heteropolar diatomic molecules.

bonding electrons ($1s^2$, $2s^2$, $2p_x^2$, $2p_y^2$) are centered on the F nucleus, the effective positive charge of the two atoms (exclusive of the bonding electrons) is e in each case. The center of positive charge is thus half-way between the nuclei, at which point we position the origin of our coordinate system (see Fig. 6.12). The dipole moment is given by the sum of the positive and negative charges multiplied by their displacements along the z axis. If $\langle z_1 \rangle$ and $\langle z_2 \rangle$ are the average values of z for the bonding electrons, we have

$$\mu = -e(\langle z_1 \rangle + \langle z_2 \rangle) + e\left(\frac{R}{2} - \frac{R}{2}\right) \qquad (6.7)$$

the term in the first parentheses arising from the electrons, and the second (which is zero) from the nuclei and the atomic electrons not involved in the bonding. Since the electrons are indistinguishable, $\langle z_1 \rangle = \langle z_2 \rangle$, so that Eq. 6.7 reduces to

$$\mu = -2e\langle z_1 \rangle \qquad (6.8)$$

To obtain the dipole moment in terms of A and B, we must calculate $\langle z_1 \rangle$; its value is given by

$$\langle z_1 \rangle = \int [\psi_{VB}(1,2)]^2 z_1 \, dv_1 \, dv_2 \qquad (6.9)$$

Substitution of Eqs. 6.3–6.6 into Eq. 6.9 gives

$$\langle z_1 \rangle = (A^2 + B^2)^{-1} \{ \tfrac{1}{2} A^2 [\int [1s_H(1)]^2 z_1 \, dv_1 + \int [2p_{zF}(1)]^2 z_1 \, dv_1] \\ + B^2 \int [2p_{zF}(1)]^2 z_1 \, dv_1 + \sqrt{2} AB \int 1s_H(1) \, 2p_{zF}(1) z_1 \, dv_1 \} \qquad (6.10)$$

In deriving Eq. 6.10, we have assumed that the overlap integral

$$S = \int 1s_H(1) \, 2p_{zF}(1) \, dv_1$$

can be neglected and have used the fact that

$$\int [1s_H(2)]^2 \, dv_2 = \int [2p_{zF}(2)]^2 \, dv_2 = 1$$

Equation 6.10 is considerably simplified if we further assume that

$$\int 1s_H(1) \, 2p_{zF}(1) z_1 \, dv_1 \simeq 0$$

This assumption is consistent with that of zero overlap.[1] The remaining integrals are average values of z for an electron in the hydrogen orbital,

$$\int [1s_H(1)]^2 z_1 \, dv_1 = -\frac{R}{2} \qquad (6.11)$$

or in the fluorine orbital,

$$\int [2p_{zF}(1)]^2 z_1 \, dv_1 = \frac{R}{2} \qquad (6.12)$$

(Problem 6.48). Equations 6.8–6.12 can be combined to give a simple, approximate formula for the molecular dipole moment:

$$\mu = \left(\frac{B^2}{A^2 + B^2}\right) eR_e = \left(\frac{B^2}{A^2 + B^2}\right) \mu_{\text{ion}} \qquad (6.13)$$

---

[1] Neglect of integrals of this type is a special case of the so-called zero differential overlap (ZDO) approximation in which all integrals that have at least one electron on more than one center are set equal to zero; for example, here the integral has electron 1 on both hydrogen and fluorine (see Reference 5 in the Additional Reading list).

almost entirely of the $1s_{\mathrm{Li}}$ atomic orbital. The 2σ orbital is dominated by the $1s_{\mathrm{H}}$ orbital, but with significant contributions from $2s_{\mathrm{Li}}$ and $2p_{z\mathrm{Li}}$ (sp hybridization, see Section 6.3). This is not surprising because $2s_{\mathrm{Li}}$ and $2p_{z\mathrm{Li}}$ have similar energies and both overlap significantly with $1s_{\mathrm{H}}$. That the wave function corresponds to a reasonable charge distribution follows from the fact that the calculated dipole moment is $\mu = 6.41\mathrm{D}\,(\mathrm{Li^+H^-})$ in approximate agreement with the experimental value of $\mu = 5.88\mathrm{D}$.

6.2.1 Electronegativity A quantity which provides information concerning the fractional ionic character of a bond is the *electronegativity difference* between the pairs of atoms (or orbitals) involved. The intuitive notion of electronegativity is the "electron attracting power" of an atom. Thus, a quantitative measure of electronegativity difference between atoms A and B should be related to the relative stabilities of the structures A^+B^- and A^-B^+. Since the ionic attraction and overlap repulsion of the two structures are approximately the same, we can follow the discussion of Section 5.2 to show that the energy difference is (ignoring polarization effects)

$$E(A^+B^-) - E(A^-B^+) = (\mathrm{IP}_A - \mathrm{EA}_B) - (\mathrm{IP}_B - \mathrm{EA}_A)$$
$$= (\mathrm{IP}_A + \mathrm{EA}_A) - (\mathrm{IP}_B + \mathrm{EA}_B)$$

Mulliken used a similar argument to define the electronegativity difference $(X_A - X_B)$ between atom A and B as the energy difference per electron between the two ionic structures,

$$X_A - X_B = \tfrac{1}{2}[(\mathrm{IP}_A + \mathrm{EA}_A) - (\mathrm{IP}_B + \mathrm{EA}_B)]$$

For the electronegativity X of an atom, we have therefore

$$X = \tfrac{1}{2}(\mathrm{IP} + \mathrm{EA}) \tag{6.19}$$

When used to calculate electronegativities the values of IP and EA given in Table 4.2 must be modified, however, because the electronic structure of the atom that is participating in a bond is often significantly different from that of the isolated system to which IP and EA correspond. As an example, we consider the F atom forming a single bond as in HF. For this case, the EA value of Table 4.2 is essentially applicable because the ground terms of $\mathrm{F}(^2P)$ and $\mathrm{F}^-(^1S)$ are appropriate for bonding; that is, in $\mathrm{F}(^2P)$, the configuration is $(1s^2\,2s^2\,2p_x^2\,2p_y^2\,2p_z)$ with one unpaired electron in $2p_z$, just the one considered in HF, and in $\mathrm{F}^-(^1S)$ the configuration is the closed shell $(1s^2\,2s^2\,2p_x^2\,2p_y^2\,2p_z^2)$ that corresponds to the ionic $\mathrm{H^+F^-}$ interaction. To obtain the proper IP we have to determine the electronic structure of F^+ appropriate for bonding (i.e., for $\mathrm{H^-F^+}$). The essential point is that it is the bonding $2p_z$ electron which has to be transferred, so that the F^+ configuration is $1s^2\,2s^2\,2p_x^2\,2p_y^2$. This has all electrons paired (i.e., it is a singlet state), in contrast to the ground term, which according to the rules of Section 4.6 is $1s^2\,2s^2\,2p^4(^3P)$. Thus, the IP value (17.4 eV) of Table 4.2

must be corrected to include the excitation energy (3.6 eV) to the F^+ bonding configuration; the resulting value is 21 eV.

From the discussion for F, it is clear that the electronegativity of an atom depends on the particular bonding scheme under consideration. An atom prepared for a given type of bond is said to be in a certain *valence state* (see also Section 6.3) so that different valence-state electronegativities can be defined. In Table 6.7 we list atomic electronegativities obtained from

Table 6.7   *Electronegativities of selected atoms in their most prevalent valence state*[a]

Atom	Bonding orbital[b]	IP(eV)	EA(eV)	X[c]	x[d]	x_P[e]
H	s	13.595	0.754	7.17	2.26	2.1
Li	s	5.390	0.6	3.0	0.95	1.0
Be	sp hybrid	7.94	1.26[f]	4.60	1.46	1.5
B	sp^2 hybrid	10.74	2.0	6.4	2.02	2.0
C	sp^3 hybrid	13.85	2.30	8.07	2.55	2.5
N	p	13.83	0.85	7.34	2.32	3.0
O	p	17.28	1.98	9.63	3.04	3.5
F	p	20.98	3.47	12.22	3.85	4.0
Na	s	5.138	0.54	2.84	0.90	0.9
Mg	sp hybrid	6.74	1.62[f]	4.18	1.32	1.2
Al	sp^2 hybrid	8.70	2.6	5.6	1.77	1.5
Si	sp^3 hybrid	11.22	3.74	7.48	2.36	1.8
P	p	10.6	1.27	5.9	1.86	2.1
S	p	12.50	2.37	7.44	2.35	2.5
Cl	p	15.09	3.65	9.37	2.96	3.0
K	s	4.339	0.49,[g] 0.75	2.41, 2.54	0.76, 0.80	0.80
Br	p	13.72	3.51	8.61	2.72	2.8
Rb	s	4.176	0.42	2.30	0.73	0.8
I	p	12.61	3.38	7.99	2.52	2.5

[a] From H. O. Pritchard and H. A. Skinner, *Chem. Rev.* **55,** 745 (1955) except for new ground-state EA values from Table 4.2.

[b] This column indicates the type of bonding orbital assumed in determining the electronegativity (see text) (i.e., *s*, *p*, or a hybrid orbital).

[c] Mulliken scale; calculated by Eq. 6.19.

[d] Mulliken scale adjusted by Eq. 6.20 to Pauling scale.

[e] Pauling thermochemical scale.

[f] Estimated by extrapolation of ionization potentials of isoelectronic ions; H. A. Skinner and H. O. Pritchard, *Trans. Faraday Soc.* **49,** 1254 (1953).

[g] These are the two values of EA listed in Table 4.2; it is not yet known which is more accurate.

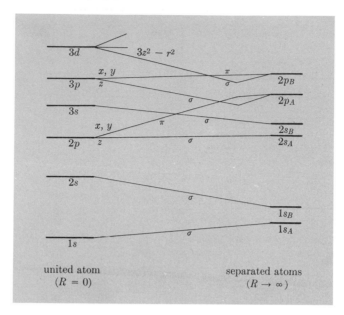

Fig. 6.14. United atom-separated atom correlation diagram for heteronuclear diatomic molecules.

Here the symbol KK is used to denote completed K shells ($n = 1$) for each atom. Comparison of these configurations with those for N_2 and O_2^+ given in Table 6.3 shows the expected correspondence between CO and N_2 and between NO and O_2^+. Since the nuclei have different charges for heteronuclear diatomic molecules like CO and NO, the MO's are not symmetric or antisymmetric with respect to inversion and the classification g, u does not apply. The configurations and term symbols for some heteronuclear diatomics are listed in Table 6.8; in this table, we designate each MO by its Λ value and its energy order (e.g., 1σ, 2σ, 3σ, . . .), rather than by its bonding or antibonding character.

The differences and similarities between the electronic structures of CO and N_2 can best be illustrated by comparing the SCF-LCAO wave functions of Tables 6.5 and 6.9. The lowest two MO's for $N_2(1\sigma_g, 1\sigma_u)$ are mainly $1s_A + 1s_B$ and $1s_A - 1s_B$, respectively; their energies are almost degenerate, so that the linear combinations

$$\tfrac{1}{2}(1\sigma_g + 1\sigma_u) = 1s_A$$
$$\tfrac{1}{2}(1\sigma_g - 1\sigma_u) = 1s_B$$

are almost equally good descriptions of these orbitals. This is another way of saying that the $1s$ orbitals of the N atoms in N_2 do not interact. In the case of CO, where the coefficients of corresponding C and O orbitals are not

Table 6.8 *Ground-state configurations and term symbols for some hetero-nuclear diatomics*

Molecule	No. of electrons	Term symbol	Configuration
LiH	4	$^1\Sigma^+$	$K(2\sigma)^2$
BeH	5	$^2\Sigma^+$	$K(2\sigma)^2\,3\sigma$
CH	7	$^2\Pi$	$K(2\sigma)^2(3\sigma)^2(1\pi)$
NH	8	$^3\Sigma^-$	$K(2\sigma)^2(3\sigma)^2(1\pi)^2$
OH	9	$^2\Pi$	$K(2\sigma)^2(3\sigma)^2(1\pi)^3$
HF	10	$^1\Sigma^+$	$K(2\sigma)^2(3\sigma)^2(1\pi)^4$
BeO, BN	12	$^1\Sigma^+$	$KK(3\sigma)^2(4\sigma)^2(1\pi)^4$
CN, BeF	13	$^2\Sigma^+$	$KK(3\sigma)^2(4\sigma)^2(1\pi)^4(5\sigma)$
CO	14	$^1\Sigma^+$	$KK(3\sigma)^2(4\sigma)^2(1\pi)^4(5\sigma)^2$
NO	15	$^2\Pi$	$KK(3\sigma)^2(4\sigma)^2(5\sigma)^2(1\pi)^4(2\pi)$

required by symmetry to have equal magnitudes, Table 6.9 shows 1σ to be mainly $1s_O$ and 2σ to be mainly $1s_C$, with the former considerably lower in energy than the latter. The 3σ orbital of CO is mainly $2s_O$, with some $1s_O$, $2p\sigma_O'$, $3s_O$, $2p\sigma_C'$, $3s_C$, and $1s_C$ mixed in. Thus, 3σ can be described as an O orbital that is only slightly bonding. In N_2, the corresponding orbital ($2\sigma_g$) is also mainly $2s$ ($2s\sigma_g$ in this case) with some contribution from other orbitals. The 4σ orbital is primarily an O orbital with $2s$ and $2p\sigma$ contributions ($-2s_O + 2p\sigma_O + 2p\sigma_O'$) and has a slightly higher degree of mixture of C orbitals than in the case of 3σ; the corresponding orbital in N_2 is $2\sigma_u$. The 5σ orbital is mainly $2s_C$ with some $2p\sigma_C'$, $3s_C$, and $2p\sigma_O$, $2p\sigma_O'$ mixed in; it is related to $3\sigma_g$ of N_2, which has more $2s$, $2p$ mixing, however. The 1π orbital is a mixture of $2p\pi_O$, $2p\pi_O'$, $2p\pi_C$ in that order, but the contributions from the three AO's are comparable; the $1\pi_u$ orbital in N_2 is similarly composed primarily of $2p\pi$ and $2p\pi'$.

The charge density of each MO in CO has its center displaced toward one or the other atom. This contrasts with N_2 in which each of the orbitals contributes a symmetric charge distribution. Table 6.10 gives the average $\langle z \rangle$ calculated for each CO orbital. The atom towards which the charge distribution of each MO is displaced is also indicated. The shift of four of the six electron distributions toward O compensates for its excess of nuclear charge over that of C, and can be taken as a manifestation of the greater electronegativity of O. The 5σ electrons have their centers of charge beyond the C atom; these essentially "lone-pair" electrons help to balance the shift of 3σ, 4σ, and 1π towards O, so that the average position of the

Table 6.10 $\langle z \rangle$ for CO[a]

MO	$\langle z \rangle$ (a.u.)	n[b]	Atom
1σ	-1.0657	2	O
2σ	1.0653	2	C
3σ	-0.4841	2	$\sim$O
4σ	-1.0910	2	O
5σ	1.5700	2	C
1π	-0.5572	4	$\sim$O
av.	-0.1600		

[a] W. M. Huo, *J. Chem. Phys.* **43,** 624 (1965). $R_e = 2.132$ a.u.; the origin is at the midpoint between C and O, and C is in the $+z$ direction.
[b] Number of electrons occupying each orbital.

over-all electron cloud is near the origin (-0.1600 a.u.). Accordingly, the dipole moment calculated from both the electronic and nuclear charge distribution turns out to be small. Experimentally, $\mu = 0.112$D corresponding to C$^-$O$^+$, whereas the SCF calculation illustrated here gives C$^+$O$^-$ (see Problem 6.11). The difference in sign of μ for the experimental and SCF values is somewhat unusual. It arises because the magnitude of μ is so small that a not unreasonable error (± 0.2D) in the calculated value can change the sign. In HCl, for example, $\mu_{\text{expt}} = 1.081$D and $\mu_{\text{SCF}} = 1.97$D. A configuration-interaction calculation for CO [F. Grimaldi, A. Lecourt, and C. Moser, *Intern. J. Quantum Chem.* **1S,** 153 (1967)] removes the sign discrepancy and gives $\mu = 0.17$D(C$^-$O$^+$).

The N$_2$ and CO orbital energies can be seen to have a similar pattern as shown in Table 6.11. There are the inner-shell pairs ($1\sigma_g$, $1\sigma_u$ in N$_2$; 1σ, 2σ in CO) which have very low energies and the valence orbitals ($2\sigma_g$, $2\sigma_u$, $3\sigma_g$, $1\pi_u$ in N$_2$; 3σ, 4σ, 5σ, 1π in CO) which are comparable in energy. However, in contrast to N$_2$, 1π for CO has a lower calculated energy than 5σ. The latter is also the case for BF, which is isoelectronic with N$_2$ and CO (see Table 6.11).

For CO, the lowest excited state is formed by transferring an electron from 5σ to 2π. The 2π has a distribution with its center displaced towards the O atom. Thus, in the excited state with the configuration and term

$$\text{CO} \quad (1\sigma)^2(2\sigma)^2(3\sigma)^2(4\sigma)^2(5\sigma)(1\pi)^4(2\pi) \quad {}^3\Pi$$

a large dipole moment is to be expected, since the balance of nuclear and electronic charges of the ground state is now destroyed. R. S. Freund and

Table 6.11 *Energies of occupied orbitals in N₂, CO, BF*[a]

N₂		CO		BF	
Orbital	Energy	Orbital	Energy	Orbital	Energy
$1\sigma_g$	-15.68195	1σ	-20.66123	1σ	-26.37504
$1\sigma_u$	-15.67833	2σ	-11.35927	2σ	$-\ 7.70897$
$2\sigma_g$	$-\ 1.47360$	3σ	$-\ 1.51920$	3σ	$-\ 1.69759$
$2\sigma_u$	$-\ 0.77796$	4σ	$-\ 0.80235$	4σ	$-\ 0.85373$
$3\sigma_g$	$-\ 0.63495$	5σ	$-\ 0.55304$	5σ	$-\ 0.40424$
$1\pi_u$	$-\ 0.61544$	1π	$-\ 0.63771$	1π	$-\ 0.74447$

[a] From the SCF-LCAO-MO calculations referred to in Tables 6.5 and 6.9. All values in a.u. are for the equilibrium internuclear distance.

W. Klemperer [*J. Chem. Phys.* **43**, 2422 (1965)] determined μ to be 1.38D in the lowest ³Π state which is to be compared with an SCF calculation of 2.46D [W. Huo, *J. Chem. Phys.* **45**, 1554 (1966)]. This doubly bonded excited state is expected to bear more resemblance to the carbonyl group (>C=O) than to the ground state; the carbonyl dipole moment is approximately 2.3D.

6.3 POLYATOMIC MOLECULES

Although polyatomic molecules appear to be very complicated because they contain many electrons distributed over several nuclei, in most cases the bonding can be understood by means of the approaches used for diatomic molecules. The reason for this is that most chemical bonds are localized; that is, the bonding electrons can be divided up into electron pairs that bond only two atoms, in a manner analogous to the bonding in diatomics. Furthermore, the bond between a given pair of atoms often has very similar properties in different molecules. For example, the O—H bond energy in H_2O, as estimated from the energy absorbed in the reaction $H_2O \rightarrow H + OH$, is 118 kcal mole⁻¹; in H_2O_2, the O—H bond energy given by the energy absorbed in the reaction $H_2O_2 \rightarrow H + HO_2$ is 110 kcal mole⁻¹. These values of the O—H bond energy agree to within 20% with the dissociation energy of the OH radical, 100 kcal mole⁻¹. Thus, to a first approximation, one can add bond energies to obtain total dissociation energies for polyatomic molecules. Similarly, the C—H bond length is close to 1.1 Å whether in methane (1.09 Å), ethane (1.11 Å), ethylene (1.07 Å), or some other molecule. The fact that a certain vibrational frequency corresponds to a given type of bond (e.g., the CH frequency is ∼2900 cm⁻¹ and the OH frequency is ~ 3600 cm^{-1}) is often used to identify that group in a complex organic molecule (see Chapter 7). Thus, more generally, the potential-energy function associated with a particular bond, as determined by R_e, D_e, ν_e, is fairly constant in different molecules. Even dipole moments can be identified with bonds (the so-called *bond dipole*) and the total molecular dipole can be estimated by adding vectorially the individual bond moments. However, the charge distribution in a bond is affected by the molecular environment more than the potential function, so that larger deviations from additivity occur in bond moments than in dissociation energies. To illustrate how the concept of localized bonds is applied to polyatomic molecules, we consider some of the hydrides of the second-row elements.

6.3.1 Water   Proceeding in much the same way as we did in discussing HF, we write the ground-state configurations of the isolated H and O atoms as

$$\text{H} \quad (1s)$$
$$\text{O} \quad (1s^2\, 2s^2\, 2p_z^{\,2}\, 2p_x\, 2p_y)$$

Two localized MO's can be formed from the singly occupied $2p_{xO}$ and $2p_{yO}$ orbitals, each in linear combination with a $1s_H$ orbital. The configurational MO wave function can therefore be represented

$$\psi = 1s_O^2\, 2s_O^2\, 2p_{zO}^2\, \phi_x^2\, \phi_y^2$$

where

$$\phi_x = (a^2 + b^2)^{-1/2}\, [a\, 1s_{Hx} + b\, 2p_{xO}]$$
$$\phi_y = (a^2 + b^2)^{-1/2}\, [a\, 1s_{Hy} + b\, 2p_{yO}] \tag{6.21}$$

In Eq. 6.21, the same coefficients are used for ϕ_x and ϕ_y since the same types of AO's are used in the two bonds. For maximum bond stability, the $1s_H$ and $2p_O$ orbitals should overlap as much as possible, since this leads to large resonance integrals. The maximum overlap is obtained when the hydrogen-atom nuclei lie along the x and y axes, as in Fig. 6.15a. According to the bonding picture given by Eq. 6.21, H_2O has two O—H bonds at a $90°$ angle. The experimental result is that the O—H bond length is 0.96 Å and the bond angle is $104.5°$. The larger bond angle is due to a number of factors neglected in the simple model, including the mutual repulsion of the two protons. The H_2O molecule is found to have an electric dipole moment of 1.8D pointing midway between the O—H bonds and away from the O atom (i.e., the O atom is at the negative end of the dipole), indicating that the two O—H bonds are polar, with $b/a > 1$ in Eq. 6.21. This agrees with the expectation from the electronegativity difference between O and H (Table 6.7). Thus, the protons have a net positive charge which gives rise to some repulsion between them. In H_2S, the S—H bond length is 1.34 Å

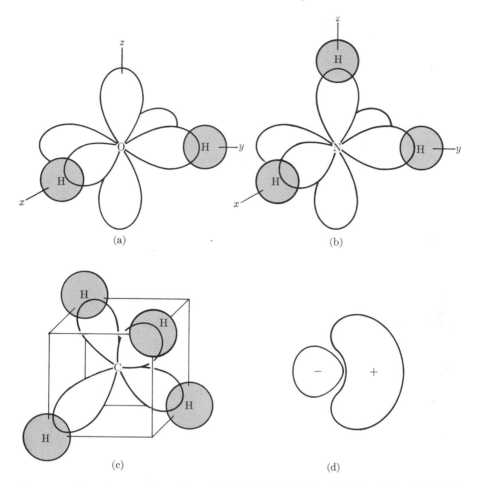

Fig. 6.15. Structure and bonding of (a) H_2O, (b) NH_3, (c) CH_4, (d) sp^3 hybrid orbital.

and the dipole moment is only 1.1D. Consequently, the mutual proton (or bond-dipole) repulsion is less than in H_2O, and the experimental bond angle of 92° is much closer to 90°. Other factors [e.g., hybridization (see Section 6.3.5) and an antibonding interaction between the protons] also contribute to the widening of the bond angle.

The localized VB function for the water molecule is obtained by a corresponding extension of the diatomic results. We have

$$1s_O^2 2s_O^2 2p_{zO}^2 \{A[2p_{xO}(1)1s_{Hx}(2) + 1s_{Hx}(1)2p_{xO}(2)] + B2p_{xO}(1)2p_{xO}(2)\}$$
$$\times \{A[2p_{yO}(3)1s_{Hy}(4) + 1s_{Hy}(3)2p_{yO}(4)] + B2p_{yO}(3)2p_{yO}(4)\} \quad (6.22)$$

where A, B can be determined from the experimental dipole moment or the electronegativity difference (Problems 6.12 and 6.13).

Since the two electrons in each bond are expected to have paired spins, as in the single bond of a diatomic molecule (e.g., H_2, Section 5.6), the spins of the electrons in the $2p_{x0}$ and $2p_{y0}$ orbitals are no longer coupled together; that is, the effective or valence state of the O atom with two OH bonds is not the lowest triplet term (^3P) but rather an excited form (0.5 eV above ^3P) with almost random orientation of the spins.

6.3.2 Ammonia   In the ground-state electron configuration of the nitrogen atom $N(1s^2 2s^2 2p_x 2p_y 2p_z)$, the three $2p$ orbitals are each singly occupied (see Section 4.5). Thus, we expect to form equivalent bonds to three hydrogen atoms. The configurational MO wave function can be written

$$\psi = 1s_N^2 2s_N^2 \phi_x^2 \phi_y^2 \phi_z^2$$

where

$$\phi_x = (a^2 + b^2)^{-1/2}(a\, 1s_{Hx} + b\, 2p_{xN})$$
$$\phi_y = (a^2 + b^2)^{-1/2}(a\, 1s_{Hy} + b\, 2p_{yN}) \qquad (6.23)$$
$$\phi_z = (a^2 + b^2)^{-1/2}(a\, 1s_{Hz} + b\, 2p_{zN})$$

Maximum overlap between the $1s_H$ and the $2p_{xN}$, $2p_{yN}$, and $2p_{zN}$ orbitals requires the H atoms to be located along the x, y, and z axes, corresponding to N—H bond angles of 90° as shown in Fig. 6.15b. The experimental angle is 107°, indicating that factors corresponding to those present in H_2O occur in NH_3 as well. For PH_3, in which the P—H bond lengths are 1.43 Å (compared with 1.01 Å for N—H bonds in NH_3) and the dipole moment is 0.58D (compared with 1.47D for NH_3), the P—H bond angles are 93°.

6.3.3 Methane and sp hybridization   The ground-state configuration of carbon $C(1s^2 2s^2 2p_x 2p_y)$ suggests that carbon should behave like oxygen and form a CH_2 molecule with the C—H bonds at right angles to one another. While the CH_2 molecule exists and has a bond angle close to that of H_2O in its lowest singlet state (Problem 6.37), it is well known that CH_4 is a very stable "carbon hydride." This demonstrates in a striking manner that by the neglect of filled shells or subshells in our discussion of bonding, we have oversimplified the problem; that is, we have generally ignored the possibility (though we mentioned it for electronically excited He_2 in Section 6.1.4) that excited atomic states can be involved in bond formation. If one of the $2s$ electrons in carbon is excited to the $2p$ level, we obtain a configuration with the four singly occupied orbitals ($2s$, $2p_x$, $2p_y$, $2p_z$). In order that the four electrons in these orbitals can form four "independent" electron-pair bonds, their spins must be random; that is, the carbon atom must be further excited from the lowest ($1s^2 2s 2p_x 2p_y 2p_z$) term (^5S) to the

$(1s^2 2s 2p_x 2p_y 2p_z)$ valence state, which is about 8.26 eV (190 kcal mole^{-1}) above the ground state. Since each C—H bond stabilizes the molecule by about 100 kcal mole^{-1}, the $(2s \rightarrow 2p)$ *promotion energy* is "returned with interest" when the four bonds are formed (see Fig. 6.16). It is this excess of bond stabilization energy over the required promotion energy that leads to the stability of CH_4.

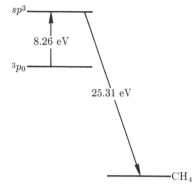

Fig. 6.16. Excitation energy of the carbon atom to the sp^3 valence state and the stabilization energy due to the formation of four C—H bonds.

From the nature of the carbon valence state, we might expect that methane would have three C—H bonds formed from $1s_H$ and $2p_C$ orbitals lying along each of the Cartesian axes, as in NH_3, and an additional weaker bond formed from a fourth $1s_H$ orbital and the $2s_C$ orbital. Experimentally, all four bonds are found to be equivalent, with a common 109°28′ H—C—H bond angle. Four carbon AO's pointing in the required tetrahedral directions can be constructed by taking linear combinations of the $2s$, $2p_x$, $2p_y$, and $2p_z$ orbitals. Such linear combinations are called *hybrid orbitals*, or simply *hybrids*; since they consist of one s orbital and three p orbitals in the present case, they are referred to as sp^3 hybrids. A form for these sp^3 hybrid orbitals is

$$\chi_1 = \frac{1}{\sqrt{4}} (2s + 2p_x + 2p_y + 2p_z)$$

$$\chi_2 = \frac{1}{\sqrt{4}} (2s - 2p_x - 2p_y + 2p_z)$$

$$\chi_3 = \frac{1}{\sqrt{4}} (2s + 2p_x - 2p_y - 2p_z) \tag{6.24}$$

$$\chi_4 = \frac{1}{\sqrt{4}} (2s - 2p_x + 2p_y - 2p_z)$$

Table 6.12   *Hybridization of carbon orbitals and bond angles*

Bond	Hybrid AO	R_e(Å)	D_e(kcal)	Expected H—C—H bond angle	Experimental H—C—H bond angle
C—C	sp^3	1.54	88	109.47°	(C_2H_6):109.3°
C—C	sp^2	1.34	167	120°	(C_2H_4):117°
C—C	sp	1.20	230		
C—H	sp^3	1.11	98		
C—H	sp^2	1.07	104		
C—H	sp	1.06	113		

bonds for these hydrocarbon systems. It is interesting to note the change in the CH bond with hybridization. In spite of the fact that in all cases one has only a single bond, its length decreases with increasing amount of s in the hybrid and its strength increases. These results can be understood in terms of the degree of overlap between the carbon hybrid and the $1s$ hydrogen orbital.

6.3.5   Hybridization in other molecules   According to the variation principle, the correct combination of orbitals is the one that minimizes the energy. Thus, the extent of hybridization is determined by the molecular energy. Since the energy is always lowered when the $2s$ and $2p_z$ orbitals are mixed, some sp hybridization, though not necessarily a large amount, takes place in all diatomic molecules.

The mixing of several σ molecular orbitals in the SCF-LCAO treatment of diatomic molecules (Section 6.1.7) may be regarded as hybridization. Physically, it makes no difference whether we take linear combinations of atomic orbitals first and form molecular orbitals from such hybrid orbitals, or if we form the molecular symmetry orbitals first (as we did in Section 6.1.7) and then take linear combinations of them; it is only a matter of mathematical convenience, and the final form of the orbitals is the same.

For nonlinear molecules like H_2O and NH_3, hybridization is of more interest because, as in C_2H_4, the bond angles are affected by the amount of s character in the hybrid orbitals. To see how hybridization is introduced, we consider H_2O and look at the oxygen ground state ($1s^22s^22p_z^22p_x2p_y$), which leads to pure p bonds (p_x and p_y), and the excited states ($1s^22s2p_z^22p_x^22p_y$), ($1s^22s2p_z^22p_x2p_y^2$), each of which can have an s and a p orbital involved in bonding. In contrast to C, the promotion $s \rightarrow p$ does not increase the number of bonds that can be formed, because in every case there are only two unpaired electrons. This means that the only stabilizing effect of these excited states is due to the increased bond strength resulting from the greater overlap of the hybrid orbitals. Consequently, the degree of hybridization is a balance between this stabilization and the destabilization from the promotion-energy requirements.

The basic assumption relating the bond angle to hybridization is that the bond is formed in the direction of maximum overlap. If there is no s character at all (i.e., if the orbitals are $2p_x$, $2p_y$, and $2p_z$) the bond angle is 90°, whereas sp^3 hybridization gives the tetrahedral bond angle of 109.47°. Hybridization thus provides a contribution leading to bond angles larger than 90°, in addition to the proton repulsion discussed earlier. The exact importance of each effect is hard to determine; for example, the best SCF-LCAO calculation for the H_2O molecule gives an equilibrium bond angle of 106.5° in comparison with the experimental value of 104.5°. Empirically from the known bond angles and the "criterion of maximum overlap," the "s character" in the bonds of H_2O (104.5°) is 0.200, while in NH_3(106.8°) the s character is 0.224, as compared with the tetrahedral value of 0.250 (Problems 6.18 and 6.19). It should be pointed out that the maximum-overlap description is only an approximate one and that deviations can occur. Moreover, a direct experimental measure of hybridization is not available. Thus, hybridization is to be regarded as a valid concept only because it is useful for correlating a variety of data concerning molecular structure (e.g., bond angles) and properties of the electron distribution (e.g., quadrupole coupling constants; see Section 7.14).

6.4 NONLOCALIZED BONDS

Although the bonding in many polyatomic molecules can be understood in terms of localized bonds, there are systems for which such a description is not adequate. In this section, we discuss a number of examples in which delocalization is important and also consider the relationship between localized and nonlocalized bonds.

6.4.1 Benzene Since benzene is a classic molecule of this type, we examine the problem of its electronic structure from both the VB and MO points of view. Following our development for planar hydrocarbons like ethylene, we assume that the six carbon atoms of the benzene ring form localized sp^2 hybrid σ bonds between adjacent atoms. There are also six σ bonds formed from the remaining sp^2 hybrid carbon orbitals and the $1s_H$ orbitals of the six hydrogen atoms (Fig. 6.19a). The singly occupied $2p_z$ carbon orbitals (taking the plane of the molecule to be the xy plane), which are called π orbitals, can combine to form localized bonds in two different ways as

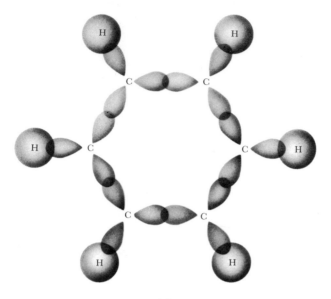

(a)

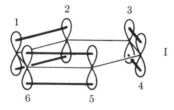

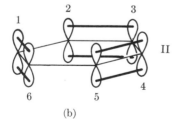

(b)

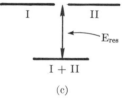

(c)

Fig. 6.19. Structure and bonding of benzene; **(a)** σ bonds, **(b)** π bonds, **(c)** relation between the resonance energy, the Kekulé energy (I, II) and the true energy (I + II) of the benzene molecule.

shown in Fig. 6.19b. These two structures (I and II) are what are usually called the Kekulé structures of the benzene molecule.

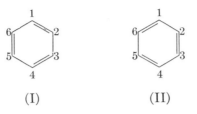

$$(I) \qquad\qquad\qquad (II)$$

VB Theory. The VB wave function for the two structures can be written

$$\psi_I = (1, 2)(3, 4)(5, 6)$$
$$\psi_{II} = (2, 3)(4, 5)(6, 1)$$

where $(1, 2) = [2p_{z1}(1)\, 2p_{z2}(2) + 2p_{z2}(1)\, 2p_{z1}(2)]$, etc. We have included only the $2p_z$ electrons in writing the wave function, the σ bonds being treated as a fixed core. Such a treatment of benzene is often referred to as the π-electron model. If the bonds C_1—C_2, C_3—C_4, C_5—C_6 were shorter than the bonds C_2—C_3, C_4—C_5, C_6—C_1, it is clear that structure I would be more stable than structure II. But since all of the C—C bond lengths in benzene are known to be equal (1.397 Å), the electronic energies of the two structures are the same. Thus the two structures are "equivalent" and there is no way to choose between them. The situation is similar to the one that arose in the VB treatment of H_2. There, the identity of the electrons prevented the use of either $1s_A(1)\, 1s_B(2)$ or $1s_B(1)\, 1s_A(2)$ for the approximate wave function. The difficulty was resolved by introducing a linear combination of the two functions, which resulted in a sizeable increase in the calculated bond energy. For benzene, the problem of two equivalent structures is solved correspondingly by taking the normalized linear combination

$$\psi = N(\psi_I + \psi_{II})$$

which gives a calculated energy for the π electrons lower than the energy of either of the structures, as illustrated schematically in Fig. 6.19c. The actual structure is said to be a "resonance hybrid" of the two Kekulé structures and the *resonance energy* is defined as the difference between the energy of the resonance hybrid and a single structure of lowest energy. Although we do not go through the details of the VB calculation, we can write down the result. The π-electron energy can be expressed in terms of a Coulomb integral Q and an exchange integral J for each electron-pair bond between nearest-neighbor atoms, analogous to the integrals used in the VB discussion of H_2 in Section 5.6.1; integrals between the nonadjacent atoms are small and can be neglected in the simple treatment. Because of the hexagonal symmetry of the molecule, all Q and J values are the same.

For a single Kekulé structure (e.g., ψ_I), the energy expression for the $2p_z$ electrons in the field of the σ-electron core is

$$E_I = E_{II} = 6Q + 1.5J$$

The coefficient 1.5 of J arises from the electron-pair bonds between $(1, 2)$, $(3, 4)$, and $(5, 6)$, which give $3J$ and the nonbonding interactions between $(2, 3)$, $(4, 5)$, and $(6, 1)$, which give $-\frac{3}{2}J$. In fact, a general VB energy formula for a set of localized bonds, such as in ψ_I or ψ_{II}, can be written down; it is

$$E = \sum_{\substack{i>j \\ \text{(all)}}} Q_{ij} + \sum_{\substack{i>j \\ \text{(bonding)}}} J_{ij} - \frac{1}{2} \sum_{\substack{i>j \\ \text{(nonbonding)}}} J_{ij} \tag{6.28}$$

where i, j go over all pairs of neighboring orbitals for the Coulomb integrals Q_{ij}, over the bonded pairs (i, j) in the first exchange sum, and over the nonbonded pairs (i, j) in the second exchange sum. If Eq. 6.28 is applied to benzene, for which all Q_{ij} and J_{ij} are equal, the expression E_I results.

The function $N(\psi_I + \psi_{II})$ yields an energy[2]

$$E_{(I+II)} = 6Q + 2.4J$$

Comparing this result with E_I, we see that the resonance energy is

$$E_{\text{res}} = \Delta E = E_{(I+II)} - E_I = 0.9J \tag{6.29}$$

If the three Dewar structures

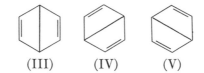

| (III) | (IV) | (V) |

are also included in the calculation, the calculated π-electron energy is

$$E_{(I+II+III+IV+V)} = 6Q + 2.6J$$

and the resonance energy is $1.1J$. The Dewar structures are thus seen to make a small contribution.

MO Theory. The MO theory can also be used to describe the bonding in benzene; moreover, as mentioned before, it is particularly suited to the study of excited states and electronic transitions (see Chapter 7). As in the VB description, we assume that the in-plane orbitals combine to form localized σ bonds. For the six $2p_z$ or π electrons, there is no unique way of forming localized bonds; that is, we can construct localized MO's involving AO's $(1, 2)$, $(3, 4)$, $(5, 6)$ or $(2, 3)$, $(4, 5)$, $(6, 1)$. One way of proceeding would be to take all possible bonding and antibonding configurations that

---

[2] For the method used in calculating the energy of a VB function composed of more than one structure, see References 1, 5, and 6 in the Additional Reading list.

can be formed with these localized orbitals and to carry out a configuration-interaction calculation so as to determine the linear combination that minimizes the energy. The alternative approach is to introduce delocalized MO's; that is, MO's that are not localized between a pair of orbitals but extend over the entire molecule. Since there are six degenerate atomic $2p_z$ or π orbitals, we can combine them to form six delocalized orthogonal and normalized MO's; that is, we have

$$\phi_i = N_i[C_{i1}\, 2p_{z1} + C_{i2}\, 2p_{z2} + C_{i3}\, 2p_{z3}$$
$$+ C_{i4}\, 2p_{z4} + C_{i5}\, 2p_{z5} + C_{i6}\, 2p_{z6}]$$

where $C_{i1}, C_{i2}, \ldots, C_{i6}$ are the coefficients of $2p_{z1}, 2p_{z2}, \ldots, 2p_{z6}$ in the MO ϕ_i and N_i is the appropriate normalizing factor. Although these coefficients can be determined by a variational calculation, we can justify their

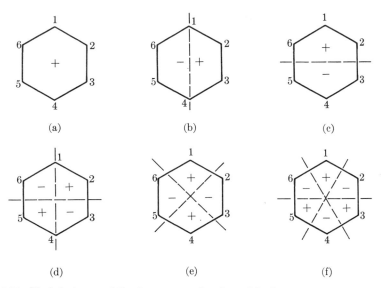

Fig. 6.20. Nodal planes of the benzene molecular orbitals.

values by simpler arguments that are based on the hexagonal symmetry of the benzene molecules. For the MO of lowest energy, no nodes in addition to that in the benzene plane (due to the form of the $2p_z$ orbitals) should be introduced. This means that the coefficients of $2p_z$ orbitals all have the same sign (Fig. 6.20a). Since the carbon atoms are all equivalent, each $2p_z$ orbital is weighted the same. The result is

$$\phi_1 = \frac{1}{\sqrt{6}}\,(2p_{z1} + 2p_{z2} + 2p_{z3} + 2p_{z4} + 2p_{z5} + 2p_{z6}) \qquad (6.30)$$

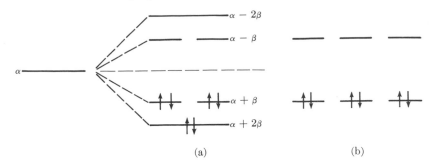

(a)   (b)

Fig. 6.21. Orbital energies from simple MO theory; (**a**) benzene, (**b**) three ethylene molecules.

$$\phi_b = \frac{1}{\sqrt{2}}\,(2p_{zA} + 2p_{zB})$$

$$\phi_a = \frac{1}{\sqrt{2}}\,(2p_{zA} - 2p_{zB}) \tag{6.39}$$

The corresponding bonding and antibonding π electron energies for the stretched ethylene molecule are[3]

$$\epsilon_b = \alpha + \beta$$
$$\epsilon_a = \alpha - \beta \tag{6.40}$$

and the two π electrons of each ground-state ethylene molecule are paired in the bonding orbital. Thus, the Kekulé structure has the total π-electron energy $6\alpha + 6\beta$, giving the energy difference (the delocalization energy)

$$E_{\mathrm{deloc}} = \Delta E = E_{\mathrm{benzene}} - E_{\mathrm{Kekulé}} = 2\beta \tag{6.41}$$

The delocalization energy of MO theory corresponds to the resonance energy of VB theory (Eq. 6.29).

6.4.2 *Naphthalene* For molecules without symmetry, the LCAO-MO's are determined by means of a variational calculation; that is, the coefficients of the AO's comprising each MO are chosen so as to minimize the energy of the molecule. This results in the SCF equations discussed in Section 6.1.7 for which an iterative solution is required. In applications to π-electron systems such as that of benzene (Section 6.4.1), the SCF treatment is simplified by neglecting the dependence of the effective Hamiltonian on the form of the orbitals and introducing constant integral parameters (e.g.,

[3] Since the normal ethylene C—C bond length (1.339 Å) is significantly shorter than that of benzene, we expect β to be somewhat larger for ethylene than for benzene; consequently, the β value appearing in Eq. 6.40 has to be adjusted to take account of the bond-length change.

the α and β integrals). This approximation, which is called Hückel MO theory, reduces the calculation to an application of the linear variation principle discussed in Section 5.6.4, generalized to MO's formed from n AO's. If the ith MO is given by

$$\phi_i = C_{1i}\,\chi_1 + C_{2i}\,\chi_2 + \cdots + C_{ni}\,\chi_n \tag{6.42}$$

the set of homogenous linear equations for the coefficients C_{ji} are

$$(H_{11} - \epsilon_i S_{11})C_{1i} + (H_{12} - \epsilon_i S_{12})C_{2i} + \cdots + (H_{1n} - \epsilon_i S_{1n})C_{ni} = 0$$
$$(H_{21} - \epsilon_i S_{21})C_{1i} + (H_{22} - \epsilon_i S_{22})C_{2i} + \cdots + (H_{2n} - \epsilon_i S_{2n})C_{ni} = 0$$

$$(H_{n1} - \epsilon_i S_{n1})C_{1i} + (H_{n2} - \epsilon_i S_{n2})C_{2i} + \cdots + (H_{nn} - \epsilon_i S_{nn})C_{ni} = 0 \tag{6.43}$$

and the secular equation for the orbital energies is

$$\begin{vmatrix} H_{11} - \epsilon S_{11} & H_{12} - \epsilon S_{12} & \cdots & H_{1n} - \epsilon S_{1n} \\ H_{21} - \epsilon S_{21} & H_{22} - \epsilon S_{22} & \cdots & H_{2n} - \epsilon S_{2n} \\ & & \cdot & \\ & & \cdot & \\ H_{n1} - \epsilon S_{n1} & H_{n2} - \epsilon S_{n2} & \cdots & H_{nn} - \epsilon S_{nn} \end{vmatrix} = 0 \tag{6.44}$$

The integrals H_{jk} and S_{jk} appearing in Eqs. 6.43 and 6.44 are

$$H_{jk} = \int \chi_j h_e \chi_k \, dv$$
$$S_{jk} = \int \chi_k \chi_k \, dv$$

while the energy ϵ_i is given by

$$\epsilon_i = \int \phi_i h_e \phi_i \, dv$$

In Eq. 6.44, the energy ϵ appears without a subscript because there are n values of ϵ such that the determinant is equal to zero; these values ϵ_1, ϵ_2, . . . , ϵ_n are just the energies associated with the orbitals $\phi_1, \phi_2, \ldots, \phi_n$.

If the AO's are normalized, we have

$$S_{jj} = 1$$

Furthermore, in the simple Hückel treatment (see the discussion of benzene), the overlap integrals are neglected, so that

$$S_{jk} = 0 \quad k \neq j$$

Another simplifying assumption in the Hückel treatment is that the integrals H_{jk} ($k \neq j$) are nonzero only for adjacent atoms, and that for ad-

Table 6.13   *Reflection symmetry of the naphthalene orbitals of Eq. 6.45*

Orbital	σ_h	σ_v'	σ_v''
		Symmetry planes[a]	
A_1	−	+	+
A_2	−	+	−
A_3	−	−	+
A_4	−	−	−
B_1	−	+	+
B_2	−	+	−
B_3	−	−	+
B_4	−	−	−
C_1	−	+	+
C_2	−	+	−

[a] See Fig. 6.22 for definitions.

AO's to set up the secular determinant for the orbital energies and the secular equations for the coefficients. Whichever set is used, the final answers are the same; however, as we find below, the secular determinant is considerably simplified when symmetry orbitals are employed. First, the integrals of the type

$$\int A_1 \, h_e \, B_2 \, dv \equiv \langle A_1 | h_e | B_2 \rangle$$

must be evaluated. Using the Hückel assumptions, one obtains (Problem 6.27)

$$
\begin{aligned}
&\langle A_j | h_e | A_j \rangle = \alpha && j = 1, 2, 3, 4 \\
&\langle A_j | h_e | A_k \rangle = 0 && j, k = 1, 2, 3, 4 \quad j \neq k \\
&\langle B_j | h_e | B_j \rangle = \alpha + (-1)^{j+1}\beta && j = 1, 2, 3, 4 \\
&\langle B_j | h_e | B_k \rangle = 0 && j, k = 1, 2, 3, 4 \quad j \neq k \\
&\langle C_j | h_e | C_j \rangle = \alpha + (-1)^{j+1}\beta && j = 1, 2 \\
&\langle C_1 | h_e | C_2 \rangle = 0 && \\
&\langle A_j | h_e | B_j \rangle = \beta && j = 1, 2, 3, 4 \\
&\langle A_j | h_e | B_k \rangle = 0 && j, k = 1, 2, 3, 4 \quad j \neq k \\
&\langle A_j | h_e | C_j \rangle = \sqrt{2}\,\beta && j = 1, 2 \\
&\langle A_j | h_e | C_k \rangle = 0 && j = 1, 2, 3, 4 \quad k = 1, 2 \quad j \neq k \\
&\langle B_j | h_e | C_k \rangle = 0 && j = 1, 2, 3, 4 \quad k = 1, 2
\end{aligned}
\tag{6.46}
$$

From these results, we see that nondiagonal elements in the secular determinant occur only for the sets

$$(A_1, B_1, C_1) \quad (A_2, B_2, C_2) \quad (A_3, B_3) \quad (A_4, B_4)$$

that is, only for orbitals with the same symmetry behavior on reflection in the two planes. By grouping these sets together in the secular determinant we can decompose the 10×10 determinant into two 3×3 and two 2×2 determinants. The resulting determinant is shown in Table 6.14. The determinant is easily factored to give

$$\begin{vmatrix} \alpha - \epsilon & \beta & \sqrt{2}\,\beta \\ \beta & \alpha + \beta - \epsilon & 0 \\ \sqrt{2}\,\beta & 0 & \alpha + \beta - \epsilon \end{vmatrix} \times \begin{vmatrix} \alpha - \epsilon & \beta & \sqrt{2}\,\beta \\ \beta & \alpha - \beta - \epsilon & 0 \\ \sqrt{2}\,\beta & 0 & \alpha - \beta - \epsilon \end{vmatrix}$$

$$\begin{aligned} \epsilon_4 &= \alpha + \beta & \epsilon_7 &= \alpha - \beta \\ \epsilon_1 &= \alpha + 2.303\beta & \epsilon_{10} &= \alpha - 2.303\beta \\ \epsilon_8 &= \alpha - 1.303\beta & \epsilon_3 &= \alpha + 1.303\beta \end{aligned} \qquad (6.47)$$

$$\times \begin{vmatrix} \alpha - \epsilon & \beta \\ \beta & \alpha + \beta - \epsilon \end{vmatrix} \times \begin{vmatrix} \alpha - \epsilon & \beta \\ \beta & \alpha - \beta - \epsilon \end{vmatrix} = 0$$

$$\begin{aligned} \epsilon_2 &= \alpha + 1.618\beta & \epsilon_9 &= \alpha - 1.618\beta \\ \epsilon_6 &= \alpha - 0.618\beta & \epsilon_5 &= \alpha + 0.618\beta \end{aligned}$$

Each of the determinants in the product can be set individually equal to zero to find its roots. They are listed, numbered in order of increasing energy, under the determinant from which they arise (Problem 6.28). The coefficients of the MO's corresponding to the energy ϵ_i can be found by substituting ϵ_i into the appropriate set of homogenous equations. For example, since ϵ_1 derives from the first secular determinant, ϕ_1 is a linear combination of A_1, B_1, and C_1; that is

$$\phi_1 = a_1 A_1 + b_1 B_1 + c_1 C_1$$

and the coefficients a_1, b_1, and c_1 are given by the equations, corresponding to Eq. 6.43 for this symmetry type,

$$\begin{aligned} (\alpha - \epsilon_1)a_1 + \beta b_1 + \sqrt{2}\,\beta c_1 &= 0 \\ \beta a_1 + (\alpha + \beta - \epsilon_1)b_1 &= 0 \\ \sqrt{2}\,\beta a_1 + (\alpha + \beta - \epsilon_1)c_1 &= 0 \end{aligned}$$

From the second and third equations

$$\begin{aligned} b_1 &= -\left(\frac{\beta}{\alpha + \beta - \epsilon_1}\right)a_1 \\ c_1 &= -\left(\frac{\sqrt{2}\,\beta}{\alpha + \beta - \epsilon_1}\right)a_1 \end{aligned} \qquad (6.48)$$

The coefficient a_1 is obtained by requiring ϕ_1 to be normalized; that is,

$$a_1{}^2 + b_1{}^2 + c_1{}^2 = a_1{}^2\left[1 + \left(\frac{\beta}{\alpha + \beta - \epsilon_1}\right)^2 + \left(\frac{\sqrt{2}\,\beta}{\alpha + \beta - \epsilon_1}\right)^2\right] = 1 \qquad (6.49)$$

Table 6.14 Secular determinant for the π-electron system of naphthalene in the Hückel approximation

	A_1	B_1	C_1	A_2	B_2	C_2	A_3	B_3	A_4	B_4
A_1	$\alpha - \epsilon$	β	$\sqrt{2}\,\beta$	0	0	0	0	0	0	0
B_1	β	$\alpha + \beta - \epsilon$	0	0	0	0	0	0	0	0
C_1	$\sqrt{2}\,\beta$	0	$\alpha + \beta - \epsilon$	0	0	0	0	0	0	0
A_2	0	0	0	$\alpha - \epsilon$	β	$\sqrt{2}\,\beta$	0	0	0	0
B_2	0	0	0	β	$\alpha - \beta - \epsilon$	0	0	0	0	0
C_2	0	0	0	$\sqrt{2}\,\beta$	0	$\alpha - \beta - \epsilon$	0	0	0	0
A_3	0	0	0	0	0	0	$\alpha - \epsilon$	β	0	0
B_3	0	0	0	0	0	0	β	$\alpha + \beta - \epsilon$	0	0
A_4	0	0	0	0	0	0	0	0	$\alpha - \epsilon$	β
B_4	0	0	0	0	0	0	0	0	β	$\alpha - \beta - \epsilon$

The numerical results are (Problem 6.29)

$$\phi_1 = 0.6011A_1 + 0.4641B_1 + 0.6525C_1 \tag{6.50}$$

or, in terms of the AO's,

$$\phi_1 = 0.3005[(1) + (4) + (5) + (8)] + 0.2307[(2) + (3) + (6) + (7)] \\ + 0.4614[(9) + (10)] \tag{6.51}$$

Corresponding calculations give all of the MO's:

$$
\begin{aligned}
\phi_1 &= 0.6011A_1 + 0.4641B_1 + 0.6525C_1 \\
\phi_2 &= 0.5257A_3 + 0.8507B_3 \\
\phi_3 &= 0.7992A_2 + 0.3470B_2 + 0.4908C_2 \\
\phi_4 &= 0.8165B_1 - 0.5773C_1 \\
\phi_5 &= 0.8507A_4 + 0.5257B_4 \\
\phi_6 &= 0.8507A_3 - 0.5257B_3 \\
\phi_7 &= 0.8165B_2 - 0.5773C_2 \\
\phi_8 &= 0.7992A_1 - 0.3470B_1 - 0.4908C_1 \\
\phi_9 &= 0.5257A_4 - 0.8507B_4 \\
\phi_{10} &= 0.6011A_2 - 0.4614B_2 - 0.6525C_2
\end{aligned}
\tag{6.52}
$$

The nodes of these orbitals are shown by dotted lines in Fig. 6.23; as can be seen by comparing with Eq. 6.47, the orbital energies bear a direct relation to the number of nodes.

In the ground state, the ten π electrons of naphthalene doubly occupy each of the five lowest orbitals $\phi_1, \ldots, \phi_5$. The total ground-state energy in the Hückel approximation is thus

$$E_{\text{naph}} = 2(\epsilon_1 + \epsilon_2 + \cdots + \epsilon_5) = 10\alpha + 13.68\beta$$

The delocalization energy is found by subtracting from E_{naph} the energy of five stretched ethylene molecules,

$$5E_{\text{ethylene}} = 5[2(\alpha + \beta)] = 10\alpha + 10\beta$$

that is,

$$E_{\text{deloc}} = E_{\text{naph}} - 5E_{\text{ethylene}} = 3.68\beta$$

Corresponding treatments of other aromatic molecules yield the values listed in Table 6.15.

An experimental estimate of the resonance or delocalization energy can be made by taking the difference of the experimental heat of combustion of benzene and the heat of combustion of a Kekulé molecule as calculated from the bond energies for C—C, C=C, and C—H. Fitting this result (36 kcal mole^{-1}) to the calculated VB resonance energy ($1.1J$), we have $J = -33$ kcal mole^{-1}; correspondingly, the calculated MO delocalization energy (2β) gives $\beta = -18$ kcal mole^{-1}. To determine if these values are con-

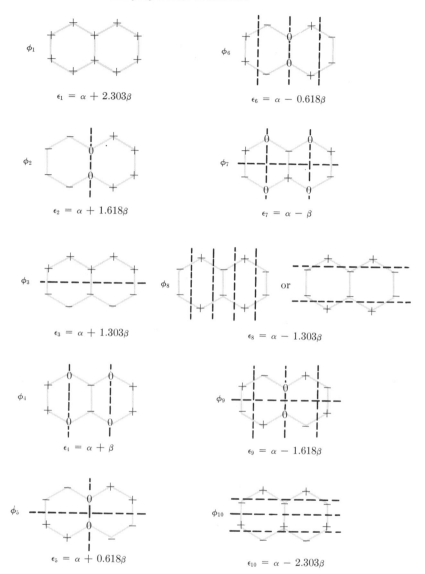

Fig. 6.23. Molecular orbitals for naphthalene.

sistent, we apply them to the VB and MO results for other aromatic hydrocarbons. A comparison of the calculated resonance or delocalization energy
and the experimental data is made in Table 6.15. It can be seen that the VB
and MO methods provide values of J and of β, respectively, which are in
satisfactory agreement for the different molecules, indicating that the two

simple theories of the π-electron delocalization in these systems are consistent. Inclusion of overlap in Hückel calculations gives qualitatively similar results, but slightly different empirical values of β as determined from experimental resonance energies.

A spectroscopic estimate of the resonance integral β can also be made by measuring the benzene transitions corresponding to the degenerate configuration changes

$$\phi_1{}^2\ \phi_2{}^2\ \phi_3{}^2 \rightarrow \phi_1{}^2\ \phi_2{}^2\ \phi_3\ \phi_4, \qquad \phi_1{}^2\ \phi_2{}^2\ \phi_3\ \phi_5,$$
$$\phi_1{}^2\ \phi_2\ \phi_3{}^2\ \phi_4, \qquad \phi_1{}^2\ \phi_2\ \phi_3{}^2\ \phi_5$$

all of which have an energy difference of $2|\beta|$ according to the simple theory (see Section 7.11.4). Since the transition energy is about 120 kcal mole^{-1}, the spectroscopic value of $|\beta|$ is about 60 kcal mole^{-1}, in contrast to the thermodynamic value of 18 kcal mole^{-1}. This inconsistency is a consequence

Table 6.15 *Resonance energy of aromatic hydrocarbons (kcal mole^{-1})*[a]

Molecule	VB resonance energy	MO delocalization energy	Expt'l integral $-J$	$-\beta$
Benzene	$-1.106J = 36$	$-2\beta = 36$	33	18
Naphthalene	$-2.040J = 61$	$-3.683\beta = 61$	30	17
Anthracene	$-2.951J = 84$	$-5.314\beta = 84$	28	16
Phenanthrene	$-3.019J = 92$	$-5.449\beta = 92$	30	17
			av. 30	17

[a] K. Higasi, H. Baba, and A. Rembaum, *Quantum Organic Chemistry* (Wiley-Interscience, New York, 1965). Kekulé and Dewar structures are included in the calculations.

of the many simplifications introduced in the development of the Hückel MO model (e.g., neglect of overlap, electron repulsion, sigma electrons, and interactions between nonadjacent atoms.) Most applications of quantum theory to chemical problems have such limitations; that is, the drastic approximations which must be introduced to obtain tractable formulas limit the theory to correlating closely related phenomena or interpreting the variation of properties in a series of similar compounds. It is possible to introduce more refined treatments that eliminate some of the approximations of the Hückel theory. Calculations of this type are considerably more complicated but do tend to yield more consistent results (see Reference 9 in

and so on. Table 6.16 lists the polynomials $D_n(x)$ for $1 \leq n \leq 10$. For even values of n, the form of $D_n(x)$ is that of a polynomial in x^2; for odd values of n, the form of $D_n(x)$ is that of a product of x times a polynomial in x^2. Thus, the roots x_i of Eq. 6.55 [i.e., $D_n(x_i) = 0$] are symmetric about $x = 0$ for both odd and even n, and include zero if n is odd. For $1 \leq n \leq 5$ the roots are easily found by solving at most a quadratic equation (Problem 6.41); the results for $1 \leq n \leq 4$ are given in Table 6.17. Also included in the table are the corresponding $\epsilon_i(\epsilon_i = \alpha - x_i\beta)$ and the coefficient ratios for the atomic p_π orbitals, which are found by substituting ϵ_i in Eq. 6.43 and

Table 6.16 *Explicit expressions for the polynomials $D_n(x)$*

n	$D_n(x)$
1	x
2	$x^2 - 1$
3	$x(x^2 - 2)$
4	$x^4 - 3x^2 + 1$
5	$x(x^4 - 4x^2 + 3)$
6	$x^6 - 5x^4 + 6x^2 - 1$
7	$x(x^6 - 6x^4 + 10x^2 - 4)$
8	$x^8 - 7x^6 + 15x^4 - 10x^2 + 1$
9	$x(x^8 - 8x^6 + 21x^4 - 20x^2 + 5)$
10	$x^{10} - 9x^8 + 28x^6 - 35x^4 + 15x^2 - 1$

solving for C_{ji}/C_{1i}; the coefficients themselves are obtained from Table 6.17 by normalizing the MO's.

It is evident from Table 6.17 that the orbital energies increase monotonically with the number of orbital nodes. Since the resonance integral β is negative, MO's for which $x < 0$ are more stable (have lower energy) than the individual p_π AO's (whose energies are approximated by the atomic integral α). Similarly, MO's for which $x > 0$ are less stable than the AO's. Thus, it is convenient to classify MO's as bonding if $x < 0$, antibonding if $x > 0$, and nonbonding if $x = 0$. Moreover, as expected from the symmetry of the x_i, the energies are distributed symmetrically about the nonbonding value α; the odd-n systems have a nonbonding orbital ($\epsilon_i = \alpha$), while the even-n orbitals do not.

It can also be seen from the table, by inspection of the coefficients C_{1i} and C_{ni}, that each MO is symmetric or antisymmetric about the symmetry plane located at the midpoint of the molecule and perpendicular to the chain. For odd-n systems, this plane goes through the central atom so that

Table 6.17 Values of x_i and ϵ_i and the coefficients of atomic p_π orbitals in the MO's for linear π systems

n	i	x_i	ϵ_i	C_{2i}/C_{1i}	C_{3i}/C_{1i}	C_{4i}/C_{1i}	Nodes
1	1	0	α				0
2	1	-1	$\alpha + \beta$	$+1$			0
	2	$+1$	$\alpha - \beta$	-1			1
3	1	$-\sqrt{2}$	$\alpha + \sqrt{2}\,\beta$	$+\sqrt{2}$	$+1$		0
	2	0	α	0	-1		1
	3	$+\sqrt{2}$	$\alpha - \sqrt{2}\,\beta$	$-\sqrt{2}$	$+1$		2
4	1	$-\frac{1}{2}(\sqrt{5}+1)$	$\alpha + \frac{1}{2}(\sqrt{5}+1)\beta$	$+\frac{1}{2}(\sqrt{5}+1)$	$+\frac{1}{2}(\sqrt{5}+1)$	$+1$	0
	2	$-\frac{1}{2}(\sqrt{5}-1)$	$\alpha + \frac{1}{2}(\sqrt{5}-1)\beta$	$+\frac{1}{2}(\sqrt{5}-1)$	$-\frac{1}{2}(\sqrt{5}-1)$	-1	1
	3	$+\frac{1}{2}(\sqrt{5}-1)$	$\alpha - \frac{1}{2}(\sqrt{5}-1)\beta$	$-\frac{1}{2}(\sqrt{5}-1)$	$-\frac{1}{2}(\sqrt{5}-1)$	$+1$	2
	4	$+\frac{1}{2}(\sqrt{5}+1)$	$\alpha - \frac{1}{2}(\sqrt{5}+1)\beta$	$-\frac{1}{2}(\sqrt{5}+1)$	$+\frac{1}{2}(\sqrt{5}+1)$	-1	3

its p_π orbital has zero coefficient in the antisymmetric MO (e.g., C_{22} for $n = 3$). The values of the first (C_{1i}) and last (C_{ni}) coefficients of each orbital i have the same magnitude and alternate in sign. In the lowest-lying MO $(i = 1)$, the two end atomic p_π orbitals have the same sign; in the next MO $(i = 2)$, they have opposite signs, and so on. Thus, for i odd the end atomic p_π orbitals have the same sign, while for i even they have opposite sign. This implies that the MO's alternate in symmetry, the lowest MO being symmetric (S), the next being antisymmetric (A), and so on. This result is of importance for the formulation of rules for certain types of chemical reactions (the Woodward–Hoffmann rules; see Section 6.4.4).

As the first example, we consider butadiene, which has four π orbitals and four π electrons. According to the *Aufbau* principle, the MO's ϕ_1 and ϕ_2 are doubly occupied in the ground state, giving a total π-electron energy of $4\alpha + 2\sqrt{5}\beta$. To calculate the delocalization energy, we subtract from this value the π-electron energy of two ethylene molecules, namely, $4\alpha + 4\beta$, obtaining

$$E_{\text{deloc}} = (2\sqrt{5} - 4)\beta = 0.472\beta$$

The allyl cation has three π orbitals and two π electrons. Thus ϕ_1 is doubly occupied and the π-electron energy is $2\alpha + 2\sqrt{2}\beta$. Subtraction of the π-electron energy of ethylene gives the delocalization energy

$$E_{\text{deloc}} = (2\sqrt{2} - 2)\beta = 0.828\beta$$

The allyl radical with three π electrons has the electron configuration $\phi_1^2\phi_2$ in the ground state. The additional electron makes no contribution to the stability of the molecule in the Hückel approximation, since the orbital ϕ_2 is nonbonding $(\epsilon_2 = \alpha)$. Similarly, the allyl anion with ground-state configuration $\phi_1^2\phi_2^2$ has two bonding and two nonbonding π electrons so that its π-electron bond energy is the same as that of the allyl cation and the allyl radical. More refined calculations show that these three systems have similar, but not identical, bond energies.

A VB treatment of linear unsaturated hydrocarbons in which each carbon atom contributes a single π electron leads to resonance between two or more structures, each with alternating single and double bonds. For butadiene, the principal nonionic structures are

$$\text{CH}_2\text{=CH—CH=CH}_2 \qquad \overline{\text{CH}_2\text{—CH=CH—CH}_2}$$
$$\text{(I)} \qquad\qquad\qquad \text{(II)}$$

Structure II has a higher energy than does structure I, since the "long bond" between the end atoms involves a negligible exchange integral. Thus structure I is most important, and the mixing of the two structures contributes a relatively small amount of resonance energy. Molecules such as butadiene with alternating single and double bonds in their principal VB structure are called *conjugated* systems.

6.4.4 Woodward–Hoffmann rules for cycloaddition reactions

In addition to providing a measure of increased molecular stability (delocalization energy) and the ordering of MO energies, the Hückel treatment of conjugated molecules discussed in Section 6.4.3 also allows the formulation of a very useful set of rules followed by certain types of chemical reactions. The rules that we discuss in this section were first formulated by R. B. Woodward and Roald Hoffmann in 1965. They provide an outstanding illustration of the utility in chemistry of relatively simple theoretical ideas based on symmetry. We consider only one type of reaction, the cycloaddition reaction in which two conjugated molecules add to form a single ring. For a more complete discussion of rules and for their application to a wider variety of reactions, the reader is referred to Reference 14 in the Additional Reading list.

The simplest example of a cycloaddition reaction is the cycloaddition of two ethylene molecules to form cyclobutane:

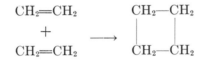

In Fig. 6.24a are shown the π orbitals of the two ethylenes which bond together to form the two new σ bonds of cyclobutene. Following Woodward and Hoffmann, we note that there is a vertical symmetry plane (σ_1) and a horizontal symmetry plane (σ_2) for two ethylene molecules oriented as shown in Fig. 6.24b. The wave functions for the two molecules can be classified as S or A with respect to reflection in each of these planes. A function symmetric with respect to both planes is designated as SS, a function symmetric with respect to σ_1 and antisymmetric with respect to σ_2 is designated SA, and similarly for the two other possible combinations. The linear combinations of the π bonding and antibonding MO's for the two reactant ethylenes and their symmetry classifications are shown on the left-hand side of Fig. 6.25. It is assumed that the two ethylenes are far apart so that the interaction between them is small relative to the difference between the bonding and antibonding energy of the orbitals within a molecule. Thus, ϕ_1 and ϕ_2, which are bonding for both ethylenes, are lower in energy than ϕ_3 and ϕ_4, which are antibonding; furthermore, since the overlap between the two molecules is positive in ϕ_1, ϕ_3, while it is negative in ϕ_2, ϕ_4, orbital ϕ_1 is lower in energy than ϕ_2 and ϕ_3 is lower than ϕ_4. On the right-hand side of Fig. 6.25 are the corresponding combinations of bonding (σ) and antibonding (σ^*) orbitals in the product cyclobutane. Here it is assumed that

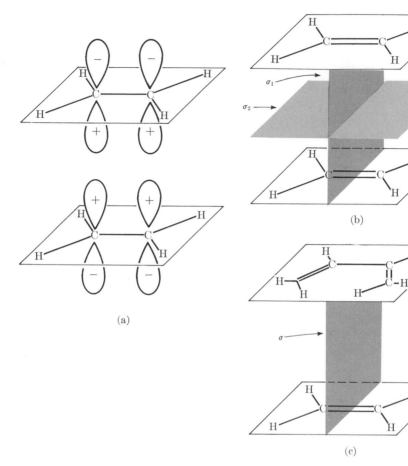

Fig. 6.24. Cycloaddition **(a)** The bonding π orbitals of the two ethylenes oriented for cycloaddition. Symmetry planes for **(b)** ethylene plus ethylene and **(c)** butadiene plus ethylene. After R. Hoffmann and R. B. Woodward, *J. Am. Chem. Soc.* **87,** 2046 (1965).

the newly formed σ bonds in cyclobutane are stronger than the interaction between them. Thus, ψ_1 and ψ_2, which are bonding, are lower in energy than ψ_3 and ψ_4, which are antibonding; furthermore, ψ_1 is lower than ψ_2 and ψ_3 is lower than ψ_4 because the overlap between the σ orbitals in the different bonds is positive in ψ_1, ψ_3 and negative in ψ_2, ψ_4.

The orbitals of reactant and product molecules can now be connected to form a correlation diagram analogous to that introduced previously for diatomic molecules (Section 6.1.3). To draw the correlation diagram, we connect orbitals of the same symmetry; the result is shown in Fig. 6.25.

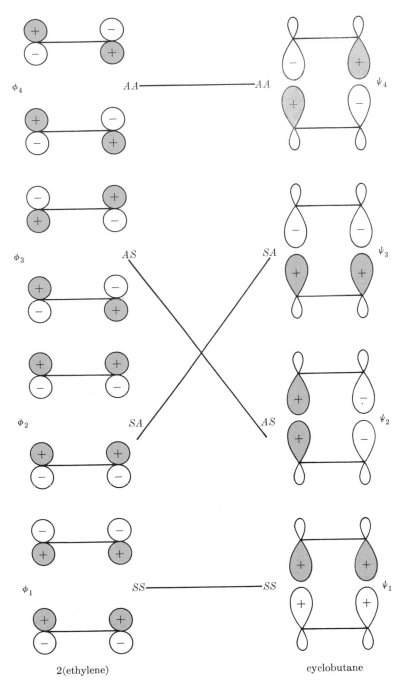

2(ethylene)   cyclobutane

Fig. 6.25. Orbital correlation diagram for two ethylene molecules and cyclobutane.

It is assumed that the *thermal* cycloaddition reaction involves two ground-state ethylene molecules; their configuration in terms of the present notation can be written $\phi_1^2\phi_2^2$. From Fig. 6.25, it is evident that this reactant configuration correlates with the excited product configuration $\psi_1^2\psi_3^2$, which suggests that there will be a high activation barrier. By contrast, for a *photochemical* reaction in which the reactants are excited to the configuration $\phi_1^2\phi_2\phi_3$ by absorption of light, the system can proceed smoothly (there is no energy barrier) to the product configuration $\psi_1^2\psi_2\psi_3$. This contrast between the thermal and photochemical cycloaddition is in agreement with experiment.

For the somewhat more complicated case of the cycloaddition of butadiene and ethylene, the MO's can be classified according to their symmetry with respect to reflection in the single vertical plane shown in Fig. 6.24c, and one can again construct an orbital correlation diagram (Fig. 6.26). The situation is the reverse of that for two ethylenes; that is, the thermal cycloaddition of butadiene and ethylene is seen to proceed smoothly in the ground state ($\phi_1^2\phi_2^2\phi_3^2$ connects with $\psi_1^2\psi_2^2\psi_3^2$), while the photochemical reaction is inhibited by an energy barrier ($\phi_1^2\phi_2^2\phi_3\phi_4$ connects with $\psi_1^2\psi_3^2\psi_2\psi_5$).

We note that the thermally "forbidden" but photochemically "allowed" cycloaddition of two ethylene molecules is characterized by the correlation of bonding orbitals of reactants with antibonding orbitals of products; on the other hand, all bonding orbitals of reactants are correlated with bonding orbitals of products for the thermally allowed cycloaddition of butadiene and ethylene. If we assume for two arbitrary conjugated systems that correlation of bonding and antibonding orbitals is characteristic of photochemical cycloadditions and that correlation of bonding orbitals only with bonding orbitals is characteristic of thermal cycloadditions, general rules can be derived by which one can determine which mechanism applies to a given reaction.

Let the two conjugated systems contribute m and n p_π AO's, respectively, and, for simplicity, let m and n both be even and let the two molecules be neutral. Then in the ground configurations of the reactants there are $m/2$ and $n/2$ doubly occupied bonding MO's in the two molecules. According to Table 6.17 and the discussion in Section 6.4.3, the bonding MO's alternate in symmetry starting with S for the lowest orbital, A for the next lowest, and so on. Thus, if $m/2$ is even, there are $m/4$ occupied bonding S orbitals and $m/4$ occupied bonding A orbitals in the first molecule; if $m/2$ is odd, there are $(m/2 - 1)/2 + 1 = m/4 + \frac{1}{2}$ occupied bonding S orbitals and $(m/2 - 1)/2 = m/4 - \frac{1}{2}$ occupied bonding A orbitals in the first molecule. Similar statements with m replaced by n apply for the second molecule. In the product molecule

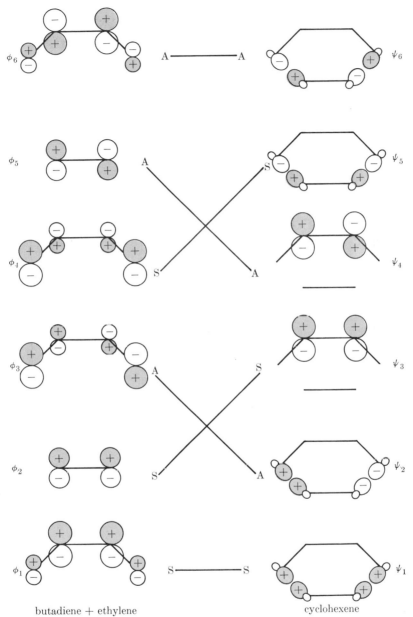

butadiene + ethylene   cyclohexene

Fig. 6.26. Orbital correlation diagram for butadiene plus ethylene and cyclohexene.

potential-energy surface connecting reactants with products would require calculations of the SCF type including configuration interaction. The latter is often needed to obtain the proper behavior of the wave function in both the reactant and product limits (see Section 5.6.5).

6.4.5 *Delocalization in other molecules*   The necessity for including delocalization in the description of the π electrons of aromatic and conjugated systems raises the question of its more general occurrence in molecules; that is, how do we know that a localized theory is adequate for the polyatomic molecules that we considered earlier. To obtain some understanding of this point, we look once again at the water molecule. We assume as we did in Section 6.3.1 that for the O atom only the ground-state configuration $(1s^2 2s^2 2p_z^2 2p_x 2p_y)$ is involved in bonding, so that we can treat the H_2O molecule as a system consisting of four electrons in the four orbitals $2p_{xO}$, $2p_{yO}$, $1s_{Hx}$, $1s_{Hy}$. In the VB method, we can form the two structures shown in Fig. 6.27, one with $(2p_{xO}, 1s_{Hx})$, $(2p_{yO}, 1s_{Hy})$ bonds (Fig. 6.27a), and the other with $(2p_{xO}, 1s_{Hy})$, $(2p_{yO}, 1s_{Hx})$ bonds (Fig. 6.27b). The first structure (ψ_I), which is the one considered earlier, has two strong bonds between the directed, overlapping orbitals. The second structure (ψ_{II}) has its two bonds between orbitals that overlap only weakly and so is much less stable than the first. Writing the ground-state wave function as a linear combination of the two structures, we have

$$\psi = c_I \psi_I + c_{II} \psi_{II}$$

In contrast to benzene, where the two equivalent structures make equal contributions, the coefficient c_{II} is expected to be much smaller than c_I; that is, because of the difference in stability between the two structures, ψ_I makes the dominant contribution and ψ_{II} can be neglected. An approximate variational calculation shows that including both ψ_I and ψ_{II} stabilizes the molecule by less than 0.1 eV relative to ψ_I alone. This argument justifies the localized VB model for systems in which strongly directional orbitals make one structure much more important than all others that correspond to different bonding schemes.

To introduce delocalized orbitals into the MO description of the H_2O molecule, we proceed to set up linear combinations of AO's by analogy with the benzene calculation. From the four AO's $2p_{xO}$, $2p_{yO}$, $1s_{Hx}$, and $1s_{Hy}$, we can form four MO's

$$\phi_i = [c_{i1}\, 2p_{xO} + c_{i2}\, 2p_{yO} + c_{i3}\, 1s_{Hx} + c_{i4}\, 1s_{Hy}]$$
$$i = 1, 2, 3, 4$$

To simplify the determination of the coefficients c_{i1}, c_{i2}, c_{i3}, c_{i4}, we can use symmetry arguments similar to those made for naphthalene. Since the

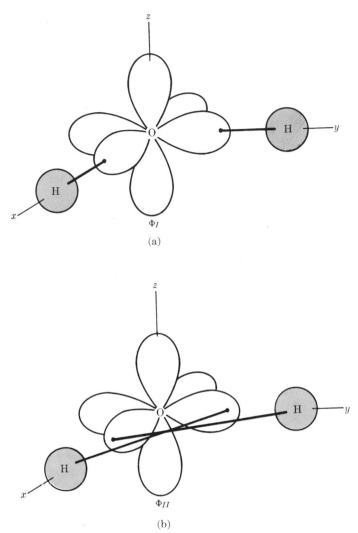

Fig. 6.27. Two structures for the water molecule.

orbitals $2p_{x\text{O}}$ and $2p_{y\text{O}}$ are equivalent and the orbitals $1s_{\text{H}x}$ and $1s_{\text{H}y}$ are equivalent, we form the linear combinations

$$2p_s = \frac{1}{\sqrt{2}}(2p_{x\text{O}} + 2p_{y\text{O}}) \quad 2p_a = \frac{1}{\sqrt{2}}(2p_{x\text{O}} - 2p_{y\text{O}})$$

$$\text{H}_s = \frac{1}{\sqrt{2}}(1s_{\text{H}x} + 1s_{\text{H}y}) \quad \text{H}_a = \frac{1}{\sqrt{2}}(1s_{\text{H}x} - 1s_{\text{H}y})$$

If we reflect the orbitals $2p_{x\mathrm{O}}$ and $2p_{y\mathrm{O}}$ in a plane perpendicular to that of the molecule and bisecting the HOH angle (Fig. 6.28), we find that they simply interchange; that is,

$$2p_{x\mathrm{O}} \to 2p_{y\mathrm{O}} \qquad 2p_{y\mathrm{O}} \to 2p_{x\mathrm{O}}$$

and similarly

$$1s_{\mathrm{H}x} \to 1s_{\mathrm{H}y} \qquad 1s_{\mathrm{H}y} \to 1s_{\mathrm{H}x}$$

Thus, on reflection,

$$2p_s \to 2p_s \qquad 2p_a \to -2p_a$$
$$\mathrm{H}_s \to \mathrm{H}_s \qquad \mathrm{H}_a \to -\mathrm{H}_a$$

that is, the s-type orbitals are symmetric with reflection in the bisecting plane, while the a-type orbitals are antisymmetric with respect to that plane. Since the interaction between orbitals with different symmetries is zero (Problems 6.25 and 6.26), we can simplify the expressions for the MO's by combining only symmetric or antisymmetric orbitals with each other; that is,

$$\phi_{s1} = N_{s1}(2p_s + \lambda_1 \mathrm{H}_s) \qquad \phi_{s2} = N_{s2}(2p_s - \lambda_2 \mathrm{H}_s)$$
$$\phi_{a1} = N_{a1}(2p_a + \mu_1 \mathrm{H}_a) \qquad \phi_{a2} = N_{a2}(2p_a - \mu_2 \mathrm{H}_a) \tag{6.59}$$

Here $N_{s1} = (1 + \lambda_1^2)^{-1/2}, \ldots, N_{a2} = (1 + \mu_2^2)^{-1/2}$ and $\lambda_1, \lambda_2, \mu_1,$ and μ_2 are positive coefficients whose values can be determined by the variation principle. If we look at the nodes of the four orbitals (Fig. 6.29), we see that ϕ_{s1} and ϕ_{a1} have no nodes in the bonding region, while ϕ_{s2} and ϕ_{a2} both have

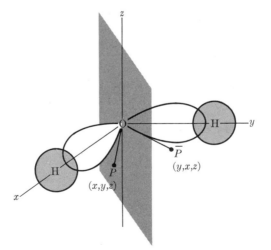

Fig. 6.28. Reflection in the plane bisecting the HOH angle; the point $P(x, y, z)$ is taken into the point $\bar{P}$ (y, x, z).

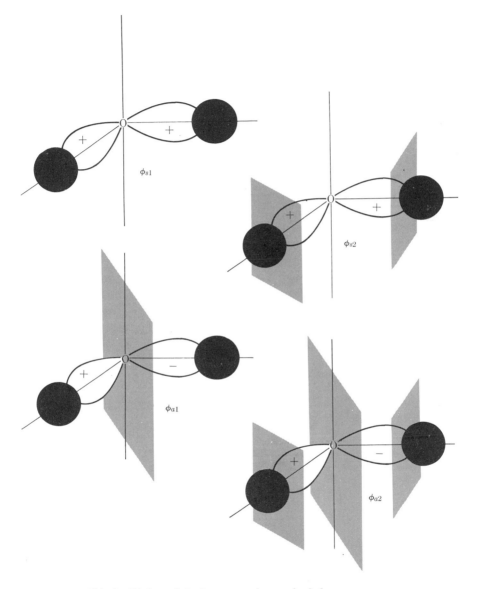

Fig. 6.29. MO's for H_2O; nodal planes are shown shaded.

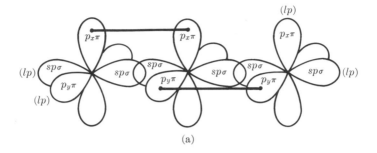

(a)

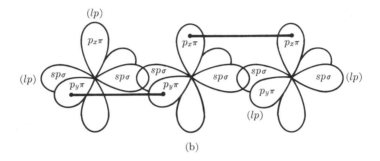

(b)

Fig. 6.32. Bonding in CO_2 (lone-pair orbitals are marked lp); **(a)** and **(b)** show the two equivalent π structures.

electrons make a negligible contribution to bonding and consider only the $2s$, $2p$ valence electrons of carbon and oxygen. Since CO_2 is known to be linear in the ground electronic state, we assume that the C atom forms σ bonds to the oxygen atoms with two sp hybrid orbitals oriented in opposite directions like those of acetylene (Figs. 6.17d and 6.17e). If we take the molecular axis to be in the z direction, the C atom has one electron in the $2p_x$ orbital and one in the $2p_y$ orbital available for π bonding. Each of the two O atoms also can be expected to form sp hybrid orbitals, one of which forms a σ bond with an sp hybrid of the C atom and one of which is occupied by two lone-pair electrons. Since the bonding and lone-pair orbitals are not equivalent, the two hybrids need not have exactly the same form (i.e., the bonding hybrid has more p character and the lone pair has more s character). Of the two remaining O-atom orbitals, $2p_x$ and $2p_y$, one is doubly occupied (a second lone pair) and the other is singly occupied and so can form a π bond with the corresponding $2p\pi$ orbital of the C atom. The resulting bonding scheme is shown in Fig. 6.32. There are clearly two possible bonding arrangements, as shown in Figs. 6.32a and 6.32b. In Fig. 6.32a the p_x orbitals·form the π bond on the left while the p_y orbitals form the π bond on the right, and conversely for Fig. 6.32b. Since these two bonding structures are equivalent, resonance between them makes the molecule symmetric about the axis and enhances the stability of the molecule relative to that of either (a) or (b) alone. The C—O bond length in CO_2 is 1.16Å, which is between the triple-bond value of 1.13 Å in CO and the double-bond value of 1.21 Å in $H_2C{=}O$.

In the simplest MO picture of CO_2, the $1s$ orbitals of the three atoms can be assumed not to participate in the bonding, as in the VB treatment. Thus, the MO's for CO_2 are constructed by taking linear combinations of the $2s$ and $2p$ orbitals of the C atom and the two O atoms. Since CO_2 is linear and has a center of symmetry, the MO's are classified in the same way as for homonuclear diatomic molecules. The MO's with no nodal planes containing the molecular axis are designated σ_g or σ_u, depending upon whether they are symmetric or antisymmetric, respectively, under inversion (see Section 6.1.6); correspondingly, the MO's with one nodal plane through the molecular axis are designated π_g or π_u. Each of the AO's of carbon can also be classified according to these symmetry properties, but appropriate linear combinations of the oxygen orbitals must first be formed before they can be so classified. The classifications and pictorial representations of the C orbitals and the required linear combinations of the O orbitals are shown in Fig. 6.33; verification of these results is left to the reader.

After classification of the AO's, the next step is to form MO's by taking linear combinations of all the orbitals of each type, with the coefficients determined by the variation principle according to the SCF-LCAO-MO method. The results of such a SCF calculation for CO_2 by A. D. McLean are given in Table 6.19. McLean extended the calculation to include d_σ, d_σ' orbitals on the carbon atom and two sets of $2p$ orbitals ($2p$ and $2p'$) on both carbon and oxygen; the prime on one orbital indicates that it has a different orbital exponent from the unprimed orbital. Figure 6.34a shows a schematic representation of the MO's of CO_2 as given by the SCF-LCAO results of Table 6.19. The two lowest MO's, $1\sigma_u$ and $1\sigma_g$, are essentially the antisymmetric and symmetric combinations of the oxygen $1s$ orbitals. The next MO, $2\sigma_g$, lies 9 a.u. higher and is essentially the carbon $1s$ orbital. Thus $1\sigma_u$, $1\sigma_g$, and $2\sigma_g$ are nonbonding, verifying the validity of our earlier assumption that the $1s$ orbitals can be ignored in the bonding scheme. The $3\sigma_g$ orbital is a σ bonding MO composed mainly of the $2s$ orbitals from each of the three atoms. The $2\sigma_u$ orbital, which has about the same energy as $3\sigma_g$, represents bonding between the oxygen $2s$ orbitals and the carbon $2p_\sigma$ orbital. The coefficients of the AO's in the $4\sigma_g$ orbital clearly show sp hybridization of the oxygen $2s$ and $2p$ orbitals; these sp hybrids are combined with the carbon $2s$ orbital, but since the main lobes of the hybrids point outward ($2s$ and $2p_\sigma$ have coefficients of opposite sign), the bonding in $4\sigma_g$ is very

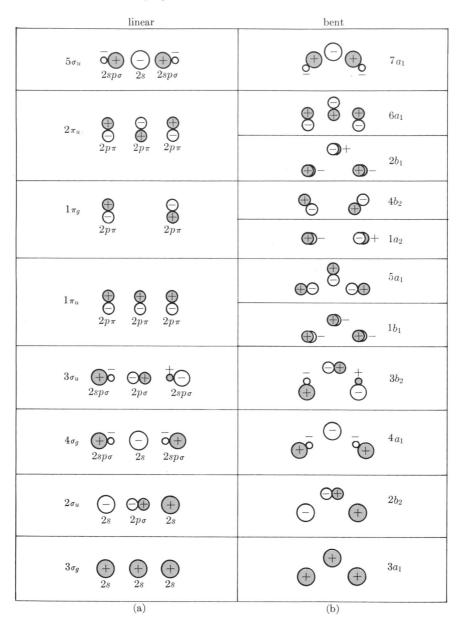

Fig. 6.34. Schematic representation of CO_2 valence orbitals in **(a)** linear and **(b)** bent conformations.

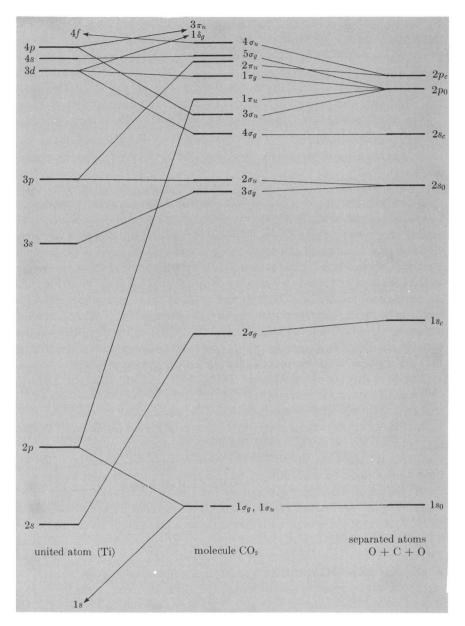

Fig. 6.35. MO correlation diagram for CO_2.

below that of the antibonding $2\pi_u$ orbital. The highest of the four new a_1 orbitals is strongly C—O antibonding, as was the $5\sigma_u$ orbital. Figure 6.37 also shows that a $1\pi_g$ orbital is expected to increase in energy as the two O atoms are brought closer together by bending, so that the two $1\pi_g$ MO's go from nonbonding orbitals in the linear case to O—O antibonding orbitals in the bent case. The energies of the remaining orbitals change only slightly with bending.

From this qualitative consideration of the consequences of bending the

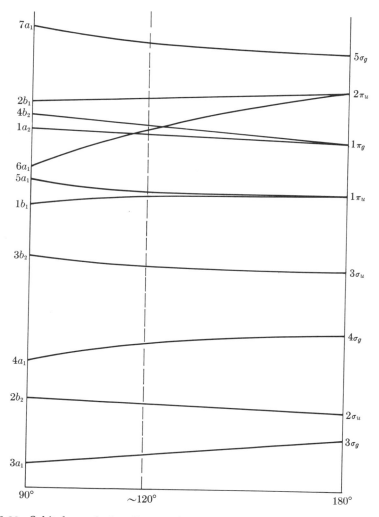

Fig. 6.38. Orbital correlation diagram for linear (180°) and bent (90°) geometries of CO_2; ordering of MO levels at 90° is only approximate.

CO_2 molecule, we can construct the correlation diagram in Fig. 6.38, which shows the relative MO energy levels in the linear and in the bent geometry [the diagram is similar to one discussed by A. D. Walsh, *J. Chem. Soc.* **1953**, 2266]; the order for the linear case is given on the right and that for the bent case on the left. In the ground-state configuration of CO_2, the orbitals through $1\pi_g$ are occupied. Bending produces a lowering of the energies of the two $4\sigma_g$ electrons, but this is more than compensated by the increase in energy of the four $1\pi_g$ electrons. Thus, CO_2 is linear in the ground state. In the excited configuration $(1\pi_g)^3(2\pi_u)$, $(6a_1)$ the $2\pi_u$ orbital whose energy decreases strongly upon bending, is occupied. This leads to a bent molecule; the lowest state of this configuration is observed to be bent at an O—C—O angle of about 122°. A schematic correlation diagram for several low-lying configurations and terms of CO_2 is presented in Fig. 6.39. It shows the linear ground state and three low-lying bent excited states. The energy ordering of the orbitals at ~120° is that given by the dashed line in Fig. 6.38.

The orbital correlation diagram shown in Fig. 6.38 is typical of the valence MO's of small triatomic molecules. The critical orbital is $2\pi_u$; that is, if $2\pi_u$ is empty, the molecule is most stable in the linear geometry, and if $2\pi_u$ is occupied the molecule is more stable in a bent geometry. The triatomic molecules with 16 or fewer valence electrons (e.g., C_3, C_2N, CN_2, NCO, CO_2^+, CO_2) have an empty $2\pi_u$ orbital and are linear, while those with 17, 18, 19, or 20 valence electrons (e.g., NO_2, O_3, ClO_2, Cl_2O) begin to fill the $2\pi_u$ orbital and are bent. Such predictions, known as *Walsh's rules*, though for the most part based originally upon qualitative arguments of the type presented here, are generally confirmed for small triatomic molecules by available observations and by more quantitative MO calculations that have recently been carried out for a few representative systems. Ap-

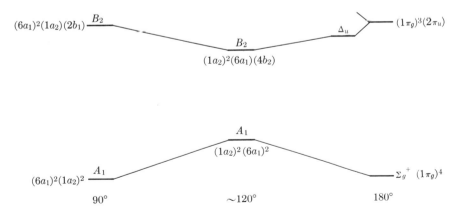

Fig. 6.39. Schematic configuration correlation diagram for linear and bent geometries of CO_2.

parent exceptions to Walsh's rules for XY_2 molecules occur for cases in which d orbitals may contribute to the bonding. For example, $CaCl_2$ and $CaBr_2$, with 16 valence electrons, are linear as predicted, but CaF_2, $BaCl_2$, and $BaBr_2$, with the same number of valence electrons, are bent. The electronic structure of the trihalide ions $BrIBr^-$, $ClICl^-$, I_3^-, $ClBrCl^-$, and Br_3^- with 22 valence electrons may be complicated by the contribution of d orbitals to the filled MO's, but the experimental evidence indicates that these molecules are linear, as predicted by Walsh's rules.

6.4.8 MO treatment of transition-metal complexes

In Section 5.5 we saw that certain aspects of the physical properties of transition-metal complexes can be understood in terms of the ionic interaction between closed-shell ligands and the open-shell metal ion. According to this simple electrostatic model, the ligands produce a field which removes the fivefold degeneracy of the metal-ion $3d$ orbitals. Application of the *Aufbau* principle and Hund's rules gives a ligand-field stabilization energy (LFSE) and spin degeneracy that correlates qualitatively with measurements of heats of hydration, paramagnetism, and several other properties. However, experiments that probe the electron distribution functions of these compounds indicate that the simple ionic model cannot be exact because significant covalency is present. For example, the electron spin resonance spectrum (see Section 7.13) of compounds like $[MnF_6]^{4-}$, which have one unpaired d electron in the configuration $(t_{2g}^5 e_g^0)$, can be used to estimate the density of the unpaired electron at the ligand nuclei. The magnitude of the observed (electron–spin, nuclear–spin) hyperfine interaction indicates that the delocalization of the t_{2g} electron varies between 10 and 30%, depending upon the complex. Also, the electrostatic picture based on undistorted metal-ion d orbitals predicts that term splittings which are due to the electron-electron repulsion between the d electrons should be the same in the free ion and the complex; experimental term splittings obtained from the spectra of complexes are often lower than those of the free ion. A corresponding result is found for the spin-orbit interaction. All of these deviations from the ionic model for transition-metal complexes are given a satisfactory interpretation by a MO treatment in which metal-ion orbitals and combinations of ligand orbitals of the same symmetry are mixed to form delocalized MO's.

Octahedral Complexes. Molecular orbitals for the octahedral MX_6 molecule are constructed by methods corresponding to those described in previous sections for other polyatomic molecules. The metal-ion orbitals and combinations of ligand orbitals can be classified according to their symmetry properties in the same way as were the CO_2 orbitals. In an octahedral molecule the orbitals can be nondegenerate (designated a), doubly degenerate (designated e), or triply degenerate (designated t). Since an octahedral molecule (see Fig. 5.7) has a center of symmetry, the orbitals can be classified also as g or u according to whether they are symmetric or antisymmetric, respectively, under inversion of the coordinates. An additional property of an octahedral complex with six identical ligands is that it is symmetric to rotation by 90°, 180°, 270°, or 360° about any of the ligand-metal bonds (the x, y, or z axis in Fig. 5.7); that is, such a rotation interchanges only equivalent ligands. The a and t orbitals can be further classified by their symmetry properties under 90° rotation about one of

Table 6.21   *Classification of octahedral orbitals*[a]

Designation	Degeneracy	Inversion	90° rotation about fourfold axis
a_{1g}	1	a_{1g}	a_{1g}
a_{2g}	1	a_{2g}	$-a_{2g}$
$e_g^{(1)}, e_g^{(2)}$	2	$e_g^{(1)}, e_g^{(2)}$	$e_g^{(1)}, -e_g^{(2)}$
$t_{1g}^{(1)}, t_{1g}^{(2)}, t_{1g}^{(3)}$	3	$t_{1g}^{(1)}, t_{1g}^{(2)}, t_{1g}^{(3)}$	$-t_{1g}^{(2)}, t_{1g}^{(1)}, t_{1g}^{(3)}$
$t_{2g}^{(1)}, t_{2g}^{(2)}, t_{2g}^{(3)}$	3	$t_{2g}^{(1)}, t_{2g}^{(2)}, t_{2g}^{(3)}$	$-t_{2g}^{(2)}, t_{2g}^{(1)}, -t_{2g}^{(3)}$
a_{1u}	1	$-a_{1u}$	a_{1u}
a_{2u}	1	$-a_{2u}$	$-a_{2u}$
$e_u^{(1)}, e_u^{(2)}$	2	$-e_u^{(1)}, -e_u^{(2)}$	$e_u^{(1)}, -e_u^{(2)}$
$t_{1u}^{(1)}, t_{1u}^{(2)}, t_{1u}^{(3)}$	3	$-t_{1u}^{(1)}, -{}^{(1)}t_{1u}^{(2)}, -t_{1u}^{(3)}$	$-t_{1u}^{(2)}, t_{1u}^{(1)}, t_{1u}^{(3)}$
$t_{2u}^{(1)}, t_{2u}^{(2)}, t_{2u}^{(3)}$	3	$-t_{2u}^{(1)}, -{}^{(1)}t_{2u}^{(2)}, -t_{2u}^{(3)}$	$-t_{2u}^{(2)}, t_{2u}^{(1)}, -t_{2u}^{(3)}$

[a] The sign attached to the symmetry designation in columns 3 and 4 indicates the behavior of the orbital under the symmetry operation. The superscripts on e and t orbitals identify the members of the degenerate set; the order of the t orbitals in the last column specifies the transformation of the orbitals (e.g., $t_{1u}^{(1)} \rightarrow -t_{1u}^{(2)}, t_{1u}^{(2)} \rightarrow t_{1u}^{(1)}$ by a 90° rotation).

these fourfold symmetry axes. If an a orbital is unchanged by the rotation through 90°, it is given the subscript 1; if it changes sign, it is given the subscript 2. When the three degenerate t orbitals are rotated by 90°, two of them are transformed into other members of the set or their negatives. If the third t orbital remains unchanged by the rotation, the set of degenerate t orbitals is designated by the subscript 1; otherwise, it changes sign under the rotation and the set is assigned the subscript 2. The classifications of the orbitals of octahedral complexes are summarized in Table 6.21. (For a discussion of more general methods that make use of orbital symmetry, see Reference 12 in the Additional Reading list.)

Once the symmetry orbitals (i.e., orbitals of $a_{1g}, a_{2g}, \ldots, t_{2u}$ symmetry) have been formed by taking appropriate linear combinations of the ligand orbitals, the ligand and metal-ion orbitals are combined into MO's with coefficients evaluated from the variation principle by means of a set of linear equations analogous to Eq. 6.43 or the more general SCF equa-

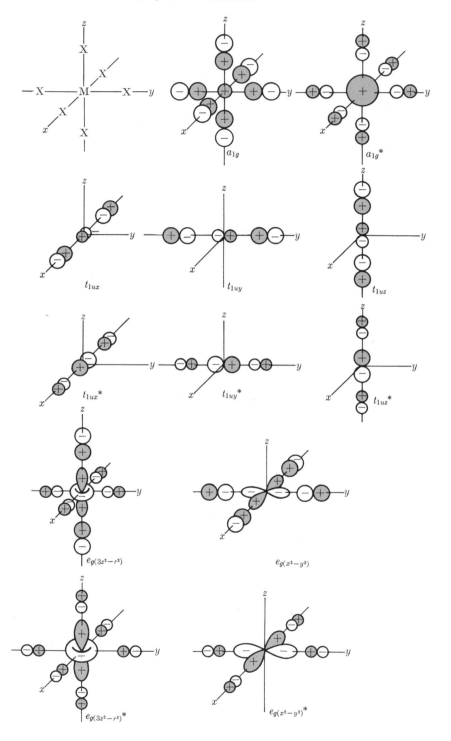

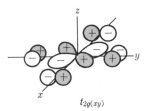

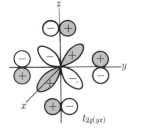

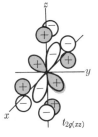

$t_{2g(xy)}$ — $t_{2g(yz)}$ — $t_{2g(xz)}$

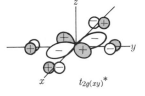

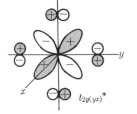

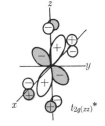

$t_{2g(xy)}^*$ — $t_{2g(yz)}^*$ — $t_{2g(xz)}^*$

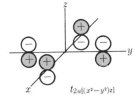

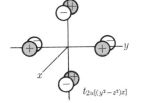

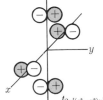

$t_{2u[(x^2-y^2)z]}$ — $t_{2u[(y^2-z^2)x]}$ — $t_{2u[(z^2-x^2)y]}$

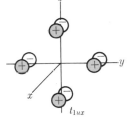

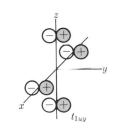

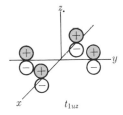

t_{1ux} — t_{1uy} — t_{1uz}

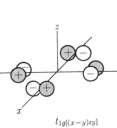

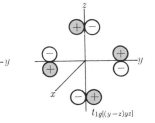

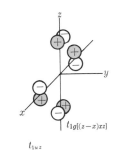

$t_{1g[(x-y)zy]}$ — $t_{1g[(y-z)yz]}$ — $t_{1g[(z-x)xz]}$

t_{1uz}

Fig. 6.40. *Left and above:* Combinations of ligand orbitals and metal-ion orbitals of the same symmetry type in octahedral MX_6

tions. As in the case of naphthalene (Section 6.4.2), it is important to use symmetry orbitals here because if two orbitals ϕ_i and ϕ_j have different symmetry (e.g., if one is symmetric and the other antisymmetric under reflection in a symmetry plane), the integral H_{ij} is zero and the two orbitals do not mix; that is, only a_{1g} ligand orbitals can mix with a_{1g} metal orbitals, e_g with e_g, t_{2g} with t_{2g}, and so on. The possible combinations of metal-ion $3d$, $4s$, and $4p$ orbitals with ligand $2p$ orbitals are pictured with their symmetry designations in Fig. 6.40. The relative sizes of the AO's are meant to indicate roughly the relative values of their coefficients in the MO's obtained by an approximate SCF-LCAO-MO calculation for $Cr\,F_6{}^{3-}$.

A typical MO energy-level diagram for MX_6 is given in Fig. 6.41, where the level order corresponds to the results of a calculation for $CrF_6{}^{3-}$. The $1a_{1g}$ orbital in Fig. 6.41 is the σ-bonding a_{1g} orbital shown in Fig. 6.40; it is formed mainly from the $4s$ metal-ion orbital and the $2p_\sigma$ ligand orbitals. The $1e_g$ orbitals in Fig. 6.41 are doubly degenerate σ-bonding orbitals formed from the metal-ion $d_{3z^2-r^2}$ and $d_{x^2-y^2}$ orbitals and linear combinations of ligand p_σ orbitals that have large overlap with these d orbitals, as shown in Fig. 6.40. Next in energy are the triply degenerate $1t_{2g}$ π-bonding orbitals composed of d_{xy}, d_{yz}, and d_{xz} and the linear combinations of ligand $2p_\pi$ orbitals that overlap strongly with them. The $1t_{1u}$ MO's are triply degenerate σ-bonding orbitals formed from the $4p$ metal-ion orbitals and the appropriate pairs of $2p_\sigma$ ligand orbitals. The $2t_{1u}$, t_{2u}, and t_{1g} orbitals are essentially nonbonding combinations of four $2p_\pi$ ligand orbitals whose energy is very close to that of the unperturbed $2p_\pi$ orbitals. Of these MO's the $2t_{1u}$ lies lowest in energy, as expected, since it can mix slightly with the $4p$ metal-ion orbitals. The orbitals described so far are 18 in number and are filled by the 36 $2p$ ligand electrons. The remaining orbitals are all antibonding combinations of metal-ion and ligand orbitals of the various symmetries. The first two, $2t_{2g}$ and $2e_g$, are occupied by the metal-ion $3d$ electrons, and correspond in the simpler ionic model to the t_{2g} and e_g levels into which the $3d$ level is split by the ligand field. The energy separation between $2t_{2g}$ and $2e_g$ can thus be identified with 10δ, although δ has to be obtained from the MO energies since it is not given by the quantity corresponding to the ionic model ($\frac{1}{6}(Q/R^5)\langle r^4 \rangle$) (Section 5.5). In $CrF_6{}^{3-}$, there are three metal-ion $3d$ electrons and the ground-state configuration is

$$(1a_{1g})^2(1e_g)^4(1t_{2g})^6(1t_{1u})^6(2t_{1u})^6(t_{2u})^6(t_{1g})^6(2t_{2g})^3$$

For different metal ions and different ligands, the ordering of the lower-lying bonding and nonbonding orbitals can vary, but the antibonding t_{2g} and e_g levels generally contain the same number of electrons as were in the $3d$ level of the free metal ion. It is for this reason that the simple ionic picture, which ignores mixing of ligand and metal-ion orbitals, is in qualitative accord with experimental thermodynamic and spectroscopic data. However, the more sophisticated MO treatment, which involves mixing of the

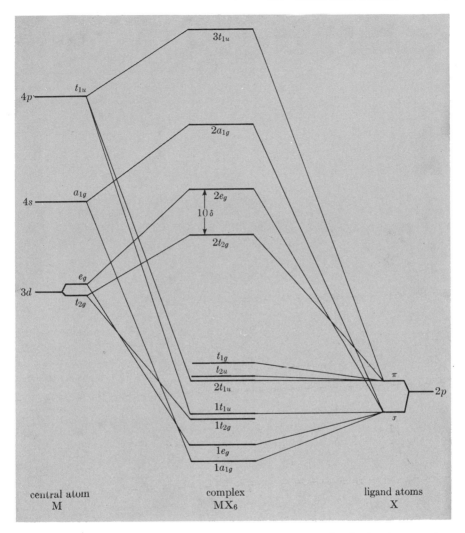

Fig. 6.41. Energy-level diagram for an octahedral MX_6 molecule showing correlations with atomic orbitals. The connection of levels shows the major contribution of the M and X_6 orbitals to the MX_6 orbitals, as indicated in Fig. 6.40. MO level order corresponds to semiempirical calculations for CrF_6^{3-} [C. J. Ballhausen and H. B. Gray, *Molecular Orbital Theory* (Benjamin, New York, 1964), pp. 128–131].

metal-ion orbitals with those of the ligands, is also able to explain the deviations from the simple model listed at the beginning of this section. For the spin-orbit coupling and electron-electron repulsion terms, which depend on the values of $\langle 1/r^3 \rangle$ and $\langle 1/r_{12} \rangle$, respectively, the required reduction of these averages is achieved by shifting some of the metal electrons to the ligands, since this increases their average distance from the metal ion and from each other, as well.

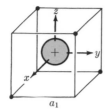

a_1

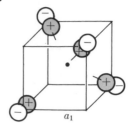

a_1

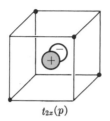

$t_{2x}(p)$

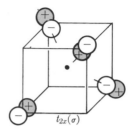

$t_{2x}(\sigma)$

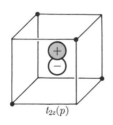

$t_{2y}(p)$

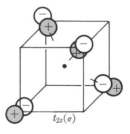

$t_{2y}(\sigma)$

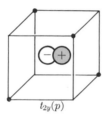

$t_{2z}(p)$

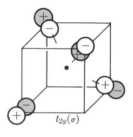

$t_{2z}(\sigma)$

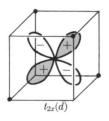

$t_{2x}(d)$

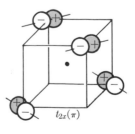

$t_{2x}(\pi)$

central atom M

ligand atoms X_4

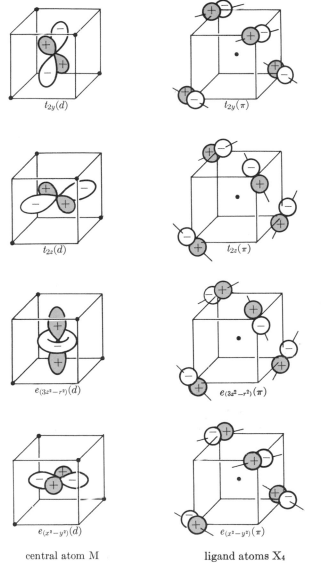

central atom M   ligand atoms X_4

Fig. 6.42. *Left and above:* Metal-ion orbitals and combinations of ligand orbitals of various symmetry types in tetrahedral MX_4. The $\sigma(2s)$ and the nonbonding $t_1(\pi)$ orbitals are not shown.

Tetrahedral Complexes. For MX_4 complexes in which the ligands are tetrahedrally arranged (Fig. 5.14), the procedure for obtaining the MO's is similar to that outlined for octahedral complexes. Again, the MO's can be singly, doubly, or triply degenerate and are designated a, e, and t, respectively. However, the g, u designation is not applicable since there is no center of symmetry. The orbitals can be further classified by their properties under 90° rotation about the z axis followed by reflection in the central plane perpendicular to the axis (i.e., the xy plane, see Fig. 5.14). Table 6.22 summarizes the classification of the orbitals for tetrahedral symmetry. The $3d$, $4s$, and $4p$ metal-ion orbitals and the various linear combinations of ligand $2p$ orbitals are shown in Fig. 6.42, together with their symmetry classifications. An appropriate SCF-LCAO-MO level diagram which results from combining metal-ion and ligand orbitals of the same symmetry subject to the variation principle is shown in Fig. 6.43, where the level ordering is that obtained from calculations for MnO_4^{2-}. In addition to the

Table 6.22 *Classification of tetrahedral orbitals*[a]

Designation	Degeneracy	90° rotation followed by reflection
a_1	1	a_1
a_2	1	$-a_2$
$e^{(1)}, e^{(2)}$	2	$e^{(1)}, -e^{(2)}$
$t_1^{(1)}, t_1^{(2)}, t_1^{(3)}$	3	$-t_1^{(2)}, t_1^{(1)}, t_1^{(3)}$
$t_2^{(1)}, t_2^{(2)}, t_2^{(3)}$	3	$-t_2^{(2)}, t_2^{(1)}, -t_2^{(3)}$

[a] See footnote to Table 6.21.

ligand orbitals shown in Fig. 6.42, it is found that the $2s$ orbitals must be included to give good agreement with experiment; that is, for MnO_4^{2-}, ligand sp hybridization makes an important contribution. The $1a_1$ and $1t_2$ orbitals are strongly bonding, the $1e$ and $2t_2$ orbitals are weakly bonding, and the $2a_1$, $3t_2$, and t_1 orbitals are nonbonding and are mainly composed of linear combinations of ligand $2p$ orbitals. Up to this point there are 16 orbitals, which are filled by the $2s$ and $2p$ ligand electrons. The remaining MO's are antibonding; the first two of these, $2e$ and $4t_2$, are occupied primarily by the metal-ion $3d$ electrons. These orbitals thus correspond to the e and t_2 levels into which the $3d$ metal-ion level is split by the ligand field in the simple ionic model, and their energy separation corresponds to 10δ.

For MnO_4^{2-}, there is only one metal-ion $3d$ electron in addition to the electrons of the closed-shell O^{2-} ions, so that the ground-state configuration is

$$MnO_4^{2-}(1a_1)^2(1t_2)^6(1e)^4(2t_2)^6(2a_1)^2(3t_2)^6(t_1)^6(2e)^1$$

A variety of electronic transitions (e.g., $t_1 \rightarrow 2e$, $3t_2 \rightarrow 2e$, $t_1 \rightarrow 4t_2$) have been observed, in agreement with the energy-level scheme shown in Fig. 6.43. From the Hückel-type MO calculations for MnO_4^{2-} it is clear that,

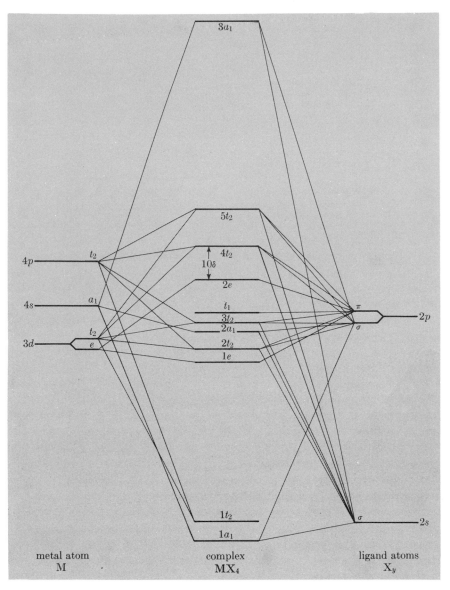

Fig. 6.43. MO diagram for tetrahedral MX_4 showing correlation with AO's. All contributions of the M and X_4 orbitals to the MX_4 orbitals are shown. MO level order corresponds to semiempirical calculations for MnO_4^{2-} [C. J. Ballhausen and H. B. Gray, *Molecular Orbital Theory* (Benjamin, New York, 1964), pp. 123–128].

except for $1a_1$ and $1t_2$ (which are composed almost entirely of ligand s orbitals), all the occupied MO's include significant covalent mixing between metal-ion and ligand orbitals; for example, the $1e$ orbital is found to be composed of approximately 60% metal-ion and 40% ligand orbitals. Thus, the description of transition-metal complexes that is provided by MO theory (often called *ligand-field theory* in this application) is considerably more complicated than the simple ionic (or *crystal-field theory*) picture.

SUMMARY

This chapter has been concerned with the quantum-theoretical principles by which bonding in diatomic and polyatomic molecules can be classified and understood. Parallel treatments of molecules by the VB and MO methods show that if delocalization and resonance can be ignored (most diatomics, H_2O, CH_4, NH_3, C_2H_6, C_2H_4, C_2H_2, etc.), the VB picture is simpler; otherwise (O_2, C_6H_6, CO_2, etc.) the MO treatment often gives a useful first approximation with a minimum of calculational difficulty. Improvements on the simple qualitative treatments are provided by self-consistent-field calculations which have been carried out for many diatomic and a few polyatomic molecules. An approximate method of this type (Hückel MO model) is applied to a variety of aromatic and conjugated systems. It is used also to introduce the Woodward–Hoffmann rules for cycloaddition and other organic reactions. The behavior of MO orbitals and their energies as a function of molecular geometry serves to predict the shapes of small triatomic molecules. A refinement of the crystal-field model for transition-metal complexes is provided by MO calculations for these systems.

In Chapter 7 we discuss several spectroscopic methods for measuring many of the experimental molecular data that have made possible the development of a theory of the chemical bond.

ADDITIONAL READING

[1] C. A. COULSON, *Valence* (Oxford University Press, London, 1961), 2nd ed.

[2] L. PAULING, *The Nature of the Chemical Bond* (Cornell University Press, Ithaca, N.Y., 1960), 2nd ed.

[3] C. J. BALLHAUSEN and H. B. GRAY, *Molecular Orbital Theory* (Benjamin, New York, 1964).

[4] J. N. MURREL, S. F. A. KETTLE, and J. M. TEDDER, *Valence Theory* (Wiley, New York, 1965).

[5] R. DAUDEL, R. LEFEBVRE, and C. MOSER, *Quantum Chemistry* (Wiley-Interscience, New York, 1959).

[6] K. Higasi, H. Baba, and A. Rembaum, *Quantum Organic Chemistry* (Wiley-Interscience, New York, 1965).

[7] R. G. Parr, *The Quantum Theory of Molecular Structure* (Benjamin, New York, 1963).

[8] J. D. Roberts, *Notes on Molecular Orbital Calculations* (Benjamin, New York, 1961).

[9] L. Salem, *Molecular Orbital Theory of Conjugated Systems* (Benjamin, New York, 1966).

[10] A. Streitweiser, Jr., *Molecular Orbital Theory for Organic Chemists* (Wiley, New York, 1961).

[11] C. A. Coulson and A. Streitweiser, *Dictionary of π-Electron Calculations* (Freeman, San Francisco, 1965).

[12] F. A. Cotton, *Chemical Applications of Group Theory* (Wiley-Interscience, New York, 1963).

[13] A. Liberles, *Introduction to Molecular-Orbital Theory* (Holt, Rinehart and Winston, New York, 1966).

[14] R. B. Woodward and R. Hoffmann, *The Conservation of Orbital Symmetry* (Academic Press, New York, 1970).

[15] G. Herzberg, *Spectra of Diatomic Molecules* (Van Nostrand, New York 1950).

PROBLEMS

6.1 Express the dissociation energies D_e of the third-row diatomic molecules Na_2, Mg_2, . . . , Ar_2 in terms of resonance integrals β involving atomic orbitals with $n = 3$. Give the bond order of each molecule as predicted by the simple MO theory.

6.2 Write simple VB wave functions for each of the neutral second-row homonuclear diatomic molecules. Give the bond order of each molecule as predicted by the simple VB theory and compare with the MO results of Table 6.2.

6.3 Consider a C_2 molecule in the ground state electronic configuration $(2p\pi)^2$; assume that there is an electron in each of the degenerate orbitals, and that their spins are opposed (i.e., a singlet state). What is the term symbol, including the symmetry with respect to reflection in a plane through the nuclei?

6.4 One of the low-lying excited configurations of O_2 is $(2p\pi)^3(2p\pi^*)^3$. Find term symbols for all the possible states arising from this configuration. (*Hint:* The problem can be simplified by working with the corresponding hole configuration.)

$$Ans. \ {}^1\Sigma_u{}^+, \ {}^3\Sigma_u{}^+, \ {}^1\Sigma_u{}^-, \ {}^3\Sigma_u{}^-, \ {}^1\Delta_u, \ {}^3\Delta_u.$$

6.5 Let ϕ_1' and ϕ_2' be two approximate real orthonormal MO's of the same symmetry species (e.g. σ_g, π_u, a_{1g}, e_u) and let h_e be the effective one-electron Hamiltonian for the system. The orbital energies are thus given by

$$\epsilon_1' = \int \phi_1' h_e \phi_1' \, dv = H_{11}'$$
$$\epsilon_2' = \int \phi_2' h_e \phi_2' \, dv = H_{22}'$$

If ϕ_1' and ϕ_2' are only approximate eigenfunctions of h_e, then $h_e \phi_1'$ is not a constant times ϕ_1' and will therefore not in general be orthogonal to ϕ_2'. Thus the integrals

$$H_{12}' = H_{21}' = \int \phi_2' h_e \phi_1' \, dv$$

are not equal to zero. Orthonormal linear combinations of ϕ_1' and ϕ_2' can be found so that the corresponding integrals vanish when ϕ_1' and ϕ_2' are replaced by

$$\phi_1 = a_{11}\phi_1' + a_{12}\phi_2'$$
$$\phi_2 = a_{21}\phi_1' + a_{22}\phi_2'$$

where the a_{ij} are constants.

(a) If the a_{ij} satisfy the linear equations ($i = 1, 2$)

$$a_{i1}(\epsilon_1' - \epsilon_i) + a_{i2}H_{12}' = 0$$
$$a_{i1}H_{12}' + a_{i2}(\epsilon_2' - \epsilon_i) = 0$$

where

$$\epsilon_1 = \int \phi_1 h_e \phi_1 \, dv$$
$$\epsilon_2 = \int \phi_2 h_e \phi_2 \, dv$$

show that

$$\int \phi_1 \phi_2 \, dv = 0$$
$$\int \phi_1 h_e \phi_2 \, dv = 0$$

(*Hint:* Solve the secular equation for ϵ_1 and ϵ_2 and show that $\epsilon_2 - \epsilon_2' = \epsilon_1' - \epsilon_1$; then express the two integrals in terms of a_{12}, a_{22}, ϵ_1, ϵ_1', ϵ_2, ϵ_2' and substitute the result from the secular equation.)

(b) When two approximate MO's of the same symmetry combine to form two orthogonal linear combinations with $H_{12} = 0$, show that the energy of the lower MO is lowered (i.e., if $\epsilon_1 < \epsilon_2$, show that $\epsilon_1 < \epsilon_1'$) and that the energy of the higher MO is raised (i.e., that $\epsilon_2 > \epsilon_2'$).

6.6   The IP's of N are

i	IP$_i$ (eV)
1	14.549
2	29.612
3	47.438
4	77.470
5	97.887
6	552.063
7	667.000

An SCF-LCAO ground-state calculation of N_2 yields a total electronic energy (including the nuclear repulsion Z^2e^2/R_e) of -108.9928 a.u. at $R_e = 2.0134$ a.u. The experimental dissociation energy D_e of N_2 is 9.902 eV. (a) What is the percent error in the calculated total electronic energy? (b) What is the sign of the calculated dissociation energy? (c) Theoretical dissociation energies often are estimated from the difference between the calculated SCF energies of the molecule and that of its component atoms. Given that the SCF energy of N is -54.40095 a.u., estimate the theoretical D_e for N_2.

6.7 From the experimental values $R_e = 1.5853$ Å and $\mu = 5.882$D for LiH, determine the coefficient ratios B/A in the VB wave function (Eq. 6.3) and b/a in the MO (Eq. 6.14).

6.8 Evaluate the integral

$$\int 1s_H(1)\, 2p_{zF}(1)\, z_1\, dv_1$$

where $1s_H$ is the ground-state hydrogen wave function centered at $(0, 0, -R/2)$ and $2p_{zF}$ is a Slater orbital (Chapter 4) centered at $(0, 0, R/2)$. By what amount does inclusion of this integral change the value of μ for HF if B/A in Eq. 6.3 is taken to be 0.83? (*Hint:* Use the confocal elliptic coordinates ξ, η, and ϕ of Problem 5.25 to evaluate the integral; note that $z = \frac{1}{2}R\xi\eta$.)

6.9 Estimate the fraction of ionic character of HF from the electronegativity difference of the two atoms. Compare your result with the value estimated from the dipole moment.

6.10 If electrons did not repel one another, but were only attracted to positive nuclei without any shielding effects, what would be the values of IP and EA for C and H atoms? Would CH_4 be covalent or ionic?

6.11 From the data given in Table 6.10, calculate the SCF dipole moment of CO.

6.12 From the experimental dipole moment for H_2O (1.8D) and the experimental bond angle (104.5°), find the bond dipole moment for OH in H_2O. From your result and the experimental bond length (0.96 Å) estimate the ratio B/A in Eq. 6.22.

6.13 Estimate the fraction of ionic character of the O—H bond in H_2O from the electronegativity difference between O and H. Use the result to obtain the ratio B/A in Eq. 6.22, and compare with the result for Problem 6.12.

6.14 Consider a hybrid orbital of the type

$$\chi_1 = N[2s + a(2p_x + 2p_y + 2p_z)]$$

where N and a are constants.
(a) Find an expression for N that normalizes χ_1 for all values of a. (b) By substituting expressions for $2s$, $2p_x$, $2p_y$, and $2p_z$ from Table 3.1 show that for a given value of r, the function χ_1 has its maximum at the point $P_1(x = y = z = r/\sqrt{3})$. What are the

polar angles θ_1 and ϕ_1 that give the direction of the orbital; (i.e., the direction of $\overrightarrow{OP_1}$)? (c) Construct a second orbital χ_2 that has the same form as χ_1 but with the signs of $2p_x$, $2p_y$, and $2p_z$ chosen so that the point P_1 at which χ_2 is a maximum for a given r is in the direction $(\theta_2 = \theta_1,\ \phi_2 = \pi + \phi_1)$. What is the angle $\overrightarrow{OP_1}$, $\overrightarrow{OP_2}$? Show that χ_2 is normalized. (d) What value of a makes the orbitals χ_1 and χ_2 orthogonal? (e) Construct orbitals χ_3 and χ_4 by changing only the signs of $2p_x$, $2p_y$, and $2p_z$ in χ_1 so that $\theta_3 = \theta_4 = \pi - \theta_1$, $\phi_3 = \phi_1 - \pi/2$, $\phi_4 = \phi_1 + \pi/2$ and compare with Eq. 6.24. Locate the points P_1, P_2, P_3, and P_4 for $r = \sqrt{3}$ in a Cartesian coordinate system and compare with Fig. 6.15c. (f) Show that each pair of orbitals χ_i, χ_j is orthogonal.

6.15   Consider a hybrid orbital of the form

$$\chi_2 = N_2[2s - b(2p_x - c2p_y)]$$

where N_2, b, and c are positive constants.
(a) Find an expression for N_2 that normalizes χ_2 for any value of b and c. (b) Show that χ_2 can be written

$$\chi_2 = N_2[f_1(r) - bf_2(r)(\cos \phi_2 - c \sin \phi_2)]$$

where $x = r \cos \phi_2$ and $y = r \sin \phi_2$. (c) Find the angle ϕ_2 (in terms of c) that maximizes χ_2 for a given value of r. (d) Construct χ_3 with the same form as χ_2 but with the sign of the coefficient of $2p_y$ changed. What is ϕ_3 in terms of c? (e) Find the value of c for which $\phi_3 - \phi_2 = 2\pi/3$. (f) Find the value of b that makes χ_2 and χ_3 orthogonal. (g) Find the values of N_1, d, and e for which the orbital

$$\chi_1 = N_1[2s + d2p_x + e2p_y]$$

is normalized and is orthogonal to both χ_2 and χ_3. (h) Locate the directions ϕ_1, ϕ_2, and ϕ_3 in a sketch of the xy plane. What are the angles $\phi_2 - \phi_1$ and $\phi_1 - \phi_3$? (i) Compare the forms of χ_1, χ_2, and χ_3 with Eq. 6.25.

6.16   Repeat Problem 6.15 with $\phi_3 - \phi_2 = 117°$, as in ethylene. Find the normalized orthogonal orbitals χ_1, χ_2, and χ_3 and compare with Eq. 6.26.

6.17   (a) If S_s and S_p are the overlap integrals between a hydrogen $1s_H$ orbital centered at the origin and a $2s$ and $2p_\sigma$ orbital, respectively, centered at $z = R$, write an expression for the overlap integral S between the $1s_H$ orbital and a normalized hybrid orbital centered at $z = R$ and of the form

$$N(2s + \lambda 2p_\sigma)$$

where N and λ are constants. Determine the value of N required to normalize the hybrid orbital.
(b) Assuming that $S_s = S_p$, plot S as a function of the fraction of s character, $1/(1 + \lambda^2)$, and find the fraction at which S is a maximum.
(c) For Slater orbitals with $\zeta_{2s} = \zeta_{2p} = 1.5$ and $R > 3a_0$, the overlap S_p is greater than S_s. At smaller separations, S_s becomes larger than S_p. At $R = 2.08a_0$ (approxi-

mately the C—H bond length), $S_p = 0.495$ and $S_s = 0.591$. Find the fraction of s character that makes S a maximum for these values of S_s and S_p and compare with your answer to part (b). Does your result indicate that the degree of hybridization changes with internuclear distance?

6.18 (a) Derive general expressions for a set of four orthonormal hybrid orbitals $(\chi_1, \chi_2, \chi_3, \chi_4)$ that are linear combinations of $2s$, $2p_x$, $2p_y$, and $2p_z$ under the restriction that χ_1 and χ_2 are equivalent and that χ_3 and χ_4 are equivalent (i.e., that they have the same amount of s character).

Ans.

$$\chi_1 = \lambda(2s) + \tfrac{1}{2}(2p_x) + \tfrac{1}{2}(2p_y) + (\tfrac{1}{2} - \lambda^2)^{1/2}(2p_z)$$

$$\chi_2 = \lambda(2s) - \tfrac{1}{2}(2p_x) - \tfrac{1}{2}(2p_y) + (\tfrac{1}{2} - \lambda^2)^{1/2}(2p_z)$$

$$\chi_3 = (\tfrac{1}{2} - \lambda^2)^{1/2}(2s) + \tfrac{1}{2}(2p_x) - \tfrac{1}{2}(2p_y) - \lambda(2p_z)$$

$$\chi_4 = (\tfrac{1}{2} - \lambda^2)^{1/2}(2s) - \tfrac{1}{2}(2p_x) + \tfrac{1}{2}(2p_y) - \lambda(2p_z)$$

(b) Show that λ^2 is the fraction of s character in χ_1 and χ_2.
(c) Show that for $\lambda^2 = \tfrac{1}{4}$, the orbitals reduce to tetrahedral sp^3 hybrids, and for $\lambda^2 = \tfrac{1}{2}$ to the digonal sp hybrids plus $2p$ orbitals. Determine the directions of the maxima (at constant r) of each of the four orbitals.
(d) Show that the angle $\phi_1 - \phi_2$ between the maxima (at constant r) for χ_1 and χ_2 is related to λ^2 by the expression

$$\cos(\phi_1 - \phi_2) = -\frac{\lambda^2}{1 - \lambda^2}$$

(e) Find the value of λ^2 for which $\phi_1 - \phi_2$ equals $104.5°$, the bond angle in H_2O.

6.19 (a) Derive a general expression for a set of four orthonormal hybrid orbitals $(\chi_1, \chi_2, \chi_3, \chi_4)$ that are linear combinations of $2s$, $2p_x$, $2p_y$, and $2p_z$ under the restriction that χ_2, χ_3, and χ_4 have the same amount of s character.

Ans.

$$\chi_1 = (1 - 3\lambda^2)^{1/2}(2s) + \sqrt{2}\,\lambda(2p_z)$$

$$\chi_2 = \lambda(2s) + \left(\frac{2}{3}\right)^{1/2}(2p_x) - \left(\frac{1}{3} - \lambda^2\right)^{1/2}(2p_z)$$

$$\chi_3 = \lambda(2s) - \frac{1}{\sqrt{6}}(2p_x) + \frac{1}{\sqrt{2}}(2p_y) - \left(\frac{1}{3} - \lambda^2\right)^{1/2}(2p_z)$$

$$\chi_4 = \lambda(2s) - \frac{1}{\sqrt{6}}(2p_x) - \frac{1}{\sqrt{2}}(2p_y) - \left(\frac{1}{3} - \lambda^2\right)^{1/2}(2p_z)$$

(b) Show that λ^2 is the fraction of s character in χ_2, χ_3, and χ_4
(c) Show that for $\lambda^2 = \tfrac{1}{3}$, the orbitals reduce to trigonal sp^2 hybrids plus a p orbital. To what do the orbitals reduce for $\lambda^2 = 0$?

(d) Show that the angles between maxima (at constant r) of pairs of the three equivalent orbitals are equal and that

$$\cos(\phi_2 - \phi_3) = \cos(\phi_3 - \phi_4) = \cos(\phi_4 - \phi_2) = -\frac{\lambda^2}{1 - \lambda^2}$$

(e) Find the value of λ^2 for which $\phi_2 - \phi_3$ equals $106.8°$, the bond angle in NH_3.

6.20   Use Eq. 6.28 to calculate the energy of the Kekulé and Dewar structures of benzene. Ignore Coulomb and exchange integrals involving AO's associated with nonadjacent atoms.

Ans. Kekulé: $6Q + 1.5J$; Dewar: $6Q$.

6.21   Find a benzene MO that contains a single node passing through atoms 2 and 5 of Fig. 6.20 (*Hint:* Take a linear combination of ϕ_2 and ϕ_3 of Eqs. 6.31 and 6.32.)

6.22   Use Eq. 6.36 to find the ratios a/b and c/d in Eqs. 6.32 and 6.34, respectively.

6.23   Find the energies of the bonding and antibonding π orbitals for benzene and ethylene in terms of α and β.

Ans. Eqs. 6.38 and 6.40.

6.24   Sketch the orbitals of Eq. 6.45 on naphthalene carbon frameworks in perspective similar to that of Fig. 6.22. Indicate the positive and negative lobes of the p_z orbitals and verify the results of reflection in σ_h, σ_v', and σ_v'' given in Table 6.13.

6.25   Consider two orbitals ϕ_s and ϕ_a that are symmetric and antisymmetric, respectively, with respect to replacement of z by $-z$; that is,

$$\phi_s(x, y, -z) = \phi_s(x, y, z)$$
$$\phi_a(x, y, -z) = -\phi_a(x, y, z)$$

Show that the overlap integral is zero, i.e., that

$$\int \phi_s \, \phi_a \, dv = 0$$

6.26   Let ϕ_+ be an orbital that is symmetric upon reflection in a given plane and let ϕ_- be orbital that is antisymmetric upon reflection in the same plane. (a) Show that

$$\int \phi_+ \, \phi_- \, dv = 0$$

(*Hint:* Define the coordinate system so that the symmetry plane corresponds to the xy plane; reflection in the plane thus corresponds to $z \rightarrow -z$. Is $\phi_+\phi_-$ a symmetric or antisymmetric function of z?)
(b) If h_e is an effective one-electron Hamiltonian that is invariant upon reflection in the plane (i.e., the nuclear arrangement in the molecule after reflection is indistinguishable from the arrangement before reflection), show that

$$H_{+-} = \int \phi_+ h_e \phi_- \, dv = 0$$

6.27 Verify Eq. 6.46 for $\langle A_1|h_e|A_1\rangle$, $\langle A_1|h_e|A_2\rangle$, $\langle B_2|h_e|B_2\rangle$, $\langle B_2|h_e|B_3\rangle$, $\langle A_1|h_e|B_1\rangle$, and $\langle C_1|h_e|C_1\rangle$.

6.28 Find the roots of the factored secular determinant in Eq. 6.47.

6.29 Substitute the value of ϵ_1 from Eq. 6.47 into Eqs. 6.48 and 6.49 to find the numerical values of a_1, b_1, and c_1. Compare with Eq. 6.50 and verify the result for ϕ_1 given in Eq. 6.51.

6.30 Show how to introduce the orbitals ϕ_x and ϕ_y given in Eq. 6.62 into Eq. 6.60 by adding and subtracting columns 7 and 9 and columns 8 and 10 after multiplying by the appropriate constants. Put your result into the form of Eq. 6.61.

6.31 Write down the four MO's that can be formed from the $1s$ AO's of four H atoms arranged at the corners of a square (Fig. 6.31). Label the orbitals s or a if symmetric or antisymmetric, respectively, with respect to reflection in plane v and prime the orbitals that are antisymmetric with respect to reflection in plane v'. Show that ϕ_a and $\phi_s{}'$ are equivalent and that they therefore are degenerate. Calculate the orbital energies in the Hückel approximation.

6.32 Classify each of the $2s$ and $2p_x$, $2p_y$, $2p_z$ orbitals of the atom X in Fig. 6.36 according to its symmetry or antisymmetry with respect to 180° rotation about the z axis and reflection in the σ_1 plane. Give the designation of each of the orbitals.

$$Ans.\ 2s : a_1, \quad 2p_x : b_1, \quad 2p_y : b_2, \quad 2p_z : a_1.$$

6.33 Construct sums and differences of the $2s$, $2p_x$, $2p_y$, and $2p_z$ orbitals of the atoms Y_1 and Y_2 in Fig. 6.36 that are symmetric or antisymmetric with respect to 180° rotation about the z axis and reflection in the σ_1 plane.

Ans.

$$2s_1 + 2s_2 : a_1 \qquad 2s_1 - 2s_2 : b_2 \qquad 2p_{x1} + 2p_{x2} : b_1$$
$$2p_{x1} - 2p_{x2} : a_2 \qquad 2p_{y1} + 2p_{y2} : b_2 \qquad 2p_{y1} - 2p_{y2} : a_1$$
$$2p_{z1} + 2p_{z2} : a_1 \qquad 2p_{z1} - 2p_{z2} : b_2$$

6.34 Use Walsh's rules to predict the shapes of NOCl, $N_3{}^-$, BO_2, N_2O, NF_2, and SO_2 in their ground electronic states.

6.35 The electric dipole moment of gaseous Li_2O is found to be zero, while that of Cs_2O is found to be nonzero. Are these results consistent with Walsh's rules as discussed in the text? Is one of the molecules an exception to the rules? What feature of the electronic structures of the two molecules might suggest that one of them obeys the rules while the other does not?

6.36 Construct a valence-orbital angular correlation diagram for dihydride molecules XH_2, using qualitative reasoning similar to that used to obtain Fig. 6.38. Discuss the shapes of H_2O, NH_2, CH_2, BeH_2, and HgH_2 in terms of your diagram. (*Hint:* Construct MO's from combinations of $1s_{H1} + 1s_{H2}$ and $1s_{H1} - 1s_{H2}$ with the valence s and p orbitals of atom X having the appropriate symmetry. Classify the MO's by the designation σ_g, σ_u, π_u for the linear geometry and a_1, b_1, b_2 for the bent geometry. Determine which orbitals are bonding, antibonding, and non-bonding, and arrange the orbitals in order of increasing energy in each geometry. Connect the orbitals of the appropriate symmetry without allowing correlation lines for orbitals of the same symmetry to cross.)

Ans.

   The valence orbital σ_u is expected to increase in energy when the molecule is bent and the higher-energy π_u orbital is expected to stabilize with bending; thus XH_2 molecules with four or fewer valence electrons are linear and those with more than four valence electrons are bent. By this analysis BeH_2 and HgH_2 are predicted to be linear.

6.37   (a) Use the results of Problem 6.36 to determine the ground electron configuration and term symbol of CH_2 and to predict whether CH_2 is linear or bent in the ground state.

$$Ans.\ (1\sigma_g)^2(2\sigma_g)^2(1\sigma_u)^2(1\pi_u)^2 \quad {}^3\Sigma_g^- \quad \text{(linear)}$$

(b) Do you expect CH_2 to be linear or bent in the excited state ${}^1\Delta_g$ with the same configuration as the ground state, except that one of the degenerate $1\pi_u$ orbitals is doubly occupied?

Ans. Bent (bond angle is 102.4°).

6.38   Make a table which gives the transformation properties of the metal-ion $3d$, $4s$, and $4p$ orbitals in a tetrahedral complex when they are reflected in a plane passing through two ligands and bisecting the line between the other two (for example, the plane $x = y$ in Fig. 6.42).

6.39   Give the ground-state electron configuration of $MnO_4{}^{2-}$, using Fig. 6.43 for the MO level order.

6.40   Derive Eq. 6.56 by expressing the determinant of Eq. 6.55 in terms of a Laplace expansion.

Hint: In a Laplace expansion of a $n \times n$ determinant, the determinant is replaced by the sum of the product of each element in the first row (or column) multiplied by the $(n-1) \times (n-1)$ determinant obtained by striking out the row and column of the element. Each product is multiplied by $(-1)^{i+j}$, where i and j are the row and column numbers of the element, respectively. For example,

$$\begin{vmatrix} a & b & c \\ d & e & f \\ g & h & i \end{vmatrix} = a\begin{vmatrix} e & f \\ h & i \end{vmatrix} - b\begin{vmatrix} d & f \\ g & i \end{vmatrix} + c\begin{vmatrix} d & e \\ g & h \end{vmatrix}$$

6.41   (a) Verify each entry $(n > 2)$ in Table 6.16 by means of Eq. 6.56. (b) Find the roots of $D_n(x)$ for $1 \leq n \leq 4$ and compare your results with those listed in

Table 6.17. (c) Calculate the ratios C_{ji}/C_{1i} for $n = 4$; $i, j = 1, \ldots, 4$ and compare with the ratios given in Table 6.17.

6.42 Sketch the MO's of butadiene ($n = 4$), using the coefficient ratios given in Table 6.17 as a guide to the relative signs and sizes of the atomic p_π orbitals in each MO.

6.43 Calculate $D_5(x)$, x_i, and the delocalization energy of the pentadienyl cation, radical, and anion in terms of β. What relative bond strengths do you predict for these three species compared with one another, and compared with that of butadiene?

6.44 Show that if m is the number of p_π AO's in a conjugated molecule, the number of bonding π MO's with S symmetry is $m/4$ if $m/2$ is even and $m/4 + \frac{1}{2}$ if $m/2$ is odd; show that the number of bonding π MO's with A symmetry is $m/4$ if $m/2$ is even and $m/4 - \frac{1}{2}$ if $m/2$ is odd. Check your results by counting the number of bonding π MO's of each symmetry in conjugated molecules with $m = 2, 4, 6, 8$. Choose all possible pairs of these four molecules and determine the number of bonding MO's of each symmetry in the cycloaddition reaction of each pair. Compare your results with Table 6.18.

6.45 Show that if $n/2$ and $m/2$ are both even or both odd, then $m + n = 4k$, where k is an integer; show that if one of the pair of numbers $n/2$, $m/2$ is even and the other is odd, then $m + n = 4k + 2$, where k is an integer. Use these results and Table 6.18 to write down the Woodward–Hoffmann rules for cycloaddition.

6.46 Predict whether the cycloaddition of butadiene and hexatriene proceeds thermally or photochemically.

6.47 Construct orbital and configuration correlation diagrams for the cycloaddition of two butadiene molecules. Predict whether the reaction proceeds thermally or photochemically. Do your predictions based upon the correlation diagrams agree with those of the Woodward–Hoffmann rules?

6.48 Evaluate the integrals in Eqs. 6.11 and 6.12, using the exact hydrogenic function for $1s_H$ (Chapter 3) and the Slater orbital for $2p_{zF}$ (Chapter 4). The hydrogen nucleus is located at $(0, 0, -R/2)$ and the fluorine nucleus is located at $(0, 0, R/2)$. (See Fig. 6.12.)

6.49 (a) A bonding and an antibonding diatomic MO are to be constructed from the AO's χ_A and χ_B. If overlap is neglected, show that the MO energies ϵ_b and ϵ_a for the bonding and antibonding combinations are given in the simple MO theory by the solutions to the secular equation

$$\begin{vmatrix} \alpha_A - \epsilon & \beta \\ \beta & \alpha_B - \epsilon \end{vmatrix} = 0$$

where

$$\alpha_A = \int \chi_A \, h_e \, \chi_A \, dv$$
$$\alpha_B = \int \chi_B \, h_e \, \chi_B \, dv$$
$$\beta = \int \chi_A \, h_e \, \chi_B \, dv = \int \chi_B \, h_e \, \chi_A \, dv$$

where h_e is the effective one-electron Hamiltonian. (b) Write out the explicit expressions for ϵ_b and ϵ_a in terms of α_A, α_B, and β, and show that when $\alpha_A = \alpha_B = \alpha$ they reduce to $\epsilon_b = \alpha + \beta$, $\epsilon_a = \alpha - \beta$.

(c) For the case $\alpha_A > \alpha_B$ and $\beta < (\alpha_A - \alpha_B)$, show that

$$\epsilon_b = \alpha_B - \frac{\beta^2}{\alpha_A - \alpha_B} + \frac{\beta^4}{(\alpha_A - \alpha_B)^3} - \cdots$$

$$\epsilon_a = \alpha_A + \frac{\beta^2}{\alpha_A - \alpha_B} - \frac{\beta^4}{(\alpha_A - \alpha_B)^3} + \cdots$$

Hint: Use the approximation

$$(x + y)^{1/2} = x^{1/2} + \tfrac{1}{2}x^{-1/2}\, y - \tfrac{1}{8}x^{-3/2}\, y^2 + \cdots \qquad y^2 < x^2$$

(d) From your answers to parts (b) and (c), do you expect more or less stabilization (energy lowering) of the bonding MO and destabilization (energy raising) of the antibonding MO as the difference between the AO energies of χ_A and χ_B increases?

For the following problems a programmable desk calculator or digital computer will be useful.

C6.1   One-electron two-center integrals such as the overlap integrals

$$\int \phi_A(1) \, \phi_B(1) \, dv_1$$

when transformed into confocal elliptic coordinates (see Problem 5.25) can be reduced to sums of products of integrals of the form

$$A_n(\alpha) \equiv \int_1^\infty x^n e^{-\alpha x} \, dx$$

$$B_n(\alpha) \equiv \int_{-1}^{+1} x^n e^{-\alpha x} \, dx$$

The values of these integrals are given in the general case by the formulas

$$A_n(\alpha) = e^{-\alpha} \sum_{k=1}^{n+1} \frac{n!}{\alpha^k (n - k + 1)!}$$

$$B_n(\alpha) = -e^{-\alpha} \sum_{k=1}^{n+1} \frac{n!}{\alpha^k (n - k + 1)!} - e^{\alpha} \sum_{k=1}^{n+1} \frac{(-1)^{n-k} n!}{\alpha^k (n - k + 1)!}$$

(a) Verify the formulas for $n = 0$ and $n = 1$ by working out the integrals and comparing with the general formulas.

(b) Express the overlap integrals

$$\int 1s_A(1)\ 1s_B(1)\ dv_1 \qquad\qquad \int 1s_A(1)\ 2s_B(1)\ dv_1$$
$$\int 1s_A(1)\ 2p_{\sigma B}(1)\ dv_1 \qquad\qquad \int 2s_A(1)\ 2s_B(1)\ dv_1$$
$$\int 2s_A(1)\ 2p_{\sigma B}(1)\ dv_1 \qquad\qquad \int 2p_{\sigma A}(1)\ 2p_{\sigma B}(1)\ dv_1$$

in terms of the integrals $A_n(\alpha)$, $B_n(\alpha)$, taking the AO's to be Slater orbitals with $\zeta_A = \zeta_B = 1$ and the AB distance R to be arbitrary.

(c) Using a computer program for evaluating the integrals, calculate the overlap integrals for $0 \le R \le 6a_0$ in steps of $0.1a_0$ and plot the overlap integrals versus R. Compare your results with those listed by R. S. Mulliken, C. A. Rieke, D. Orloff, and H. Orloff, *J. Chem. Phys.* **17,** 1248 (1949) (This paper is reprinted in Reference 3 in the Additional Reading list.)

C6.2 For the sp^3 hybrid orbitals of Problems 6.18, in which χ_1 and χ_2 are equivalent and χ_3 and χ_4 are equivalent, plot the angle between equivalent pairs ($|\phi_1 - \phi_2|$ and $|\phi_3 - \phi_4|$, respectively) as functions of the s character (λ^2) of χ_1 and χ_2 and the s character $[(\frac{1}{2} - \lambda^2)^{1/2}]$ of χ_3 and χ_4. Allow the s characters to range between 0 and 1 in steps of 0.01. In CH_2Cl_2, $\measuredangle$ (H—C—H) = $112 \pm 0.3°$ and $\measuredangle$ (Cl—C—Cl) = $111.8°$; is sp^3 hybridization alone adequate to describe the bonds in this molecule?

C6.3 Using a computer program for solving up to 10 homogeneous linear equations, find the MO energies ϵ_i and the coefficients of the AO's C_{ij} ($i, j = 1, \ldots,$ 10) for the π electrons of napthalene in the Hückel approximation. Set up a secular determinant analogous to Eq. 6.44 in terms of the $2p_z$ AO's rather than using symmetry orbitals as was done in the text. Compare your results with those given in Eqs. 6.47 and 6.52. [*Hint:* Let $x = (\alpha - \epsilon)/\beta$.]

C6.4 Calculate the MO energies ϵ_i ($i = 1, \ldots, 10$) in the Hückel approximation for an open-chain conjugated molecule with 10 π electrons (see Section 6.4.3) by finding the zeros of the polynomial $D_{10}(x)$ of Table 6.16. Use a standard computer program for polynomial zeros (e.g., Newton–Raphson iteration).

7

Molecular spectra

The line spectra of atoms, which were discussed in Chapters 2 and 3, provided the impetus for the development of the quantum theory and served as a standard by which the success of the theory could be measured. We now turn our attention to molecular spectra. As we have seen in Chapters 5 and 6, the electronic structure of molecules is more complex and varied than that of atoms. In addition, the rotational and vibrational motion of the nuclei contributes to the molecular energies. It is not surprising, then, that molecular spectra are more complicated than the relatively simple line spectra of atoms. Nevertheless, their interpretation is a very worthwhile endeavor because of the wealth of information that can be obtained. Most of our experimental knowledge of the electronic structure, internuclear spacings, vibrational frequencies, and other important properties of molecules is derived from spectral data. In the present chapter, we examine certain aspects of molecular spectra and show what can be learned from them.

7.1 THE ELECTROMAGNETIC SPECTRUM

The total energy of a molecule is made up of many different contributions (electronic, vibrational, rotational, electron spin, nuclear spin, etc.); the energy level spacings are therefore such that radiation is emitted and absorbed in many regions of the electromagnetic spectrum (Fig. 7.1). The observed wavelengths range from 10^{-1} Å for electronic x-ray levels to 10^5 cm for nuclear spin levels. To obtain a convenient numerical scale for particular

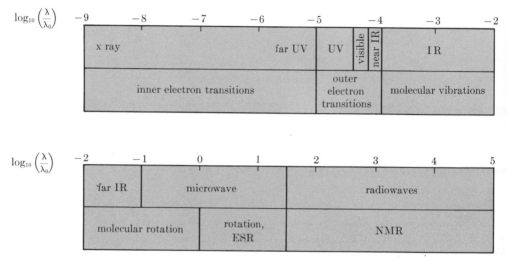

Fig. 7.1. The electromagnetic spectrum; the logarithmic scale is used because the range of the spectrum depicted here covers 14 orders of magnitude ($\lambda_0 = 1$ cm).

transitions, it has become customary to use different units to specify the wavelength or energy of the radiation in various spectral regions. Depending on the region, the wavelength is given in terms of meters (m), centimeters (cm), millimeters (mm), microns (μ), or nanometers (nm). The corresponding transition energy may be expressed in terms of electron volts (eV), Rydbergs (i.e., $e^2/2a_0$), or atomic units (a.u., i.e., e^2/a_0). Instead of the energy, the wave number ($\bar{\nu}$) is often given in cm^{-1} or the frequency (ν) in cycles per second or hertz (Hz), kilohertz (kHz), megahertz (MHz), or kilomegahertz (kMHz). A convenient set of conversion factors is given in Table 7.1.

As shown in Fig. 7.1, historical names are in common use to describe the various spectral regions; for example, the "ultraviolet (uv)" region (10^3 to 4×10^3 Å), the "x-ray" region (10^{-1} to 10^3 Å), the "visible" region (4×10^3 to 7×10^3 A), the "infrared (ir)" region (1 to 100 μ), the "microwave"

Table 7.1   *Conversion factors*[a, b]

$$\tilde{\nu} = 1/\lambda = \nu/c \qquad\qquad\qquad E = h\nu$$

$$1 \text{ eV} = (4.80298 \times 10^{-10} \text{ esu}) \times [(299.7925)^{-1} \text{ statvolt}] = 1.60210 \times 10^{-12} \text{ erg}$$

$$\sim \frac{1.60210 \times 10^{-12} \text{ erg}}{6.6256 \times 10^{-27} \text{ erg sec}} = 2.41804 \times 10^{14} \text{ sec}^{-1}$$

$$\sim \frac{2.41804 \times 10^{14} \text{ sec}^{-1}}{2.997925 \times 10^{10} \text{ cm sec}^{-1}} = 8065.73 \text{ cm}^{-1}$$

$$1 \text{ erg} = (1.60210 \times 10^{-12} \text{ eV}^{-1})^{-1} = 6.24181 \times 10^{11} \text{ eV}$$
$$1 \text{ sec}^{-1} \sim (2.41804 \times 10^{14} \text{ eV}^{-1})^{-1} = 4.13558 \times 10^{-15} \text{ eV}$$
$$1 \text{ cm}^{-1} \sim (8065.73 \text{ eV}^{-1})^{-1} = 1.23981 \times 10^{-4} \text{ eV}$$

10^2 cm = 1 m

10^{-1} cm = 1 mm

10^{-4} cm = 1 μ

10^{-7} cm = 1 nm

10^{-8} cm = 1 Å

1 sec^{-1} = 1 Hz

10^3 sec^{-1} = 1 kHz

10^6 sec^{-1} = 1 MHz

10^9 sec^{-1} = 1 kMHz

10^5 eV = 1 keV

10^6 eV = 1 MeV

10^9 eV = 1 BeV

[a] Also, see the Appendix.

[b] The symbol ($\sim$) indicates that two numbers correspond to each other, although they have different units (as is evident from the table).

region (0.1 mm to 30 cm), and the "radiofrequency (rf)" region (1kMHz to 1kHz).

A survey of the electromagnetic spectrum is presented in Table 7.2.

7.2   GENERAL FEATURES OF MOLECULAR SPECTRA

The general features of molecular spectra can be best inferred by looking at a diagram such as that shown in Fig. 7.2. Here we have drawn the potential-energy functions for two electronic states of the hydrogen molecule. Both curves correspond to bound states since they have minima relative to the separated atoms. The lower curve is for the ground state (designated $X^1\Sigma_g^+$, where X is the standard symbol for the ground state), which dissociates into two ground-state H atoms [H(1s) + H(1s)]; the upper curve is for an excited state (designated $B^1\Sigma_u^+$ because it is the second excited state with the same spin multiplicity, the standard symbols in order of excitation being $A, B, C, \ldots$), which dissociates into one ground state and one excited state H atom [H(1s) + H(2p)]. As we remember from Chapter

Table 7.2 *Survey of the electromagnetic spectrum*

λ	Energy, wave number, or frequency	Origin	Name
10^{-1} to 10^3 Å	124 keV to 12.4 eV	atomic and molecular inner-electron transitions	x ray, far uv
10^3 to 10^4 Å	12.4 to 1.24 eV 10^5 to 10^4 cm^{-1}	atomic and molecular outer-electron transitions	uv (10^3 to 4×10^3 Å) visible (4×10^3 Å to 7×10^3 Å) near ir ($>7 \times 10^3$ Å)
10^4 to 10^6 Å 1 to 10^2 μ	1.24 to 0.0124 eV 10^4 to 100 cm^{-1}	molecular vibrations	ir
10^2 to 10^5 μ 0.1 to 10^2 mm	100 to 0.1 cm^{-1}	molecular rotation	far ir (10^2 to 10^3 μ) microwave (10^3 μ to 0.1 mm)
10 to 30 cm	0.1 to 0.033 cm^{-1} 3 to 1 kMHz	some rotations electron spin resonance	microwave
0.3 to 300 m	1 kMHz to 1 MHz	nuclear magnetic resonance	radiowaves

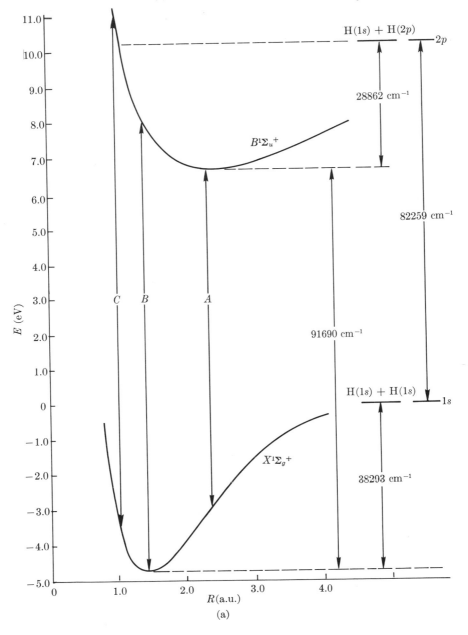

Fig. 7.2. Potential-energy curves for the ground state $(X \Sigma_g^+)$ and an excited state $(B^1\Sigma_u^+)$ of H_2; the energy of two $H(1s)$ atoms is taken as zero. **(a)** Without vibration-rotation.   (*Continued on next page.*)

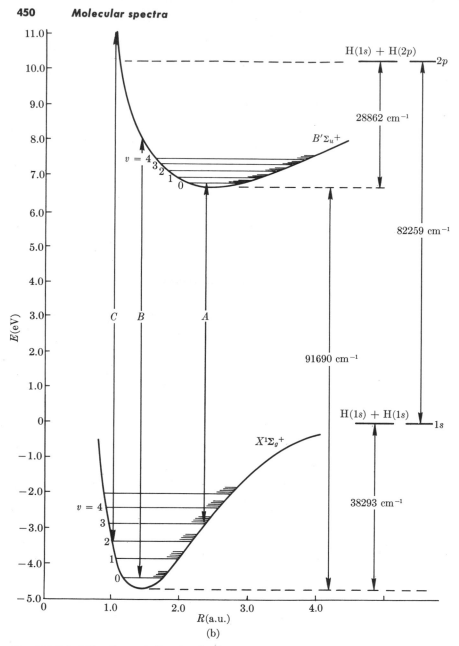

Fig. 7.2 (b). Vibration-rotation levels shown.,

5, these potential curves are obtained by assuming that the nuclei are infinitely heavy and that their motion does not contribute to the energy of the system. If this were true, we would expect molecular spectra to be composed of a continuous range of frequencies corresponding to the continuous range of internuclear distances. Thus, for the emission spectrum from the excited ($B^1\Sigma_u^+$) state to the ground state ($X^1\Sigma_g^+$), Fig. 7.2 shows three hypothetical transitions, each of which would have a different frequency since the energy difference between the two states depends on the distance. We have drawn all lines vertically, corresponding to the assumption that the infinitely heavy nuclei are stationary so that the internuclear distance does not change during the transition.

At very low resolution, (i.e., with a spectrometer that does not resolve wavelengths to better than $\sim$0.1 Å), molecular spectra do have the appearance expected from the above argument; that is, "continuous" bands are observed, each one associated with a pair of electronic states. An example is given in Fig. 7.3a, which shows the emission spectrum of a number of states

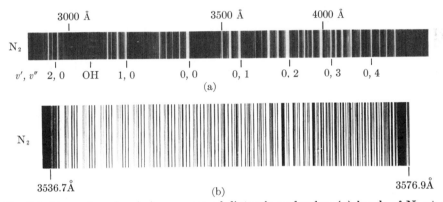

Fig. 7.3. Examples of emission spectra of diatomic molecules; (**a**) bands of N_2 at low resolution [redrawn from R. W. B. Pease and A. G. Gaydon, *The Identification of Molecular Spectra* (Chapman and Hall, London, 1963), 3rd ed., plate 3]; the band heads are indicated beneath the spectrum. (**b**) Bands of N_2 resolved into individual lines corresponding to transitions between rotation-vibration sublevels of the electronic states [from G. Herzberg, *Spectra of Diatomic Molecules* (Van Nostrand, New York, 1950), 2nd ed., Fig. 22].

of N_2. At higher resolution, one finds that instead of the apparent continuum, the spectrum is composed of sharp lines. These lines arise from the fact that the nuclei are not infinitely heavy and that nuclear motion, like that of the electrons, is quantized. Thus, the actual molecular energy levels do not correspond to the stationary-nucleus potential curves, but rather to the energy levels superimposed on these curves by the nuclear vibrational and rotational motion. Correspondingly, the spectral transitions take place between a vibration-rotation level of the excited electronic state and a vibration-rotation level of the ground electronic state. Since there are many such levels, a large number of individual lines are possible for a single electronic transition. Such a high resolution spectrum is shown in Fig. 7.3b. As illustrated in Fig. 7.4, the electronic spectra of molecules at intermediate resolution often have the appearance of "bands," each of which is composed of many vibration-rotation transitions.

The transition marked C in Fig. 7.2b is somewhat different in that the upper level is above the energy of the dissociation products for the molecule in the excited electronic state. Unlike the discrete or quantized energy

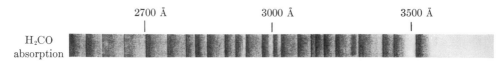

Fig. 7.4. Band structure of H_2CO (redrawn from R. W. B. Pease and A. G. Gaydon, *op. cit.*, plate 14). The spectra of some of the diatomic fragments also appear.

values corresponding to the region below the dissociation limit, the allowed values of the energy above the dissociation limit are continuous; that is, just as for the ionized H atom (proton plus unbound electron), there is no quantum restriction on the energy of two unbound atoms. Thus, the electronic spectrum of H_2 in the wavelength region of transition C is found to be continuous at even the highest resolving power (see Fig. 7.5).

The complexity of the molecular spectrum resulting from the possible transitions between two electronic states (Figs. 7.3–7.5) contrasts with that of the separated atoms into which the two states dissociate. As indicated

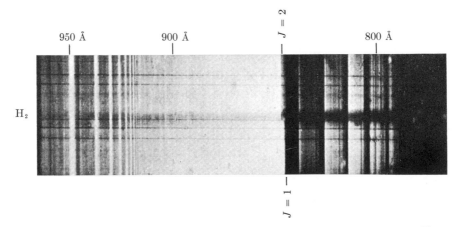

Fig. 7.5. Absorption spectrum of H_2 showing the onset of the continuum [from H. Beutler, *Z. Physik. Chem.*, **B29,** 315 (1935), Fig. 2]. The dark portion of the spectrum corresponds to the continuum. The onset of the continuum for two different rotational states is indicated by the respective rotational quantum numbers J.

on the right-hand side of Fig. 7.2, which neglects relativistic fine structure, there is only a single sharp atomic line corresponding to the $H(2p) \rightarrow H(1s)$ transition; that is, the first line of the H-atom Lyman series.

Instead of studying the structure of transitions between different electronic states, it is often useful to measure the emission or absorption spectrum of a *single* electronic state (Fig. 7.6). Such transitions, which are

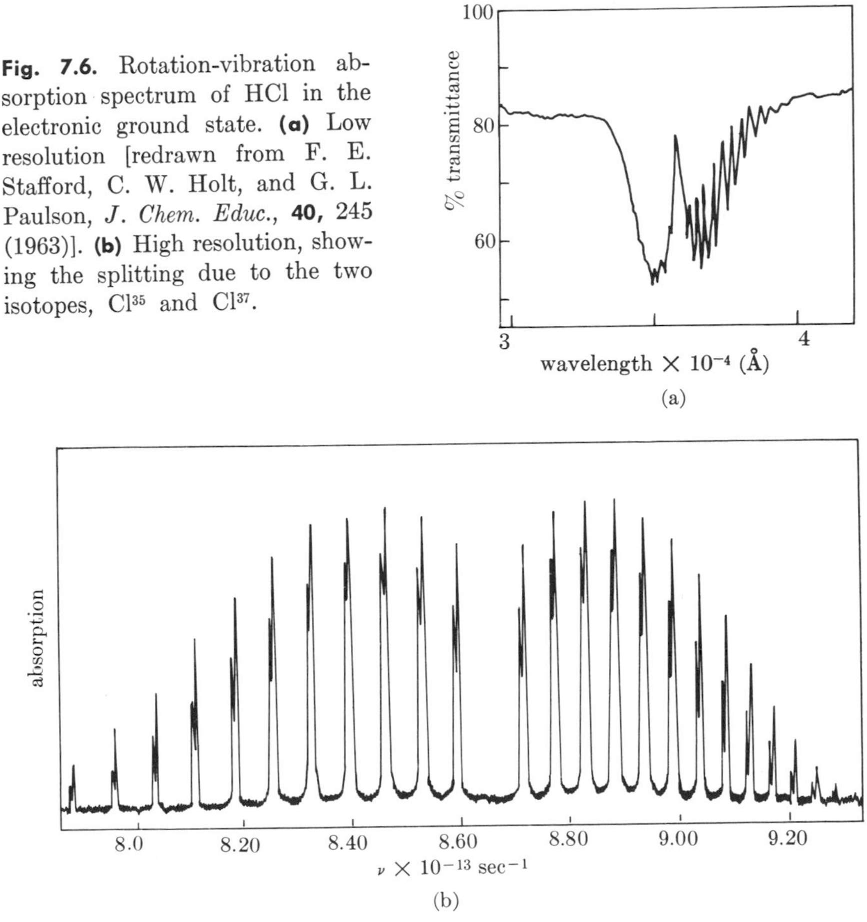

Fig. 7.6. Rotation-vibration absorption spectrum of HCl in the electronic ground state. **(a)** Low resolution [redrawn from F. E. Stafford, C. W. Holt, and G. L. Paulson, *J. Chem. Educ.*, **40,** 245 (1963)]. **(b)** High resolution, showing the splitting due to the two isotopes, Cl^{35} and Cl^{37}.

(a)

(b)

generally at longer wavelengths than those involving different electronic states, arise from pairs of states of different nuclear motion corresponding to the same electronic level. They can be pure rotational transitions or transitions involving changes in both the vibrational and rotational state.

As is evident from the foregoing discussion, the unraveling of a molecular electronic spectrum may require considerable skill. We begin our detailed

Having obtained the electronic wave function $\Psi(\mathbf{r}, R)$, we complete the determination of the molecular wave function by writing it in the form

$$\Psi(\mathbf{r}, \mathbf{R}) = \psi(\mathbf{r}, R)\, \chi(\mathbf{R}) \tag{7.7}$$

where $\chi(\mathbf{R})$ is the nuclear wave function which depends on the nuclear coordinate $\mathbf{R}$ alone. To determine the equation satisfied by $\chi(\mathbf{R})$, we substitute Eq. 7.7 into Eq. 7.1 and obtain

$$\left(-\frac{\hbar^2}{2\mu}\nabla^2 - \frac{\hbar^2}{2m}\sum_{i=1}^{n}\nabla_i^2 + V(\mathbf{r}, R)\right)\psi(\mathbf{r}, R)\,\chi(\mathbf{R})$$

$$= \left(-\frac{\hbar^2}{2m}\sum_{i=1}^{n}\nabla_i^2 + V(\mathbf{r}, R)\right)\psi(\mathbf{r}, R)\,\chi(\mathbf{R}) \tag{7.8}$$

$$+ \left(-\frac{\hbar^2}{2\mu}\nabla^2\right)\psi(\mathbf{r}, R)\,\chi(\mathbf{R}) = E\psi(\mathbf{r}, R)\,\chi(\mathbf{R}).$$

We note in Eq. 7.8 that since the ∇_i differentiate only electronic coordinates, they have no effect on $\chi(\mathbf{R})$, which depends only on the nuclear coordinates; by contrast ∇^2, which differentiates with respect to the nuclear coordinates, acts on both $\psi(\mathbf{r}, R)$ and $\chi(\mathbf{R})$. Making use of Eq. 7.6 to substitute for the electronic term, we find

$$\left(-\frac{\hbar^2}{2\mu}\nabla^2 + E(R)\right)\psi(\mathbf{r}, R)\,\chi(\mathbf{R}) = E\psi(\mathbf{r}, R)\,\chi(\mathbf{R}) \tag{7.9}$$

Since the electronic wave function varies slowly with the internuclear distance, we assume that

$$\nabla^2\psi(\mathbf{r}, R)\,\chi(\mathbf{R}) \simeq \psi(\mathbf{r}, R)\,\nabla^2\chi(\mathbf{R}) \tag{7.10}$$

neglecting any contributions from the derivatives of $\psi(\mathbf{r}, R)$ with respect to R; this is the essential element in the Born–Oppenheimer approximation. With the approximation in Eq. 7.10, we can write Eq. 7.9 in the form

$$\psi(\mathbf{r}, R)\left(-\frac{\hbar^2}{2\mu}\nabla^2 + E(R)\right)\chi(\mathbf{R}) = E\psi(\mathbf{r}, R)\,\chi(\mathbf{R}) \tag{7.11}$$

Dividing both sides by $\psi(\mathbf{r}, R)$, we finally obtain

$$\left(-\frac{\hbar^2}{2\mu}\nabla^2 + E(R)\right)\chi(\mathbf{R}) = E\chi(\mathbf{R}) \tag{7.12}$$

Equation 7.12 is the Schrödinger equation for the nuclear motion of a diatomic molecule. It is independent of the electronic motion whose effect appears only through the term $E(R)$, which is the electronic energy as a function of R and acts as the potential-energy for the motion of the nuclei.

If we introduce the spherical polar coordinates (R, θ, ϕ) of one nucleus with respect to the other as origin and write out ∇^2 in full (see Section 4.1.1), Eq. 7.12 becomes

$$\left\{ -\frac{\hbar^2}{2\mu R^2} \left[\frac{\partial}{\partial R}\left(R^2 \frac{\partial}{\partial R} \right) + \frac{1}{\sin\theta}\frac{\partial}{\partial\theta}\left(\sin\theta \frac{\partial}{\partial\theta} \right) + \frac{1}{\sin^2\theta}\frac{\partial^2}{\partial\phi^2} \right] \right.$$
$$\left. + E(R) \right\} \chi(R, \theta, \phi) = E\chi(R, \theta, \phi) \qquad (7.13)$$

Equation 7.13 has exactly the same form as the H-atom Schrödinger equation (Eq. 3.9), except that the Coulomb potential $-Ze^2/r$ is replaced by the radial potential $E(R)$ from the electronic problem for fixed nuclei (Eq. 7.6). Following Eqs. 3.9–3.13, we can thus separate the wave function for nuclear motion into a radial part $\mathcal{R}(R)$, which is different from the H-atom radial function because the potential is different, and an angular part $S(\theta, \phi)$, which is the same as the angular part of the H-atom solutions because in both cases the potential energy is independent of angle. We have

$$\chi(R, \theta, \phi) = \mathcal{R}(R) \, S(\theta, \phi) \qquad (7.14)$$

where the functions $S(\theta, \phi) = S_{JM}(\theta, \phi)$ are specified by the molecular total angular-momentum quantum number J and the z-component angular-momentum quantum number M, analogous to the H-atom quantum numbers l and m, respectively. Thus, for a molecule with angular function $S_{JM}(\theta, \phi)$, the square of the total angular momentum is $J(J+1)\hbar^2$ and the z component is $M\hbar$. Although we do not prove this here (for specific cases, see Problem 7.9), one can show that the angular part of the nuclear kinetic-energy operator obeys the equation

$$-\frac{\hbar^2}{2\mu R^2} \left(\frac{1}{\sin\theta}\frac{\partial}{\partial\theta}\sin\theta\frac{\partial}{\partial\theta} + \frac{1}{\sin^2\theta}\frac{\partial^2}{\partial\phi^2} \right) S_{JM}(\theta, \phi)$$
$$= \frac{J(J+1)\hbar^2}{2\mu R^2} S_{JM}(\theta, \phi) \qquad (7.15)$$

Upon multiplication by $2\mu R^2$, Eq. 7.15 becomes

$$-\hbar^2 \left(\frac{1}{\sin\theta}\frac{\partial}{\partial\theta}\sin\theta\frac{\partial}{\partial\theta} + \frac{1}{\sin^2\theta}\frac{\partial^2}{\partial\phi^2} \right) S_{JM}(\theta, \phi) = J(J+1)\hbar^2 S_{JM}(\theta, \phi) \qquad (7.16)$$

The functions $S_{JM}(\theta, \phi)$ are thus seen to be eigenfunctions of the operator

$$\mathbf{M}^2 = -\hbar^2 \left(\frac{1}{\sin\theta}\frac{\partial}{\partial\theta}\sin\theta\frac{\partial}{\partial\theta} + \frac{1}{\sin^2\theta}\frac{\partial^2}{\partial\phi^2} \right) \qquad (7.17)$$

with eigenvalue $J(J+1)\hbar^2$. Since $J(J+1)\hbar^2$ is the square of the total angular momentum, the operator $\mathbf{M}^2$ in Eq. 7.17 is the operator for the square of the *angular momentum*. Substituting from Eqs. 7.14 and 7.15 into Eq. 7.13, we find

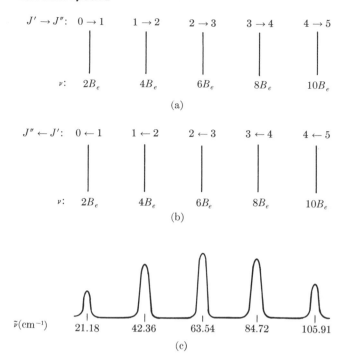

Fig. 7.7. Pure rotational spectrum of a diatomic molecule; (**a**) absorption ($J'' = J' + 1$), (**b**) emission ($J'' = J' - 1$), (**c**) pure rotational lines in the absorption spectrum of HCl. (Rigid rotator model.)

take place, we calculate the order of magnitude of $2B_e$ from Eq. 7.26 by using

$$\hbar \simeq 10^{-27} \text{ erg sec} \qquad \mu \simeq 10^{-24} \text{ g}$$
$$h \simeq 10^{-26} \text{ erg sec} \qquad R_e \simeq 10^{-8} \text{ cm}$$

The result is

$$2B_e \simeq \frac{(10^{-27})^2}{(10^{-24})(10^{-8})^2(10^{-26})} = 10^{12} \text{ sec}^{-1} = 1000 \text{ kMHz} \sim 33 \text{ cm}^{-1}$$

Comparing this result with Table 7.2, we see that the rotational spectrum is found in the far infrared or microwave region. A typical spectrum, that obtained for the HCl molecule, is shown in Fig. 7.7c.

Measurement of the line spacing in the rotational spectrum of a diatomic molecule gives B_e, from which I_e and R_e can be determined by Eqs. 7.26 and 7.21. As an example, we consider the molecule $N^{14}O^{16}$, whose rotational lines in the rigid-rotator model would be separated by $\Delta\nu = 102.1690$ kMHz; it is clear that the particular isotopes have to be specified since the reduced mass and therefore the rotational constant and spectrum depend on them (Problem 7.20 and Section 7.6.4). From this value for $\Delta\nu$, we obtain

$$B_e = \tfrac{1}{2}(\Delta\nu) = 51.0845 \text{ kMHz}$$

$$
\begin{aligned}
I_e &= \frac{\hbar}{4\pi B_e} = \frac{1.05450 \times 10^{-27}}{6.283185} \frac{1}{2B_e} \text{ g cm}^2 \\
&= \frac{1.67829 \times 10^{-28}}{2B_e} \text{ g cm}^2 \\
&= \frac{(1.67829 \times 10^{-28})(6.02252 \times 10^{39})}{2B_e} \text{ a.m.u. } \text{\AA}^2 \\
&= \frac{1.01075 \times 10^{12}}{2B_e} = 9.8929 \text{ a.m.u. } \text{\AA}^2
\end{aligned}
$$

$$
R_e = \left(\frac{I_e}{\mu}\right)^{1/2} = \left(\frac{(9.8929)(14.00307 + 15.99491)}{(14.00307)(15.99491)}\right)^{1/2} \text{\AA}
$$

$$= 1.15108 \text{ \AA}$$

Since microwave frequencies can be measured very precisely, the accuracy with which R_e can be determined for molecules with permanent dipole moments is limited not by the measurement but rather by the approximations in the rigid-rotator theory (see Section 7.6).

For molecules without permanent moments, such as H_2, the electric dipole pure rotation transition probability is zero. In such systems, the moment of inertia and the internuclear distance can often be found from analysis of the rotational fine structure of electronic absorption bands (see Section 7.7.1) and from rotational Raman spectra (see Section 7.10). In some cases rotation spectra of molecules without a permanent dipole moment have been induced by collisions in the pure gas or with other atoms or molecules in a mixture. The effect of such collisions is to induce a dipole moment in the nonpolar molecule during the collision; for example, the overlap of the charge clouds can so distort them as to produce a dipole moment. If the collision rate is large, the number of nonpolar molecules with induced moments at a given time may be sufficient to allow the observation of a microwave absorption spectrum with characteristic rotational lines. Since the intensity of the rotational lines is proportional to the collision rate, it depends on the square of the pressure, rather than the pressure itself as does the rotational spectrum intensity of polar molecules.

7.4.1 The rotational Stark effect

When an electric field is applied to a sample of molecules whose rotational spectrum is being recorded, shifts and splittings of the lines are observed (the rotational Stark effect). The effects on molecular energy levels are analogous to those already discussed for the

H atom in an external magnetic field, which makes the energy of the system depend upon orientation (i.e., on the quantum number m, Section 3.12). In the present case, the electric dipole moment of the molecule ($\mathbf{u}_{el}$) interacts with the electric field and contributes a perturbation term of the form

$$\Delta E_{el} = -(\mathbf{u}_{el})_z \mathcal{E} = (\mathbf{u}_0 \cos \theta) \mathcal{E} \tag{7.28}$$

where the electric field $\mathcal{E}$ is taken to be in the z direction. A first-order perturbation calculation (Section 4.15) gives no contribution since the expectation value of $(\mathbf{u}_{el})_z$ is zero for all the rotator states [Problem 7.10(b)]. Thus, one has to take account of the change induced in the rotator wave function by the electric field to determine the level shifts. This corresponds to the classical result that rotating molecules speed up slightly when turning towards the field direction and slow down when turning away from it. In quantum-mechanical terms, the electric field mixes the two states $J \pm 1$ (corresponding to acceleration and retardation of rotation) with the state J, the degree of mixing depending upon the product of the permanent dipole moment μ_0 and the electric field $\mathcal{E}$ (Problem 7.50). For the perturbation corresponding to Eq. 7.28, the energy levels for $J \neq 0$ are found to be

$$E_{JM} = \frac{J(J+1)\hbar^2}{2I_e} - \frac{I_e \mu_0^2 \mathcal{E}^2}{\hbar^2}\left(\frac{3M^2 - J(J+1)}{J(J+1)(2J-1)(2J+3)}\right) \tag{7.29a}$$

For $J = 0$, the energy level E_{00} can be obtained from Eq. 7.29a by first setting $M = 0$ and then taking the limit as $J \to 0$; the result is

$$E_{00} = -\frac{I_e \mu_0^2 \mathcal{E}^2}{3\hbar^2} \tag{7.29b}$$

Since state J contains a mixture of the unperturbed states J, $J + 1$, and $J - 1$ (the perturbed ground state contains a mixture of $J = 0$ and $J = 1$), there are now weakly allowed lines corresponding to $\Delta J = \pm 2$ in addition to the strongly allowed lines corresponding to $\Delta J = \pm 1$. The permanent electric dipole moment of the molecule can often be determined to four or five significant figures from the spacing between the Stark perturbed rotational lines (see Problem 7.11).

7.5 VIBRATIONAL SPECTRA OF DIATOMIC MOLECULES

In the rigid-rotator model for the nuclear motion of a diatomic molecule, the internuclear distance is assumed to be fixed at R_e, so that the radial Schrödinger equation (Eq. 7.19) for the nuclei can be ignored. To look at the radial (vibrational) motion by itself, we assume that the rotational energy term in Eq. 7.19 can be approximated by its rigid-rotator value; that is, we substitute

$$\frac{J(J+1)\hbar^2}{2\mu R^2} \simeq \frac{J(J+1)\hbar^2}{2\mu R_e^2} = E_{\rm rot} \qquad (7.30)$$

and obtain

$$-\frac{\hbar^2}{2\mu R^2}\frac{d}{dR}\left(R^2\frac{d\Re}{dR}\right) + E_{\rm rot}\,\Re(R) + E(R)\,\Re(R) = E_{v,J}\Re(R) \qquad (7.31)$$

where $E_{v,J}$ is the total energy (i.e., electronic energy plus energy of nuclear motion). By taking the zero of energy to be the minimum in the potential-energy function $E(R)$, we separate out the electronic energy and can write the energy of nuclear motion $E_{\rm nucl}$ as

$$E_{v,J} \equiv E_{\rm nucl} = E_{\rm vib} + E_{\rm rot} \qquad (7.32)$$

Cancelling the term $E_{\rm rot}\Re(R)$ on both sides of Eq. 7.31, we have

$$-\frac{\hbar}{2\mu R^2}\frac{d}{dR}\left(R^2\frac{d\Re}{dR}\right) + E(R)\,\Re(R) = E_{\rm vib}\Re(R) \qquad (7.33)$$

Equation 7.33 is the Schrödinger equation for the vibrational motion of the molecule. The essential approximation in deriving it from the complete nuclear motion equation was the separation of rotation and vibration by the rigid-rotator assumption (Eq. 7.30). This is a satisfactory first approximation though refinements to include the rotation-vibration interaction can be introduced without great difficulty (see Sections 7.6.2 and 7.6.3).

The kinetic-energy term in Eq. 7.33 can be simplified by writing

$$\Re(R) = \frac{1}{R}\,\chi(R) \qquad (7.34)$$

and substituting Eq. 7.34 into Eq. 7.33 to obtain

$$-\frac{\hbar^2}{2\mu}\frac{d^2}{dR^2}\,\chi(R) + E(R)\,\chi(R) = E_{\rm vib}\,\chi(R) \qquad (7.35)$$

For most stable molecules the potential-energy function $E(R)$ has a shape similar to the curves for H_2 in Fig. 7.2. We can expand this function in a power series about $R = R_e$ to obtain

$$E(R) = E(R_e) + \left(\frac{dE}{dR}\right)_{R_e}(R - R_e)$$

$$+ \frac{1}{2}\left(\frac{d^2E}{dR^2}\right)_{R_e}(R - R_e)^2 + \cdots \qquad (7.36)$$

Since we chose the zero of energy at the minimum of the curve, $E(R_e) = 0$; moreover, because this point is a minimum, the first derivative is zero $[(dE/dR)_{R_e} = 0]$. Thus, we have

$$E(R) = \frac{1}{2}\left(\frac{d^2E}{dR^2}\right)_{R_e}(R - R_e)^2 + \cdots$$

If we introduce the variable $\rho = (R - R_e)$, we can write

$$E(\rho) = \tfrac{1}{2}k_e\rho^2 + \cdots$$

where the force constant k_e is given by

$$k_e = \left(\frac{d^2E}{dR^2}\right)_{R_e}$$

Neglect of terms involving powers of $R - R_e$ (or ρ) higher than the second in the series for $E(R)$ corresponds to replacing the exact expression by a simple parabola, with the same minimum and curvature at the minimum as the exact curve, as shown in Fig. 7.8. This is a good approximation for R close to R_e.

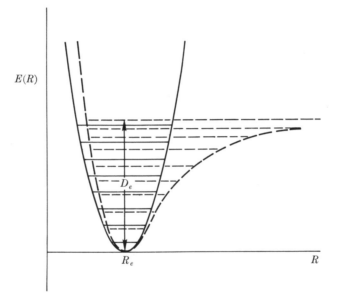

Fig. 7.8. Parabolic (or harmonic) approximation (solid curve) to a diatomic potential-energy curve $E(R)$ (dashed curve).

The resulting radial Schrödinger equation has the form

$$-\frac{\hbar^2}{2\mu}\frac{d^2}{d\rho^2}\chi(\rho) + \frac{1}{2}k_e\rho^2\chi(\rho) = E_{\text{vib}}\chi(\rho) \qquad (7.37)$$

which is identical to the harmonic-oscillator equation studied in Chapter 2 (Eq. 2.116). It was pointed out there that the allowed energies are

$$E_{\text{vib}} = E_v = (v + \tfrac{1}{2})h\nu_e \quad v = 0, 1, 2, \ldots \qquad (7.38)$$

where ν_e is the classical frequency given by

$$\nu_e = \frac{1}{2\pi}\left(\frac{k_e}{\mu}\right)^{1/2} \tag{7.39}$$

The harmonic-oscillator wave functions are shown in Fig. 7.9. The ground-state wave function is a bell-shaped curve with its maximum at $\rho = 0$ (i.e., at $R = R_e$), and the excited-state wave functions have nodes

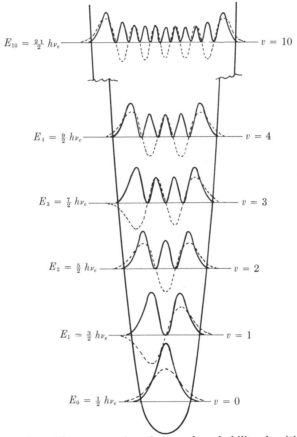

$E_{10} = \frac{21}{2} h\nu_e$ —— $v = 10$

$E_4 = \frac{9}{2} h\nu_e$ —— $v = 4$

$E_3 = \frac{7}{2} h\nu_e$ —— $v = 3$

$E_2 = \frac{5}{2} h\nu_e$ —— $v = 2$

$E_1 = \frac{3}{2} h\nu_e$ —— $v = 1$

$E_0 = \frac{1}{2} h\nu_e$ —— $v = 0$

Fig. 7.9. Harmonic-oscillator wave functions and probability densities. (Redrawn from G. Herzberg, *op. cit.*, Fig. 42, p. 77.)

which increase in number with increasing energy. For the harmonic-oscillator potential, the most probable value of R in the ground state is R_e, while for the excited states, values of R nearer the points at which $E_v = E(R)$ become more probable. At these points the kinetic energy is zero, so that classically the motion of the oscillator would stop and reverse direction. These are therefore the "classical turning points"; values of ρ greater than this value correspond to "tunneling" into the classically forbidden region (see Section 3.4). At very high values of v, the regions in the neighborhood of the turning points become most probable. This result is in accord with the correspondence principle, since the classical oscillator moves more slowly in the vicinity of the turning points and consequently spends more time there than in regions near R_e, where the velocity is a maximum.

To determine the selection rules for vibrational transitions we again make use of the time-dependent discussion of Section 3.11. Thus, for a transition from v' to v'' to be electric-dipole allowed, the transition moment intregals

$$
\begin{aligned}
\mu_{v'',v'} &= \int_0^\infty \mathcal{R}_{v''}(R)\mu(R - R_e)\,\mathcal{R}_{v'}(R)\,R^2\,dR \\
&= \int_{-\infty}^{+\infty} \chi_{v''}(\rho)\mu(\rho)\chi_{v'}(\rho)\,d\rho
\end{aligned}
\tag{7.40}
$$

must be nonzero.[1] The function $\mu(\rho)$ is the molecular dipole moment expressed as a function of ρ. If $\mu(\rho)$ is independent of ρ (e.g., zero for all internuclear distances as in a homonuclear diatomic like H_2), the integral $\mu_{v'',v'}$ is zero for all $v'' \neq v'$ because of the orthogonality of the vibrational wave functions. Thus, only if the dipole moment varies with distance can vibrational transitions occur, the intensity of the transition being dependent upon the rate of change of the function $\mu(\rho)$. Figure 7.10 shows the approximate variation of μ with R for the molecule CO. In the simplest approximation for harmonic-oscillator wave functions (Problem 7.13), the selection rule is

$$
v'' = v' \pm 1 \qquad \Delta v = \pm 1
\tag{7.41}
$$

Thus, for absorption, the wave number of the transition as given by the Bohr relation is

$$
\begin{aligned}
\bar{\nu} &= \frac{1}{hc}(E_{v''} - E_{v'}) = (v'' - v')\frac{\nu_e}{c} = (v'' - v')\omega_e \\
&= (v' + 1 - v')\omega_e = \omega_e
\end{aligned}
\tag{7.42}
$$

where we have adopted the standard notation ω_e for ν_e/c; a corresponding result holds for emission. Consequently, if the vibration of the molecule

---

[1] In using the lower limit $-\infty$ for the second integral, we have made an approximation that is commonly used to simplify difficult integrals. Since R has limits $0 \leq R < \infty$, the limits of $\rho(= R - R_e)$ are $-R_e \leq \rho < \infty$; however, for $\rho < -R_e$, the oscillator wave functions $\chi_v(\rho)$ are generally sufficiently small that extending the limit from $-R_e$ to $-\infty$ introduces only a negligibly small error into the integral. Since the integral from $-\infty$ to ∞ is much easier to evaluate, so that extension of the lower limit is well justified.

were accurately approximated by the harmonic model, only one vibrational transition frequency would be allowed, namely, the classical frequency of the oscillator, $\bar{\nu} = \omega_e$. The vibrational absorption spectrum of a sample of such "harmonic-oscillator" molecules would be expected to show a single intense line.

The actual potential function does not have the simple harmonic form of Eq. 7.37, but has additional terms

$$E(R) = \frac{1}{2} k_e (R - R_e)^2 + \frac{1}{3!} \left(\frac{d^3 E}{dR^3} \right)_{R_e} (R - R_e)^3$$

$$+ \frac{1}{4!} \left(\frac{d^4 E}{dR^4} \right)_{R=R_e} (R - R_e)^4 + \cdots$$

These *anharmonic* contributions to the potential change the wave function and lead to small nonzero transition moment integrals for transitions with

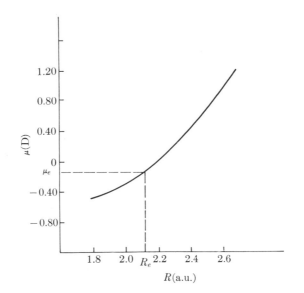

Fig. 7.10. An estimate of the dependence of the electric dipole moment on internuclear distance, derived from intensities of the vibrational transitions of CO. [From the results of R. C. Herman and K. E. Shuler, *J. Chem. Phys.*, **22**, 481 (1954).]

wave numbers roughly equal to $2\omega_e$, $3\omega_e$, etc. Such "overtones" correspond to transitions with $\Delta v = \pm 2, \pm 3$, etc. However, because the effect of anharmonicity is small for the lowest vibrational states of most molecules, the relative intensity drops rather sharply as Δv increases in magnitude.

To estimate the order of magnitude expected for ω_e, we need to know the force constant k_e. For the simplest estimate, we can assume that a molecule is "dissociated" when the internuclear distance is about twice its equilibrium value. With this assumption we have

$$D_e \approx \tfrac{1}{2}k_e(2R_e - R_e)^2 = \tfrac{1}{2}k_eR_e^2$$

Using values corresponding to HCl, namely,

$$D_e = 4.614 \text{ eV} \quad R_e = 1.27460 \text{ Å} \quad \mu = 0.9795925 \text{ a.m.u.}$$

we obtain

$$k_e \simeq \frac{2D_e}{R_e^2} = \frac{(2)(4.614)}{(1.275)^2} \simeq 6 \text{ eV Å}^{-2}$$

$$\simeq 9 \times 10^4 \text{ erg cm}^{-2}$$

$$\simeq 1 \text{ millidyne Å}^{-1}$$

and

$$\omega_e = \frac{1}{2\pi c}\left(\frac{k_e}{\mu}\right)^{1/2} \simeq \frac{1}{2\pi(3 \times 10^{10})} \times \left(\frac{(9 \times 10^4)(6 \times 10^{23})^{1/2}}{0.98}\right)$$

$$\simeq 1300 \text{ cm}^{-1}$$

This result suggests that we can expect vibration frequencies to be of the order of 1000 cm^{-1} for diatomic molecules. The observed range for ω_e is of the order of a few hundred cm^{-1} for weakly bound and/or heavy molecules to a few thousand cm^{-1} for strongly bound and/or light molecules such as H$_2$ and the lighter hydrides (ω_e for HCl is found to be 2989.74 cm^{-1}). Vibration spectra are therefore normally found in the infrared region. Some typical values for ω_e are listed in Table 7.3.

The infrared vibrational spectra of polyatomic molecules are more complicated than diatomic vibrational spectra, since lines at several frequencies generally appear, each corresponding to a different mode of motion of the nuclei that produces a change in the dipole moment of the molecule; the observed lines may correspond to bending as well as stretching of bonds. Some simple cases are discussed in Section 7.9. Force constants for stretching and bending of a given type of bond are observed to be relatively invariant from molecule to molecule; thus the characteristic frequencies of the various groups in an unknown molecule often can be extracted from its ir spectrum as an aid in the identification of the molecule. Several characteristic bond frequencies are listed in Table 7.4.

For the reason just discussed, the ir spectrometer is an important tool for chemical analysis. Relatively simple standard techniques have been developed for recording absorption spectra of gases, liquids, solutions, and solids. Because ordinary glass absorbs ir radiation, the optical systems of ir spectrometers are commonly made of ground salt crystals such as NaCl or

Table 7.3 Spectroscopically determined properties of selected diatomic molecules in the ground electronic state.[a,b]

Molecule	$\omega_e(\text{cm}^{-1})$	$\omega_e x_e(\text{cm}^{-1})$	$\tilde{B}_e(\text{cm}^{-1})$	$\tilde{a}_e(\text{cm}^{-1})$	$\tilde{D}_e(\text{cm}^{-1})$	$R_e(\text{Å})$	$D_0(\text{eV})$	Ground state
$C^{12}H^1$	2861.6	64.3	14.457	0.534	14.5×10^{-4}	1.1198	3.47	$^2\Pi$
$Cl^{35}Cl^{35}$	564.9	4.0	0.2438	0.0017	...	1.988	2.475	$^1\Sigma_g^+$
$C^{12}O^{16}$	2170.21	13.461	1.9314	0.01749	6.43×10^{-6}	1.1282	11.108	$^1\Sigma^+$
H^1H^1	4395.2	117.91	60.81	2.993	0.0464_8	0.7417	4.4763	$^1\Sigma_g^+$
H^1Cl^{35}	2989.74	52.05	10.5909	0.3019	5.32×10^{-4}	1.27460	4.430	$^1\Sigma^+$
Li^7Li^7	351.44	2.592	0.6727	0.00704	$9.8_6 \times 10^{-6}$	2.673	1.03	$^1\Sigma_g^+$
$N^{14}N^{14}$	2359.61	14.456	2.010	0.0187	5.8×10^{-6}	1.094	9.756	$^1\Sigma_g^+$
$N^{14}O^{16}$	1904.03	13.97	1.7046	0.0178	$\sim 5 \times 10^{-6}$	1.1508	6.49	$^2\Pi$
$O^{16}O^{16}$	1580.361	12.0730	1.44567	0.01579	$4.95_7 \times 10^{-6}$	1.20740	5.080	$^3\Sigma_g^-$

[a] From G. Herzberg, Spectra of Diatomic Molecules (Van Nostrand, Princeton, N.J., 1950); the data for many other molecules are also listed in this volume.
[b] The spectroscopic constants listed in this table are defined in Section 7.4 through 7.6.

Table 7.4 *Characteristic group frequencies*[a]

Group	Bond-stretching vibration (cm^{-1})	Group	Bond-bending vibration (cm^{-1})
≡C—H	3300	C≡C—H (with H)	700
C—H (vinyl)	3020	=C (with H H)	1100
—C—H	2960	—C—H (with H H)	1000
—O—H	3680 (gas)[b] 3400 (liquid)[b]	C (with H H)	1450
N—H	3350	—C—H (with H)	1450
C=O	1700	C—C≡C	300
—C≡N	2100		
—C≡C—	2050		
C=C	1650		
—C—C—	900		
—C—F	1100		
—C—Cl	650		
—C—Br	560		
—C—I	500		

[a] From G. Herzberg, *Infrared and Raman Spectra* (Van Nostrand, Princeton, N.J., 1945), p. 195.

[b] The difference is largely due to hydrogen bonding in the liquid phase.

KBr. More recently, interferometers (see Section 2.2) have been employed for very sensitive recording of ir emission spectra from extraterrestrial sources. To avoid virtually complete absorption of the ir radiation by the earth's atmosphere, these compact instruments can be mounted in rockets or satellites, along with computers for analyzing the spectra.

At low resolution, only broad vibrational peaks are observed. In the next section, the rotational fine structure of these peaks is discussed.

7.6 DIATOMIC ROTATION-VIBRATION SPECTRA

Under high resolution, the peak due to a vibrational transition is seen to be composed of a number of individual lines. These lines correspond to changes in the rotational quantum number J simultaneous with the change in the vibrational quantum number v. If we make the rigid-rotator, harmonic-oscillator approximation (i.e., if we continue to neglect rotation-vibration interaction due to the distortion caused by the centrifugal force and the dependence of the moment of inertia upon the vibrational state), the energy levels for nuclear motion are given by a combination of Eqs. 7.23 and 7.38; that is,

$$E_{\text{nucl}} = E_{v,J} = E_{\text{rot}} + E_{\text{vib}} = (v + \tfrac{1}{2}) hc\omega_e + J(J + 1) \, hc\tilde{B}_e \quad (7.43)$$

where $\tilde{B}_e$ is the rotational constant of Eq. 7.26 expressed in cm^{-1},

$$\tilde{B}_e = \frac{B_e}{c} = \frac{\hbar^2}{2I_e hc} \quad (7.44)$$

The wave number corresponding to a simultaneous vibrational and rotational transition is therefore

absorption:

$$\tilde{\nu} = \frac{1}{hc} (E_{v'',J''} - E_{v',J'}) = (v'' - v')\omega_e + [J''(J'' + 1) - J'(J' + 1)]\tilde{B}_e$$
$$(7.45a)$$

emission:

$$\tilde{\nu} = \frac{1}{hc} (E_{v',J'} - E_{v'',J''}) = (v' - v'')\omega_e + [J'(J' + 1) - J''(J'' + 1)]\tilde{B}_e$$
$$(7.45b)$$

where we have used a prime to refer to the initial state and a double prime to refer to the final state.

The same selection rules hold for simultaneous vibration-rotation transitions as for the separate transitions. Making use of Eqs. 7.24 and 7.41, we can distinguish four cases:

(i) $\Delta v = +1$ $\Delta J = +1$
(ii) $\Delta v = +1$ $\Delta J = -1$
(iii) $\Delta v = -1$ $\Delta J = +1$
(iv) $\Delta v = -1$ $\Delta J = -1$

Since ω_e is generally one or two orders of magnitude larger than $\tilde{B}_e$, cases (i) and (ii) correspond to absorption of radiation, while cases (iii) and (iv) correspond to emission. Each set of absorption or emission lines associated with a single vibrational transition in the vibration-rotation spectrum consists of two groups of lines, called *branches*, due to different rotational transitions; the so-called "*P* branch" involves transitions in which $\Delta J = -1$ and corresponds to decreasing rotational energy, and the so-called "*R* branch" involves transitions in which $\Delta J = +1$ and corresponds to increasing rotational energy. The designation "*Q* branch" is used to refer to transitions in which $\Delta J = 0$, a case which does not occur in vibration-rotation spectra of diatomic molecules with Σ electronic ground states, but can occur for polyatomic molecules and in the vibration-rotation spectra of diatomic molecules with nonzero electronic angular momentum about the molecular axis.

Substitution of the selection rules for case (i) into Eq. 7.45a gives the absorption lines in the *R* branch:

$$\tilde{\nu}_R = \omega_e + 2\tilde{B}_e(J' + 1) \quad J' = 0, 1, 2, \ldots \qquad (7.46a)$$

Use of the case (ii) selection rules gives the absorption lines in the *P* branch:

$$\tilde{\nu}_P = \omega_e - 2\tilde{B}_e J' \quad J' = 1, 2, 3, \ldots \qquad (7.46b)$$

According to Eqs. 7.46 the lines in the *P* branch are spaced $2B_e$ apart and lie on the low-frequency (or long wavelength) side of ω_e, while the lines of the *R* branch are also spaced $2B_e$ apart and lie on the high-frequency (or short wavelength) side of ω_e.

In Eqs. 7.46 J' is the rotational quantum number for the initial state. This implies that J' cannot be zero in the *P* branch, since the condition [see case (ii)] that $\Delta J = J'' - J' = -1$ would give the final-state rotational quantum number J'' a negative value; for the *R* branch, all values of J', including $J' = 0$, are allowed. Consequently, the first line of the *R* branch (Eq. 7.46a) is $\tilde{\nu}_R = \omega_e + 2\tilde{B}_e$ and the first line of the *P* branch (Eq. 7.46b) is $\tilde{\nu}_P = \omega_e - 2\tilde{B}_e$; that is, the line with wave number ω_e is missing. The absence of a line at the band center is useful for finding the origin of the spectrum. The allowed transitions in the vibration-rotation absorption spectrum of a diatomic molecule are given in Fig. 7.11 and a typical spectrum is shown in Fig. 7.12. As in the pure rotation spectrum (Section 7.4), the line spacings can be used to obtain values of $\tilde{B}_e$ and from $\tilde{B}_e$ to evaluate I_e and R_e for the molecule.

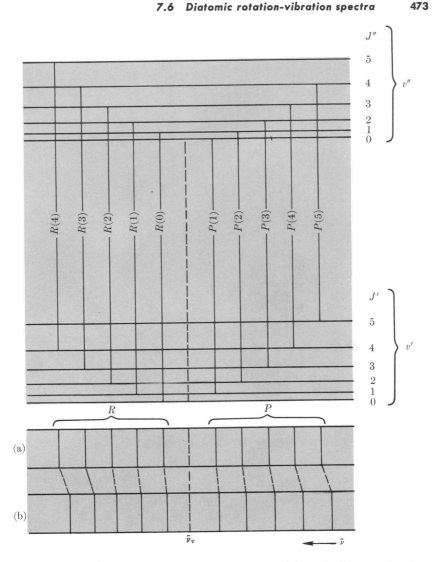

Fig. 7.11. Rotational fine structure for a $v' \leftrightarrow v''$ transition **(a)** without vibration-rotation interaction, **(b)** with vibration-rotation interaction. [After G. Herzberg, Spectra of Diatomic Molecules (Van Nostrand, Princeton, N.J., 1950), p. 112.]

A corresponding discussion applies to the emission spectrum (see Problem 7.14).

The spectrum in Fig. 7.12 shows a clear variation of intensity with J'. This can be understood by realizing that the intensity of an absorption line is proportional to the number of molecules in the initial state. The relative

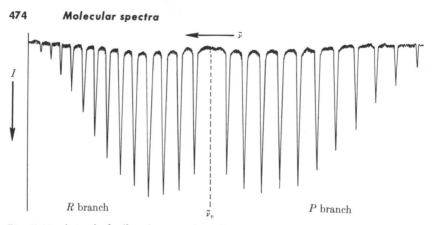

R branch $\tilde{\nu}_v$ P branch

Fig. 7.12. A typical vibration-rotation absorption spectrum for a diatomic molecule showing the R and P branches and the location of the absent line ($\tilde{\nu}_v$).

number of molecules in any given (v', J') initial state with energy $E_{v',J'}$ can be shown to be proportional to the Boltzmann factor

$$N_{v',J'} \propto g_{v',J'} \exp\left(-\frac{E_{v',J'}}{kT}\right) \tag{7.47}$$

where k is the Boltzmann constant and $g_{v',J'}$ is the degeneracy of the state. Since the vibrational states are nondegenerate and the rotational degeneracy is $2J' + 1$, corresponding to the different M' states ($M' = -J'$, $-J' + 1, \ldots, J'$), the vibrational-rotational state degeneracy is $g_{v',J'} = 2J' + 1$. Because the energy $E_{v',J'}$ increases with increasing J', the factor $\exp(-E_{v',J'}/kT)$ decreases. However, the degeneracy increases with J', so that the relative population

$$(2J' + 1) \exp\left[-J'(J' + 1)\frac{B_e h}{kT}\right] \tag{7.48}$$

has a maximum. Differentiating Eq. 7.48, we find that the maximum occurs near

$$J' \simeq \frac{1}{2}\left[\left(\frac{2kT}{B_e h}\right)^{1/2} - 1\right] \tag{7.49}$$

and that Eq. 7.48 satisfactorily accounts for the observed variation in intensity in the individual branches (Problem 7.16).

7.6.1 Anharmonicity and the dissociation limit Having developed a simple theory for the vibrational-rotational spectra of diatomic molecules, we are ready to look at some of the corrections that have to be introduced in more accurate studies. All of them follow from the complete solution of the Schrödinger equation with the exact potential $E(R)$. However, it is convenient to consider the various corrections successively so as to make clear their role in the observed spectrum.

In the discussion of the harmonic-oscillator model for diatomic molecules, we pointed out that the presence of overtones demonstrates that anharmonic terms contribute to $E(R)$. Moreover, if one examines the overtones, one finds that their spacing is somewhat less than $2\omega_e$, $3\omega_e$, etc. Correspondingly, when transitions are observed between different neighboring pairs of vibrational levels, $\bar{\nu}$ is found to decrease somewhat as v' increases; this also indicates that the level spacing is not constant. To account for these deviations from the harmonic model, we must use more terms in the potential-energy-function expansion given in Eq. 7.36 or, equivalently, with $\rho = R - R_e$,

$$E(\rho) = \frac{1}{2} k_e \rho^2 + \frac{1}{3!} \left(\frac{d^3 E}{d\rho^3} \right)_0 \rho^3 + \frac{1}{4!} \left(\frac{d^4 E}{d\rho^4} \right)_0 \rho^4 + \cdots \qquad (7.50)$$

where the subscript (0) on the derivatives indicates that they are to be evaluated at $\rho = 0$. Although it is possible to determine each of the derivatives in Eq. 7.50 directly from the observed spectrum, it is often convenient to introduce a relatively simple mathematical function for $E(\rho)$ that has the general shape of the dashed curve in Fig. 7.8 and includes some parameters to be evaluated from the spectral data. One commonly used function, called the *Morse function* or Morse potential after P. M. Morse, is

$$E(\rho) = D_e(1 - e^{-\beta\rho})^2 = D_e(1 - e^{-\beta(R-R_e)})^2 \qquad (7.51)$$

where D_e, the equilibrium dissociation energy discussed in Chapters 5 and 6, is the difference between the minimum of the curve at R_e and the asymptotic energy at $R \to \infty$. The constant β is positive, so that for $R \ll R_e$, where ρ is large and negative, $E(\rho)$ is large and positive, corresponding to the short-range overlap repulsion. Inspection of Eq. 7.51 shows that the minimum occurs at $\rho = 0$ (at $R = R_e$) and that as $\rho \to \infty$ the function $E(\rho)$ approaches D_e (i.e., the zero of energy is chosen at the minimum of the curve).

The constant β can be related to the force constant k_e by expanding $E(\rho)$ about $\rho = 0$ in a Maclaurin series to give

$$E(\rho) = D_e(\beta^2\rho^2 - \beta^3\rho^3 + \tfrac{7}{12} \beta^4\rho^4 - \cdots) \qquad (7.52)$$

Comparing Eqs. 7.50 and 7.52, we see that

$$D_e \beta^2 = \tfrac{1}{2} k_e \qquad (7.53)$$

It is clear from Eqs. 7.51 and 7.53 that once the three parameters R_e, D_e, and k_e are known, the Morse curve is completely determined. This means that all the anharmonic corrections can be evaluated by comparing the

Morse-function expansion (Eq. 7.52) with that general expression (Eq. 7.50). For the first two anharmonic corrections to the potential energy function, we have

$$\frac{1}{3!}\left(\frac{d^3E}{d\rho^3}\right)_0 = -\beta^3 D_e \quad \frac{1}{4!}\left(\frac{d^4E}{d\rho^4}\right)_0 = \frac{7}{12}\beta^4 D_e \tag{7.54}$$

Conversely, assuming the Morse function to have the correct form, we can use the experimental data to find D_e for a molecule from energy expressions including anharmonicity by means of Eqs. 7.53 and 7.54. The vibration energies can be calculated for the Morse function either by substituting Eq. 7.51 into Eq. 7.33 and solving the resulting differential equation, or by treating the anharmonic terms of Eq. 7.52 as "perturbations" of the harmonic states and finding the energy corrections by the techniques of perturbation theory. The result is[2]

$$E_v = (v + \tfrac{1}{2})\omega_e hc - (v + \tfrac{1}{2})^2 \omega_e x_e hc \tag{7.55}$$

The positive constant $\omega_e x_e$ is called the first anharmonicity constant, and is given by

$$\omega_e x_e = \frac{h\nu_e}{4D_e}\omega_e \tag{7.56}$$

Values of the first anharmonic correction $\omega_e x_e$ for some molecules are listed in Table 7.3.

The selection rule for an anharmonic oscillator is

$$\Delta v = \pm 1, \pm 2, \pm 3, \ldots$$

that is, all transitions are allowed. However, the "fundamental" lines corresponding to $\Delta v = \pm 1$ are still the most intense, and the intensity decreases very rapidly as $|\Delta v|$ increases.

From Eq. 7.55, we see that in the ground vibrational state a molecule has zero-point vibrational energy

$$E_0 = \tfrac{1}{2}hc\omega_e - \tfrac{1}{4}hc\omega_e x_e \tag{7.57}$$

The bond dissociation energy at $0°K$, symbolized by D_0, is the difference between the energy for infinite separation of the atoms and the lowest vibration level. From Eq. 7.57 and the definition of D_e, we have

$$D_0 = D_e - \tfrac{1}{2}hc\omega_e + \tfrac{1}{4}hc\omega_e x_e \tag{7.58}$$

In the harmonic-oscillator approximation, the third term on the right-hand side of Eq. 7.58 is omitted and D_0 is written

$$D_0 = D_e - \tfrac{1}{2}hc\omega_e \tag{7.59}$$

---

[2] The agreement between the accurate and second-order perturbation-theory calculations of the spectrum of the Morse oscillator is accidental. In the general case, the perturbation results differ from the accurate values by a truncation error.

Applying the Bohr frequency rule and the selection rules to Eq. 7.55, we obtain the following expressions for the vibrational absorption lines from the ground state ($v = 0$) to the excited states ($v = 1, 2, \ldots$)

fundamental: $\qquad (v = 1) \quad \tilde{\nu}_1 = \omega_e - 2\omega_e x_e$

first overtone: $\qquad (v = 2) \quad \tilde{\nu}_2 = 2\omega_e - 6\omega_e x_e$

second overtone: $\qquad (v = 3) \quad \tilde{\nu}_3 = 3\omega_e - 12\omega_e x_e$

In general, for absorption from the vibrational ground state (i.e., $v = 0$) to the level v,

$$\tilde{\nu}_v = v\omega_e - v(v + 1)\omega_e x_e \qquad (7.60)$$

One way to obtain an estimate of D_e is to use the vibrational overtone spacings to find $\omega_e x_e$ and then to use the Morse relation (Eq. 7.56) in the form

$$D_e = \frac{hc\omega_e^2}{4\omega_e x_e} \qquad (7.61)$$

The overtone spacing $\Delta\tilde{\nu}_v$ as a function of v is found from Eq. 7.60 to be

$$\Delta\tilde{\nu}_v = \tilde{\nu}_v - \tilde{\nu}_{v-1} = \omega_e - 2v\omega_e x_e \qquad (7.62)$$

We see that as v increases, the value of $\Delta\tilde{\nu}_v$ decreases. The second difference is

$$\Delta^2\tilde{\nu}_v = \Delta\tilde{\nu}_v - \Delta\tilde{\nu}_{v-1} = -2\omega_e x_e \qquad (7.63)$$

Fundamental and overtone frequencies of HCl and their first and second differences are given in Table 7.5. From the last column and Eq. 7.63, $\omega_e x_e$ is seen to be about 52 cm^{-1}. Insertion of this value into Eq. 7.62 allows ω_e to be obtained from the values of $\Delta\tilde{\nu}_v$ given in Table 7.5 (Problem 7.52); or, equivalently, ω_e can be obtained by extrapolation of $\Delta\tilde{\nu}_v$ to $v = 0$. Substitu-

Table 7.5 *Vibrational absorption lines of H^1Cl^{35} in the ground electronic state.*[a]

v[b]	$\tilde{\nu}_v(\mathrm{cm}^{-1})$	$\Delta\tilde{\nu}_v(\mathrm{cm}^{-1})$	$\Delta^2\tilde{\nu}_v(\mathrm{cm}^{-1})$
1	2885.9_0	2885.9_0	
2	5668.0_5	2782.1_5	-103.7_5
3	8346.9_8	2678.9_3	-103.2_2
4	10923.1_1	2576.1_3	-102.8_0
5	13396.5_5	2473.4_4	-102.6_9

[a] G. Herzberg, *Spectra of Diatomic Molecules* (Van Nostrand, Princeton, N.J., 1950), p. 55.
[b] Upper-state vibrational quantum number; transitions correspond to $0 \rightarrow v$.

Problem 7.18). For good results, one must have values of ω_v for large values of v. Although these are hard to obtain from pure vibrational spectra because the intensities of the bands corresponding to large v are too small to observe in most cases, they can often be determined from the vibrational fine structure of electronic emission spectra, which we discuss in Section 7.7. First, however, we consider the corrections to the energy of nuclear motion arising from the incomplete separation of rotation and vibration.

7.6.2   Centrifugal distortion   In using Eq. 7.43 for the vibration-rotation levels, we assumed that rotation was adequately described by the rigid-rotator model and neglected the fact that a rotating molecule experiences a centrifugal force which stretches the molecule so that the effective equilibrium internuclear distance increases with J. One way of introducing this centrifugal distortion is to include the rotational term in the potential-energy function from which the R_e value is determined as a function of J. For a rotating harmonic oscillator we see from Eq. 7.18 with $E(R)$ replaced by the harmonic oscillator potential that there are three contributions to the energy; that is, the radial kinetic energy (expressed as derivatives with respect to R), the rotational energy, and $E(R)$. The second and third terms, which together are often regarded as an effective potential for the radial motion, can be written as

$$E_J(R) = \frac{1}{2} k_e (R - R_e)^2 + \frac{J(J+1)\hbar^2}{2\mu R^2} \tag{7.71}$$

where the second term in Eq. 7.71 is called the *centrifugal potential*.

If we let R_e' be the value of R for which $E_J(R)$ is a minimum, we have

$$\left. \frac{dE_J(R)}{dR} \right|_{R=R_e'} = k_e(R_e' - R_e) - \frac{J(J+1)\hbar^2}{\mu(R_e')^3} = 0 \tag{7.72}$$

from which R_e' may be determined as a function of J. Since the correction is small, we can obtain it by solving Eq. 7.72 by successive approximation. Little error is introduced if we use the first-order correction, which results from setting $R_e' = R_e$ in the rotation term of Eq. 7.72; we find

$$R_e' = R_e + \frac{J(J+1)hB_e}{2\pi^2 \nu_e^2 \mu R_e} \tag{7.73}$$

where we have introduced B_e and ν_e by Eqs. 7.26 and 7.39, respectively. If we replace R_e by R_e' in the rotational energy equation (Eq. 7.23), we have

$$E_J = \frac{J(J+1)\hbar^2}{2\mu(R_e')^2} \tag{7.74}$$

Substituting for R_e' by Eq. 7.73 in Eq. 7.74, expanding the result in powers of $J(J + 1)$, and keeping only the first two terms, we obtain (Problem 7.19)

$$E_J = J(J + 1)hc\tilde{B}_e - [J(J + 1)]^2\, hc\tilde{D}_e \qquad (7.75)$$

where the centrifugal distortion constant $\tilde{D}_e$ is given by

$$\tilde{D}_e = \frac{4\tilde{B}_e{}^3}{\omega_e{}^2} \qquad (7.76)$$

The corrections due to $\tilde{D}_e$ are relatively small except for high J values in most molecules; some representative $\tilde{D}_e$ values for diatomics are given in Table 7.3.

7.6.3 Vibration-rotation interaction For very precise work an additional vibration-rotation interaction term has to be introduced. Clearly a molecule that is vibrating while rotating has contributions from values of the internuclear distance other than the equilibrium value. In fact, since the inverse of the moment of inertia appears in the rotational energy the average of $1/R^2$ over the probability distribution for the particular vibrational state v is required; that is, $\langle 1/R^2 \rangle_v$ should be used. For a *harmonic* oscillator, the values of R near the turning points become increasingly probable as v increases (Fig. 7.9). Since $1/R^2$ is small for the outer turning point, the inner turning point makes the greater contribution, and the result is that $\langle 1/R^2 \rangle_v$ increases with v for a harmonic oscillator. Just the reverse is true for a *Morse* oscillator, however. From the form of the potential (see Fig. 7.8), it is evident that the particle samples larger R values as v increases. Moreover, since the potential energy is high in this region the classical Morse oscillator slows down and spends more time there so that $\langle 1/R^2 \rangle_v$ decreases with increasing v. The corresponding quantum-mechanical description shows that the form of the wave function is such that these larger values of R become more probable relative to those near the inner turning point. If the calculations are carried out for the rotating Morse oscillator, it is found that the rotational energy is

$$E_{\text{rot}} = J(J + 1)hc\tilde{B}_v \qquad (7.77)$$

where the rotation constant $\tilde{B}_v$ depends upon the vibrational quantum number. To a first approximation,

$$\tilde{B}_v = \tilde{B}_e - \tilde{\alpha}_e(v + \tfrac{1}{2}) \qquad (7.78)$$

where $\tilde{\alpha}_e$ is given by

$$\tilde{\alpha}_e = \frac{6\tilde{B}_e{}^2}{\omega_e}\left[\left(\frac{\omega_e x_e}{\tilde{B}_e}\right)^{1/2} - 1\right] \qquad (7.79)$$

Thus, we see that the vibration-rotation interaction constant $\tilde{\alpha}_e$ is negative for the harmonic oscillator ($\omega_e x_e = 0$), and positive for oscillators that are sufficiently anharmonic, in accordance with our qualitative discussion. Values of $\tilde{\alpha}_e$ for several molecules are given in Table 7.3. The shifts in the vibration-rotation spectrum of a diatomic molecule in the presence of non-zero $\tilde{\alpha}_e$ is indicated in Fig. 7.11b.

In summary, the expression for $E_{v,J}$ of a rotating anharmonic oscillator in which first-order corrections are made for centrifugal distortion and rotational interaction is

$$
\begin{aligned}
E_{v,J} = & \; hc\omega_e(v + \tfrac{1}{2}) - hc\omega_e x_e(v + \tfrac{1}{2})^2 + hc\tilde{B}_e J(J + 1) \\
& - hc\tilde{D}_e J^2(J + 1)^2 - hc\tilde{\alpha}_e(v + \tfrac{1}{2})J(J + 1)
\end{aligned}
\tag{7.80}
$$

Equation 7.80 can be regarded as the first few terms of a power-series expansion of the rotation vibration energy of a diatomic molecule in terms of the variables $(v + \tfrac{1}{2})$ and $J(J + 1)$. Although higher power terms have been introduced in some cases, Eq. 7.80 is an adequate approximation for most applications. Values of the spectroscopic constants in Eq. 7.80 for some representative diatomic molecules are given in Table 7.3.

7.6.4 Isotope effects
To the extent that the Born–Oppenheimer approximation is valid (see Section 7.3), the potential-energy function of a molecule is not altered by the substitution of one isotope for another (e.g., D for H in H_2). Thus, the nuclear motion is changed only through the dependence of the momenta and kinetic energies of the nuclei on their masses. This means that by comparing the spectra of different isotopic species of the same molecule, we can obtain a check on our conclusions concerning the form of the potential function. In analyzing the results for isotopic molecules, we consider the changes expected for the spectroscopic constants ω_e, $\omega_e x_e$, $\tilde{B}_e$, $\tilde{D}_e$, and $\tilde{\alpha}_e$, appearing in the expression for $E_{v,J}$ (Eq. 7.80). Since the force constant k_e, the equilibrium internuclear distance R_e, and the bond dissociation energy D_e remain the same, the constants are altered only by the different values of the reduced mass μ for isotopic molecules. To indicate the explicit dependence upon μ, we collect the formulas obtained in the earlier sections for the spectroscopic constants of a Morse oscillator; they are

$$
\omega_e = \frac{1}{2\pi c}\left(\frac{k_e}{\mu}\right)^{1/2}
\tag{7.81a}
$$

$$
\omega_e x_e = \frac{hc\omega_e^2}{4D_e}
\tag{7.81b}
$$

$$
\tilde{B}_e = \frac{\hbar^2}{2\mu r_e^2 hc}
\tag{7.81c}
$$

$$
\tilde{D}_e = \frac{4\tilde{B}_e^3}{\omega_e^2}
\tag{7.81d}
$$

$$\tilde{\alpha}_e = \frac{6\tilde{B}_e^{\,2}}{\omega_e}\left[\left(\frac{\omega_e x_e}{\tilde{B}_e}\right)^{1/2} - 1\right] \tag{7.81e}$$

From Eqs. 7.81, the reader can verify that ratios of the constants for the isotopic analog (i) to those of the reference molecule (r) are

$$\frac{(\omega_e)_i}{(\omega_e)_r} = \left(\frac{\mu_r}{\mu_i}\right)^{1/2} \tag{7.82a}$$

$$\frac{(\omega_e x_e)_i}{(\omega_e x_e)_r} = \frac{\mu_r}{\mu_i} \tag{7.82b}$$

$$\frac{(\tilde{B}_e)_i}{(\tilde{B}_e)_r} = \frac{\mu_r}{\mu_i} \tag{7.82c}$$

$$\frac{(\tilde{D}_e)_i}{(\tilde{D}_e)_r} = \left(\frac{\mu_r}{\mu_i}\right)^{2} \tag{7.82d}$$

$$\frac{(\tilde{\alpha}_e)_i}{(\tilde{\alpha}_e)_r} = \left(\frac{\mu_r}{\mu_i}\right)^{3/2} \tag{7.82e}$$

As an illustration we give experimental values of the spectroscopic constants of H_2 and D_2 in Table 7.7; we include $\omega_e y_e$, in addition to the con-

Table 7.7 Comparison of spectroscopic constants of H_2 and D_2[a,b]

Constant	H_2	D_2
ω_e	4395.34	3118.5
$\omega_e x_e$	117.90	64.10
$\omega_e y_e$	+0.29	+ 1.254
$\tilde{B}_e$	60.80	30.429
$\tilde{\alpha}_e$	2.993	1.0492
$\tilde{D}_e$	0.0465	0.01159
R_e	0.7417	0.7416

[a] G. Herzberg, *Spectra of Diatomic Molecules* (Van Nostrand, Princeton, N.J., 1950).
[b] All values are given in units of cm^{-1}.

stants in Eq. 7.82. As can be shown by use of Eq. 7.82, the deviations from the expected ratios are small (Problem 7.20). This result is strong evidence for the adequacy of the vibrating rotator model for diatomic molecules.

An important consequence of the isotope effect on the spectroscopic constants is that the zero-point energy of a molecule decreases with increasing reduced mass. To a first approximation, we have

$$\frac{(E_0)_i}{(E_0)_r} = \frac{\frac{1}{2}hc(\omega_e)_i}{\frac{1}{2}hc(\omega_e)_r} = \left(\frac{\mu_r}{\mu_i}\right)^{1/2} \tag{7.83}$$

The difference in E_0 is experimentally manifested as a corresponding difference in the measured bond energy $D_0 = D_e - E_0$. For example, for HCl, $D_0 = 4.430$ eV, while for DCl, $D_0 = 4.481$ eV. The reader can verify that these values are in agreement with Eq. 7.83.

7.7   ELECTRONIC TRANSITIONS

In our discussion of the electronic structure of diatomic molecules (Sections 6.1 and 6.2), we showed how the MO method can serve to determine the energy levels and the ground-state configuration and term symbol. The MO method can be used also to obtain excited-state configurations and to examine the nature of the electronic transitions. As an example, we consider the O_2 molecule for which experimental results and theoretical analyses are available. The O_2 spectrum has been the subject of intensive study because of its importance in atmospheric phenomena.

In Section 6.1.6, we saw that the ground-state configuration of O_2 can be written

$$(1\sigma_g)^2(1\sigma_u)^2(2\sigma_g)^2(2\sigma_u)^2(1\pi_u)^4(3\sigma_g)^2(1\pi_g)^2$$

or, specifying the orbitals in more detail (Table 6.3),

$$(1s\sigma_g)^2(1s\sigma_u{}^*)^2(2s\sigma_g)^2(2s\sigma_u{}^*)^2(2p\pi_u)^4(2p\sigma_g)^2(2p\pi_g{}^*)^2$$

Since the electron configuration for the highest occupied orbital $(1\pi_g)^2$ does not correspond to a closed shell, several terms are associated with this configuration. As shown in Section 6.1.6, these terms correspond to different orbital and spin arrangements of the two $1\pi_g$ electrons. They are the ground-state term $(^3\Sigma_g{}^-)$ and the two excited state terms $(^1\Sigma_g{}^+, {}^1\Delta_g)$. By Hund's rules the triplet has the lowest energy; of the two singlets, the one with the larger value of Λ has the lower energy. The order of the states is therefore

$$^3\Sigma_g{}^- < {}^1\Delta_g < {}^1\Sigma_g{}^+$$

with experimental energy differences

$$E(^1\Delta_g) - E(^3\Sigma_g{}^-) = 0.98 \text{ eV} \quad E(^1\Sigma_g{}^+) - E(^3\Sigma_g{}^-) = 1.63 \text{ eV}$$

The potential curves for these three states, all of which dissociate to ground-state atoms, are shown as the lowest curves in Fig. 7.13. Since several different states with the same term symbol can arise, letters preceding the term symbol are often used to distinguish them. The ground state is usually assigned the letter X, and other states with the same multiplicity as the ground state are assigned capital letters A, B, C, etc., in order of increasing energy or in order of their discovery. States with multiplicity different from that of the ground state are often assigned lower-case letters a, b, c, etc. As indicated in Fig. 7.13, the three terms arising from the ground configuration of O_2 are designated

$$X^3\Sigma_g^- \qquad a^1\Delta_g \qquad b^1\Sigma_g^+$$

In addition to the three states corresponding to the ground configuration, many states associated with excited configurations of O_2 have been

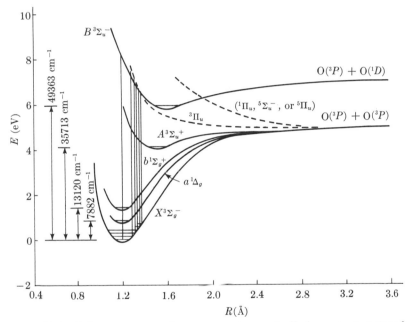

Fig. 7.13. Potential-energy curves for some spectroscopically important states of O_2; vibrational levels and quantum numbers are indicated for most of the states. [Adapted from R. F. Gilmore, *J. Quant. Spectry. Radiative Transfer*, **5**, 369 (1965).]

studied. Several of these, with the configurations from which they derive, are given in Table 7.8. The energy of each of these states is, of course, a function of the internuclear distance R, and each bound state has in general a different R_e, k_e, and D_e and a unique set of rotational-vibrational levels associated with it. A potential curve can be constructed for each electronic state by assuming a functional form, such as the Morse curve, and fitting its parameters to the measured constants. Alternatively, if a preted as an extra uncertainty in the energy of the upper ($B\ ^3\Sigma_u^-$) state, which arises from a shortened lifetime. The origin of this appears to be the dissociation of the excited state into two ground-state oxygen atoms due to a nonradiative transition to one of the repulsive states shown by dotted lines in Fig. 7.13. Such a dissociation from a bound state by means of a non-

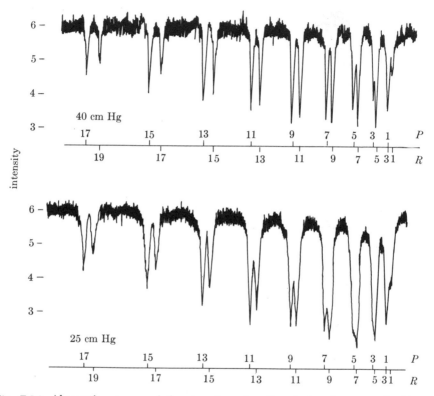

Fig. 7.14. Absorption traces of the $v' = 2 \rightarrow v'' = 0$ and $v' = 3 \rightarrow v'' = 0$ vibrational Schumann–Runge bands at pressures of 40 and 25 cm Hg, respectively. [From R. D. Hudson and V. L. Carter, *J. Opt. Soc. Am.*, **58,** 1621 (1968).]

radiative transition to a repulsive state is called *predissociation*. The process is analogous to auto-ionization in atoms (see Section 4.3.2).

The spin-forbidden atmospheric $a \rightleftarrows X$ and $b \rightleftarrows X$ bands absorb strongly in the infrared (12680 Å) and red (7600 Å) regions, respectively. The Herzberg $A \rightleftarrows X$ bands account for much of the ultraviolet emission observed in the night sky. In addition to the listed bands of O_2, transitions to higher upper electronic states have been observed. These states appear to involve a single electron very loosely bound to an O_2^+ core, in that the transitions resemble those of one-electron atoms, with the transition frequency $\bar{\nu}$ roughly proportional to $1/n^2$. The values of $\bar{\nu}$ converge to a limit that corresponds to the IP for the lower state. These series, called *Rydberg series*, are observed in many molecules and are often used for determining IP's (see Sections 6.4.6 and 7.11.2).

One important application of electronic spectroscopy is to the accurate evaluation of dissociation energies. In O_2, this is achieved by making use of the fact that the Schumann–Runge bands $[B^3\Sigma_u^- \leftarrow X^3\Sigma_g^-]$ form a progression which converges to a well-defined limit; that is, as the band frequency increases, the spacings get smaller and smaller, until they reach an observed dissociation continuum. The so-called *convergence limit* is at 1759 Å, which corresponds to 7.047 eV. This is the energy required to go from the lowest vibrational state of the ground electronic state ($X^3\Sigma_g^-$) to a pair of oxygen atoms. At least one of these must be in an excited state, because the ground-state dissociation energy D_0 is known to be less than 6.3 eV from application of Eq. 7.61 with the equilibrium values ω_e, $\omega_e x_e$ replaced by the ground vibration state values ω_0, $\omega_0 x_0$.[4] If one can identify the O-atom states, their known excitation energies can be subtracted from 7.047 eV to obtain the desired dissociation energy of $X^3\Sigma_u^-$ into *ground-state* atoms $[O(^3P)]$. To identify the dissociation product of the $B\ ^3\Sigma_u^-$, we make use of the fact that only ground-configuration O atoms (^3P, ^1S, ^1D) could be involved because the energy 7.047 eV is not large enough to produce excited configurations; the lowest excited configuration state is 9.1 eV above the ground state. Thus, we have to consider $^3P + {}^1S$, $^3P + {}^1D$, $^1S + {}^1S$, $^1S + {}^1D$, $^1D + {}^1D$. Of these, $^1S + {}^1S$, $^1S + {}^1D$, and $^1D + {}^1D$ can be eliminated immediately because two singlet atoms ($S = 0$) cannot combine to give a triplet molecule ($S = 1$). The two remaining possibilities, $^3P + {}^1S$ and $^3P + {}^1D$, have energies of 1.967 and 4.188 eV, respectively, above the $^3P + {}^3P$ ground state. Thus, the value of D_0 for $X^3\Sigma_g^-$ would be 5.080 eV if the dissociation products of the upper state are $^3P + {}^1S$ and 2.859 eV if they are $^3P + {}^1D$. Since vibrational levels of $X^3\Sigma_g^-$ up to 3.4 eV above $v = 0$ have been observed, D_0 must be greater than that; that is, the value 2.859 eV is excluded. This leaves 5.080 eV as the accurate value for D_0. Figure 7.13 is drawn to be consistent with this conclusion.

7.7.1 Electronic band structure
Having discussed the general features and energetics of electronic transitions, we now concern ourselves with their rotational-vibrational fine structure, which has already been illustrated in the Schumann–Runge absorption spectrum shown in Fig. 7.14. We consider only the least complicated case, namely, transitions in which neither

---

[4] The two ground-state atoms $O(^3P)$ are known not to be able to form an O_2 molecule in a $^3\Sigma_u^-$ state from general symmetry arguments often called the Wigner-Witmer rules (see Reference 4 in the Additional Reading list).

to the coordinate system shown in the figure. This coordinate system is introduced as a reference system to which the molecule and its rotation can be referred and so is to be regarded as fixed in space (so-called *space-fixed coordinates*). Suppose that one of the atoms is at point P, which has the perpendicular distance a_P from the axis $\hat{\omega}$. As the molecule rotates about $\hat{\omega}$, the atom rotates with it and the components (x_P, y_P, z_P) of the position

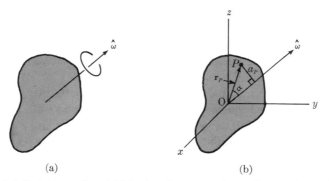

Fig. 7.16. (a) Rotation of a rigid body about rotation axis $\hat{\omega}$; **(b)** coordinates of point P in the rigid body relative to rotation axis $\hat{\omega}$.

vector $\mathbf{r}_P$ relative to the space-fixed coordinate axes change. However, the length r_P, the distance a_P, and the angle α between $\mathbf{r}_P$ and $\hat{\omega}$ remain constant since they are determined by the bond angles and distances in the molecule and not by its orientation in space. From Fig. 7.16 one can write

$$a_P{}^2 = r_P{}^2 - (r_P \cos \alpha)^2 \qquad (7.87)$$

The linear speed v_P of the atom at point P is also constant and is given by

$$v_P = a_P \omega \qquad (7.88)$$

If the atom at P has mass m_P, its rotational kinetic energy T_P is

$$T_P = \tfrac{1}{2} m_P \, v_P{}^2 \qquad (7.89)$$

Substituting from Eqs. 7.87 and 7.88 into Eq. 7.89, we find

$$T_P = \tfrac{1}{2} m_P [r_P{}^2 - (r_P \cos \alpha)^2]\omega^2 = \tfrac{1}{2} m_P [r_P{}^2\omega^2 - (r_P \, \omega \cos \alpha)^2] \quad (7.90)$$

It is convenient here to introduce the product of the square of the length of the position vector $\mathbf{r}_P$ and that of the angular velocity vector $\boldsymbol{\omega}$, which is defined as the vector of length ω in the direction $\hat{\omega}$. In terms of the space-fixed coordinate system (Fig. 7.16b), $\boldsymbol{\omega}$ has components $\omega_x, \omega_y, \omega_z$, which do not vary with time, and the square of its length is $\omega^2 = \omega_x{}^2 + \omega_y{}^2 + \omega_z{}^2$. Similarly writing $r_P{}^2 = x_P{}^2 + y_P{}^2 + z_P{}^2$, where x_P, y_P, z_P vary with time although $r_P{}^2$ is a constant, we can express the product $r_P{}^2\omega^2$ in the form

$$r_P{}^2\omega^2 = (x_P{}^2 + y_P{}^2 + z_P{}^2)(\omega_x{}^2 + \omega_y{}^2 + \omega_z{}^2) \tag{7.91}$$

The quantity $r_P\,\omega\cos\alpha$, which is also a constant, can be written as the scalar product of the two vectors $\mathbf{r}_P$ and $\boldsymbol{\omega}$,

$$r_P\omega\,\cos\alpha = \mathbf{r}_P \cdot \boldsymbol{\omega} = x_P\omega_x + y_P\omega_y + z_P\omega_z \tag{7.92}$$

Inserting Eqs. 7.91 and 7.92 into Eq. 7.90 we obtain for the kinetic energy of atom P

$$T_P = \tfrac{1}{2}m_P(y_P{}^2 + z_P{}^2)\omega_x{}^2 + \tfrac{1}{2}m_P(x_P{}^2 + z_P{}^2)\omega_y{}^2 + \tfrac{1}{2}m_P(x_P{}^2 + y_P{}^2)\omega_z{}^2$$
$$- \, m_P x_P y_P \omega_x\omega_y - m_P y_P z_P \omega_y\omega_z - m_P x_P z_P \omega_x\omega_z \tag{7.93}$$

The rotational energy of the molecule consists only of kinetic energy, since there is no angle-dependent potential. Since the total rotational energy E_{rot} is the sum of the rotational kinetic energies of the individual atoms, we have

$$E_{\text{rot}} = \sum_P T_P = \frac{1}{2}\left[\sum_P m_P(y_P{}^2 + z_P{}^2)\right]\omega_x{}^2 + \frac{1}{2}\left[\sum_P m_P(x_P{}^2 + z_P{}^2)\right]\omega_y{}^2$$
$$+ \frac{1}{2}\left[\sum_P m_P(x_P{}^2 + y_P{}^2)\right]\omega_z{}^2 - \left[\sum_P m_P \, x_P \, y_P\right]\omega_x\omega_y \tag{7.94}$$
$$- \left[\sum_P m_P \, y_P \, z_P\right]\omega_y\omega_z - \left[\sum_P m_P \, x_P \, z_P\right]\omega_x\omega_z$$

To simplify Eq. 7.94, we introduce certain properties of the rigid body which we are using to approximate the rotating molecule. This is done most simply by defining a second Cartesian coordinate system that is fixed in the molecule rather than in space. This *molecule-fixed* coordinate system rotates with the molecule and at each time its relation to the space fixed system can be specified by three angles; these can be chosen to be the two polar angles (θ_z, ϕ_z) which locate the molecule-fixed z axis in space and a third angle (β_x) which determines the orientation of the molecule-fixed x axis in the plane perpendicular to the z axis (Fig. 7.17). The orientation of the molecular-axis system within the molecule is arbitrary and can be chosen to obtain E_{rot} in a convenient form. It can be shown that there exists an orientation, with the origin at the center of mass of the molecule, such that the equations

$$\sum_P m_P \, x_P{}' \, y_P{}' = \sum_P m_P \, x_P{}' \, z_P{}' = \sum_P m_P \, y_P{}' \, z_P{}' = 0 \tag{7.95}$$

are satisfied, where $x_P{}'$, $y_P{}'$, $z_P{}'$ are the coordinates of atom P in terms of the *molecule-fixed* coordinate system. The molecule-fixed axes determined in this way are called the *principal axes* of the molecule. For molecules with sufficient symmetry, the principal axes are easy to find. One principal axis is the axis of highest symmetry; a second axis can be chosen perpendicular to the first and to a plane of symmetry (if one exists); the third axis is then fixed by the requirement that it be perpendicular to the other two. Examples of principal axes are given in Fig. 7.18 for the molecules CO_2, H_2O, NH_3, and CH_4.

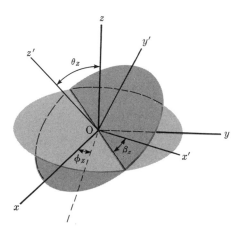

Fig. 7.17. Orientation of molecule-fixed coordinate system (x', y', z') relative to space fixed coordinate system (x, y, z); θ_z and ϕ_z are the angles specifying the z' axis and β_x is the angle specifying the x' axis relative to the intersection of the $x'y'$ and the xy planes.

We assume in what follows that the coordinates of the atoms in a molecule are referred to its principal axes with the center of mass at the origin. The principal moments of inertia of the molecule I_x, I_y, and I_z are then defined by the equations

$$I_x = \sum_P m_P(y_P'^2 + z_P'^2)$$

$$I_y = \sum_P m_P(x_P'^2 + z_P'^2) \qquad (7.96)$$

$$I_z = \sum_P m_P(x_P'^2 + y_P'^2)$$

To introduce the molecular coordinate system into Eq. 7.94, we choose the space-fixed axes so that, at a particular instant of time, they coincide with the principal axes of the molecule. At that time, Eqs. 7.95 and 7.96 are valid for the space-fixed axes, as well as for the molecule-fixed axes. These

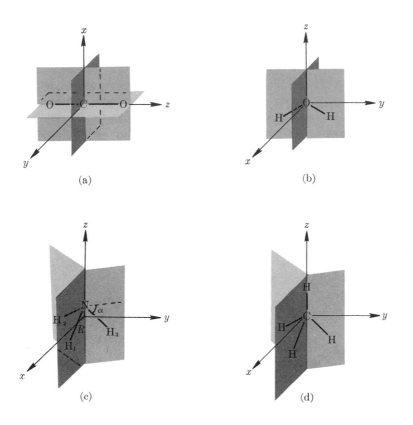

Fig. 7.18. Principal axes of **(a)** CO_2, **(b)** H_2O, **(c)** NH_3, **(d)** CH_4.

expressions with x'_P, y'_P, z'_P replaced by x_P, y_P, z_P into Eq. 7.94, we can write it in the very simple form

$$E_{\text{rot}} = \tfrac{1}{2}I_x\,\omega_x{}^2 + \tfrac{1}{2}I_y\,\omega_y{}^2 + \tfrac{1}{2}I_z\,\omega_z{}^2 \tag{7.97}$$

where ω_x, ω_y, ω_z correspond to rotation about the principal axes of the molecule. It turns out, although we do not prove this essential result, that Eq. 7.97 is valid even when the molecule- and space-fixed axes do not coincide, as long as I_x, I_y, I_z and the rotation components ω_x, ω_y, ω_z are defined with respect to the molecular coordinate system. Thus, Eq. 7.97 can be used to describe the rotational energy of the molecule without need to concern ourselves with the details of the time variation of the angles θ_z, ϕ_z, and β_x.

M_z. Demonstration of the selection rules for J requires consideration of the symmetric-top wave functions. These consist of sums of products of the angular functions used for the hydrogen atom; they are discussed in Reference 5 of the Additional Reading list. From Eq. 7.107 and the selection rules, the absorption and emission wave numbers (cm^{-1}) for a rigid symmetric top are given by Eq. 7.103, where

$$\tilde{B} = \frac{\hbar}{4\pi c I_B} \qquad (7.108)$$

Asymmetric-Top Molecules. Nonlinear planar molecules such as H_2O and H_2CO and nonplanar molecules without a threefold or higher symmetry axis, such as CH_2Cl_2, can only "accidentally" have two principal moments of inertia that coincide. Otherwise, all three moments of inertia are different, and the molecule is called an *asymmetric top*. In this case the expression for the rotational energy (Eq. 7.99) cannot be reduced further and no simple formula for the rotational energy levels can be written in terms of the quantum numbers J and K. It is possible, however, to obtain a qualitative picture of the asymmetric-top levels by relating them to those of the prolate and oblate symmetric top. We label the three moments of inertia A, B, C, in order of increasing magnitude

$$I_A < I_B < I_C$$

Since I_A, I_B, and I_C each appear in one denominator in Eq. 7.99, decreasing a moment of inertia increases the energy of each level. Thus we expect the asymmetric top to have levels that lie *below* the corresponding levels of the *prolate* symmetric top obtained by setting I_C equal to I_B. Similarly, we expect the asymmetric top to have levels that lie *above* the corresponding levels of the *oblate* symmetric top obtained by setting I_A equal to I_B. The levels for these two limiting cases are given by

$$\text{prolate:} \quad E_{JK} = \frac{\hbar^2}{2I_B} J(J+1) + \frac{\hbar^2}{2}\left(\frac{1}{I_A} - \frac{1}{I_B}\right) K^2$$

$$\text{oblate:} \quad E_{JK} = \frac{\hbar^2}{2I_B} J(J+1) + \frac{\hbar^2}{2}\left(\frac{1}{I_C} - \frac{1}{I_B}\right) K^2$$

The limiting prolate and oblate levels appropriate for the asymmetric top molecule H_2O are plotted in Fig. 7.20. By gradually increasing I_C from the value I_B in the prolate system to its value in H_2O, then increasing I_A from its value in H_2O to the value I_B in the oblate system, we perturb the energy levels in the manner qualitatively indicated in the diagram. When the equality of two of the moments of inertia is removed, levels with the same J and $|K|$ ($\neq 0$) split into two levels, one with $K = |K|$ and one with $K = -|K|$, as shown in Fig. 7.20. Since the total angular momentum remains constant throughout the distortion process, levels with a given value of J in the prolate system must be connected to levels with the same value of J in the oblate system. There are thus always $2J + 1$ levels for each J in the asymmetric top. Although these levels do not correspond to any quantum number, they are conveniently numbered J_τ, where $\tau = -J$, $-J + 1, \ldots, 0, \ldots, J$ in order of increasing energy.

The asymmetric-top analysis provides an example of an approach that is often useful for gaining an understanding of a complicated physico-chemical system or phenomenon. One constructs simple limiting models that are known to bracket the properties of the system. A qualitative understanding of the real system is obtained by "interpolating" between the limiting model results.

Computer programs are now available for calculating the rotational spectra of asymmetric tops for assumed values of I_A, I_B, and I_C. The observed spectra are then fitted by varying I_A, I_B, and I_C until agreement is obtained. Corresponding programs are used also to determine the moments of inertia from other types of spectra (e.g., rotational fine structure in electronic spectra). In recent years, such computerization has facilitated the accurate determination of molecular structures from spectral data. Table 7.10 lists the results obtained for a few of the most common systems; some of these structures were obtained from microwave spectra (e.g., HCN), others from electronic spectra (e.g., CH_2), and others from ir and Raman spectra (e.g., C_2H_6). In many cases (e.g., HCOOH), the spectra of several isotopic species must be analyzed to obtain enough information to determine values for the individual bond lengths and bond angles.

Spherical-Top Molecules. Some molecules (e.g., CH_4) with a high degree of symmetry have all three moments of inertia equal (Problem 7.27). The rotational energy of these *spherical-top* molecules can be easily shown to be given by Eq. 7.102 (Problem 7.27). Such molecules cannot exhibit a pure rotational spectrum since they have no permanent electric dipole moment. If the values of the three moments of inertia of a dipolar molecule coincide accidentally, the molecule is called an *accidental spherical top*. The pure rotation frequencies of this type of molecule are determined from Eq. 7.103 and the selection rule

$$\Delta J = 0, \pm 1$$

7.9 VIBRATIONS OF POLYATOMIC MOLECULES

Having discussed the rotation of polyatomic molecules in the rigid body model, we turn to the vibrational motion of the component atoms. Whereas diatomic molecules have only a single vibrational coordinate, which corresponds to the motion of one atom relative to the other, polyatomic molecules

Table 7.10 Rotational constants and structures of some common polyatomic molecules[a]

Molecule	State[b]	A_0 (cm^{-1})[c]	B_0 (cm^{-1})[c]	C_0 (cm^{-1})[c]	Bond distances (Å)	Bond angles
CH_2	ground $^3\Sigma_g^-$	20.1	~7.9	7.1	~1.03	180°
	excited $^1A_1$		11.2		1.11	102.4°
NH_2	$^2B_1$	23.73	12.94	8.17	1.024	103.4°
H_2O	$^1A_1$	27.877	14.512	9.285	0.956	104.5°
H_2S	$^1A_1$	10.374	8.991	4.732	1.328	92.2°
HCN	$^1\Sigma^+$		1.47822		$\{$H—C:1.064, C—N:1.156$\}$	180°
N_2O	$^1\Sigma^+$		0.419011		$\{$N—N:1.128, N—O:1.184$\}$	180°
CO_2	$^1\Sigma_g^+$		0.39021		1.1621	180°
NO_2	$^2A_1$	8.0012	0.43364	0.41040	1.1934	134.1°
O_3	$^1A_1$	3.5535	0.44525	0.39479	1.278	116.8°
SO_2	$^1A_1$	2.02736	0.34417	0.293535	1.4308	119.3°
NH_3	$^1A_1$		9.4443	6.196	1.0124	106.67°
C_2H_2	$^1\Sigma_g^+$		1.17660		$\{$C—H:1.060, C—C:1.203$\}$	180°
H_2CO (planar)	$^1A_1$	9.4053	1.2954	1.1343	$\{$C—H:1.102, C—O:1.210$\}$	$\angle$HCH:121.1°
CH_4	$^1A_1$		5.2412		1.0940	109°27'

Table 7.10 (Continued)

Molecule	State[b]	A_0 (cm^{-1})[c]	B_0 (cm^{-1})[c]	C_0 (cm^{-1})[c]	Bond distances (Å)	Bond angles
HCOOH (planar)	$^1A'$	2.58548	0.402112	0.347447	C=O:1.202 C—O:1.343 C—H:1.097 O—H:0.972	∠OC=O:124.9° ∠HC=O:124.1° ∠COH:106.3°
C$_2$H$_4$ (planar)	^1A_g	4.828	1.0012	0.8282	C—H:1.086 C—C:1.339	∠HCH:117.6°
C$_6$H$_6$ (planar)	$^1A_{1g}$		0.1896		C—C:1.397 C—H:1.084	∠HCC:120° ∠CCC:120°

[a] From G. Herzberg, *Electronic Spectra of Polyatomic Molecules* (Van Nostrand, Princeton, N.J., 1967).

[b] Ground electronic state, unless otherwise specified.

[c] The constants A_0, B_0, and C_0 are defined in terms of the principal moments of inertia by $A_0 = h/(4\pi c I_A{}^0)$, $B_0 = h/(4\pi c I_B{}^0)$, and $C_0 = h/(4\pi c I_C{}^0)$, where the superscript and subscript (0) refer to the ground vibrational state.

always have more than one vibrational coordinate. A triatomic molecule, for example, requires three coordinates per atom or a total of nine coordinates to specify its position in space. Three of these coordinates can be taken as the coordinates of the center of mass of the molecule. If the molecule is nonlinear, we have seen (Section 7.8) that three angular coordinates are needed to specify its orientation. There remain three vibrational coordinates associated with the relative motion of the nuclei. If the triatomic molecule is linear, only two angles (e.g., the two polar angles) are used to specify the orientation of the molecular axis. This leaves four coordinates for expressing the vibrational motion of the molecule. Correspondingly, nonlinear molecules with N atoms have $3N - 6$ vibrational coordinates, while linear molecules have $3N - 5$. In this section we are concerned with the identification of the vibrational coordinates, with useful expressions for them, and with the quantization of the vibrational motion required to determine the allowed molecular energies.

7.9.1 *Normal-mode coordinates*   Although the internuclear distance is the natural vibration coordinate for a diatomic molecule (Section 7.5), it is not self-evident that the proper generalization is to use all of the internuclear distances in a polyatomic system. In fact, for a nonlinear molecule of five atoms, there are ten internuclear distances and only $3N - 6 = 9$ independent vibrational coordinates; thus, for this case, as for molecules with a greater number of atoms, some restriction on the choice of coordinates is required to obtain an independent set. Moreover, for many molecules a combination of certain internuclear distances and bond angles is a reasonable choice to try instead. We begin, therefore, by introducing a very general set of coordinates; they correspond to the Cartesian displacements of each of the atoms from its equilibrium position. In Fig. 7.21a is sketched a nonlinear triatomic molecule ABC with its bond lengths and bond angles at their equilibrium values; that is, each of the atoms is located relative to the others such that the total potential energy of the molecule is a minimum. If a coordinate system is fixed relative to the molecule (Fig. 7.21a); the Cartesian displacement coordinates Δx_A, Δy_A, Δz_A of atom A and similarly for B and C have the form shown in the figure. For the atoms at their equilibrium positions, the displacement coordinates are all equal to zero.

The $3N - 6$ vibrational coordinates for a molecule can always be written as a linear combination of the Cartesian atomic displacement coordinates. For a diatomic molecule located on the x axis for convenience (Fig. 7.21b), the vibrational coordinate $R - R_e$, which is the change in internuclear distance from the equilibrium value, is

$$R - R_e = \Delta x_A - \Delta x_B$$

with Δx_A and Δx_B related such that

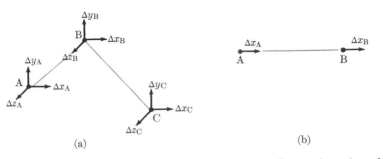

(a) (b)

Fig. 7.21. Cartesian displacement coordinates for **(a)** a nonlinear triatomic molecule ABC, and **(b)** a diatomic molecule.

$$m_A \, \Delta x_A + m_B \, \Delta x_B = 0$$

that is, the atomic displacements for a vibration are in such a ratio that the center of mass of the molecule remains fixed. The appropriate generalization of the diatomic result to specific polyatomic molecules is described below. However, we consider first the general behavior of the molecular vibrations of polyatomics.

Let $\xi_1, \xi_2, \ldots, \xi_s$ be the linear combinations of the atomic displacement coordinates that form vibrational coordinates of a polyatomic molecule. In terms of ξ_i, the vibrational kinetic energy T_{vib} of the molecule has the form

$$\begin{aligned}
2T_{\text{vib}} = {} & d_{11} \, \dot{\xi}_1^2 + d_{22} \, \dot{\xi}_2^2 + \cdots + d_{ss} \, \dot{\xi}_s^2 \\
& + 2(d_{12} \, \dot{\xi}_1 \, \dot{\xi}_2 + d_{13} \, \dot{\xi}_1 \, \dot{\xi}_3 + \cdots + d_{s-1,s} \, \dot{\xi}_{s-1} \, \dot{\xi}_s)
\end{aligned} \tag{7.109}$$

where $\dot{\xi}_i = d\xi_i/dt$ and the d_{ij} are constants with the dimension of mass (see Section 7.9.3); that is, the kinetic energy is composed of sums of products of the velocities $\dot{\xi}_i$ associated with the vibrational coordinates ξ_i. The vibrational potential energy V_{vib} can be a more general function of the vibrational coordinates which we write as

$$V_{\text{vib}} = V(\xi_1, \xi_2, \ldots, \xi_s)$$

If V is expanded in a Maclaurin series about the equilibrium positions of the atoms (i.e., about $\xi_1 = \xi_2 = \cdots = \xi_s = 0$), one obtains

$$\begin{aligned}
V_{\text{vib}} = {} & V_0 + c_1 \, \xi_1 + c_2 \, \xi_2 + \cdots + c_s \, \xi_s \\
& + \tfrac{1}{2}(c_{11} \, \xi_1^2 + c_{22} \, \xi_2^2 + \cdots + c_{ss} \, \xi_s^2) \\
& + c_{12} \, \xi_1 \, \xi_2 + c_{13} \, \xi_1 \, \xi_3 + \cdots + c_{s-1,s} \, \xi_{s-1} \, \xi_s \\
& + \tfrac{1}{6} c_{111} \, \xi_1^3 + \cdots
\end{aligned} \tag{7.110}$$

The coefficient c_i of ξ_i in Eq. 7.110 is the derivative of V_{vib} with respect to ξ_i, evaluated at the equilibrium position of the molecule; that is,

$$c_i = \left(\frac{\partial V_{\text{vib}}}{\partial \xi_i}\right)_0 = -(F_i)_0 = 0 \tag{7.111}$$

where $(F_i)_0$ is the force associated with the coordinate ξ_i at equilibrium. By the definition of the equilibrium geometry, the forces $(F_1)_0$, $(F_2)_0$, $\ldots$, $(F_s)_0$ are zero. If we take the zero of potential energy to be that of the equilibrium geometry, the constant term V_0 is zero. If, further, it is possible to assume that the displacements from equilibrium are so small that powers of ξ_i greater than the second can be ignored in the expression for V_{vib}, Eq. 7.110 becomes

$$\begin{aligned} 2V_{\text{vib}} = {} & c_{11}\,\xi_1{}^2 + c_{22}\,\xi_2{}^2 + \cdots + c_{ss}\,\xi_s{}^2 \\ & + 2(c_{12}\,\xi_1\,\xi_2 + c_{13}\,\xi_1\,\xi_3 + \cdots + c_{s-1,s}\,\xi_{s-1}\,\xi_s) \end{aligned} \tag{7.112}$$

In the analysis of the rotation of a rigid molecule (Section 7.8), we pointed out that it was possible to choose the coordinate axes such that off-diagonal terms in the kinetic energy vanished. By a corresponding procedure, we can select the vibrational coordinates so that the total vibrational energy has the simple form

$$E_{\text{vib}} = T_{\text{vib}} + V_{\text{vib}} = \frac{1}{2}\sum_{i=1}^{s}(d_{ii}\,\dot{\xi}_i{}^2 + c_{ii}\,\xi_i{}^2) \tag{7.113}$$

that is, the coefficients d_{ij} and c_{ij} vanish for $i \neq j$; the diagonal coefficients d_{ii} and c_{ii} are identified as the effective mass and force constant, respectively, associated with the coordinate ξ_i. Since we have assumed that ξ_i is a linear combination of Cartesian coordinate displacements, the momentum p_i is

$$p_i = d_{ii}\,\dot{\xi}_i$$

and E_{vib} can be written in the Hamiltonian form (i.e., in terms of coordinates and momenta)

$$E_{\text{vib}} = \frac{1}{2}\sum_{i=1}^{s}\left(\frac{1}{d_{ii}}\,p_i{}^2 + c_{ii}\,\xi_i{}^2\right) \tag{7.114}$$

The classical equations of motion are easily derived from Newton's second law, which relates the force and acceleration associated with each vibrational coordinate; that is,

$$d_{ii}\,\ddot{\xi}_i = F_i = -\frac{\partial V_{\text{vib}}}{\partial \xi_i} \tag{7.115}$$

From Eqs. 7.115 and 7.113 we obtain

$$\ddot{\xi}_i = -\frac{c_{ii}}{d_{ii}}\,\xi_i \tag{7.116}$$

Comparison of Eq. 7.116 with Eq. 2.111 shows that it has the same form as that of a harmonic oscillator; the solution to Eq. 7.116 is thus (see Section 2.10)

$$\xi_i = \xi_i^0 \cos (2\pi \nu_i t + \phi_i) \tag{7.117}$$

where ξ_i^0 is the maximum displacement (i.e., the amplitude), ν_i is the frequency, given by

$$\nu_i = \frac{1}{2\pi} \left(\frac{c_{ii}}{d_{ii}} \right)^{1/2} \tag{7.118}$$

and ϕ_i is the phase angle such that $\xi_i^0 \cos \phi_i$ is the displacement at $t = 0$. When the vibrations of a polyatomic molecule are described by a set of coordinates ξ_i that undergo the simple harmonic motion given by Eq. 7.117, the coordinates ξ_i are the so-called *normal-mode coordinates* for the molecule. In accord with the standard convention, we designate a set of normal-mode coordinates by the symbols $Q_1, Q_2, \ldots, Q_s$. The Q_i can be constructed as a linear combination of the more general displacements ξ_i by requiring that the kinetic energy and the potential energy satisfy Eq. 7.113; that is, (through second powers of the coordinates) they contain no cross terms such as $\dot{Q}_i \dot{Q}_j$ or $Q_i Q_j$ for $i \neq j$. Equivalently, the Q_i can be constructed by requiring that their equations of motion be of the form of Eq. 7.116, namely,

$$D_i \ddot{Q}_i + C_i Q_i = 0 \tag{7.119}$$

Before illustrating the construction of normal-mode coordinates for particular cases, we outline the quantum-mechanical description of the vibrations of polyatomic molecules.

7.9.2 Quantization of vibrational energy

The quantum-mechanical vibrational Hamiltonian H_{vib} can be obtained from Eq. 7.119 by substituting $-(\hbar/i)\partial/\partial Q_i$ for the momentum associated with the normal-mode coordinate Q_i (see Section 2.7). The result is

$$H_{\text{vib}} = \sum_{i=1}^{s} h_i \tag{7.120}$$

where

$$h_i = -\frac{\hbar^2}{2D_i} \frac{\partial^2}{\partial Q_i^2} + \frac{C_i}{2} Q_i^2 \tag{7.121}$$

The vibrational Schrödinger equation is

$$H_{\text{vib}} \psi_{\text{vib}}(Q_1, Q_2, \ldots, Q_s) = E_{\text{vib}} \psi_{\text{vib}}(Q_1, Q_2, \ldots, Q_s) \tag{7.122}$$

We saw in Chapter 2 that when the Hamiltonian operator has the form of Eq. 7.120, namely, a sum of operators h_i, each of which operates on a single coordinate Q_i, the solution to Eq. 7.122 is a simple product of wave functions, each of which is associated with one coordinate; that is,

$$\psi_{\text{vib}}(Q_1, Q_2, \ldots, Q_s) = \psi_1(Q_1)\, \psi_2(Q_2) \cdots \psi_s(Q_s) \qquad (7.123)$$

Substitution of Eq. 7.123 into Eq. 7.122 gives a set of s equations of the form (Problem 7.53)

$$h_i\, \psi_i(Q_i) = E_i\, \psi_i(Q_i) \qquad i = 1, \ldots, s \qquad (7.124)$$

Since Eq. 7.124 has the same form as Eq. 2.118, the normal-mode energies E_i are

$$E_i = E_{v_i} = (v_i + \tfrac{1}{2})h\nu_i \qquad (7.125)$$

where ν_i is the classical normal-mode frequency given by

$$\nu_i = \frac{1}{2\pi}\left(\frac{D_i}{C_i}\right)^{1/2} \qquad (7.126)$$

The total vibrational energy is the sum of the individual vibrational energies associated with the normal mode coordinates; that is,

$$E_{\text{vib}} = E_{v_1, v_2, \ldots, v_s} = \sum_{i=1}^{s} E_{v_i} = \sum_{i=1}^{s} \left(v_i + \tfrac{1}{2}\right) h\nu_i \qquad (7.127)$$

From Eq. 7.127, the zero-point energy of a polyatomic molecule is obtained by setting $v_i = 0$ for each of the s normal modes; that is,

$$E_0 = E_{0,0,\ldots,0} = \tfrac{1}{2}h\sum_{i=1}^{s}\nu_i \qquad (7.128)$$

The pure vibrational spectrum of a polyatomic molecule is given by the Bohr frequency rule

$$\nu = \frac{1}{h}\left(E_{v_1', v_2', \ldots, v_s'} - E_{v_1'', v_2'', \ldots, v_s''}\right) \qquad (7.129)$$

for the transition between the upper state v_1', v_2', $\ldots$, v_s' and the lower state v_1'', v_2'', $\ldots$, v_s''. In the harmonic-oscillator approximation, the vibrational selection rule for the ith normal mode is

$$\Delta v_i = \pm 1 \quad (i = 1, 2 \ldots, s) \qquad \Delta v_j = 0 \quad (j \neq i)$$

that is, the allowed transitions arise when the energy of a *single* normal mode changes by one quantum. As in the diatomic case, anharmonic terms in the potential introduce transitions involving more than one quantum for a given mode or the simultaneous change of several modes. Such transitions are frequently observed in polyatomic species, although their intensities are generally lower than those with $\Delta v_i = \pm 1$. Only the frequencies corresponding to normal-mode vibrations that change the electric dipole moment appear in ir absorption and emission spectra. Transitions involving other normal modes can be observed in the Raman spectrum of the molecule if they correspond to changes in the electric polarizability (see Section 7.10). For more details about vibrational spectra, the reader is referred to References 1 and 5 in the Additional Reading list.

We now illustrate how to determine normal-mode coordinates and the relationship of observed frequencies to the potential-energy parameters for triatomic molecules.

7.9.3 Nonlinear symmetric triatomic molecules

The molecule Y—X—Y with bond angle $\alpha(\neq 180°)$ has three vibrational coordinates. Since the molecule is planar, motion of its atoms out of the plane correspond to either translation or rotation rather than to vibration. Thus, a treatment of the vibrational motion can be restricted to the plane of the molecule and only the displacement coordinates $\Delta x_1, \Delta y_1, \Delta x_2, \Delta y_2, \Delta x_3, \Delta y_3$ for the three atoms have to be considered. Rather than attempting to use these displacement coordinates directly to determine the vibrational motion, it is convenient to introduce combinations of them that are appropriate to the molecular symmetry; that is, since the molecule YXY has a plane of symmetry (the yz plane), *symmetry coordinates* that are symmetric or antisymmetric with respect to reflection can be constructed. Two such vibrational displacements are shown in Figs. 7.22a and 7.22b. From these figures, it is evident that the two vibrational displacements are both unchanged by reflection in the yz plane, since upon reflection

$$\Delta x_1 \rightarrow -\Delta x_1 \qquad \Delta y_1 \rightarrow \Delta y_1$$
$$\Delta x_2 \rightarrow -\Delta x_3 \qquad \Delta y_2 \rightarrow \Delta y_2$$
$$\Delta x_3 \rightarrow -\Delta x_2 \qquad \Delta y_3 \rightarrow \Delta y_3$$

As indicated in the figures, these displacements are obtained by taking the linear combinations of the atomic Cartesian displacements that are symmetric with respect to reflection in the yz plane in the case of S_1 (Fig. 7.22a) and antisymmetric in the case of S_2 (Fig. 7.22b). In both cases, the position of the center of mass remains fixed, since motion of the center of mass corresponds to translation rather than to vibration. This implies that the values of the atomic displacements must be related by the equations

$$m_X \Delta x_1 + m_Y(\Delta x_2 + \Delta x_3) = 0$$
$$m_X \Delta y_1 + m_Y(\Delta y_2 + \Delta y_3) = 0$$

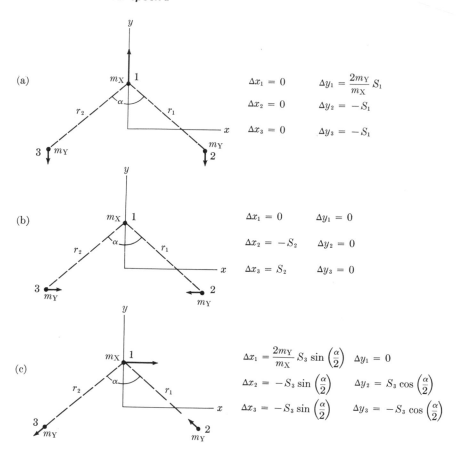

Fig. 7.22. Vibrational displacement coordinates for a nonlinear symmetric molecule XY₂.

In terms of the displacement coordinates S_1 and S_2 corresponding to the vibrations of Figs. 7.22a and 7.22b, respectively, we find that the Cartesian displacements for the first vibration are

$$\Delta x_1 = \Delta x_2 = \Delta x_3 = 0 \quad \Delta y_1 = \frac{2m_Y}{m_X} S_1 \quad \Delta y_2 = -S_1 \quad \Delta y_3 = -S_1 \quad (7.130a)$$

and for the second vibration,

$$\Delta x_1 = \Delta y_1 = \Delta y_2 = \Delta y_3 = 0 \quad \Delta x_2 = -S_2 \quad \Delta x_3 = S_2 \quad (7.130b)$$

Thus, the vibrational motion can be described by specifying the value of S_1 and S_2 as a function of time.

For the third vibrational displacement, we attempt to construct one that is antisymmetric with respect to reflection. It is clear that Δx_2 and Δx_3 for this displacement must have the same magnitude and sign, while Δy_2 and Δy_3 must have opposite signs; thus the third displacement has the form (Fig. 7.22c)

$$\Delta x_1 = \frac{2m_Y}{m_X} aS_3 \qquad \Delta y_1 = 0$$

$$\Delta x_2 = -aS_3 \qquad \Delta y_2 = bS_3 \qquad (7.131)$$

$$\Delta x_3 = -aS_3 \qquad \Delta y_3 = -bS_3$$

The factor $2m_Y/m_X$ in Δx_1 insures that the center of mass is stationary. We must now choose a and b so that there is also no rotation associated with this displacement. This can be done by requiring that the angular momentum about the z axis be zero for small displacements. From the definition of the z component of angular momentum, the result for the three-atom system is (see Problem 7.54)

$$m_X(x_1 \dot{y}_1 - y_1 \dot{x}_1) + m_Y(x_2 \dot{y}_2 - y_2 \dot{x}_2 + x_3 \dot{y}_3 - y_3 \dot{x}_3) = 0$$

For small displacements $x_1 \simeq x_{10}$, $\dot{y}_1 \simeq \Delta y_1/\Delta t$, etc., where x_{10}, y_{10}, . . . , y_{30} are the equilibrium values of the coordinates. Thus, we require that

$$m_X(x_{10} \Delta y_1 - y_{10} \Delta x_1) + m_Y(x_{20} \Delta y_2 - y_{20} \Delta x_2 + x_{30} \Delta y_3 - y_{30} \Delta x_3) = 0 \qquad (7.132)$$

Substitution of Eqs. 7.131 into Eq. 7.132 gives ($y_{20} = y_{30}$, $x_{20} = -x_{30}$)

$$\frac{b}{a} = \frac{y_{10} - y_{20}}{x_{20}} = \frac{\cos{(\alpha/2)}}{\sin{(\alpha/2)}}$$

where α is the equilibrium bond angle; this relation is most simply satisfied by choosing a and b to have the values

$$a = \sin\frac{\alpha}{2} \qquad b = \cos\frac{\alpha}{2}$$

The displacements of the Y atoms are thus seen to be collinear with the X– Y bonds, as shown in Fig. 7.22c.

Having introduced the three independent vibrational coordinates and expressed them in terms of the symmetry displacements S_1, S_2, and S_3, we can write an arbitrary vibration of the molecule as a linear combination of S_1, S_2, and S_3. The corresponding Cartesian coordinate displacements are

$$\Delta x_1 = \frac{2m_Y}{m_X} S_3 \sin\frac{\alpha}{2} \qquad \Delta y_1 = \frac{2m_Y}{m_X} S_1$$

$$\Delta x_2 = -S_2 - S_3 \sin\frac{\alpha}{2} \qquad \Delta y_2 = -S_1 + S_3 \cos\frac{\alpha}{2} \qquad (7.133)$$

$$\Delta x_3 = S_2 - S_3 \sin\frac{\alpha}{2} \qquad \Delta y_3 = -S_1 - S_3 \cos\frac{\alpha}{2}$$

The vibrational kinetic energy in Cartesian coordinates is

$$2T_{\text{vib}} = m_X(\dot{x}_1^2 + \dot{y}_1^2) + m_Y(\dot{x}_2^2 + \dot{y}_2^2 + \dot{x}_3^2 + \dot{y}_3^2) \qquad (7.134)$$

By making use of the fact that we can write $\dot{x}_1$ in the form

$$\dot{x}_1 = \frac{dx_1}{dt} = \frac{d\Delta x_1}{dt} = \frac{2m_Y}{m_X}\dot{S}_3 \sin\frac{\alpha}{2}$$

(since α, which is fixed at the equilibrium value, does not vary with time) and that corresponding equations exist for the other velocities, one can show (Problem 7.32) that Eq. 7.134 can be written as

$$2T_{\text{vib}} = 2m_Y(p\dot{S}_1^2 + \dot{S}_2^2 + r\dot{S}_3^2) \qquad (7.135)$$

where

$$p = 1 + \frac{2m_Y}{m_X}$$

$$r = 1 + \frac{2m_Y}{m_X}\sin^2\frac{\alpha}{2}$$

Thus, the kinetic energy contains no cross terms. This is a result of the orthogonality of the three displacements; that is, S_1 and S_2 are displacements in the y and x directions, respectively, and are symmetric with respect to reflection in the yz plane, while S_3 is antisymmetric with respect to that plane. Using Eq. 7.112 for the potential-energy function in terms of the coordinates S_1, S_2, S_3, we have

$$2V_{\text{vib}} = C_{11} S_1^2 + C_{22} S_2^2 + C_{33} S_3^2$$
$$+ 2(C_{12} S_1 S_2 + C_{23} S_2 S_3 + C_{13} S_1 S_3) \qquad (7.136)$$

Reflection in the yz plane replaces S_3 by $-S_3$, but leaves S_1 and S_2 unchanged. The potential-energy function itself must remain unchanged by a reflection that interchanges the two identical Y atoms, since the molecule itself is unaltered by such a transformation. From Eq. 7.136, we obtain on reflection in the yz plane

$$2V_{\text{vib}} = C_{11} S_1^2 + C_{22} S_2^2 + C_{33} S_3^2$$
$$+ 2(C_{12} S_1 S_2 - C_{23} S_2 S_3 - C_{13} S_1 S_3) \qquad (7.137)$$

Adding Eqs. 7.136 and 7.137 and dividing the sum by two, we find

$$2V_{\text{vib}} = C_{11} S_1^2 + 2C_{12} S_1 S_2 + C_{22} S_2^2 + C_{33} S_3^2 \qquad (7.138)$$

That is, the terms in the potential energy that change sign on reflection (i.e., those that involve products of coordinates with different symmetries) must have zero coefficients ($C_{23} = C_{13} = 0$).

Newton's equations of motion (Eq. 7.115) obtained from Eqs. 7.135 and
7.138 are

$$2m_Y p \ddot{S}_1 + C_{11} S_1 + C_{12} S_2 = 0$$
$$2m_Y \ddot{S}_2 + C_{12} S_1 + C_{22} S_2 = 0 \qquad (7.139)$$
$$2m_Y r \ddot{S}_3 + C_{33} S_3 = 0$$

Of these, only the equation for S_3 has the form of Eq. 7.119. Thus, S_3 is a
normal-mode coordinate, and henceforth we designate it as Q_3. Associated
with Q_3 is the vibrational frequency ν_3,

$$\nu_3 = \frac{1}{2\pi}\left(\frac{c_{33}}{2m_Y r}\right)^{1/2}$$

To find the other normal-mode coordinates we write

$$Q = g_1 S_1 + g_2 S_2 \qquad (7.140)$$

where the constants g_1 and g_2 are to be chosen so that Q obeys an equation
of the form of Eq. 7.119, namely,

$$\ddot{Q} + \lambda Q = 0 \qquad (7.141)$$

where we have written λ for $4\pi^2\nu^2$ (see Eq. 7.126). Substitution of Eqs.
7.140 and 7.139 into Eq. 7.141 and rearranging gives (Problem 7.33)

$$\left[g_1\left(\frac{C_{11}}{2m_Y p} - \lambda\right) + g_2\frac{C_{12}}{2m_Y}\right]S_1 + \left[g_1\frac{C_{12}}{2m_Y p} + g_2\left(\frac{C_{22}}{2m_Y} - \lambda\right)\right]S_2 = 0 \qquad (7.142)$$

Since the vibrational displacements S_1 and S_2 are independent, they can
have arbitrary values. A particular choice is $S_2 = 0$ and $S_1 \neq 0$, which gives
the equation

$$g_1\left(\frac{C_{11}}{2m_Y p} - \lambda\right) + g_2\frac{C_{12}}{2m_Y} = 0 \qquad (7.143a)$$

correspondingly, $S_1 \neq 0$, $S_2 = 0$ gives

$$g_1\frac{C_{12}}{2m_Y p} + g_2\left(\frac{C_{22}}{2m_Y} - \lambda\right) = 0 \qquad (7.143b)$$

Equations 7.143 are linear homogeneous equations for the coefficients g_1
and g_2 which can be simultaneously satisfied only if the (secular) determinant of the coefficients vanishes (see Section 5.6.4); that is, if

$$\begin{vmatrix} \dfrac{C_{11}}{2m_Y p} - \lambda & \dfrac{C_{12}}{2m_Y} \\ \dfrac{C_{12}}{2m_Y p} & \dfrac{C_{22}}{2m_Y} - \lambda \end{vmatrix} = 0 \qquad (7.144)$$

From Eq. 7.144 we obtain

$$\lambda^2 - \left(\frac{C_{11} + C_{22}p}{2m_Yp}\right)\lambda + \left(\frac{C_{11}C_{22} - C_{12}{}^2}{4m_Y{}^2p}\right) = 0 \qquad (7.145)$$

and from Eq. 7.143,

$$\frac{g_2}{g_1} = -\frac{C_{11} - 2m_Yp\lambda}{pC_{12}} \qquad (7.146)$$

Since Eq. 7.145 is quadratic, it has two roots, which we label λ_1 and λ_2. Substitution of each of these roots into Eq. 7.146 yields a ratio g_2/g_1 for the normal coordinate Q in Eq. 7.40; the normal coordinates associated with λ_1 and λ_2 are called Q_1 and Q_2, respectively. It can also be shown (Problem 7.33) that the two solutions to Eq. 7.145 obey the relations

$$\lambda_1 + \lambda_2 = 4\pi^2(\nu_1{}^2 + \nu_2{}^2) = \frac{C_{11} + C_{22}p}{2m_Yp}$$

$$\lambda_1\lambda_2 = 16\pi^4\nu_1{}^2\nu_2{}^2 = \frac{C_{11}C_{22} - C_{12}{}^2}{4m_Y{}^2p} \qquad (7.147)$$

It is evident from Eqs. 7.146 and 7.147 that the two frequencies ν_1 and ν_2 and the ratios of S_1 and S_2 in the normal-mode coordinates Q_1 and Q_2 depend upon the magnitudes of C_{11}, C_{22}, and C_{12}, which in turn are determined by the forces between atoms in a given molecule. A common practice is to assume that the potential-energy function can be expressed in terms of stretching of the bonds and changes in the bond angles. If Δr_1, Δr_2, and Δr_3 are the amounts by which the bonds X_1—Y_2, X_1—Y_3, and Y_2—Y_3 are lengthened and δ is the increase in the bond angle, then in the limit of small displacements it can be shown (Problem 7.34) that

$$\Delta r_1 = p\, S_1\cos\frac{\alpha}{2} - S_2\sin\frac{\alpha}{2} - rS_3$$

$$\Delta r_2 = p\, S_1\cos\frac{\alpha}{2} - S_2\sin\frac{\alpha}{2} + rS_3 \qquad (7.148)$$

$$\Delta r_3 = -2S_2$$

$$r_0\delta = -2p\, S_1\sin\frac{\alpha}{2} - 2S_2\cos\frac{\alpha}{2}$$

where r_0 is the equilibrium X—Y bond length and the constants p and r are defined after Eq. 7.135.

The simplest model for the potential-energy function is the *central-force field*. In this model V_{vib} is given by

$$2V_{\text{vib}} = k_1(\Delta r_1)^2 + k_2(\Delta r_2)^2 + k_3(\Delta r_3)^2 \qquad (7.149)$$

In the present case of a Y—X—Y molecule, $k_2 = k_1$. Thus there are two constants, k_1 and k_3, to be fitted to three normal-mode frequencies. An additional parameter can be added by modifying the central-force field to include interaction between the bonds, that is, by use of the potential-energy function

$$2V_{\text{vib}} = k_1[(\Delta r_1)^2 + (\Delta r_2)^2] + 2k_{12}(\Delta r_1)(\Delta r_2) + k_3(\Delta r_3)^2 \quad (7.150)$$

From Eqs. 7.150 and 7.148, it can be shown that (Problem 7.35) the coefficients in the potential function (Eq. 7.138) are

$$C_{11} = 2p^2(k_1 + k_{12}) \cos^2 \frac{\alpha}{2}$$

$$C_{22} = 2(k_1 + k_{12}) \sin^2 \frac{\alpha}{2} + 4k_3$$

$$\quad (7.151)$$

$$C_{33} = 2r^2(k_1 - k_{12})$$

$$C_{12} = -2p(k_1 + k_{12}) \sin \frac{\alpha}{2} \cos \frac{\alpha}{2}$$

For many molecules, even this modified central-force-field model is unsatisfactory, since imaginary values of k_1 and k_{12} are obtained when Eqs. 7.151 are substituted into Eq. 7.147 together with λ_1 and λ_2 determined from the observed frequencies (see Problem 7.36).

A generally more satisfactory three-parameter function is the *modified valence-force-field model,* namely,

$$2V_{\text{vib}} = k_1[(\Delta r_1)^2 + (\Delta r_2)^2] + 2k_{12}(\Delta r_1)(\Delta r_2) + k_\delta r_0^2 \delta^2 \quad (7.152)$$

In this case

$$C_{11} = 2p^2(k_1 + k_{12}) \cos^2 \frac{\alpha}{2} + 4k_\delta p^2 \sin^2 \frac{\alpha}{2}$$

$$C_{22} = 2(k_1 + k_{12}) \sin^2 \frac{\alpha}{2} + 4k_\delta \cos^2 \frac{\alpha}{2}$$

$$\quad (7.153)$$

$$C_{33} = 2r^2(k_1 - k_{12})$$

$$C_{12} = -2p(k_1 + k_2 - 2k_\delta) \cos \frac{\alpha}{2} \sin \frac{\alpha}{2}$$

The value of k_{12} is often found to be small. The "pure valence-force field" with k_{12} set equal to zero then gives a satisfactory two-parameter potential-energy function.

7.9.4   Equilateral symmetric triatomic molecule

In the special case for which $m_X = m_Y$ and the three equilibrium interatomic distances are the same, the equations of the previous section are greatly simplified, and the normal-mode coordinates Q_1 and Q_2 can be obtained without making use of the vibrational frequencies. To demonstrate this result, we use the simple central-force-field model for the potential energy (Eq. 7.149), although the same conclusions are reached from Eq. 7.150, provided interaction terms for all three pairs of bonds are included. For an equilateral molecule, $\alpha = 60°$ and $\sin(\alpha/2) = \frac{1}{2}$, $\cos(\alpha/2) = \sqrt{3}/2$. Furthermore, if all three masses are equal, $p = 2r = 3$. Since the three bonds are equivalent, $k_1 = k_2 = k_3$. Using these values and setting $k_{12} = 0$, we obtain from Eq. 7.151

$$C_{11} = \frac{27}{2} k_1$$

$$C_{22} = C_{33} = \frac{9}{2} k_1 \qquad (7.154)$$

$$C_{12} = -\frac{3\sqrt{3}}{2} k_1$$

Substitution of Eqs. 7.154 into Eq. 7.145 gives the solutions (Problem 7.39)

$$\lambda_1 = \frac{3k_1}{m_X}$$

$$\lambda_2 = \frac{3k_1}{2m_X}$$

From the expression for ν_3 given for the Y—X—Y molecule, which applies to X_3 with use of the appropriate values of C_{33} and r, we have

$$\lambda_3 = \frac{3k_1}{2m_X}$$

Two of the vibrational frequencies (λ_2 and λ_3) are thus degenerate for an equilateral X_3 molecule. Substitution of the expressions for λ_1 and λ_2 into Eq. 7.146 gives the ratio $g_2/g_1 = -1/\sqrt{3}$ and $1/\sqrt{3}$, respectively. The normal-mode coordinates are therefore

$$Q_1 = S_1 - \frac{1}{\sqrt{3}} S_2$$

$$Q_2 = S_1 + \frac{1}{\sqrt{3}} S_2 \qquad (7.155)$$

$$Q_3 = S_3$$

When Eqs. 7.155 are solved for S_1, S_2, and S_3 in terms of Q_1, Q_2, and Q_3, and the results substituted into Eq. 7.133 with $\alpha = 60°$, we obtain (Problem 7.39)

$$
\begin{aligned}
\Delta x_1 &= Q_3 & \Delta y_1 &= Q_1 + Q_2 \\
\Delta x_2 &= \frac{\sqrt{3}}{2}(Q_1 - Q_2) - \frac{1}{2} Q_3 & \Delta y_2 &= -\frac{1}{2}(Q_1 + Q_2) + \frac{\sqrt{3}}{2} Q_3 \quad (7.156) \\
\Delta x_3 &= -\frac{\sqrt{3}}{2}(Q_1 - Q_2) - \frac{1}{2} Q_3 & \Delta y_3 &= -\frac{1}{2}(Q_1 + Q_2) - \frac{\sqrt{3}}{2} Q_3
\end{aligned}
$$

The normal-mode displacements can be obtained from Eq. 7.156 by successively setting all but one of the Q_i equal to zero. The resulting expressions are shown in Fig. 7.23. An example of a molecule predicted to have this geometry is the ground electronic state of the trihydrogen ion H_3^+. Since the displacement Q_1 does not change the electric dipole moment of the ion from its equilibrium value of zero, only the frequency ν_2 is observable in

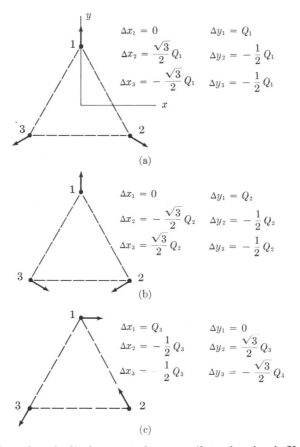

(a)

$$\Delta x_1 = 0 \qquad \Delta y_1 = Q_1$$
$$\Delta x_2 = \frac{\sqrt{3}}{2} Q_1 \qquad \Delta y_2 = -\frac{1}{2} Q_1$$
$$\Delta x_3 = -\frac{\sqrt{3}}{2} Q_1 \qquad \Delta y_3 = -\frac{1}{2} Q_1$$

(b)

$$\Delta x_1 = 0 \qquad \Delta y_1 = Q_2$$
$$\Delta x_2 = -\frac{\sqrt{3}}{2} Q_2 \qquad \Delta y_2 = -\frac{1}{2} Q_2$$
$$\Delta x_3 = \frac{\sqrt{3}}{2} Q_2 \qquad \Delta y_3 = -\frac{1}{2} Q_2$$

(c)

$$\Delta x_1 = Q_3 \qquad \Delta y_1 = 0$$
$$\Delta x_2 = -\frac{1}{2} Q_3 \qquad \Delta y_2 = \frac{\sqrt{3}}{2} Q_3$$
$$\Delta x_3 = -\frac{1}{2} Q_3 \qquad \Delta y_3 = -\frac{\sqrt{3}}{2} Q_3$$

Fig. 7.23. Normal-mode displacements for an equilateral molecule X_3.

an ir spectrum, although indirect information about both ν_1 and ν_2 can be obtained [F. Petty and T. F. Moran, Chem. Phys. Letters **5**, 64 (1970)].

7.9.5   Linear symmetric triatomic molecules   Another special case in which the normal modes can be determined easily is that of the linear symmetric triatomic molecule ($\alpha = 180°$). Examples are CO_2 and SO_2, which have been extensively studied. Setting $\sin (\alpha/2) = 1$ and $\cos (\alpha/2) = 0$ in Eqs. 7.151 for the modified central-force field, we obtain

$$C_{11} = 0$$
$$C_{22} = 2(k_1 + k_{12} + 2k_3)$$
$$C_{33} = 2r^2(k_1 - k_{12})$$
$$C_{12} = 0$$

Thus, the restoring force for the S_1 displacement that corresponds to bending of the molecule is zero under the central-force assumption. Since this is unsatisfactory, we use the modified valence-force field (Eqs. 7.153), which gives

$$C_{11} = 4k_\delta p^2$$
$$C_{22} = 2(k_1 + k_{12})$$
$$C_{33} = 2r^2(k_1 - k_{12}) \qquad p - r = 1 + \frac{2m_Y}{m_X} \qquad (7.157)$$
$$C_{12} = 0$$

Since C_{12} vanishes for the linear symmetric triatomic molecule, the symmetry coordinates S_1 and S_2, as well as S_3, are normal-mode coordinates (see Eq. 7.139). Labeling these Q_2, Q_1, and Q_3, respectively, where we have interchanged subscripts 1 and 2 to conform to standard conventions, we find that the Cartesian displacement coordinates can be written (Eq. 7.133)

$$\Delta x_1 = \frac{2m_Y}{m_X} Q_3 \qquad \Delta y_1 = \frac{2m_Y}{m_X} Q_2$$
$$\Delta x_2 = -Q_1 - Q_3 \qquad \Delta y_2 = -Q_2$$
$$\Delta x_3 = Q_1 - Q_3 \qquad \Delta y_3 = -Q_2$$

The corresponding frequencies are given by (Problem 7.41)

$$\lambda_1 = \frac{1}{m_Y} (k_1 + k_{12})$$

$$\lambda_2 = \frac{2p}{m_Y} k_\delta$$

$$\lambda_3 = \frac{p}{m_Y} (k_1 - k_{12})$$

The normal-mode displacements for linear Y—X—Y are shown in Fig. 7.24.

In the introduction to this section, we remarked that a linear triatomic molecule has four vibrational coordinates, although here we have considered only three. The fourth coordinate is identical with Q_2, but corresponds to bending at right-angles to Q_2, that is, in the z direction. We can designate the four normal-mode coordinates Q_1, Q_{2y}, Q_{2z}, and Q_3. No additional vibrational frequency is obtained by adding Q_{2z}, since the frequencies ν_{2z} and ν_{2y} are obviously equal. However, the two degenerate coordinates must be

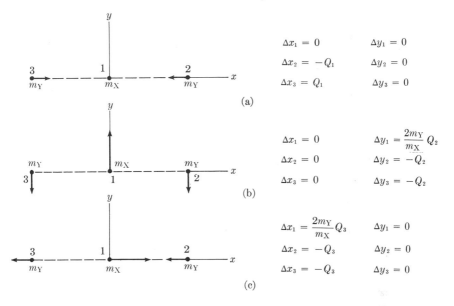

$$\Delta x_1 = 0 \qquad \Delta y_1 = 0$$
$$\Delta x_2 = -Q_1 \qquad \Delta y_2 = 0$$
$$\Delta x_3 = Q_1 \qquad \Delta y_3 = 0$$

(a)

$$\Delta x_1 = 0 \qquad \Delta y_1 = \frac{2m_Y}{m_X} Q_2$$
$$\Delta x_2 = 0 \qquad \Delta y_2 = -Q_2$$
$$\Delta x_3 = 0 \qquad \Delta y_3 = -Q_2$$

(b)

$$\Delta x_1 = \frac{2m_Y}{m_X} Q_3 \qquad \Delta y_1 = 0$$
$$\Delta x_2 = -Q_3 \qquad \Delta y_2 = 0$$
$$\Delta x_3 = -Q_3 \qquad \Delta y_3 = 0$$

(c)

Fig. 7.24. Normal-mode displacements for a linear symmetric molecule XY_2. Not shown is the bending vibration identical with Q_2 but in the zx plane.

taken into account when the number of vibrational degrees of freedom are counted in partitioning the total internal energy among vibrational and rotational modes, for example.

Vibrations of asymmetric triatomic molecules (see Problem 7.42) and molecules with more than three atoms can be treated by procedures corresponding to those that we have outlined, but in the latter case one must in general find the roots of larger secular determinants than the 2×2 determinant in Eq. 7.144. Often the determinants can be factored by the use of symmetry principles. For further details, see Reference 5 in the Additional Reading list.

7.10 RAMAN SPECTRA

In 1928, while investigating the scattering of light by liquids, C. V. Raman made a discovery that has proved to be of great importance for the investigation of molecular structure. He noted that when a substance is irradiated with light of a certain frequency, the scattered beam contains not only light of the same frequency as the incident beam (the Rayleigh line), but also lines with slightly lower frequencies (Stokes lines) and slightly higher frequencies (anti-Stokes lines). Raman made a thorough study of these shifted spectral lines and in 1930 was awarded the Nobel prize for his work.

A simple quantum-mechanical picture of Raman scattering can be given in terms of the visible-light analog of the Compton effect (Section 2.3). In the Compton effect, the frequencies of the high-energy photons are shifted as a result of the transfer of energy and momentum to the ejected electron; that is, the collision can be considered essentially elastic because the photon energy is so high that the electron behaves as if it were free. Corresponding elastic collisions occur between visible photons and molecules. However, here the frequency of the scattered light is effectively unchanged because the molecular mass is so large (see Eq. 2.27). This type of scattering, called Rayleigh scattering, leads to an unshifted line and is the dominant process. An inelastic scattering process can also occur in which the frequency of the photon is shifted because it alters the motion of the molecular nuclei. Since the nuclear motion is quantized, only certain excitations and de-excitations are allowed, namely, those corresponding to the rotational-vibrational energy levels. If ν is the frequency of the incident light, ν' is the frequency of the scattered light, E_n is the energy of the initial state of the molecule, and E_m the energy of the final state, we have

$$h\nu' = h\nu + (E_n - E_m)$$

Thus if $E_m > E_n$ (excitation),

$$\nu' = \nu - \nu_{nm} \qquad \text{(Stokes line)}$$

where ν_{nm} is the Bohr frequency for the excitation. If, on the other hand, $E_m < E_n$ (de-excitation)

$$\nu' = \nu + \nu_{nm} \qquad \text{(anti-Stokes line)}$$

The intensities of the lines are proportional to the number of molecules in the initial state. Among the Stokes lines are those for which n is the ground state; for the anti-Stokes lines, n must be an excited state. Since at ordinary temperatures the populations of vibrational-rotational excited states is relatively small, the anti-Stokes lines are fainter than the Stokes lines.

To discuss further the nature of the interaction between the scattered photon and the molecule, we use a classical approach. The complete quantum treatment, which is rather complicated, makes use of second-order perturbation theory to describe the absorption of a photon of frequency ν and the simultaneous emission of a photon of frequency ν' (see Reference 5 of the Additional Reading list). We consider a molecule irradiated with light that contains a component with frequency equal to one of the normal-mode frequencies ν_i of the molecule. The electric field $\mathcal{E}_i$ associated with that component oscillates in time with frequency ν_i,

$$\mathcal{E}_i = \mathcal{E}_i^0 \cos (2\pi \nu_i t)$$

(see Eq. 2.3). Suppose that the electric dipole moment μ of the molecule depends on the normal-mode displacement Q_i. If we expand μ as a power series in Q_i and keep only the linear term, we have

$$\mu = \mu_0 + \mu_i' Q_i \qquad (7.158)$$

where μ_i' is the slope of the curve $\mu(Q_i)$ at $Q_i = 0$; that is,

$$\mu_i' = \left(\frac{\partial \mu}{\partial Q_i}\right)_{Q_i = 0}$$

According to the classical picture of the interaction of radiation with a molecule, the normal-mode vibration is driven by the oscillating field at the frequency ν_i and the intensity of the ν_i component of the radiation is decreased by an amount equivalent to the energy required to drive the Q_i vibration. Since the vibration frequencies of most molecules are in the ir region, observation of the intensity of the ir radiation transmitted through a sample as a function of frequency would yield the ir absorption spectrum of the substance. In order that a normal-mode frequency ν_i appear in the ir spectrum, it is clear from comparisons of Eq. 7.40 and Eq. 7.158 that μ_i' must be nonzero. Such a normal mode is said to be *ir active* (see Section 7.9).

An electric field interacts not only with the permanent electric dipole moment of a molecule, but also can distort the electron distribution so as to induce a dipole moment. This effect is analogous to the polarization of atoms, discussed already in Section 5.3. In the case of spherically symmetric atoms, the induced dipole moment lies in the same direction as the electric field and the constant of proportionality is the polarizability α; that is, the relation between the dipole moment vector $\mathbf{\mu}$ and the electric field vector $\mathbf{\mathcal{E}}$ is

$$\mathbf{\mu} = \alpha \mathbf{\mathcal{E}} \qquad (7.159)$$

For a molecule that does not have spherical symmetry, each component (μ_x, μ_y, μ_z) of the dipole moment $\mathbf{\mu}$ can depend upon each component ($\mathcal{E}_x$, $\mathcal{E}_y$, $\mathcal{E}_z$) of the electric field $\mathbf{\mathcal{E}}$; thus Eq. 7.159 must be replaced by the more general expression

$$\mu_x = \alpha_{xx}\mathcal{E}_x + \alpha_{xy}\mathcal{E}_y + \alpha_{xz}\mathcal{E}_z$$
$$\mu_y = \alpha_{yx}\mathcal{E}_x + \alpha_{yy}\mathcal{E}_y + \alpha_{yz}\mathcal{E}_z$$
$$\mu_z = \alpha_{zx}\mathcal{E}_x + \alpha_{zy}\mathcal{E}_y + \alpha_{zz}\mathcal{E}_z$$

The polarizability is seen to have nine components, whose values depend on the electronic structure of the molecule. Only six of them are independent, however, since it can be shown that $\alpha_{xy} = \alpha_{yx}$, $\alpha_{xz} = \alpha_{zx}$, and $\alpha_{yz} = \alpha_{zy}$. Because the various polarizability components behave like their subscripts under inversion (e.g., α_{xy} behaves like xy), they are all symmetric; that is, they do not change sign when the coordinates are transformed by replacing x, y, z with $-x$, $-y$, $-z$. Moreover, by a suitable choice of the orientation of the axes with respect to the molecule, the components α_{xy}, α_{xz}, and α_{yz} can be made to vanish. For a molecule in which the principal axes of inertia are determined by symmetry (Section 7.8), the same choice of axes leads to a diagonal polarizability (i.e., only α_{xx}, α_{yy}, and α_{zz} are non-zero). In the following discussion, we are not concerned with the directional properties of μ and $\mathcal{E}$; we therefore can simplify the notation and use the symbol α without subscripts to indicate the appropriate component of the polarizability, writing

$$\mu = \alpha \mathcal{E} \tag{7.160}$$

Suppose that a molecule is irradiated with light of frequency ν and that α depends upon one of the normal-mode displacements Q_i with frequency ν_i; then we have

$$\alpha = \alpha_0 + \alpha_i' Q_i \tag{7.161a}$$

$$Q_i = Q_{i0} \cos(2\pi\nu_i t) \tag{7.161b}$$

$$\mathcal{E} = \mathcal{E}^0 \cos(2\pi\nu t) \tag{7.161c}$$

where $\alpha_i' = (\partial\alpha/\partial Q_i)_{Q_i=0}$. According to Eqs. 7.160 and 7.161, the induced dipole moment is given by

$$\mu = (\alpha + \alpha_i' Q_i)\,\mathcal{E} = \{\alpha_0 + \alpha_i'[Q_{i0} \cos\,(2\pi\nu_i t)]\}\mathcal{E}^0 \cos\,(2\pi\nu t)$$
$$= \alpha_0\,\mathcal{E}^0 \cos(2\pi\nu t) + \alpha_i' Q_{i0}\,\mathcal{E}^0 \cos(2\pi\nu t)\cos(2\pi\nu_i t) \tag{7.162}$$

Applying the trigonometric identity

$$\cos\,(\alpha \pm \beta) = \cos\alpha\,\cos\beta \mp \sin\alpha\,\sin\beta$$

to Eq. 7.162, we obtain

$$\mu = \alpha_0\,\mathcal{E}^0 \cos(2\pi\nu t) + \tfrac{1}{2}\alpha_i' Q_{i0}\,\mathcal{E}^0 \cos[2\pi(\nu - \nu_i)t]$$
$$+ \tfrac{1}{2}\alpha_i' Q_{i0}\,\mathcal{E}^0 \cos[2\pi(\nu + \nu_i)t] \tag{7.163}$$

In the classical description, the scattered light has the three frequencies at which the induced dipole moment oscillates, namely, ν, $\nu - \nu_i$, and $\nu + \nu_i$ (see Fig. 7.25). These are the Rayleigh, Stokes, and anti-Stokes lines, re-

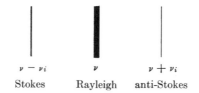

| | $\nu - \nu_i$ | ν | $\nu + \nu_i$ |
| Stokes | Rayleigh | anti-Stokes |

Fig. 7.25. The Rayleigh and Raman lines in the radiation scattered from a substance with a single Raman-active vibration of frequency ν_i.

spectively. Raman lines are observed with frequency shifts corresponding to the frequencies of each normal mode Q_i for which α_i' is nonzero. These normal modes are said to be *Raman active*. Since Rayleigh scattering occurs even when $\alpha_i' = 0$, it is universally observed. In most cases, the Rayleigh line is 10^2 or 10^3 times as intense as the Stokes and anti-Stokes lines.

It is important to note the difference between an absorption (or emission process), in which the radiation frequency must equal a resonant frequency of the system, and a scattering process, in which there is no required relationship between the two frequencies. In the latter, as discussed above, it is the change in frequency rather than the frequency itself which corresponds to a resonant frequency. If one uses a light source with a range of frequencies that includes the resonant frequency, both absorption and scattering play significant roles. One can avoid the complications that arise due to resonant absorption by choosing the incident frequency ν in the visible region, since the normal-mode frequencies lie in the ir region. Also, more intense scattering is achieved if visible rather than ir radiation is used, since the scattered intensity is proportional to ν^4.

To see how to determine which vibrations are Raman active, consider the motion of the linear symmetric molecule XY_2 (Fig. 7.22). Since the polarizability α depends on the molecular volume (Section 5.3), α decreases during one phase of the normal-mode vibration Q_1 (the Y atoms moving towards the X atom) and increases during the other phase (the Y atoms moving away from the X atom). A plot of α versus Q_1 therefore looks like that in Fig. 7.26a, and α_1' has a nonzero (negative) value; that is, since Q_1 is defined as positive for the Y atoms moving toward X, $(\partial\alpha/\partial Q_1)_{Q_1=0}$ is negative. Thus, the normal mode Q_1 is Raman active. The two other normal modes Q_2 and Q_3 are also expected to alter the polarizability. However, in contrast to Q_1, the displacements are identical in each vibrational phase for these vibrations, since both positive and negative values of Q_2 or Q_3 correspond to bending away from the linear equilibrium geometry. This means that α must either decrease during both phases (in which case α is a maximum in the equilibrium conformation) or increase during both phases (in which case α is a minimum at the equilibrium geometry). Plots of α versus Q_i for $i = 2, 3$ could be similar to Figs. 7.26b or to 7.26c. Either a maximum or a minimum requires that the first derivatives α_2' and α_3' be zero. Thus Q_2 and Q_3 are Raman inactive.

To compare the Raman result with the ir spectrum of linear symmetric XY_2, we must consider the behavior of the field-independent dipole moment μ (Eq. 7.158). Although μ is zero by symmetry for XY_2 in the equilibrium geometry, vibrations such as Q_2 and Q_3 produce a nonzero moment due to the distortion of the electron distribution by the asymmetric nuclear positions. Thus, the normal-mode displacements Q_2 and Q_3 lead to nonzero values of μ_2' and μ_3', respectively, and are ir active. Since the displacement

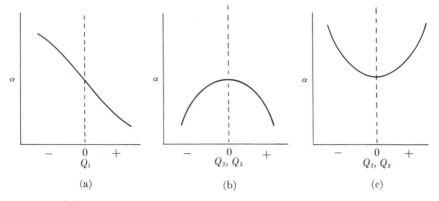

α

α

α

$-$	0	$+$
	Q_1	

(a)

$-$	0	$+$
	Q_2, Q_3	

(b)

$-$	0	$+$
	Q_2, Q_3	

(c)

Fig. 7.26. Schematic drawing of the dependence of the polarizability α on the normal-model displacements of a linear symmetric molecule XY_2.

Q_1 does not destroy the symmetry of the molecule, the dipole moment remains zero for both phases of Q_1, so μ_1' is zero and Q_1 is ir inactive.

These results can be generalized to any molecule with a center of symmetry. For such a molecule, all normal-mode vibrations can be classified as g (signs of the displacements remain the same upon inversion) or u (signs

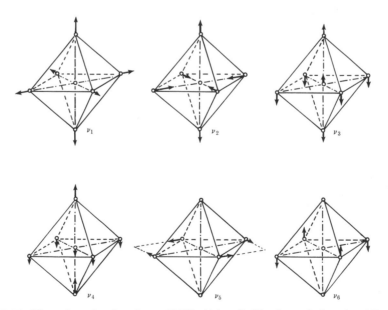

Fig. 7.27. Normal-mode vibrations of MX_6. [After G. Herzberg, *Infrared and Raman Spectra* (Van Nostrand, Princeton, N.J., 1945), Fig. 51, p. 122.]

of the displacement change upon inversion). All the ir-active modes are u, and all the Raman-active modes are g. This requires that both techniques be used if all the active normal-mode frequencies of a symmetrical molecule are to be observed. For a molecule without a center of symmetry, some modes can be both ir and Raman active. If a molecule has a large number of planes and axes of symmetry, some of its normal-mode vibrations may be both ir and Raman inactive; for example, icosahedral $B_{12}H_{12}^{2-}$ has nine ir-active modes, 22 Raman-active modes, and 35 inactive modes. Normal modes for a molecule of octahedral symmetry are shown in Fig. 7.27.

Since the change in polarizability and the field-independent dipole moment are both expressed as linear functions of the normal coordinate Q_i (Eqs. 7.158 and 7.161a), the vibrational Raman selection rules for Raman-active vibrations are the same as those obtained already for ir-active vibrations; that is,

$$\Delta v_i = \pm 1 \qquad \Delta v_j = 0 \quad (j \neq i)$$

in the harmonic approximation (see Section 7.9).

Pure rotational Raman lines are also observed in addition to the vibrational lines that we have just discussed. Although only molecules with

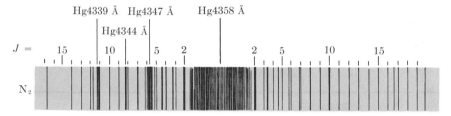

Fig. 7.28. Rotational Raman spectrum of N_2. [Redrawn with permission of the National Research Council of Canada and B. P. Stoicheff, *Can. J. Phys.*, **32,** 630 (1954), Fig. 1.]

permanent electric dipole moments can have pure rotational microwave spectra, molecules with polarizabilities that are orientation-dependent have pure rotational Raman spectra. These include all linear molecules, and in particular nonpolar molecules such as C_2H_2 and homonuclear molecules such as H_2 and N_2. The pure rotational Raman spectrum of N_2 is shown in Fig. 7.28. The Raman selection rule for a linear molecule is[7]

---

[7] More precisely, this selection rule is for a linear molecule with no angular momentum about the molecular axis due to the degenerate bending vibrations. For details, see Reference 5 in the Additional Reading list.

$$\Delta J = 0, \pm 2$$

An example of the use of Raman data to obtain the rotational constants of a homonuclear molecule is given in Problem C7.4.

Lasers provide exceptionally intense and monochromatic Raman sources. Their use has considerably extended the range of molecules for which Raman spectra can be obtained. Moreover, with the improvement in intensity, it is now easier to study the Raman spectra of gases, so that the frequencies obtained are unperturbed by the additional intermolecular forces that occur in liquids. When observed at very high intensity, the laser-induced Raman spectra consist of only the most active vibrational line (usually that of the totally symmetric vibration) and its overtones. This phenomenon is useful for producing intense coherent radiation with frequencies shifted from that of the parent laser beam. The shifted radiation, particularly at higher frequencies, can be employed as an energy source for chemical reactions and for other applications.

7.11 ELECTRONIC SPECTRA OF POLYATOMIC MOLECULES

The energy of a given electronic state of a diatomic molecule varies with the internuclear distance R and can be represented relatively simply by a plot such as is given for several states of O_2 in Fig. 7.13. For polyatomic systems, the electronic energy also depends on the geometry of the molecule. It is thus a function of the $3N - 6$ (for nonlinear molecules) or $3N - 5$ (for linear molecules) internal variables needed to specify all of the internuclear distances. Experimental or theoretical determination of the electronic energy and its dependence on the internal variables becomes increasingly difficult as the number of atoms increases. Moreover, the electronic spectra of polyatomic molecules, and particularly their vibrational and rotational fine structure, are generally more complex than those of diatomic molecules. Nevertheless, much of our knowledge of the ground state and the excited states of polyatomic molecules is based on the analysis of their electronic spectra. To provide a brief introduction to this very important subject, we first describe the main features of the electronic spectra of the two relatively small molecules, formaldehyde and ethylene, which are isoelectronic with O_2 (16 electrons). We then consider Rydberg transitions and the spectra of conjugated systems. Finally, we discuss briefly luminescence of molecules and certain of the new techniques for the study of electronic energy levels.

7.11.1 Ethylene and formaldehyde   The localized MO description of the ground state of ethylene was given in Section 6.3.4; the configuration is (see Fig. 7.29a for atom labels)

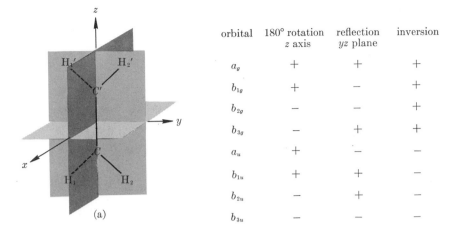

orbital	180° rotation z axis	reflection yz plane	inversion
a_g	+	+	+
b_{1g}	+	−	+
b_{2g}	−	−	+
b_{3g}	−	+	+
a_u	+	−	−
b_{1u}	+	+	−
b_{2u}	−	+	−
b_{3u}	−	−	−

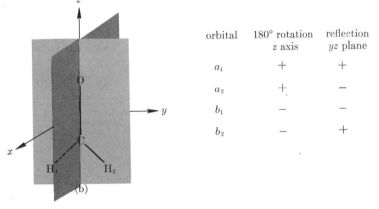

orbital	180° rotation z axis	reflection yz plane
a_1	+	+
a_2	+	−
b_1	−	−
b_2	−	+

Fig. 7.29. The symmetry axes, planes, and orbital designations of **(a)** C_2H_4 and **(b)** H_2CO.

$$(1s_C)^2(1s_{C'})^2(\sigma_{CH_1})^2(\sigma_{CH_2})^2(\sigma_{C'H_1'})^2(\sigma_{C'H_2'})^2(\sigma_{CC'})^2(\pi_{CC'})^2$$

where $1s_C$ and $1s_{C'}$ are the inner-shell carbon orbitals, which mix only slightly on molecule formation, σ_{CH_i} and $\sigma_{C'H_i'}$ $(i = 1, 2)$ are the carbon-hydrogen bonding orbitals composed of a carbon sp^2 hybrid and a hydrogen $1s$ orbital, $\sigma_{CC'}$ is the carbon σ bond composed of carbon sp^2 hybrids, and $\pi_{CC'}$ is the carbon π-bonding orbital, which is composed of carbon $2p_x$ orbitals in the coordinate system used in Fig. 7.29a. For formaldehyde, a corresponding localized MO description can be written. The ground configuration has the form (see Fig. 7.29b for atom labels)

$$(1s_O)^2(1s_C)^2(2s_O)^2(\sigma_{CH_1})^2(\sigma_{CH_2})^2(\sigma_{CO})^2(\pi_{CO})^2(n_O)^2$$

where $1s_O$ and $1s_C$ are the inner-shell orbitals, $2s_O$ is an orbital that is primarily $2s$ on oxygen though some mixing with other orbitals is expected (compare the 3σ orbital of CO in Section 6.2.2), σ_{CH_1} and σ_{CH_2} are carbon-hydrogen bonding orbitals corresponding to those in ethylene, σ_{CO} is the carbon-oxygen σ-bonding orbital composed of a carbon sp^2 hybrid and an oxygen orbital that is primarily $2p_z$, π_{CO} is the carbon-oxygen π-bonding orbital composed of $2p_x$ on the two atoms, and n_O is essentially a nonbonding $2p_y$ orbital on oxygen.

Since there is an antibonding orbital associated with each bonding orbital of C_2H_4 and H_2CO, there are a large number of valence-orbital excited configurations that can be constructed for these two systems. Many of these are observed in their electronic spectra. However, before considering the spectroscopic results, it is useful to introduce more general (delocalized) MO's and to discuss the correlation among the energy levels of O_2, C_2H_4, and H_2CO. For this purpose, we need to determine the symmetry properties of the molecules.

Ethylene (C_2H_4) has three two-fold symmetry axes (x, y, and z), three symmetry planes (xy, yz, and xz), and a center of symmetry (inversion). The symmetry axes and planes and the designations of the MO's for C_2H_4 are given in Fig. 7.29a. The $+$ and $-$ indicate symmetry and antisymmetry, respectively, of each orbital under the indicated rotation, reflection, or inversion.

Formaldehyde (H_2CO) is planar in the ground state (see Fig. 7.29b) and has the same two-fold symmetry axis (z axis) and symmetry planes (xz and yz planes) as does a bent symmetrical Y—X—Y molecule (Fig. 6.34). The orbital designations for H_2CO are given in the inset of Fig. 7.29b. The $+$ and $-$ indicate symmetry and antisymmetry, respectively, of each orbital under the indicated rotation or reflection.

To see how the orbitals of H_2CO and C_2H_4 correlate with one another and with those of O_2, we imagine the following process. Starting with O_2, we remove two protons from the nucleus of one of the O atoms and place them at the equilibrium positions for the H atoms in H_2CO. The orbitals of O_2 are thereby transformed into those of H_2CO. The σ_g and σ_u orbitals in O_2 are both symmetric with respect to $180°$ rotation about the z axis and with respect to reflection in the yz plane. These orbitals therefore become a_1 orbitals in H_2CO. The π_u and π_g orbitals of O_2 are each doubly degenerate. One member of each degenerate pair (π_{ux} and π_{gx}) has a node in the yz plane and therefore becomes a b_2 orbital in H_2CO. The other π orbitals (π_{uy} and π_{gy}) have nodes in the xz plane and are therefore antisymmetric with respect to $180°$ rotation about the z axis but symmetric with respect to reflection in the yz plane; these orbitals become b_1 orbitals in H_2CO.

Now imagine that two protons are removed from the nucleus of the O atom of H_2CO to the equilibrium positions of the H atoms in C_2H_4. The

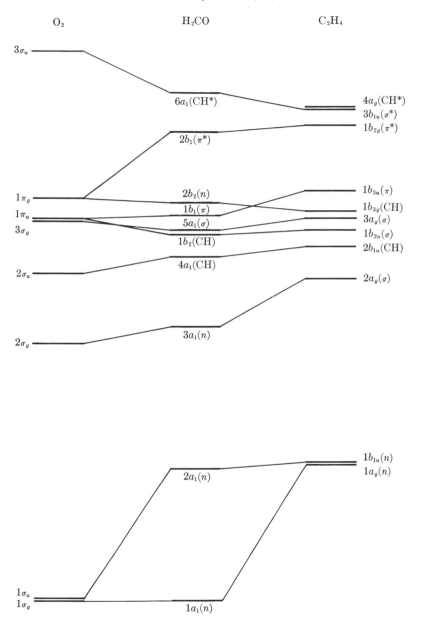

Fig. 7.30. Correlation of MO's of the isoelectronic molecules O_2, H_2CO, and C_2H_4.

H_2CO orbitals retain their symmetry with respect to the z axis and the yz plane as they are transformed into those of C_2H_4. The inversion (g, u) symmetry of the C_2H_4 orbitals can be determined from the g, u symmetry of the original O_2 orbitals. Thus a_1 becomes either a_g or b_{1u}, b_2 becomes either b_{3g} or b_{2u}, and b_1 becomes either b_{2g} or b_{3u}.

The resulting correlation diagram is shown in Fig. 7.30, in which the relative energies of the orbitals are those obtained from SCF calculations. In addition to the symmetry designation, the approximate bonding type is given for each H_2CO and C_2H_4 orbital in Fig. 7.30. The notation n, σ, π, σ^*, π^* refers to orbitals that are nonbonding, σ bonding, π bonding, σ antibonding, and π antibonding, respectively, for the two central atoms, C—O or C—C. Orbitals that are bonding and antibonding for the H atoms with the C atom to which they are attached are indicated by CH and CH*, respectively.

The relationship between the localized MO results and the delocalized orbitals of Fig. 7.30 is straightforward. Thus, in H_2CO, we have the correspondence for the occupied orbitals

$$1a_1 \leftrightarrow 1s_O \qquad\qquad 1b_2 \leftrightarrow (\sigma_{CH_1} - \sigma_{CH_2})$$
$$2a_1 \leftrightarrow 1s_C \qquad\qquad 5a_1 \leftrightarrow \sigma_{CO}$$
$$3a_1 \leftrightarrow 2s_O \qquad\qquad 1b_1 \leftrightarrow \pi_{CO}$$
$$4a_1 \leftrightarrow (\sigma_{CH_1} + \sigma_{CH_2}) \qquad\qquad 2b_2 \leftrightarrow n_O$$

and for the unoccupied orbitals

$$2b_1 \leftrightarrow \pi_{CO}^* \qquad\qquad\qquad 6a_1 \leftrightarrow (\sigma_{CH_1}^* + \sigma_{CH_2}^*)$$

where π_{CO}^*, $\sigma_{CH_1}^*$, and $\sigma_{CH_2}^*$ are the antibonding orbitals. Similar relationships hold for C_2H_4.

From the *Aufbau* principle, the ground-state configurations of these three 16-electron molecules are

$$O_2: (1\sigma_g)^2(1\sigma_u)^2(2\sigma_g)^2(2\sigma_u)^2(3\sigma_g)^2(1\pi_u)^4(1\pi_g)^2 \quad {}^3\Sigma_g^-$$

$$H_2CO: (1a_1)^2(2a_1)^2(3a_1)^2(4a_1)^2(1b_2)^2(5a_1)^2(1b_1)^2(2b_2)^2 \quad {}^1A_1$$

$$C_2H_4: (1a_g)^2(1b_{1u})^2(2a_g)^2(2b_{1u})^2(1b_{2u})^2(3a_g)^2(1b_{3g})^2(1b_{3u})^2 \quad {}^1A_g$$

Since every orbital is doubly occupied in H_2CO and C_2H_4 in the ground state, each of these molecules is a singlet and their electronic wave functions are symmetric with respect to every symmetry operation; the resulting ground-state term symbols are ${}^1A_1$ for H_2CO and ${}^1A_g$ for C_2H_4.

The most important observed transitions in H_2CO are from the ground state to the excited states with configurations

$$H_2CO: (1a_1)^2 \cdots (1b_1)^2(2b_2)(2b_1) \quad {}^1A_2$$
$$(1a_1)^2 \cdots (1b_1)(2b_2)^2(2b_1) \quad {}^1A_1$$

The first transition occurs in absorption in the wavelength range 3530–2300 Å, and corresponds to the promotion of an electron from the non-bonding $2b_2$ orbital (located mainly on the O atom) to the π antibonding $2b_1$ orbital. This so-called n → π^* transition is *forbidden* (see Problem 7.55), since it can be shown that the planar ground state ($^1A_1$) can undergo electric dipole transitions only to states of the type $^1A_1$, $^1B_1$, and $^1B_2$. However, in the excited state $^1A_2$, the molecule is nonplanar, with the C—O bond making angle of 31° with the H—C—H plane, and vibrational distortion allows the transition to occur. The analogous transition in O_2 is one in which the two electrons in the $1\pi_g$ orbitals are rearranged, namely, $^1\Delta_g \rightarrow$ $^1\Sigma_g^+$; this transition corresponds to the Noxon bands of O_2 (Table 7.9). For H_2CO, the $^1A_2$ state has one electron in each member of the pair. The spin-forbidden transition $^1A_1 \rightarrow {}^3A_2$, where $^3A_2$ is the triplet state with the same configuration as the $^1A_2$ state just described, is also weakly observed. As expected from Hund's rules, the $^3A_2$ state is found to lie lower than $^1A_2$; the energy difference is approximately 3000 cm^{-1}, although the comparison is not completely valid since the $^1A_2$ and $^3A_2$ states have slightly different geometries (the $^3A_2$ state out-of-plane angle is 35°). In addition to the states discussed here, a series of Rydberg transitions (see Section 7.11.2) involving excitation of one $2b_2$ electron have been studied.

The transition to the excited state $^1A_1$ is allowed and corresponds to the promotion of a π-bonding electron to a π-antibonding orbital. Absorption of H_2CO in the wavelength range 1750–1650 Å is generally identified with this $\pi \rightarrow \pi^*$ transition, but some spectroscopists attribute these absorption bands to a Rydberg transition (see Section 7.11.2). The analogous O_2 transition is $1\pi_u \rightarrow 1\pi_g$, which gives rise to the Schumann–Runge bands.

The most important feature of the C_2H_4 absorption spectrum is a diffuse band system beginning at 2600 Å in the liquid and at 2100 Å in the gas, with maximum intensity near 1620 Å. The upper electronic state of the transition is

$$C_2H_4: (1a_g)^2 \cdots (1b_{3g})^2(1b_{3u})(1b_{2g}) \quad {}^1B_{1u}$$

Thus the transition is of the $\pi \rightarrow \pi^*$ type (see Fig. 7.30) and corresponds roughly to the Schumann–Runge bands of O_2. The band system can be resolved into vibrational bands that correspond to twisting about the C—C axis. A careful analysis indicates that in the excited state $^1B_{1u}$, the two H—C—H planes are at right angles to one another (Fig. 7.31). The corresponding triplet state $^3B_{1u}$ is also twisted to 90°. Figure 7.32 shows a calculation of the energies of the three states of ethylene as a function of the twisting angle. When the two H—C—H planes are at 90°, the $1b_{3u}$ and $1b_{2g}$ (π and π^*) orbitals are degenerate. In this conformation Hund's rules predict that the $^3B_{1u}$ state has the lowest energy, as shown in Fig. 7.32.

Interconversion of *cis* and *trans* isomers of substituted ethylenes can

534 Molecular spectra

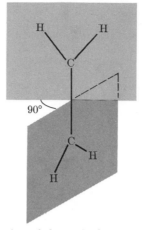

Fig. 7.31. Twisted conformation of the excited $^1B_{1u}$ and $^3B_{1u}$ states of ethylene.

occur photochemically by a mechanism in which the twisted conformation of the excited triplet state is the intermediate. These reactions are facilitated by the presence of so-called *photosensitizers* containing the $>C=O$ (carbonyl) group. The primary process is the $n \to \pi^*$ transition of the photosensitizer to the excited singlet state analogous to the $^1A_2$ state of H_2CO discussed above. The excited molecule undergoes intersystem crossing (see Section 7.11.4) to the more stable triplet $^3A_2$ state. The substituted ethylene is then excited to the twisted triplet analogous to the $^3B_{1u}$ state of C_2H_4 by collision with an excited triplet photosensitizer molecule. Subsequent de-excitation of the substituted ethylene yields a ratio of *cis* and

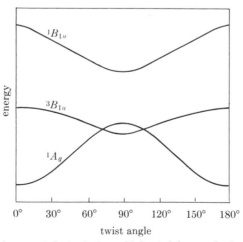

Fig. 7.32. Tortional potential of ethylene. [Adapted from calculations of V. Kaldor and I. Shavitt, *J. Chem. Phys.*, **48,** 191 (1968)].

trans isomers that depends upon the nature of the photosensitizer, in particular, on the energy of its triplet state. The process is summarized by the chemical equations

$$>C{=}O(S_0) \xrightarrow{h\nu} >C{=}O(S_1)$$

$$>C{=}O(S_1) \rightarrow >C{=}O(T_1)$$

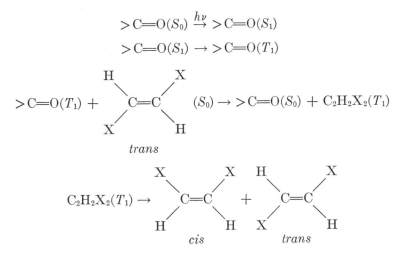

where S_0, S_1, and T_1 is the standard notation for the ground singlet, excited singlet, and lowest triplet states, respectively.

7.11.2   Rydberg transitions   The intensity of most bands in molecular electronic spectra is considerably weaker than that of atomic spectral lines. Some molecular bands are almost as intense as atomic lines, however. Many of these intense bands correspond to transitions in which an electron is excited to an orbital that lies outside the valence shells of the atoms in the molecule. The electron densities in these so-called *Rydberg orbitals* are usually very similar to those of excited atomic orbitals; thus the orbitals are designated *ns*, *np*, *nd*, etc., together with the appropriate symmetry symbol for the molecule, such as a_{1g}, b_{1u}, etc. *Rydberg states* resulting from transitions to Rydberg orbitals of a given type with successive values of the principal quantum number n can be fitted by the formula

$$E_n = E_\infty - \frac{R}{(n - \delta)^2}$$

where E_∞ is the ionization limit, R is the Rydberg constant, and δ is called the *quantum defect*, which corrects for the fact that the core is not exactly hydrogen-like. Identification of several members of a Rydberg series in molecular spectra often allows the ionization potential to be determined from the series limit (see Section 6.4.6).

In the far-uv spectrum of formaldehyde, four Rydberg series have been identified, each corresponding to excitation of an electron from the $2b_2$

orbital to a carbon Rydberg orbital. The upper state configurations have been assigned as follows:

(i) $\cdots (1b_1)^2(2b_2)^2 \rightarrow (1b_1)^2(2b_2)(3sa_1)$
$(1b_1)^2(2b_2)(4sa_1)$
$\cdots$

(ii) $\cdots (1b_1)^2(2b_2)^2 \rightarrow (1b_1)^2(2b_2)(3pb_2)$
$(1b_1)^2(2b_2)(3pb_2)$
$\cdots$

(iii) $\cdots (1b_1)^2(2b_2)^2 \rightarrow (1b_1)^2(2b_2)(3pa_1)$
$(1b_1)^2(2b_2)(3pa_1)$
$\cdots$

(iv) $\cdots (1b_1)^2(2b_2)^2 \rightarrow (1b_1)^2(2b_2)(3d)$
$(1b_1)^2(2b_2)(4d)$
$\cdots$

where the symmetry of the $3d$-type Rydberg states has not been determined. The long-wavelength members of series (ii) and (iii) are distinct because of the different symmetries (and therefore different energies) of the two sets of excited states, namely, $^1A_1$ and $^1B_2$, respectively. As n becomes large, however, this symmetry distinction becomes unimportant since the Rydberg orbital approaches more closely the character of an atomic carbon np orbital, and the two series merge. The wave number (in cm^{-1}) of the upper members of the four Rydberg series of formaldehyde are given by

(i) $\qquad \bar{\nu} = 87809 - \dfrac{R}{(n - 1.04)^2} \qquad n = 3, 4, \ldots$

(ii), (iii) $\bar{\nu} = 87710 - \dfrac{R}{(n - 0.70)^2} \qquad n = 3, 4, \ldots$

(iv) $\qquad \bar{\nu} = 87830 - \dfrac{R}{(n - 0.40)^2} \qquad n = 3, 4, \ldots$

The diffuseness of the bands does not allow determination of the series limits to the high precision obtained for atoms, but the average for the four series is 87765 cm^{-1}, corresponding to an IP of 10.88 ± 0.01 eV. High-resolution photoelectron spectroscopy (see Section 7.11.5) gives an IP of 10.88 ± 0.04 eV, in confirmation of the spectroscopic result.

7.11.3 *Spectra of conjugated hydrocarbons: free-electron model* In Section 6.4.3, we treated conjugated hydrocarbons by the Hückel MO theory. We now show that some aspects of the absorption spectra of polyenes with an even number of carbon atoms can be understood in terms of an even simpler picture, called the *free-electron model*.

Let us assume that the π electrons of an even polyene are free to move over the entire length of the molecule; that is, the electrons are confined to a one-dimensional box of length L. If $2k$ is the number of carbon atoms, one may approximate L by

$$L = (2k - 1)R$$

where R is an average C—C bond distance; that is, since there are alternating short (~ 1.34 Å) and long (~ 1.46 Å) bonds, we assume that

$$R = \tfrac{1}{2}(1.46 + 1.34) = 1.40 \text{ Å}$$

According to Eq. 2.74, the allowed energy levels are

$$E_n = \frac{n^2 h^2}{8mL^2} \qquad n = 1, 2, 3, \ldots \qquad (7.164)$$

In the ground-state, the $2k$ electrons doubly occupy each level through $n = k$ (see Fig. 7.33). The transition with the least energy difference (cor-

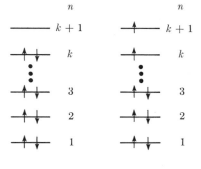

ground state   first excited state

Fig. 7.33. Free-electron model energy levels for polyenes.

responding to the band with the longest wavelength in the absorption spectrum of the molecule) is the transition of one of the electrons in the k level to the $k + 1$ level. From Eq. 7.164 and the Bohr frequency rule, $\lambda_{\max}$ is given by

$$\lambda_{\max} = \frac{8mcL^2}{(2k + 1)h} \qquad (7.165)$$

If $\lambda_{\max}$ is measured for a series of even polyenes, Eq. 7.165 may be used to calculate spectroscopic values of L, which can be compared with the result obtained from $L = (2k - 1)R$. Table 7.11 shows that the agreement is good, in view of the extreme simplicity of the free-electron model. The model gives even better results for conjugated molecules in which the two

Table 7.11

Compound	Formula	k	λ_{max}(Å)	L(Å) from spectrum	L(Å) from $(2k-1)R$
Butadiene		2	2100	5.6	4.2
Hexatriene		3	2470	7.2	7.0
Octatetraene		4	2860	8.8	9.8
Vitamin A (Retinol)		5	3060	10.1	12.6

major VB structures make equal contribution to the resonance hybrid, such as the polymethine dyes. An example is

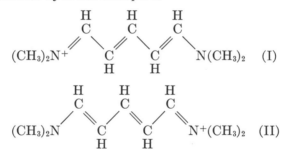

In this molecule, the C—C bond distances are all equal, and the delocalization of the π electrons is more complete than in the even polyenes.

Another series of molecules to which the free-electron model has been applied successfully are the polyacenes (benzene, naphthalene, anthracene, . . .) for which a MO description was given in Section 6.4. Here, the assumption that the π electrons are restricted to move on a circle whose perimeter is related to the size of the molecule leads to a reasonable description of the variation of the absorption wavelengths; also, two-dimensional rectangular box models have been used for the interpretation of the spectra of these systems.

7.11.4 Luminescence In addition to absorption spectroscopy, polyatomic molecules are often studied by looking at their emission after electronic excitation. To illustrate the important features of emission spectra, we use the well-studied, though not completely understood, benzene molecule as an example. When benzene is irradiated with uv light of wavelength less than about 2600 Å, it emits a characteristic spectrum in the range 3100–2590 Å. If benzene is dissolved in liquid krypton or argon and the resulting solution is frozen so that the benzene molecules are trapped in the rigid rare-gas matrix, irradiation with uv light is followed by weak emission at longer wavelengths near 3400 Å. Unlike the short-wavelength emission, the long-wavelength emission persists after the exciting radiation has been turned off. The former is called *fluorescence* and the latter *phosphorescence*. Collectively these processes are called *luminescence*.

The origin of the luminescence spectra of benzene is illustrated in Fig. 7.34, which is somewhat analogous to Fig. 4.7 for helium. The primary process is the absorption into a vibrational level of one of the excited singlet states $^1B_{2u}$, $^1B_{1u}$, or $^1E_{1u}$. All of these states, as well as the corresponding triplets ($^3B_{2u}$, $^3B_{1u}$, $^3E_{1u}$), arise from the lowest excited configurations of benzene (Section 6.4.1)

$$\phi_1^2\phi_2^2\phi_3\phi_4 \qquad \phi_1^2\phi_2^2\phi_3\phi_5 \qquad \phi_1^2\phi_2\phi_3^2\phi_4 \qquad \phi_1^2\phi_2\phi_3^2\phi_5$$

degenerate in the Hückel model. Electron-electron repulsion partly lifts the degeneracy to form linear combinations of the configurations that yield two nondegenerate ($^1B_{2u}$, $^1B_{1u}$) and one doubly-degenerate ($^1E_{1u}$) singlets and the corresponding triplets. If the absorption is into $^1B_{1u}$ or $^1E_{1u}$, transfer to a high vibrational state of $^1B_{2u}$ can occur. This process is called *internal conversion* (nonradiative transfer between states of the same multiplicity); it can be enhanced by collisions or by interaction with the lattice. Internal conversion is followed by rapid vibrational relaxation, indicated by the oscillating arrow in Fig. 7.34. Thus absorption into any of the three states $^1B_{2u}$, $^1B_{1u}$, $^1E_{1u}$ leads eventually to the population of the lowest vibrational level of the $^1B_{2u}$ electronic state. Once the molecule is in the $^1B_{2u}$ state, emission to the ground state $^1A_{1g}$ can occur. This process, by which radiation is emitted in a transition from an excited state to a lower one of the *same multiplicity*, is defined as *fluorescence*. Benzene fluorescence is illustrated by solid emission arrows in Fig. 7.34.

In addition to fluorescence, molecules in the $^1B_{2u}$ state can undergo *intersystem crossing* to one of the triplet states $^3E_{1u}$ or $^3B_{1u}$. Intersystem crossing is a process similar to internal conversion, except that the nonradiative transfer is made between states of different multiplicity. Because multiplicity changes in an isolated molecule (i.e., changes in the total electron spin of the molecule) are effected by the spin-orbit interaction (see Section 3.12.2), which is relatively weak for light atoms, intersystem crossing is generally about 10^{-6} times as fast as internal conversion. In benzene, absorption followed by intersystem crossing and vibrational relaxation leads to population of the lower vibrational levels of $^3B_{1u}$. From this state a molecule can be de-excited to the ground state $^1A_{1g}$ either by intersystem crossing to a high vibrational level of $^1A_{1g}$ followed by relaxation, or by direct

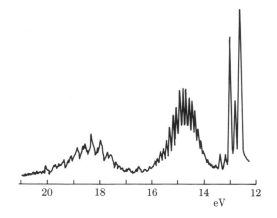

Fig. 7.35. Photoelectron spectrum of H_2O using the helium 584 Å (21.2 eV) line. Redrawn from C. R. Brundle and D. W. Turner, *Proc. Roy. Soc. (London),* **A 307,** 27 (1968).

quencies for the electronic states of ions can be obtained by photoelectron spectroscopy.

A technique for observing electronic transitions that are optically forbidden and for distinguishing between spin-forbidden and symmetry-forbidden transitions is that of *electron impact spectroscopy* [see E. N. Lassetre, *Radiation Res. Suppl.* **1,**530 (1959) and A. Kuppermann and L. M. Raff, *Discussions Faraday Soc.* **35,**30 (1963)]. Molecules to be studied are bombarded by electrons with energies in the range 20–1000 eV and the energy spectrum of the scattered electrons is measured. Energy loss of electrons with initial energy of a few hundred eV or more corresponds to inelastic scattering by molecules which undergo transitions to excited states as a result of the collision. At low initial electron energies, the optically allowed transitions become less probable and the forbidden transitions can be more easily detected. The relative intensities for allowed, spin-forbidden, and symmetry-forbidden transitions also change with scattering angle, so that in principle these different types of transitions can be distinguished and identified. Thus, electron impact spectroscopy can serve as a useful supplement to the usual optical techniques.

7.12 NUCLEAR MAGNETIC RESONANCE

Nuclear magnetic resonance (NMR) spectroscopy, which was developed to measure accurately the magnetic moments of nuclei, has become a very important tool for the identification of molecules and for the examination of their electronic structure. As the name implies, this form of spectroscopy is concerned with the energy levels of nuclei in an applied magnetic field.

For our purposes, an interaction between a magnetic field and a nucleus occurs only if the nucleus, like the electron, has a spin and an associated magnetic moment. Some nuclei (e.g., C^{12}, O^{16}; see Table 7.12) have zero spin; these nuclei are unaffected by a magnetic field and do not have an NMR spectrum. However, most nuclei have a nonzero spin characterized by a nuclear spin quantum number I (e.g., $I = \frac{1}{2}$ for H^1, F^{19}, . . . ; $I = 1$ for H^2, N^{14}). If we denote the nuclear spin angular momentum by F, we have for the square of the total nuclear spin angular momentum

$$F^2 = I(I + 1)\hbar^2 \qquad (7.166)$$

and for its z component

$$F_z = m_I \hbar \quad m_I = -I, -I + 1, . . . , I \qquad (7.167)$$

in correspondence with Eqs. 3.83 and 3.84. A nucleus with angular momentum vector $\mathbf{F}$ has a magnetic dipole moment vector (Eq. 3.85)

$$\boldsymbol{\mu}_{\text{mag}} = \gamma \mathbf{F} \qquad (7.168)$$

where the magnetogyric ratio γ, the proportionality constant relating the two, is given by

$$\gamma = g\,\frac{e}{2Mc} \qquad (7.169)$$

In Eq. 7.169, e is the electronic charge, M is the proton mass, c is the velocity of light, and the factor g, which is different for each nucleus, is a number on the order of unity. Thus, since $M/m_e = 1840$, the nuclear magnetic moment is on the order of 2000 times smaller than that of an electron. Values of gI for some nuclei are listed in Table 7.12; the product gI is often referred to as the *magnetic moment* of the nucleus (in units of $e\hbar/2Mc$; see Eq. 7.171).

If a nucleus is placed in a magnetic field with induction $\mathcal{B}$ in the z direction, the energy of the nucleus is equal to its energy in the absence of the field E_0 (which is constant and therefore can be ignored) plus the energy due to the field, which is given by

$$\begin{aligned} \Delta E_{\text{mag}} &= -(\mu_{\text{mag}})_z \mathcal{B} \\ &= -\gamma F_z \mathcal{B} \\ &= -g\left(\frac{e\hbar}{2Mc}\right) m_I \mathcal{B} \end{aligned} \qquad (7.170)$$

in correspondence with Eq. 3.93 for the electron. The quantity in parentheses in the last line of Eq. (7.170) is called the *nuclear magneton* and is designated μ_N; that is,

$$\mu_N = \frac{e\hbar}{2Mc} = 5.0505 \times 10^{-24} \text{ erg G}^{-1} \qquad (7.171)$$

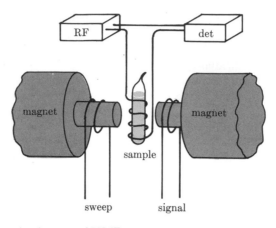

Fig. 7.37. Schematic diagram of NMR apparatus.

the nuclei which are in the state $m_I = -\frac{1}{2}$ can absorb quanta from the rf signal and undergo transitions to the $m_I = \frac{1}{2}$ state. The intensity of the rf signal therefore drops at this point, and a plot of the intensity versus $\mathcal{B}$ for a pure sample of the nuclei would give a spectrum with a single peak, such as that shown in Fig. 7.38. From a knowledge of ν_{osc} and the determination of $\mathcal{B}_{\text{res}}$, the g factor or γ can be inferred.

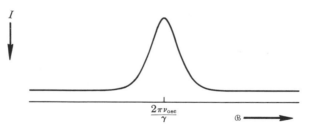

Fig. 7.38. Variation of intensity of rf signal as the magnetic field is swept through resonance in a sample of noninteracting nuclei of spin $\frac{1}{2}$.

The discussion of NMR spectroscopy up to this point makes obvious its importance to nuclear physics since it provides a tool for the measurement of nuclear g factors. However, it is not evident why it should be of such interest to chemists. The reason is that nuclei, of course, do not exist as isolated entities. Instead they are present in molecules surrounded by electrons and with other nuclei in their vicinity. Since electrons in atoms and molecules are charged particles in motion (i.e., they constitute a current), they create a magnetic field which alters the field at the nucleus and changes the resonance frequency. Moreover, if the other nuclei in the molecule have spins, their magnetic moments can interact with that of the nucleus under observation. It is these effects, which make the resonance signal of a nucleus dependent on its molecular environment, that are the source of chemical information. In what follows we indicate the features of the NMR spectrum resulting from such interactions and outline the essence of their chemical interpretation.

7.12.1   Chemical shift   The motion of the molecular electrons in the presence of the applied magnetic field sets up currents that generate a magnetic field in a direction opposed to that of the applied field. Thus the *net* field acting on the nuclear moment in a molecule is generally somewhat less than the applied field $\mathscr{B}_{app}$. The energy level splitting is reduced relative to that of the bare nucleus (see Fig. 7.39) and the resonance frequency is

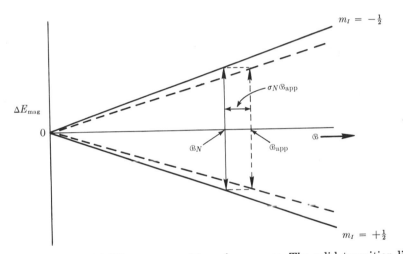

Fig. 7.39. Effect of shielding upon position of resonance. The solid transition line indicates the value of $\mathscr{B}_{res}$ for bare nuclei; the dotted transition line indicates the value of $\mathscr{B}_{res}$ when $\mathscr{B}_N$ is reduced below $\mathscr{B}_{app}$ by magnetic shielding due to the electrons in an atom or molecule.

shifted correspondingly. Since the current induced in the electrons is proportional to $\mathscr{B}_{app}$, the effective field at the nucleus ($\mathscr{B}_N$) can be written

$$\mathscr{B}_N = \mathscr{B}_{app} - \mathscr{B}_{el} = \mathscr{B}_{app}(1 - \sigma_N) \qquad (7.176)$$

where $\mathscr{B}_{el}$ is the field due to the electrons and σ_N is the *shielding factor*, defined such that $\mathscr{B}_{el} = \sigma_N \mathscr{B}_{app}$. From Eq. 7.176 and Fig. 7.39 it is evident that if a fixed external field is applied to the system, the resonance frequency for the shielded nucleus is lower than that of the bare nucleus. Correspondingly, if the NMR experiment is done by sweeping the applied field, the resonance frequency (ν_{osc}) being kept fixed as described above, the applied field must be greater for the shielded nucleus than for the bare nucleus.

At resonance, $\mathcal{B}_{app} = \mathcal{B}_{res}$, and the relation of Eq. 7.175 must hold for $\mathcal{B}_N$; that is,

$$\mathcal{B}_N = \mathcal{B}_{res}(1 - \sigma_N) = \frac{2\pi\nu_{osc}}{\gamma} \qquad (7.177)$$

or

$$\mathcal{B}_{res} = \frac{2\pi\nu_{osc}}{\gamma(1 - \sigma_N)} \qquad (7.178)$$

Since ν_{osc} and γ are fixed for a given oscillator frequency and nucleus, $\mathcal{B}_{res}$ varies only with σ_N.

The effect of shielding on the NMR spectrum is illustrated by the presence of the two peaks in the proton resonance of CH_3OH shown in Fig. 7.40a. They arise from the fact that the shielding factor (σ_{CH_3}) for the CH_3 protons is different from that (σ_{OH}) for the OH proton. The difference in the intensities of the two peaks is due to the fact that there are three CH_3 protons and only two OH protons in each molecule of the sample. For a field of 10 kG, the observed difference between the resonance peaks

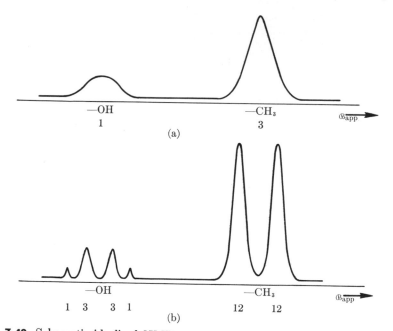

Fig. 7.40. Schematic idealized NMR spectrum of CH_3OH; **(a)** low resolution, **(b)** high resolution. The numbers represent the relative intensities of the lines. (See the footnote on p. 554.)

in Fig. 7.40a is approximately 15 mG, that is, about 1.5 parts per million (ppm) of the value of $\mathcal{B}_{app}$. Usually the shifts of $\mathcal{B}_{res}$ due to different chemical environments (*chemical shifts*) are measured relative to an internal standard such as H_2O for water-soluble samples[8] and tetramethylsilane (TMS) for samples that dissolve in organic solvents. For convenience, a dimensionless chemical shift δ_N is often defined as

$$\delta_N = \frac{\mathcal{B}_{res} - \mathcal{B}_{std}}{\mathcal{B}_{std}} \qquad (7.179)$$

where $\mathcal{B}_{std}$ is the resonance field of the standard; because of its small magnitude, δ_N is usually expressed in part per million (ppm). Although the splitting in gauss between different nuclei is dependent on the applied field, the magnitude of the chemical shift is a property of the nucleus and its chemical environment. From Eqs. 7.178 and 7.179 it follows that

$$\delta_N = \frac{\sigma_N - \sigma_{std}}{1 - \sigma_N} \simeq \sigma_N - \sigma_{std} \qquad (7.180)$$

We see that a positive value for δ_N (corresponding to an "up-field" shift relative to the standard) means that the proton is more shielded than the reference proton ($\sigma_N > \sigma_{std}$), while a negative value of δ_N (corresponding to a "down-field" shift) means that it is less shielded ($\sigma_N < \sigma_{std}$). The shielding of protons ranges over about 40 ppm as shown by the list of proton chemical shifts given in Table 7.13. The isolated H atom has a calculated shielding factor of 17.8 ppm relative to the bare proton, while the most shielded H atoms so far observed (in metal hydrides such as $[Co(CN)_5H]^{3-}$ and *trans*-$FeHCl[o-C_6H_4(PEt_2)_2]_2$ have values on the order of 65 ppm relative to the bare proton. What is most important about the shielding factor (or the δ values relative to some standard) is that they are characteristic of the chemical environment of the proton. Thus, the position of the resonance line can be used to determine the nature of the group (e.g., CH_3, NH_2, OH) in which the proton is located; typical δ values and their ranges are given in Table 7.13. When such correlations are combined with the fact that the area of the resonance peak is proportional to the number of protons of a given type, it becomes clear why NMR spectra are of considerable utility for structure determinations.

For magnetic nuclei other than protons, corresponding effects are observed, though the magnitude of the shielding and the range of chemical shifts is generally greater. Some typical results for fluorine, a nucleus which has been extensively studied, are shown in Table 7.14. Here we see that the

[8] Interaction between H_2O and the sample frequently alters the proton environment in H_2O sufficiently to render H_2O unsatisfactory as an internal standard; thus TMS or closely related compounds, which are essentially inert to interactions with the solute, are being widely used as internal standards.

Table 7.13   *Proton chemical shifts in various chemical environments*[a]

Group	δ (ppm)	Group	δ (ppm)
—SO₃H	-6.7 ± 0.3	$\equiv$C—H	$+2.4 \pm 0.4$
—CO₂H	-6.4 ± 0.8	$=\overset{\mid}{C}$—CH₃	$+3.3 \pm 0.5$
RCHO	-4.7 ± 0.3	—CH₂—	$+3.5 \pm 0.5$
RCONH₂	-2.9	RNH₃	$+3.6 \pm 0.7$
PhOH	-2.3 ± 0.3	$-\overset{\mid}{\underset{\mid}{C}}$—CH₃	$+4.1 \pm 0.6$
PhH	-1.9 ± 1.0	[Co(CN)₅H]³⁻	$+22.1$[b]
$=$CH₂	-0.6 ± 0.7	*trans*-FeHCl[o—C₆H₄(PEt₂)₂]₂	$+39.1$[b]
ROH	-0.1 ± 0.7	bare proton	-25.6
		H atom	-7.8
H₂O	0	H₂ molecule	$+1.0$
—OCH₃	$+1.6 \pm 0.3$		
—CH₂X	$+1.7 \pm 1.2$		

[a] L. H. Meyer, A. Saika, and H. S. Gutowsky, *J. Am. Chem. Soc.* **75**, 4567 (1953).
[b] J. Chatt, *Proc. Chem. Soc. (London)* **1962**, 318 (1962).

variation is much larger than that for the proton with the F atom in HF shielded by 410 ppm relative to the bare fluorine atom. By contrast, the F atom in F₂ is *less* shielded (232.8 ppm) than the bare fluorine nucleus; that is, the electronic field adds to the external field in this case (see below).

7.12.2   Interpretation of chemical shifts   Since the shielding of the nuclei is due to electronic currents in the molecule, one factor that is expected to be important is the number of electrons, or electron density, in the neighborhood of the nucleus. For an isolated atom in a ¹S state (so that no orbital or spin angular momenta with their associated magnetic moments are present), the shielding is due to the current induced in the electrons by the external field and the shielding constant σ_{at} can be shown to equal

$$\sigma_{at} = \frac{e^2}{3mc^2} \sum_i \langle r_i^{-1} \rangle \tag{7.181}$$

where the sum is over all of the electrons and $\langle r_i^{-1} \rangle$ is the average value of $1/r$ for the ith electron. From a knowledge of the ground-state configurations and the screening constants for each orbital as given in Table 4.9, the value of σ_{at}, called the atomic diamagnetic shielding constant, can be calculated. Some results for typical atoms of chemical interest are shown in Table 7.15. It is clear that the value of σ_{at} increases with the number of

Table 7.14   *Fluorine[19] chemical shifts in various chemical environments*[a]

Compound	δ (ppm)
F_2	-596.0
ClO_2F	-499.9
SF_6	-223.7
ClF_3 (pair)[b]	-313.2
(unique)[b]	-187.3
CCl_3F	-167.6
CF_4	-104.19
CHF_3	-89.0
BF_3	-36.1
SiF_4	0
HF	46.85
ClF	273.7
F^-	-25.1
bare fluorine	-363.2

[a] From D. K. Hindermann and C. D. Cornwell, *J. Chem. Phys.* **48**, 4148 (1968).

[b] The molecule ClF_3 is "T" shaped; the F atoms on either end of the horizontal bar are called here "pair F atoms," while the F atom at the bottom end of the vertical bar is called the "unique F atom."

electrons in the atom. Moreover, we see that by comparing H and F (Tables 7.13 and 7.14) that the range of molecular shielding factors is on the order of the atomic diamagnetism. Thus, in molecules, as in the isolated atoms, one important factor determining the shielding is the electron density. This interpretation can be used not only for comparing different nuclei, but more importantly for a partial understanding of the chemical shift of a single nucleus. Considering the values given in Table 7.13, we find that there is a significant correlation between the observed shielding and the expected electron density, as measured for example by the "acidity" of the proton under consideration; for example, comparing CH_3, OCH_3, OH, NH_2, and $COOH$, we see that the proton is progressively less shielded as the acidity increases. Another comparison of this type can be made by looking at the chemical shift between the CH_3 and CH_2 protons in substituted ethanes (CH_3CH_2X) as a function of the electronegativity of the X group. From the discussion in Section 6.2.1, we would expect that as the electronegativity of X increases, both sets of protons would be less shielded and, in

Table 7.15   *Atomic diamagnetic shielding constants calculated from Eq. 7.181 with screening constants from Table 4.9*

Atom	$\langle r^{-1} \rangle_{1s}$	$\langle r^{-1} \rangle_{2s}$	$\langle r^{-1} \rangle_{2p}$	$\langle r^{-1} \rangle_{3s}$	$\sum_i \langle r^{-1} \rangle_i$	σ_{at} (ppm)
H	1.0000				1.0000	17.754
He	1.6875				3.3750	59.920
Li	2.6906	0.3198			5.7010	101.22
Be	3.6848	0.4780			8.3256	147.81
B	4.6795	0.6441	0.6054		11.2526	199.78
C	5.6727	0.8042	0.7840		14.5218	257.82
N	6.6651	0.9619	0.9585		18.1295	321.87
O	7.6579	1.1229	1.1133		22.0148	390.85
F	8.6501	1.2819	1.2750		26.2390	465.85
Ne	9.6421	1.4396	1.4396		30.8010	546.84
Na	10.6259	1.6429	1.7005	0.2786	35.0192	621.73
Mg	11.6089	1.8480	1.9565	0.3675	39.3878	699.29

addition, that the chemical shift between the two types would increase, since the CH_2 protons are more affected by the X group than is the CH_3 group. The results are shown in Table 7.16; the general trend clearly supports our expectations although there are a few deviations.

The electron-density concept, which is clearly a useful one for the understanding of chemical shifts, does not provide a complete interpretation. Thus, in comparing the protons in ethane, ethylene, and acetylene, we would expect from the electron density alone that $\sigma_{ethane} > \sigma_{ethylene} > \sigma_{acetylene}$ while the experimental result is $\sigma_{ethane} > \sigma_{acetylene} > \sigma_{ethylene}$. One source of deviations of this type is the fact that electronic currents in other parts of the molecule can create magnetic fields at the nucleus under consideration. Such effects, often referred to as neighbor-anisotropy effects, are particularly important in aromatic systems where electronic currents can flow relatively freely in the plane of the ring but not perpendicular to the ring. The neighbor-anisotropy effect is believed to be the source of the anomalously high shielding in acetylene.

Consideration of an additional contribution to the shielding is essential for the analysis of the chemical shift of most atoms other than hydrogen. Since the shielding is due to electronic currents, the extent of shielding is determined not only by the electron density, but also by the degree to which the electrons are free to circulate. In particular, the asymmetry introduced into an atom by bond formation may inhibit the free circulation of the

Table 7.16   *Correlation of the difference between methyl (CH₃) and meth-ylene (CH₂) proton shielding constants in substituted ethanes with the electronegativity of the substituent group*[a]

Group	$\sigma_{CH_3} - \sigma_{CH_2}$ (Hz)	x[b]	x[c]
–SH	32	2.6	2.45
–CN	35	2.6	2.52
–COOH	37	2.6	2.57
–CHO	39	2.6	2.61
–CO–	39	2.6	2.61
–S–	40	2.6	2.64
–I	42	2.65[d]	2.68
–SCN	43	2.6	2.70
–C₆H₅	43	2.6	2.70
–Br	53	2.95[d]	2.94
–NH₂	55	3.05	2.99
–Cl	64	3.15[d]	3.19
–OH	77	3.5	3.51
–OCO–	89	3.5	3.83
–SO₄–	89	3.5	3.83
–NO₃	95	3.5	3.91
–F	96	3.90[d]	3.93

[a] B. P. Dailey and J. N. Shoolery, *J. Am. Chem. Soc.* **77**, 3977 (1955).
[b] Electronegativity of atom bonded to CH₂ group; the values used here differ slightly from those of Table 6.7.
[c] Calculated from linear fit to halogen substituted ethanes.
[d] These can be fitted by the line $x = 0.02315 (\sigma_{CH_3} - \sigma_{CH_2}) + 1.71$.

electrons, thereby reducing the diamagnetic circulation. This reduction is often referred to as a *paramagnetic* contribution to the shielding since the effect, relative to that which would result from freely circulating electrons, corresponds to a field at the magnetic nucleus that *adds* to the external field. For the fluorine nucleus, in contrast to hydrogen, the paramagnetic contribution is often the dominant term in the observed chemical shifts. In the spherically symmetric F⁻ ion, the electrons are free to circulate and the paramagnetic term is zero. In the F₂ molecule, however, the atoms have a strongly asymmetric charge distribution because the $2p$ "hole" in the valence shell is "fixed" by formation of the $2p_\sigma$ bond. This inhibits the electronic circulation and leads to a large paramagnetic contribution; thus, for F in F₂ the diamagnetic shielding due to the nine electrons is $+529$ ppm relative to the bare nucleus (somewhat larger than σ_{at}; see Table 7.15) and the paramagnetic term is -761 ppm, which leads to the net antishielding given in Table 7.14. Fluorine atoms involved in bonds that have an ionic character between that of F^- and F_2 have intermediate paramagnetic contributions and chemical shifts (see, e.g., the value for HF in Table 7.14). Thus, variation in the paramagnetic term, which dominates the fluorine chemical shifts, can give some indication of the ionic character of the bond in which the fluorine atom is involved. Similar correlations are formed in many other atoms, such as C and P.

In our discussion of the chemical shift we have assumed that the molecule can be treated as an isolated system and have ignored the effects of the surrounding medium. These effects can be very important, particularly if the medium is composed of molecules that interact with the one under consideration either chemically (e.g., by hydrogen bonding or chemical exchange) or magnetically (e.g., by creating significant magnetic fields).

7.12.3 Spin-spin coupling   Under high resolution, the NMR spectra protons (as well as other magnetic nuclei) often show that the low resolution peaks have a symmetric fine structure like that drawn for CH_3OH in Fig. 7.40b.[9] To examine the origin of this fine structure, we consider the HD molecule, whose proton and deuteron NMR spectrum are shown diagrammatically in Fig. 7.41. We see that the proton resonance consists of three

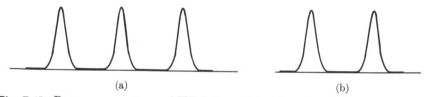

(a)   (b)

Fig. 7.41. Resonance spectrum of HD (schematic); **(a)** proton, **(b)** deuteron.

equally spaced lines of the same intensity while the deuteron spectrum has two lines of equal intensity. Moreover, the spacing of both the proton and the deuteron lines is found to be 42.8 Hz, independent of the strength of the applied external field. This fact suggests that the observed splitting is an internal molecular effect. The number of the observed lines indicates that the splitting is caused by an interaction between the proton and deuteron magnetic moments; that is, since the deuteron has a nuclear spin of 1, it can have three orientations relative to the proton, while the proton with a spin of $\frac{1}{2}$ can have two orientations relative to the deuteron.

---

[9] The spectrum of Fig. 7.40b is an idealization in that effects of rapid exchange of the hydroxyl hydrogen atom between molecules and of second-order corrections (see Section 7.12.5) are ignored.

The energy-level scheme for the proton is of the form shown in Fig. 7.42a, where the left-hand side is that for the isolated H atom (without and with shielding). On the far right side of Fig. 7.42a is shown the splitting of each of the proton levels into three levels, one for each deuteron spin orientation. Since in the proton resonance spectrum, the proton nuclear spin quantum number m_{IH} is changed by 1 while the deuteron spin orientation remains unaltered (the frequency for the deuteron resonance is very far from the proton resonance frequency), we expect the three transitions indicated in Fig. 7.42a, which correspond to the three observed resonances. The analogous transitions for the deuteron are shown in Fig. 7.42b; here the upper and lower deuteron levels with $m_{ID} = \pm 1$ are split into two by the proton, while the central deuteron level, which has $m_{ID} = 0$, is not altered by interaction with the proton.

The expression for the total magnetic energy of the proton-deuteron system in an applied magnetic field can be written

$$\Delta E_{\text{mag}} \doteq -\gamma_H \hbar \, \mathcal{B}_{\text{app}} m_{IH}(1 - \sigma_H)$$

$$- \gamma_D \hbar \, \mathcal{B}_{\text{app}} m_{ID}(1 - \sigma_D) + J_{HD} h m_{IH} m_{ID} \quad (7.182)$$

where the subscripts H and D indicate the nucleus involved and J_{HD}, which is equal to the splitting of the resonances, is called the spin-spin coupling constant (Problem 7.51). To apply Eq. 7.182 to the proton resonance, we consider the transition $m_{IH} = +\frac{1}{2} \rightarrow m_{IH} = -\frac{1}{2}$ at constant m_{ID}. The expression for the frequency analogous to Eq. 7.173 is

$$\nu_H = \frac{\gamma_H}{2\pi} \mathcal{B}_{\text{app}}(1 - \sigma_H) - J_{HD} m_{ID} \quad (7.183)$$

Since $m_{ID} = 1, 0, -1$, there are three resonant frequencies:

$$\nu_H' = \frac{\gamma_H}{2\pi} \mathcal{B}_{\text{app}}(1 - \sigma_H) + J_{HD} \quad (7.184\text{a})$$

$$\nu_H'' = \frac{\gamma_H}{2\pi} \mathcal{B}_{\text{app}}(1 - \sigma_H) \quad (7.185\text{b})$$

$$\nu_H''' = \frac{\gamma_H}{2\pi} \mathcal{B}_{\text{app}}(1 - \sigma_H) - J_{HD} \quad (7.184\text{c})$$

and the frequency difference is the coupling constant

$$\Delta \nu_H = \nu_H' - \nu_H'' = \nu_H'' - \nu_H''' = J_{HD} \quad (7.185)$$

Alternatively, with the usual experimental design (see Fig. 7.37), the oscillator frequency is fixed at ν_{osc} and the magnetic field is swept through the resonance; Eqs. 7.184 become

$$\mathcal{B}_H' = \frac{2\pi}{\gamma_H(1 - \sigma_H)} (\nu_{\text{osc}} - J_{HD}) \quad (7.186\text{a})$$

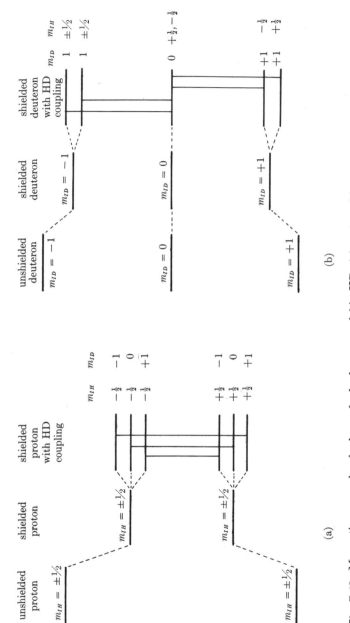

Fig. 7.42. Magnetic energy-level scheme for hydrogen nuclei in HD; **(a)** proton, **(b)** deuteron. (The shifts and splittings are not to scale.)

$$\mathcal{B}_{\mathrm{H}}{''} = \frac{2\pi}{\gamma_{\mathrm{H}}(1 - \sigma_{\mathrm{H}})} \, \nu_{\mathrm{osc}} \qquad (7.186b)$$

$$\mathcal{B}_{\mathrm{H}}{'''} = \frac{2\pi}{\gamma_{\mathrm{H}}(1 - \sigma_{\mathrm{H}})} \, (\nu_{\mathrm{osc}} + J_{\mathrm{HD}}) \qquad (7.186c)$$

from which we obtain

$$J_{\mathrm{HD}} = \frac{\gamma_{\mathrm{H}}(1 - \sigma_{\mathrm{H}})}{2\pi} (\mathcal{B}_{\mathrm{H}}{''} - \mathcal{B}_{\mathrm{H}}{'}) \simeq \frac{\gamma_{\mathrm{H}}}{2\pi} (\mathcal{B}_{\mathrm{H}}{''} - \mathcal{B}_{\mathrm{H}}{'})$$

$$\cong \frac{\gamma_{\mathrm{H}}}{2\pi} (\mathcal{B}_{\mathrm{H}}{'''} - \mathcal{B}_{\mathrm{H}}{''}) \qquad (7.187)$$

A similar treatment of the "D resonances" leads to two frequencies:

$$\nu_{\mathrm{D}}{'} = \frac{\gamma_{\mathrm{D}}}{2\pi} \mathcal{B}_{\mathrm{app}}(1 - \sigma_{\mathrm{D}}) - \frac{1}{2} J_{\mathrm{HD}}$$
$$\nu_{\mathrm{D}}{''} = \frac{\gamma_{\mathrm{D}}}{2\pi} \mathcal{B}_{\mathrm{app}}(1 - \sigma_{\mathrm{D}}) + \frac{1}{2} J_{\mathrm{HD}} \qquad (7.188)$$

since the two transitions ($m_{I\mathrm{D}} = 1 \to m_{I\mathrm{D}} = 0$ and $m_{I\mathrm{D}} = 0 \to m_{I\mathrm{D}} = -1$; $m_{I\mathrm{H}}$ constant) correspond to the same energy-level spacing. Thus

$$\Delta\nu_{\mathrm{D}} = \nu_{\mathrm{D}}{'} - \nu_{\mathrm{D}}{''} = J_{\mathrm{HD}}$$

that is, the splitting of the deuteron resonance is the same as that for the proton.

The types of splittings observed in the HD spectrum and in more complicated molecules provide structural information to supplement that obtained from the chemical shift. To relate the values of the coupling constants to the electronic structure we need an understanding of the mechanism involved. One possible source of the interaction is the direct effect of the field due to the magnetic dipole of one nucleus upon the other. This term, called the direct dipole-dipole interaction, varies as $(3 \cos^2 \theta - 1)/R^3$, where θ is the angle between the internuclear line and the external field and R is the distance between the nuclei. For rapidly tumbling molecules in a gas or liquid with all orientations equally probable, only the average value of the interaction, $\langle 3 \cos^2\theta - 1 \rangle/R^3$, contributes to the magnetic energy of the nuclei. This value is zero since

$$\langle 3 \cos^2 \theta - 1 \rangle = \int_0^\pi (3 \cos^2 \theta - 1) \sin \theta \, d\theta = \int_{-1}^{+1} (3x^2 - 1) \, dx = 0$$

Thus the direct dipole-dipole term is not observed except in oriented species obtained by the use of liquid crystals, molecular beams, or solids. In experiments where the direct dipole-dipole term does contribute, it provides information concerning the relative positions of the atoms since the magnitude of the splitting is related to the inverse cube of the internuclear distance.

In the usual liquid or solution studies, in which the direct term averages to zero, the observed splitting arises from an indirect coupling of the magnetic nuclei by the electrons of the molecule. If we consider the case of HD, we know that in its ground state the two electrons are paired to form a singlet bond (see Chapter 5); that is, the spins of the electrons are antiparallel. Since each of the electrons has a spin and associated magnetic moment (see Section 3.12), it can interact with the nuclear magnetic moments. Using a VB description, we can understand the spin-spin coupling of the nuclei in terms of Fig. 7.43. If electron 1 is on nucleus *A*, the electron

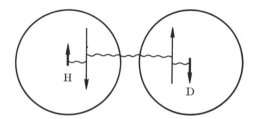

Fig. 7.43. VB description of spin-spin interaction in HD; small heavy arrows represent nuclear spins, long thin arrows represent electron spin, wavy lines represent coupling.

and nuclear spin interact and there is a tendency for their spins to be opposed; this electron-nuclear coupling is exactly that which gives rise to the hyperfine splittings in electron spin resonance (see Section 7.13). A corresponding electron spin-nuclear spin interaction takes place on atom *B*. Since the two electron spins are antiparallel in the bond, there is a coupling between the nuclei via the electrons; that is, the nuclear spin-spin interaction is transmitted along the path (proton spin)-(electron spin)-(electron spin)-(proton spin). There are two important points to be noted concerning this conclusion. First, the coupling between the electrons leading to their antiparallel spin orientation is *electrostatic* rather than magnetic, since the singlet state is bonding and the triplet is antibonding (see Section 5.6). The second point follows from the first in that for nuclei that are not directly bonded, the presence of a nonzero coupling indicates that there is a breakdown of the completely localized description of chemical bonds. Thus, in CH_3CD_3 the proton and deuteron, which are separated by two bonds, have a coupling of ~ 1.5 Hz on the average (the coupling depends on whether the H,D nuclei are *trans* or *gauche*). Comparing this with the HD coupling of 40 Hz, we can conclude there is on the order of 1/30 of a "bond" between H and D. Thus, the indirect spin-spin coupling, which is a function of the spin correlation involved in bond formation, is an extremely sensitive probe for the study of delocalized bonding. In particular it is much more sensitive than an energy criterion in which one has to attempt to evaluate the delocalization contribution as a small increment to an approximately known bond energy (see Section 6.4.5). This result illustrates the point that it is often important to find experimental techniques that permit one to make a direct measurement of the particular electronic effect of interest. The newer resonance techniques (Sections 7.12–7.14), which involve very small energy transitions, are ideally suited for this purpose.

7.12.4 Analysis of NMR spectra Although we have indicated the nature of NMR spectra in our discussion of the coupling constant in HD, it is helpful to consider a slightly more complicated case as well. For this purpose we return to examine CH_3OH under high resolution. Since there are two types of protons in the molecule, we label them a and b:

$$H_a-\overset{\overset{\displaystyle H_a}{|}}{\underset{\underset{\displaystyle H_a}{|}}{C}}-OH_b$$

Equation 7.182 applies to this case, with the subscripts H and D replaced by a and b, respectively, and with m_{Ia} and m_{Ib} replaced by M_{Ia} and M_{Ib}, the *sums* of the quantum numbers for all protons of the respective types in the molecule; that is, we have

$$\Delta E_{\text{mag}} = -\gamma\hbar\mathcal{B}_{\text{app}}M_{Ia}(1-\sigma_a) - \gamma\hbar\mathcal{B}_{\text{app}}M_{Ib}(1-\sigma_b) + J_{ab}hM_{Ia}M_{Ib}$$
$$(7.189)$$

where γ is the proton magnetogysic ratio. In writing Eq. 7.189, we have made use of the fact that equivalent atoms (i.e., the three H_a's bonded to C in a rotating CH_3 group) have the same shielding and coupling constants; also, coupling within such a group does not affect the spectrum. The possible values of M_{Ib} are $+\frac{1}{2}$, since there is only one proton of type B; but for M_{Ia} there are four possible values:

$$\tfrac{1}{2} + \tfrac{1}{2} + \tfrac{1}{2} = \tfrac{3}{2}$$
$$\tfrac{1}{2} + \tfrac{1}{2} - \tfrac{1}{2} = \tfrac{1}{2}$$
$$\tfrac{1}{2} - \tfrac{1}{2} - \tfrac{1}{2} = -\tfrac{1}{2}$$
$$-\tfrac{1}{2} - \tfrac{1}{2} - \tfrac{1}{2} = -\tfrac{3}{2}$$

In general, for n nuclei of spin $\frac{1}{2}$, the sum M_I may take on one of $n+1$ different values

$$\left(\frac{n}{2}, \frac{n}{2}-1, \ldots, -\frac{n}{2}\right)$$

$S = 1]$, electron paramagnetic resonance (EPR) or electron spin resonance (ESR) spectroscopy, is the electronic analog of NMR. We begin the description of ESR by considering the hydrogen atom in its ^2S ground state in the presence of a uniform magnetic field in the z direction. For such a one-electron atom with $S = \frac{1}{2}$ and $l = 0$, the z component of the electronic magnetic moment (see Section 3.12)

$$(\mu_{\text{mag}})_z = -g_e \left(\frac{e\hbar}{2mc} \right) m_s \tag{7.194}$$

can have two orientations with respect to the field depending on the value of $m_s (m_s = \pm \frac{1}{2})$. The magnetic energy is

$$\Delta E_{\text{mag}} = g_e \frac{e\hbar}{2mc} m_s \mathcal{B} = 2.002319 \mu_B m_s \mathcal{B} \simeq 2\mu_B m_s \mathcal{B} \tag{7.195}$$

where $\mathcal{B}$ is the applied magnetic induction, g_e is the g factor for an electron without orbital angular momentum, and μ_B is the Bohr magneton (Section 3.12). The ground-state energy level splits as indicated in Fig. 7.44.

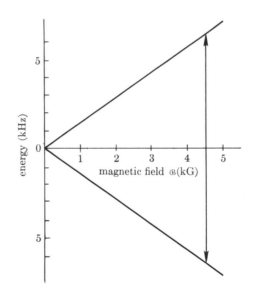

Fig. 7.44. Splitting of electronic magnetic energy levels of a ground-state H atom in a magnetic field. [Adapted from R. Beringer and M. A. Heald, *Phys. Rev.*, **95,** 1474 (1954).]

This was already described in Section 3.12, where we considered the effect of an external field on electronic transitions involving changes in the n and l quantum numbers (the Zeeman and Paschen–Back effects). As in the case of NMR, we are here concerned with transitions in which only the m_s

quantum number is altered, the values of n and l remaining unchanged. Instead of frequencies in the visible and uv region associated with the Zeeman effect, the frequency

$$\nu = \frac{\Delta E}{h} = \frac{2\mu_B \mathcal{B}}{h} \tag{7.196}$$

corresponding to the absorption process $m_s = -\frac{1}{2} \to m_s = +\frac{1}{2}$, with a commonly used $\mathcal{B}$ value of ~ 3000 G, is

$$\nu = \frac{2 \times 0.927 \times 10^{-20} \times 3000}{6.63 \times 10^{-27}} = 8.4 \times 10^9 \text{ sec}^{-1} = 8.4 \text{ kMHz}$$

This frequency is in the microwave region of the electromagnetic spectrum (Table 7.2). Thus, if radiation at this frequency is applied, we expect a single line of the form shown in Fig. 7.45a. The width of the absorption peak is due to a variety of environmental effects, which can be studied by an analysis of the so-called *line shape*. As in NMR, derivative spectra are often obtained, which for the absorption line in Fig. 7.45a would have the form

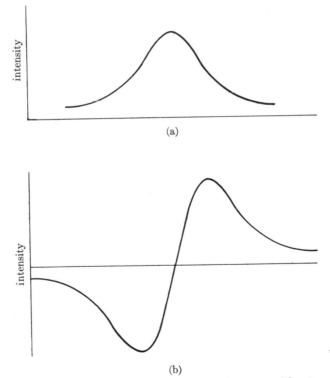

(a)

(b)

Fig. 7.45. ESR resonance line for ground-state H atoms without nuclear skin (schematic); **(a)** absorption line, **(b)** derivative spectrum.

shown in Fig. 7.45b. The basic experimental apparatus for ESR is drawn very schematically in Fig. 7.46. It consists of a source for the microwave signal, a resonant cavity containing the sample, and a signal detector. The components are connected with wave guides, rectangular metal tubes for

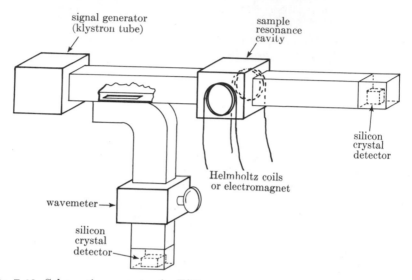

Fig. 7.46. Schematic apparatus for ESR.

conducting the microwave signal. The variable magnetic field is applied to the sample by Helmholtz coils or an electromagnet. A wavemeter (tunable cavity) between the source and a secondary detector can be adjusted to resonance with the sample cavity; it is calibrated to give the wavelength of the signal.

From the above description of the ESR measurement, it appears that nothing of chemical interest is obtained since all of the quantities determining the transition frequency (Eq. 7.196) are either fundamental constants or the known applied field. However, there are several important factors that we have not considered so far that make ESR a widely used technique in structural chemistry. One of these is that, in certain atomic and molecular systems, Eq. 7.194, and therefore Eqs. 7.195 and 7.196, are not valid; that is, the g factor may be significantly different from the pure spin value of 2.0023. This variation in g is due to the presence of orbital angular momentum ($l \neq 0$) and the coupling between the spin and orbital moments (spin-orbit coupling; see Chapter 4). As a result, g can be greater or less than the free-electron value, its exact magnitude depending on the specific system. For atoms and transition metal complexes, in particular, the observed g values have been found useful in interpreting their electronic structure. Thus, for example, in Ni^{2+} complexes the g value depends on the ligands; the g value is 2.27 in NiBr$_2$, 2.20 in NiSO$_4 \cdot$ 7H$_2$O, and 2.18 in Ni(NH$_3$)$_6$Br$_2$.

Returning to the hydrogen-atom spectrum, we remember from the discussion of NMR that for nuclei with spin, there is an interaction between the nuclear and electronic magnetic moment. In NMR this was found to be a source of the spin-spin splittings; in ESR the same interaction gives rise to so-called hyperfine splittings. Since the hydrogen atom consists of an electron with spin $S = \frac{1}{2}$ and a nucleus with spin $I = \frac{1}{2}$, we expect that the two energy levels of the electron in an external magnetic are split into four, corresponding to the four different sets of m_s and m_I values (see Fig. 7.47).

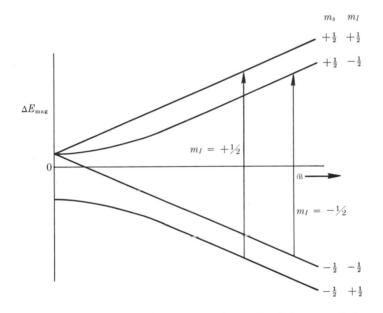

Fig. 7.47. Hyperfine splitting of H-atom ground-state levels in an applied magnetic field.

The ESR spectrum arises from electronic transitions ($m_s = -\frac{1}{2} \to m_s = \frac{1}{2}$) with the nuclear spin remaining unchanged. Consequently, for a fixed frequency of electromagnetic radiation there are two transitions at different magnetic fields, equally shifted with respect to the single transition expected in the absence of the nuclear spin splitting. The resulting spectrum is shown in Fig. 7.48.

The energy level scheme shown in Fig. 7.47 can be described by a simple extension of Eq. 7.195 to include the hyperfine interaction. For the H atom, we have

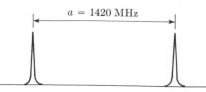

Fig. 7.48. Sketch of ESR spectrum of the H atom, showing hyperfine interaction.

$$\Delta E_{\text{mag}} = g_e \mu_B \mathcal{B} m_s + ah m_s m_I \tag{7.197}$$

where a is called the *hyperfine splitting constant* and has units of Hz. Equation 7.197 is valid for the normal case of $\mu_B \mathcal{B} \gg ah$. As can be seen in Fig. 7.47, near $\mathcal{B} = 0$ the energy levels do not vary linearly with the field. Thus Eq. 7.197 is not applicable at very small applied fields, since the quantization is more complicated when the coupling of the electron and nuclear spin due to the hyperfine interaction "competes" with the effect of the external field. From Eq. 7.197, we see that the expressions for the frequencies of the two magnetic dipole allowed ESR transitions ($\Delta m_s = \pm 1$, $\Delta m_I = 0$) are

$$\nu_1 = \frac{g_e \mu_B \mathcal{B}}{h} + \frac{1}{2} a \qquad m_I = +\frac{1}{2} \tag{7.198a}$$

$$\nu_2 = \frac{g_e \mu_B \mathcal{B}}{h} - \frac{1}{2} a \qquad m_I = -\frac{1}{2} \tag{7.198b}$$

so that the frequency difference ($\Delta \nu$) is equal to the hyperfine constant a. Alternatively, if the experiment is done at fixed frequency by varying the magnetic field, we have

$$\nu = \frac{g_e \mu_B \mathcal{B}_1}{h} + \frac{1}{2} a = \frac{g_e \mu_B \mathcal{B}_2}{h} - \frac{1}{2} a \tag{7.199}$$

where $\mathcal{B}_1$ and $\mathcal{B}_2$ are the magnetic induction values that lead to resonance. The difference in the $\mathcal{B}$ values is

$$\Delta \mathcal{B} = \mathcal{B}_2 - \mathcal{B}_1 = \left(\frac{h\nu}{g_e \mu_B} + \frac{ah}{2 g_e \mu_B} \right) - \left(\frac{h\nu}{g_e \mu_B} - \frac{ah}{2 g_e \mu_B} \right) = \frac{ah}{g_e \mu_B} \tag{7.200}$$

for $a > 0$ as it is in the hydrogen atom and many other atoms. If $\Delta \nu$ is expressed in MHz and $\mathcal{B}$ is expressed in gauss, we have

$$\Delta \nu = a = \frac{g_e \mu_B}{h} \Delta \mathcal{B} = 2.80249 \, \Delta \mathcal{B} \tag{7.201}$$

Experimental values of hyperfine splitting constants are quoted in both sets of units, the choice depending on the individual making the measurement. It should be noted that the experimental splitting depends only on the magnitude of a and not its sign; however, there exist methods (e.g.,

NMR chemical-shift measurements of systems with unpaired electrons) for determining the sign of a in appropriate cases.

For the hydrogen atom, the experimental value of a is 1420.4057 MHz or 506.84 G. Its source, already mentioned above, is the coupling between the nucleus and the electron magnetic moments. The quantitative expression for the hyperfine constant of a hydrogen atom expressed in Hz is

$$a = \frac{8\pi}{3h} g_e \mu_B g_H \mu_N |\psi(0)|^2 \tag{7.202}$$

where g_H and μ_N are the proton g factor and the nuclear magneton, respectively, and $|\psi(0)|^2$ represents the absolute value squared of the electron density at the nucleus. The coupling corresponding to Eq. 7.202 is often called the contact interaction, or *Fermi contact interaction*, after Enrico Fermi who showed (1930) that such an interaction should exist long before the ESR technique was introduced (1945–47). In addition to the Fermi contact term, there is another interaction between the electron and nuclear spins, which contributes for electrons in non-S states. This is the direct dipole-dipole term, which averages to zero in solution because of the rapid tumbling, but can be observed in solids when the molecules have fixed orientations.

From the form of the ground-state hydrogen-atom wave function (Section 3.2)

$$\psi_{1s} = (\pi a_0^3)^{-1/2} \exp\left(-\frac{r}{a_0}\right)$$

the electron density at the nucleus is

$$|\psi_{1s}(0)|^2 = \frac{1}{\pi a_0^3}$$

Substituting this result into Eq. 7.202 with the values for the various constants, we obtain 1422.75 MHz or 507.68 G, in reasonable agreement with experiment (see Problem 7.48). Correction for the finite mass of the proton [i.e., multiplication by $(1836.13/1837.13)^3$] gives the more accurate theoretical value 1420.43 MHz.

We consider now the extension of the hydrogen-atom results to more complex systems, in particular the organic radicals that are being intensively studied by ESR techniques. Such molecules contain many electrons, each of which has a spin and an associated magnetic moment. However, we can frequently neglect all of the inner-shell electrons, which form closed shells with paired spins, and restrict the discussion to the outermost electron with an unpaired spin that can interact with the external magnetic field. For the naphthalene negative ion (see Section 6.4.2), the ground-state configuration is $\phi_1{}^2 \cdots \phi_5{}^2\phi_6$ so that the unpaired electron in MO ϕ_6 gives rise to the ESR spectrum. Modifying Eq. 7.197 for this system, we have

$$\Delta E_{\mathrm{mag}} = g\mu_B \mathcal{B} m_s + ahm_s M_I + a'hm_s M_I' + \cdots \qquad (7.203)$$

As before, the first term represents the interaction of a single unpaired electron ($m_s = \pm\frac{1}{2}$) with the external field. For most organic radicals, the orbital angular momentum makes a very small contribution to the magnetic moment and g is near the free-electron value of 2.0023. The additional terms in Eq. 7.203 represent the hyperfine interactions with different types of nuclei that have hyperfine constants $a, a', \ldots$; the quantum numbers M_I correspond to the sum of the m_I values of the n nuclei with constant a, M_I' to the sum of the m_I' values of the n' nuclei with constant a', and so on. For the ordinary naphthalene negative ion with carbon C^{12} ($I = 0$), only the protons can contribute to the hyperfine

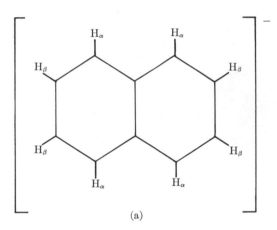

(a)

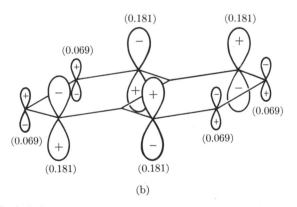

(b)

Fig. 7.49. (a) Naphthalene negative ion, showing α and β protons; **(b)** MO occupied by unpaired electron, ϕ_6; the numbers represent the squares of the coefficients of the respective AO's, according to the Hückel treatment of Section 6.4.2.

splitting. From the symmetry of the molecule we expect that the four α protons (Fig. 7.49a) have the same hyperfine constant (a_α) and the four β protons have the same hyperfine constant (a_β). The resonance frequencies depend on the values of $M_I{}^\alpha$ and $M_I{}^\beta$, each of which can have any of $n+1$ values with statistical weights given by Eqs. 7.192 and 7.193; that is, for each group of four protons we expect a splitting into five lines with intensities 1:4:6:4:1. Thus, the over-all spectrum with $a_\alpha \neq a_\beta$ should have 25 lines in all. These are shown in Fig. 7.50. In analyzing this spectrum, we can

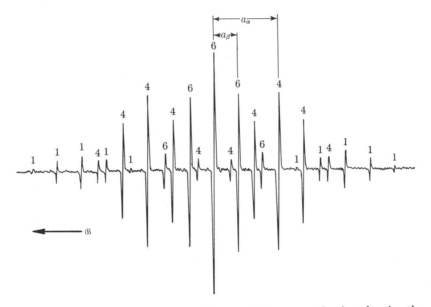

Fig. 7.50. ESR derivative spectrum of the naphthalene negative ion showing the α and β proton hyperfine splittings. [The spectrum was provided by G. Fraenkel. See text for description.]

pick out the quintet expected from the four protons with a spacing corresponding to the larger hyperfine constant; the five lines are identified by 1, 4, 6, 4, 1 in circles. Each of these lines is in turn split into a quintet with spacings corresponding to the smaller hyperfine constant; the four additional lines of each quintet are labelled by the same number as the central line (e.g., the lines of the central quintet are labelled 6). The relative intensities of the 25 lines in order from left to right are seen to be 1:4:6:4:4:16:1:24:6:16:24:4:36:4:24:16:6:24:1:16:4:4:6:4:1. Although it is clear from the spectrum that there are two types of four equivalent protons, it is not possible to know which a value corresponds to the α protons and which to the β protons without additional information. The distinction has been made by selective deuteration; for example, substituting a deuterium for an α proton alters the large splitting, which must therefore be a_α. Thus, $a_\alpha = 13.79$ MHz or 4.92 G and $a_\beta = 4.94$ MHz or 1.80 G.

The magnitude of the hyperfine splitting in the naphthalene negative ion, and for other radicals, follows from Eq. 7.202 with the reinterpretation required for a many-electron system; that is, instead of representing the total electron density at the nucleus under consideration, $|\psi(0)|^2$ is now the *net* density corresponding to the electron with unpaired spin. Moreover, since there are now several nuclei in the molecule, it is best to introduce $|\psi(\mathbf{r}_N)|^2$, which represents the density at the position of nucleus N whose hyperfine constant is being evaluated. From a comparison between the results for the hydrogen atom and those for the naphthalene negative ion, we see that in the latter the proton hyperfine constants are on the order of 1% of those in the former. This implies that in naphthalene the "unpaired spin density" at the proton is 1% that of the isolated hydrogen atom. To understand this great difference, we must look at the electronic structure of naphthalene in more detail. As was pointed out above, the unpaired electron added to naphthalene to form the negative ion is expected to be in the lowest unoccupied π orbital, namely, ϕ_6 of Fig. 7.49b. Once we realize this, we are faced with the problem of explaining not why the hyperfine constant is so small, but instead why there is any hyperfine splitting at all. Since the π orbitals have the form shown in Fig. 7.49b, the unpaired electron has a nodal plane in the plane of the molecule; that is, within the π-electron approximation, the protons are at the position of a node and $|\psi(\mathbf{r}_N)|^2$ is expected to be zero. An explanation that was first suggested for the nonzero hyperfine constants was that the protons spent part of the time in positions of finite unpaired electron density as a result of the out-of-plane vibrations; however, simple model calculations showed that such a vibrational effect was much too small. The nonzero observed splitting thus demonstrates that the assumption made for the treatment of benzene and other aromatic molecules in Section 6.4 (i.e., that electrons can be separated completely into σ electrons and π electrons) is not valid. In fact, we can use the magnitude of the splitting as a measure of the mixing of σ character into the π orbital of the unpaired electron (the so-called σ-π interaction). To do this, we look at the π-electron wave function ϕ_6 of the unpaired electron in the naphthalene negative ion (see Fig. 7.49b). The coefficient of the π orbital in the α position is $|C_\alpha| = 0.425$ and in the β position is $|C_\beta| = 0.263$. Using the square of the coefficient to obtain the electron distribution (see Section 6.4.2), we find that the fraction of an unpaired electron in an α-position π orbital is $|C_\alpha|^2 = 0.181$ and the fraction of an unpaired electron in a β-position π orbital is $|C_\beta|^2 = 0.069$. Since the σ bonds (C—H and C—C bonds) are well localized (see Section 6.4.2), it is reasonable to assume that the unpaired electron density at a given proton arises primarily from the σ-π interaction with the π orbital on the attached carbon atom. Comparing the observed a_α and a_β values (see Fig. 7.50) with the π-electron occupation numbers, we have

$$\frac{|a_\alpha|}{|C_\alpha|^2} = \frac{4.92}{0.181} = 27.2 \text{ G}$$

$$\frac{|a_\beta|}{|C_\beta|^2} = \frac{1.80}{0.069} = 26.1 \text{ G}$$

(7.204)

where we have used the absolute value of the hyperfine constant to indicate that we are only concerned with the magnitude. From Eq. 7.204 we see that the two ratios are very similar. They can be interpreted by comparison with the H atom splitting of 507 G, as implying that the σ-π interaction leads to $\approx 1/20$ of an unpaired electron at the proton if there is one unpaired electron in the π orbital. To describe the mechanism for the delocalization, we consider a simple three-electron model for the C—H fragment (Fig. 7.51a). The dominant VB structure for this system has a σ bond as shown in Fig. 7.51b and the unpaired electron in the π orbital. However, as in the case of the water molecule (Chapter 6), we can introduce the second VB structure shown in Fig. 7.51c in which there is a "bond"

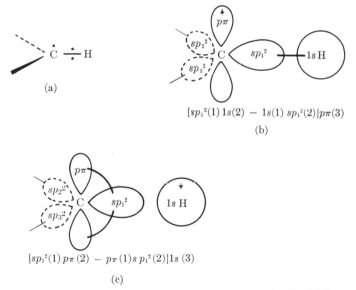

$$[sp_1{}^2(1)\,1s(2) - 1s(1)\,sp_1{}^2(2)]p\pi(3)$$

(b)

$$[sp_1{}^2(1)\,p\pi(2) - p\pi(1)s\,p_1{}^2(2)]1s(3)$$

(c)

Fig. 7.51. Three-electron model for σ-π interaction; **(a)** the C—H fragment, **(b)** principal VB structure; **(c)** VB structure with bond between the carbon σ and π orbitals and unpaired spin on the H atom. The VB functions representing structures (b) and (c) have not been antisymmetrized with respect to the unpaired electron; to do this, one should take linear combinations of permutations of electrons 1 and 3 and of electrons 2 and 3, multiplying each term derived from an odd permutation by -1 (see Section 4.2).

between the carbon σ and π orbitals and the unpaired electron is on the proton. It is the contribution of the latter structure which leads to the nonzero hyperfine interaction.

It has been found experimentally and justified in theoretical treatments by Harden McConnell and others that the constancy of the ratios (Eq. 7.204) obtained for the naphthalene negative ion holds rather generally for aromatic systems. The expression

$$a_i{}^\mathrm{H} = Q\rho_i{}^\pi$$

is often used where $a_i{}^\mathrm{H}$ is the ith proton hyperfine constant, $\rho_i{}^\pi$ is the unpaired spin density of the carbon atom attached to proton i, and Q is a "constant" whose value is generally in the narrow range -20 to -25 G, with -22.5 G being commonly used. If Q is known, a measurement of $a_i{}^\mathrm{H}$ clearly provides a value for $\rho_i{}^\pi$. Thus, one has a method for mapping out the unpaired electron distribution in aromatic radicals. This approach has been applied to a variety of systems, particularly for testing theoretical calculations, such as Hückel theory. We were, of course, doing this implicitly in Eq. 7.204 which makes use of Hückel orbitals.

Another type of delocalization that can also be studied by ESR measurements is exemplified by the results for the ethyl radical CH_3CH_2. From the simplest considerations of bonding in the radical, the unpaired electron would be expected to occupy the nonbonding $2p\pi$ orbital on the α carbon (Fig. 7.52a). If this were the case, the α hydrogens, which are sp^2 bonded to the β carbon and lie in the nodal plane of the $2p\pi$ orbital, as in aromatic systems, would be expected to yield an α triplet with $a \simeq -22.5$ G. The β hydrogens might be expected to give a $1:3:3:1$ pattern that would result from the three equivalent protons of a freely rotating methyl group, with a considerably smaller coupling constant since the β hydrogens are an extra C—C bond distance away. The experimental ESR spectrum of CH_3CH_2, as shown in Fig. 7.53, turns out to violate these predictions. The lines of intensity 2, 6, 6, and 2 form the quartet $1:3:3:1$; each of these lines in turn is the center of a triplet with relative intensities $1:2:1$. Thus, the coupling constant $a_\beta (= 26.9$ G$)$ is slightly larger than $a_\alpha (= -22.4$ G$)$, in spite of the extra C—C distance. An explanation is that there are contributions from VB structures in which the unpaired electron in the $2p\pi$ orbital forms a weak delocalized bond with an sp^3 C—H bonding hybrid orbital of the β carbon, leaving the unpaired electron on one of the β hydrogens to give the large observed hyperfine splitting. Since the three hydrogen nuclei in a freely rotating methyl group are equivalent, each C—H bonding orbital can form such a delocalized bond, and the $1:3:3:1$ pattern results. Such a bonding scheme, often referred to as *hyperconjugation*, is illustrated in Fig. 7.52b. An MO representation of hyperconjugation is indicated in Fig. 7.52c. The three H-atom $1s$ orbitals can combine as shown in the figure to form three 3-centered orbitals, one having σ

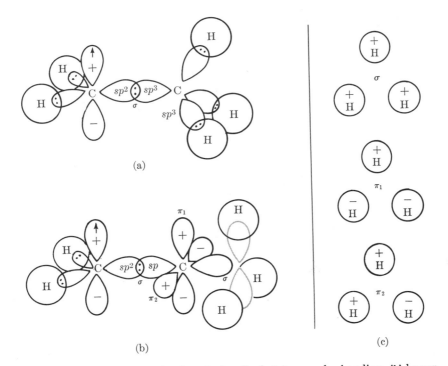

Fig. 7.52. Hyperconjugation in the ethyl radical; **(a)** normal σ bonding, **(b)** hyperconjugated π bond, **(c)** form of hydrogen σ and π orbitals.

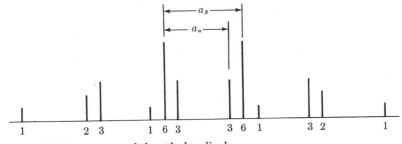

Fig. 7.53. ESR spectrum of the ethyl radical.

symmetry with respect to the "C—H$_3$ bond" and two having π symmetry. The methyl C atom, with its σ orbitals sp hybridized, can form a σ and two π bonds with the H$_3$ group. The π_1 bond has the same symmetry as the orbital containing the unpaired electron on the methylene C atom and can form a delocalized bond orbital extending over the entire C—C—H$_3$ group. This distributes the unpaired electron in such a way that it has a finite density at the methyl H-atom nuclei.

7.14 ELECTRIC QUADRUPOLE RESONANCE

In discussing the interaction between nuclei and electrons, we have considered the charge Z, which is the source of the electrostatic potential binding electrons in atoms, and the magnetic dipole moment μ_{mag}, which is minute on an absolute scale but still leads to the measurable consequences that are studied in NMR and ESR spectroscopy. We might ask whether nuclei have any higher electric or magnetic moments that affect the electron energies. The first term one thinks of is the electric dipole moment. It is believed, however, that the symmetry of the nucleus, or of any elementary particle (e.g., electrons), forbids the existence of a dipole moment. The next electrostatic term is a quadrupole moment, a simple example of which is illustrated in Fig. 7.54a. By analogy with the $(+\ -)$ point-charge model for

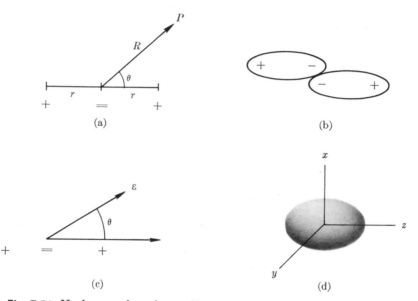

Fig. 7.54. Nuclear quadrupole coupling; **(a)** a simple quadrupole, **(b)** equivalent pair of dipoles, **(c)** relation of field direction to quadrupole axis, **(d)** ellipsoidal charge distribution of nucleus.

a dipole, a quadrupole moment can be represented by two positive charges $(+e)$ symmetrically placed with respect to a negative charge $(-2e)$ at the origin. It is clear that the dipole moment of the charge distribution shown in the figure is zero. However, there is a nonzero potential due to the separation of charge, which at large distances $(R \gg r)$ has the form (Problem 7.57)

$$\mathcal{V}(R,\theta) = \frac{er^2}{R^3} (3 \cos^2 \theta - 1) \tag{7.205}$$

and along the z axis ($\theta = 0$) yields

$$\mho(R,0) = \frac{2er^2}{R^3}$$

Thus, the interaction energy between the quadrupole and a charge e located at a distance R is

$$\Delta E = e\mho(R,0) = \frac{e^2}{R^3}(2r^2) \tag{7.206}$$

We need an expression, analogous to Eq. 7.28 for the electric dipole moment, giving the energy of a quadrupole in an electric field ε (e.g., that due to the electrons in an atom or molecule). If the field is uniform, the quadrupole energy is independent of orientation. One way of seeing this is to regard the quadrupole of Fig. 7.54a as made up two equal but opposed electric dipoles (Fig. 7.54b). Since the energy of a dipole μ in a uniform field ε is $-\mu\,\varepsilon\cos\theta$, where θ is the angle between the dipole and field directions, we have

$$\Delta E = -\mu\varepsilon[\cos\theta + \cos(\theta + \pi)] = 0$$

If the quadrupole interaction energy is to be orientation dependent, it is clear that the field must have different magnitudes at each of the two component dipoles; that is, the quadrupole must be placed in a nonuniform field (one that varies with position). For a system with cylindrical symmetry, such as a diatomic molecule, the interaction energy can be written (Problem 7.58)

$$\Delta E_Q = -\frac{e^2qQ}{4}\left[\tfrac{1}{2}(3\cos^2\theta - 1)\right] \tag{7.207}$$

where θ is the angle between the symmetry axis of the field (taken to be the z axis) and the quadrupole direction (Fig. 7.54c). In Eq. 7.207, eQ is the *quadrupole moment*, which for a set of charges e at the points $(\bar{x}_i, \bar{y}_i, \bar{z}_i)$, (Problem 7.58)

$$eQ = e\sum_i (3\bar{z}_i^2 - \bar{r}_i^2)$$

and eq is the *electric field gradient*

$$eq = \frac{\partial\varepsilon_z}{\partial z}$$

It is easy to show (Problem 7.59) that Eq. 7.207 reduces to Eq. 7.206 for the interaction of a charged particle with a quadrupole having the charge distribution of Fig. 7.54a.

For systems that do not have cylindrical symmetry, other components of the quadrupole moment and the electric field gradient have to be considered.

For the application of Eq. 7.207 to the interaction between the nuclear electric quadrupole moment and the extranuclear charge distribution in a molecule, we have to take account of the quantization of the possible orientations. In general, the nuclear quadrupole moment does not correspond to the simple distribution of point charges pictured in Fig. 7.54a. Instead, the nucleus consists of a continuous distribution with a net charge Ze. Deviations from spherical symmetry can be characterized in terms of a quadrupole moment, an octupole moment, a hexadecapole moment, and so on. For nuclei with spin quantum number $I \geq 1$, the quadrupole moment is nonzero and represents the only significant nonspherical term. It can be regarded as cylindrically symmetric and represented by an ellipsoidal charge distribution (see Fig. 7.54d) with axis coupled to the nuclear spin, whose orientation is quantized. For a diatomic molecule or other cylindrically symmetric system with fixed orientation (e.g., in a solid), the classical angular factor of Eq. 7.207 is replaced by the appropriate quantum-mechanical factor which gives the energy expression

$$\Delta E_Q = \frac{e^2 qQ}{4}\left(\frac{3m_I{}^2 - I(I+1)}{I(2I-1)}\right) \tag{7.208}$$

For a nucleus with $I = \frac{5}{2}$ (such as I^{127}), the energy level scheme resulting from Eq. 7.208 is shown in Fig. 7.55; that is, since the energy depends on $m_I{}^2$ (and not on m_I), there are three doubly degenerate levels corresponding to $m_I = \pm\frac{5}{2}, \pm\frac{3}{2}, \pm\frac{1}{2}$. The selection rule here, as in magnetic resonance, is $\Delta m_I = \pm 1$. Thus there are two allowed transitions with frequencies

$$\Delta\nu(m_I = \pm\tfrac{5}{2} \rightarrow m_I = \pm\tfrac{3}{2}) = \frac{3|e^2 qQ|}{10h}$$

$$\Delta\nu(m_I = \pm\tfrac{3}{2} \rightarrow m_I = \pm\tfrac{1}{2}) = \frac{3|e^2 qQ|}{20h}$$

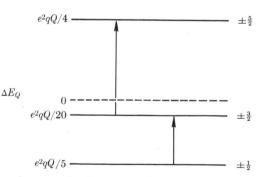

Fig. 7.55. Quadrupole energy levels and transitions of I^{127}.

From a measurement of either frequency, the magnitude $|e^2qQ|$ can be determined. Since in this type of experiment, called *pure quadrupole resonance*, the splitting is a consequence of the molecular electric field gradient, rather than an external magnetic field as in NMR, the rf coil used to generate the electromagnetic field must be of variable frequency; a number of oscillators for the measurement have been constructed. Alternatively, it is possible to do an NMR experiment in the presence of the quadrupole contribution to the energy and thereby measure $|e^2qQ|$. If the substance is a gas or can be vaporized, the quadrupole interaction can also be measured by molecular beam or microwave rotational spectroscopy. However, it should be noted that the calculation of the quadrupole splitting does not follow the simple formulas given above due to the coupling between the nuclear spin and the molecular rotational angular momentum (see Reference 11 in the Additional Reading list for further discussion).

From Eq. 7.208, it is clear that the frequency of the quadrupole resonance lines depends on the product of the nuclear quadrupole moment and the electric field gradient at the nucleus. To get an idea of the orders of magnitude involved, we make use of the fact that the form of eq is (electronic charge) $\times$ (the inverse cube of the average electronic distance). For the iodine nucleus ($Z = 53$) in a molecule

$$\frac{|e^2qQ|}{h} \simeq 53e \times (10^{-12})^2 \times e \times (10^{-8})^{-3}/h$$

$$\simeq \frac{53e^2}{h} \simeq \frac{53 \times 20 \times 10^{-20}}{6 \times 10^{-27}} \simeq 150 \times 10^7$$

$$\simeq 10^3 \, \mathrm{MHz}$$

The resulting value is of the order of that observed for the iodine atom. For lighter atoms, the values are generally smaller. Moreover, there is a very wide range for any given atom as a function of the atoms to which it is bonded in different molecules. Because the nuclear quadrupole moment is invariant with respect to the bonding, the variation in $|e^2qQ|$ is a measure of the change in the field gradient from molecule to molecule. Since this is directly related to the electron distribution, it is a significant source of chemical information. As an illustration, we consider the Cl atom in different chemical environments. Since its nuclear moment is known (see Table 7.12; for Cl^{35}, $eQ = -0.085 \times 10^{-24} \, e \, \mathrm{cm}^2$), the resonance frequency can be converted directly to the field gradient and compared with quantitative molecular calculations. However, the qualitative picture that emerges from such studies is even more important. For a Cl^- ion, which is spherically symmetric, the field gradient is zero; correspondingly, the observed value of $|e^2qQ|$ for highly ionic systems is very small (see Table 7.17). By contrast, for a covalently bonded Cl, the nuclear environment is highly asymmetric and the field gradient is large; for example, in Cl_2, $(|e^2qQ|/h) \simeq 109 \, \mathrm{MHz}$,

Table 7.17 *Quadrupole coupling constants for diatomic halides.*[a]

Molecule	$\dfrac{\|e^2qQ\|}{h}$ (MHz)	Estimated ionicity
Cl^{35} (atom)	109.74	$\cdots$
BrCl	103.6	0.056
ICl	82.5	0.115
FCl	146.0	0.259
TlCl	15.8	0.831
KCl	0.04	1.000
RbCl	0.774	0.992
CsCl	3	0.968
Br^{79} (atom)	769.76	$\cdots$
BrCl	876.8	0.110
FBr	1089.0	0.329
LiBr	37.2	0.944
NaBr	58	0.911
KBr	10.244	0.985
DBr	533	0.186
I^{127} (atom)	2292.84	$\cdots$
DI	1823	0.065
ICl	2944	0.229
LiI	198.15	0.900
NaI	259.87	0.867
KI	60	0.970

[a] From B. P. Dailey and C. H. Townes, *J. Chem. Phys.* **23**, 118 (1955).

which is near the value for an isolated Cl atom. For other compounds with intermediate ionic character, the $|e^2qQ|$ value varies between these limits. Since the primary change in going from a covalently bonded Cl to an ionic Cl^- is the addition of a $3p_z$ electron to the bonding (valence) orbital, it has been possible to interpret the observed coupling in terms of the fractional occupation number of the orbital, which in turn is related to the ionicity of the bond (see Table 7.17). Similar results have been obtained for a variety of other nuclei.

SUMMARY

This chapter has presented some of the basic concepts that are used to obtain information about molecular structure from spectroscopic data. Microwave, infrared, Raman, and electronic spectra can be employed to deduce the rotational, vibrational, and electronic energy levels. From them a variety of molecular constants (e.g., bond lengths, bond angles, dipole moments) can be determined. Many features of the electronic wave function can be obtained from electronic spectra and from photoelectron, electron impact, and magnetic and quadrupole resonance data. Resonance data also provide a sensitive probe for certain structural information.

With this chapter, we conclude the introduction to the experimental and theoretical methods used in the analysis of atomic and molecular structure. The set of principles developed in this text, when properly applied, should provide the student with a good background for statistical mechanics and chemical kinetics, which together with the quantum theory comprise the foundation of physical chemistry.

ADDITIONAL READING

[1] G. M. Barrow, *Introduction to Molecular Spectroscopy* (McGraw-Hill, New York, 1962).

[2] G. W. King, *Spectroscopy and Molecular Structure* (Holt, Rinehart, and Winston, New York, 1964).

[3] J. D. Roberts, *Nuclear Magnetic Resonance* (McGraw-Hill, New York, 1959).

[4] G. Herzberg, *Spectra of Diatomic Molecules* (Van Nostrand, Princeton, N.J., 1950), 2nd ed.

[5] G. Herzberg, *Infrared and Raman Spectra* (Van Nostrand, Princeton, N.J., 1945).

[6] G. Herzberg, *Electronic Spectra of Polyatomic Molecules* (Van Nostrand, Princeton, N.J., 1967).

[7] A. Carrington and A. D. McLachlan, *Introduction to Magnetic Resonance* (Harper and Row, New York, 1967).

[8] J. A. Pople, W. G. Schneider, and J. H. Bernstein, *High Resolution Nuclear Magnetic Resonance* (McGraw-Hill, New York, 1959).

[9] M. Bersohn and J. C. Baird, *An Introduction to Electron Paramagnetic Resonance* (Benjamin, New York, 1966).

[10] J. C. D. Brand and J. C. Speakman, *Molecular Structure* (Edward Arnold, London, 1960).

[11] C. H. Townes and A. L. Schawlow, *Microwave Spectroscopy* (McGraw-Hill, New York, 1955).

[12] J. C. Davis, Jr., *Advanced Physical Chemistry* (Ronald Press, New York, 1965), Chaps. 8, 10, and 11.

[13] C. N. Banwell, *Fundamentals of Molecular Spectroscopy* (McGraw-Hill, New York, 1968).

[14] S. P. McGlynn, T. Azumi, and M. Kinoshita, *Molecular Spectroscopy of the Triplet State* (Prentice-Hall, Englewood Cliffs, N.J., 1969).

[15] H. H. Jaffé and M. Orchin, *Theory and Applications of Ultraviolet Spectroscopy* (Wiley, New York, 1962).

PROBLEMS

7.1   The moment of inertia of the NH radical is 1.68×10^{-40} g cm². At what frequency would you expect to detect the transition $J = 2$ to $J = 3$?

7.2   In the far-ir spectrum of HBr is a series of lines having a separation of 16.94 cm⁻¹. Calculate the moment of inertia and the internuclear separation of HBr from this datum.

Ans. 3.30×10^{-40} g cm²; 1.42 Å

7.3   In the near-ir spectrum of carbon monoxide ($C^{12}O^{16}$) there is an intense band at 2144 cm⁻¹. Calculate (a) the fundamental vibration frequency of CO; (b) the period of the vibration; (c) the force constant; (d) the zero-point energy of CO in kcal mole⁻¹.

Ans. (a) 6.43×10^{13} sec⁻¹; (b) 1.56×10^{-14} sec;
(c) 1.85×10^{6} dyn cm⁻¹; (d) 2.88 kcal mole⁻¹

7.4   In the microwave spectrum of $C^{12}O^{16}$ the $J = 0 \rightarrow 1$ rotational transition has been measured at 115 271.204 MHz. Calculate the moment of inertia (in units of amu Å²) and the internuclear separation of CO.

7.5   Sketch the proton NMR spectra you would expect to find for (a) isopropyl alcohol; (b) *t*-butyl alcohol. Indicate the estimated chemical shifts, fine structure due to spin-spin splitting, and relative intensities of the lines. Discuss briefly.

7.6   (a) Which of the following chemical species would you expect to be detectable by ESR?

$$CH_2 \quad CH_3 \quad C_6H_6 \quad C_6H_6{}^+ \quad H_2 \quad He_2{}^+ \quad O_2 \quad O_2{}^+$$

(b) Sketch the ESR absorption spectrum for each of the above, if such a spectrum exists. Assume that all the nuclei are the most abundant isotope.

7.7   A compound is known to have the molecular formula C_2H_4O. Devise a method for using NMR to determine its structure.

7.8   Calculate the relative populations of the $J = 1, 2, 3, 4$ levels of CO at room temperature (25° C) and at liquid-nitrogen temperature (70° K). Make bar graphs representing the relative populations at these two temperatures. What effect will cooling the sample have upon the intensities of the ir rotational-vibrational absorption spectrum of CO?

7.9   From the discussion of the angular part of the H-atom wave function in Chapter 3, write down $S_{JM}(\theta, \phi) = \Theta_{JM}(\theta)\Phi_M(\phi)$ for $J = 0, M = 0; J = 1, M = 0;$ $J = 1, M = -1; J = 1, M = 1.$ [See Table 3.1 for $\Theta_{JM}(\theta)$ and Eq. 3.47 for $\Phi_M(\phi)$.] Show that $S_{00}, S_{10}, S_{1-1},$ and S_{11} are eigenfunctions of $\mathbf{M}^2$ with eigenvalues given by

$J(J + 1)\hbar^2$ by substituting each of these functions into Eq. 7.15. Is the real combination $S_{1-1} + S_{11}$ an eigenfunction of $\mathbf{M}^2$?

7.10 (a) Calculate the electric dipole transition moment integrals

$$\int\int S_{JM}\,\mu_{el}\,S_{J'M'}\,\sin\theta d\theta d\phi$$

where the electric dipole moment μ_{el} has components $\mu_0\cos\phi\sin\theta$, $\mu_0\sin\phi\sin\theta$, and $\mu_0\cos\theta$, and $(J, M, J', M') = (0, 0, 0, 0), (0, 0, 1, 0), (0, 0, 1, 1), (0, 0, 1, -1)$; μ_0 is a constant. Do your results agree with Eq. 7.24?

(b) Calculate the expectation (average) value of $(\mu_{el})_z = \mu_0\cos\theta$ for a diatomic molecule in the rotational states $(J, M) = (0, 0), (1, 0), (1, 1), (1, -1)$. Is the expectation value of any component μ_{el} nonzero for any rotational state of the molecule?

7.11 Using Eq. 7.29, plot the relative rotational energies for $J = 0, 1$, and 2, $M = -J, -J + 1, \ldots J$, of a polar diatomic molecule with and without an applied electric field. Draw in the strongly allowed $(\Delta J = \pm 1)$ transitions and sketch the rotation spectrum for the two cases. Express the permanent dipole moment μ_0 in terms of ε and the spacing between the two lines of longest wavelength when ε is small.

7.12 Calculate k_e in eV Å^{-2} for H_2, HCl, and Cl_2 from ω_e given in Table 7.3 and the atomic masses. Plot harmonic potentials (parabolas) having the curvature k_e for each of these molecules. Draw horizontal lines to indicate the dissociation energy D_e and the vibrational levels $(v + \frac{1}{2})hc\omega_e$ lying below D_e on the plots.

7.13 A Maclaurin expansion of the dipole moment μ of a diatomic molecule about $\rho = R - R_e = 0$, neglecting powers of ρ greater than the first, gives

$$\mu(\rho) = \mu_0 + \left(\frac{d\mu}{d\rho}\right)_0 \rho + \cdots$$

Using this approximation for $\mu(x)$, calculate transition moments

$$\int_{-\infty}^{+\infty} \psi_{v''}(\rho)\mu(\rho)\psi_{v'}(\rho)\,d\rho$$

for $(v'', v') = (0, 0), (1, 0), (1, 1)$, and $(2, 1)$. Assume the vibrational wave functions $\psi_v(\rho)$ to be solutions to the harmonic-oscillator problem:

$$\psi_0(\rho) = \left(\frac{\alpha}{\pi}\right)^{1/4} \exp(-\tfrac{1}{2}\alpha\rho^2)$$

$$\psi_1(\rho) = \left(\frac{4\alpha^3}{\pi}\right)^{1/4} \rho \exp(-\tfrac{1}{2}\alpha\rho^2)$$

$$\psi_2(\rho) = \left(\frac{\alpha}{4\pi}\right)^{1/4} (2\alpha\rho^2 - 1) \exp(-\tfrac{1}{2}\alpha\rho^2)$$

$$\alpha = \frac{2\pi\mu_A \nu_e}{\hbar} \qquad (\mu_A, \text{ reduced mass})$$

7.14   Find the allowed P- and R-branch lines in terms of ω_e and $\tilde{B}_e$ for emissson vibrational-rotational spectra, that is, for $\Delta v = -1$, $\Delta J = \pm 1$. Sketch the emission spectrum, showing the relative positions of the lines and indicate the transitions from which they arise and the spacing between them. (Ignore vibration-rotation interaction.)

7.15   Which lines in the R and P branches of the vibrational-rotational absorption spectrum of HCl have maximum intensity at 300 and at 1000° K?

7.16   Assuming equal transition probabilities for the allowed lines, use Eq. 7.46 to calculate the relative intensities of the rotational-vibrational lines in the absorption spectrum of HBr at 300° K. Plot lines with the appropriate spacing and length to represent the wave numbers and intensities of the spectrum and compare your result qualitatively with Fig. 7.12. Is the assumption of equal transition probabilities valid?

7.17   Show that ω_v as defined in Eq. 7.65 is equal to the overtone spacing $\Delta \tilde{\nu}_{v+1}$.

7.18   (a) Derive Eq. 7.69.

(b) Use the vibrational data for H_2 given in Table 7.6 to determine D_e by means of Eq. 7.70.

(c) Use the data in Table 7.6 for obtaining D_e by a Birge–Sponer extrapolation. Compare this result with that of part (b).

7.19   Derive Eq. 7.75 by substituting Eq. 7.73 for R_e' in Eq. 7.74, expanding in powers of $J(J+1)$, and keeping only the first two terms; show that $\tilde{D}_e$ is given by Eq. 7.76.

7.20   Compare the ratios of spectroscopic constants calculated from Eq. 7.82 with those obtained from the experimental values in Table 7.7.

7.21   Show the occupation of MO's and spin states of the electrons in the $^5\Pi_g$ state of O_2, the configuration of which is given in Table 7.8.

7.22   Discuss by means of a simple MO diagram how the $^3\Delta_u$, $^3\Sigma_u^+$, and $^1\Sigma_u^-$ states arise from the $(3\sigma_g)^2 (1\pi_u)^3 (1\pi_g)^3$ configuration of O_2, and use Hund's rules to determine the energy ordering of these states.

7.23   Wave numbers $\tilde{\nu}$ for the $(J' = 0, v' = 0 \rightarrow J' = 0, v'' = v)$ transitions in the Schumann–Runge O_2 bands are[1]

v	$\tilde{\nu}$	v	$\tilde{\nu}$	v	$\tilde{\nu}$
0	49357.6	2	50710.7	4	51969.8
1	50045.6	3	51352.2	5	52561.6

[1] A. G. Gaydon, *Dissociation Energies* (Chapman and Hall, London, 1968), 3rd ed., p. 72.

v	$\tilde{\nu}$	v	$\tilde{\nu}$	v	$\tilde{\nu}$
6	53122.8	12	55784.59	17	56852.41
7	53656.8	13	56085.51	18	56954.54
8	54158.9	14	56340.47	19	57030.18
9	54624.4	15	56550.54	20	57082.83
10	55053.3	16	56719.50	21	57114.77
11	55441.5				

(a) From these data determine ω_e and $\omega_e x_e$ for the upper state ($B\,^3\Sigma_u^-$).

(b) Plot the difference $\Delta\tilde{\nu}_v = \tilde{\nu}_{v+1} - \tilde{\nu}_v$ near the convergence limit and extrapolate to $\Delta\tilde{\nu} = 0$; from your result determine an accurate value for the wave number of the convergence limit. Use this value to calculate the value of D_0 for O_2 in the ground state ($X\,^3\Sigma_g^-$), and compare with the value given in Table 7.3.

(c) From the value of $\tilde{\nu}_{00}$, and results of the parts (a) and (b), determine D_0 and D_e for the $B\,^3\Sigma_u^-$ state.

7.24 The C—O bond length in linear CO_2 is 1.1621 Å. Calculate the moment of inertia of CO_2, and plot the rotational energy levels in cm^{-1} for $0 \leq J \leq 5$.

7.25 The N—H bond distance in NH_3 is 1.014 Å, and each N—H bond makes the angle 67°58′ with the threefold axis. Calculate I_A and I_B; is NH_3 prolate or oblate? Show that Eq. 7.105 is satisfied. Plot the rotational energy levels of NH_3 for $0 \leq J \leq 3$ and $0 \leq |K| \leq J$.

7.26 Ozone has a bond length of 1.278 Å and a bond angle of 116.8°. Calculate the three moments of inertia I_A, I_B, and I_C (in order of increasing magnitude). Construct a correlation diagram similar to Fig. 7.20 in which the levels of the prolate system obtained by setting $I_C = I_B$ are connected with those of the oblate system obtained by setting $I_A = I_B$.

7.27 Calculate the three principal moments of inertia of tetrahedral CH_4 (see Fig. 7.18d) in terms of the C—H bond length R. Show that the rotational energy levels are given by Eq. 7.103 and plot the levels in cm^{-1} for $0 \leq J \leq 5$.

7.28 Show that for any nonlinear planar molecule $I_C = I_A + I_B$.

7.29 The N_3H molecule is found to have the principal moments of inertia $I_A = 1.3759 \times 10^{-40}$ g cm^2, $I_B = 69.38 \times 10^{-40}$ g cm^2, and $I_C = 70.75 \times 10^{-40}$ g cm^2. Is the molecule linear? Can it have a threefold axis? Is it planar?

7.30 Formic acid (HCOOH) has the moments of inertia $I_A = 2.977 \times 10^{-40}$ g cm^2, $I_B = 21.65 \times 10^{-40}$ g cm^2, and $I_C = 24.62 \times 10^{-40}$ g cm^2. Propose a structure that is consistent with these data. (*Hint:* Assume reasonable values for the bond distances and calculate the angles.)

7.31 The pure rotational spectra of NH_3 and ND_3 consist of lines spaced 19.89 cm^{-1} apart and 10.26 cm^{-1} apart, respectively. Assuming that the two molecules have identical structures, use Eqs. 7.107 and 7.105 to calculate R and α. Compare your results with the data given in Problem 7.25.

7.32 Show that the vibrational kinetic energy for the symmetric triatomic molecule Y—X—Y with bond angle α is given by Eq. 7.135, where S_1, S_2, and S_3 are the vibrational displacement coordinates defined in Eq. 7.133 and Fig. 7.22.

7.33 (a) Show that substitution of Eqs. 7.140 and 7.139 into Eq. 7.141 yields Eq. 7.142.

(b) Show that λ_1 and λ_2 satisfy Eq. 7.145 if their sum and product are given by Eqs. 7.147.

7.34 Derive Eqs. 7.148 for the Y—X—Y molecule of Fig. 7.22. (*Hint:* Ignore powers of the displacement greater than the first.)

7.35 Derive Eqs. 7.151 by substituting Eqs. 7.148 into Eq. 7.150 and comparing the result with Eq. 7.138.

7.36 (a) Assume that a nonlinear symmetric triatomic molecule Y—X—Y, with bond angle α, has the potential-energy function given by Eq. 7.150. Show that if k_1 and k_{12} are both real, the observed vibrational frequencies must obey the inequality

$$(\lambda_1 + \lambda_2)^2 > 4\lambda_1\lambda_2 \left[1 + \left(\frac{m_X}{m_X + 2m_Y} \right) \tan^2 \frac{\alpha}{2} \right]$$

where $\lambda_i = 4\pi^2 \nu_i^2$.

(b) The vibrational frequencies of H_2O ($\alpha = 105°$) are

$$\tilde{\nu}_1 = 3642 \text{ cm}^{-1}$$
$$\tilde{\nu}_2 = 1595 \text{ cm}^{-1}$$
$$\tilde{\nu}_3 = 3756 \text{ cm}^{-1}$$

Is the potential-energy function of part (a) appropriate for this molecule?

7.37 Consider a nonlinear symmetric triatomic molecule Y—X—Y with equilibrium bond angle α and vibrational potential-energy function given by Eq. 7.152. Do the data for H_2O given in Problem 7.36 allow the use of this potential-energy function for the molecule?

7.38 Use the potential-energy function of Problem 7.37 (but set $k_{12} = 0$) and the vibrational frequencies for H_2O given in Problem 7.36 to calculate the bond angle of H_2O that satisfies Eq. 7.147.

7.39 (a) Show that an equilateral triangle molecule X_3 with the central-force-field potential

$$2V = k \left[(\Delta r_1)^2 + (\Delta r_2)^2 + (\Delta r_3)^2 \right]$$

has the frequencies

$$\nu_1 = \frac{1}{2\pi} \left(\frac{3k}{m_X} \right)^{1/2}$$
$$\nu_2 = \nu_3 = \frac{1}{2\pi} \left(\frac{3k}{2m_X} \right)^{1/2}$$

(b) Find the relation between the Cartesian displacement coordinates and the normal-mode coordinates Q_1, Q_2, and Q_3 for this system. Diagram the normal-mode displacements and compare with Fig. 7.23.

7.40 (a) The frequency of the symmetric (ir inactive) vibration of equilateral H_3^+ has been calculated to be ~ 3354 cm^{-1}. What frequency would you expect to observe in the ir spectrum of this ion if it is represented by a central-force field?

(b) What is the value of the force constant k? How does it compare with the force constant for $H_2(\bar{\nu} = 4395$ cm$^{-1})$?

7.41 (a) Show that for a modified valence-force field (Eq. 7.152) the vibrational frequencies of a linear symmetric molecule Y—X—Y are given by

$$4\pi^2 m_Y \nu_1^2 = k_1 + k_{12}$$
$$4\pi^2 m_Y \nu_2^2 = 2pk_\delta$$
$$4\pi^2 m_Y \nu_3^2 = p(k_1 - k_{12})$$

where $p = 1 + \dfrac{2m_Y}{m_X}$.

(b) The observed vibrational frequencies of CO_2 are 1337, 667, and 2349 cm^{-1}. Calculate the values of k_1, k_{12}, and k_δ in dyn cm^{-1}.

(c) Can the bond interaction constant k_{12} be neglected? What is the percent error in the calculated value of ν_1 if k_{12} is set equal to zero and k_1 is determined by ν_3 alone?

7.42 (a) Consider the linear asymmetric molecule XYZ with equilibrium bond lengths r_{10}(X—Y) and r_{20}(Y—Z). Find the values of a, b, c, and d (in terms of m_X, m_Y, and m_Z, r_{10}, and r_{20}) that make the following symmetry coordinates pure vibrations (i.e., containing no translation or rotation for small displacements), where x is the molecular axis:

$$S_1: \quad \Delta x_X = \Delta x_Y = \Delta x_Z = 0$$
$$\Delta y_X = -aS_1 \qquad \Delta y_Y - S_1 \qquad \Delta y_Z = -bS_1$$
$$S_2: \quad \Delta x_X = S_2 \qquad \Delta x_Y = 0 \qquad \Delta x_Z = -cS_2$$
$$\Delta y_X = \Delta y_Y = \Delta y_Z = 0$$
$$S_3: \quad \Delta x_X = -S_3 \qquad \Delta x_Y = dS_3 \qquad \Delta x_Z = -S_3$$
$$\Delta y_X = \Delta y_Y = \Delta y_Z = 0$$

Ans.

$$a = \frac{m_Y}{m_X}\left(\frac{r_{10}}{r_{10}+r_{20}}\right) \quad b = \frac{m_Y}{m_Z}\left(\frac{r_{10}}{r_{10}+r_{20}}\right) \quad c = \frac{m_X}{m_Z} \quad d = \frac{m_X + m_Z}{m_Y}$$

(b) Which of the vibrations of part (a) is doubly degenerate?

(c) Find the constants d_{ij} in the kinetic-energy expression

$$2T = d_{11}\dot{S}_1^2 + d_{22}\dot{S}_2^2 + d_{33}\dot{S}_3^2 + 2(d_{12}\dot{S}_1\dot{S}_2 + d_{23}\dot{S}_2\dot{S}_3 + d_{13}\dot{S}_1\dot{S}_3)$$

(d) Using the simple valence-force field

$$2V = k_1(r_1 - r_{10})^2 + k_2(r_2 - r_{20})^2 + k_\delta \left(\frac{r_{10}r_{20}}{r_{10} + r_{20}}\right)^2 \delta^2$$

where δ is the change in bond angle, find the constants c_{ij} in the potential-energy expression

$$2V = C_{11}S_1^2 + c_{22}S_2^2 + c_{33}S_3^2 + 2(c_{12}S_1S_2 + c_{23}S_2S_3 + c_{13}S_1S_3)$$

Hint: Show that

$$r_1 - r_{10} = \left(\frac{m_X + m_Y + m_Z}{m_Y}\right)S_3 - S_2$$

$$r_2 - r_{20} = -\left(\frac{m_X + m_Y + m_Z}{m_Y}\right)S_3 - \frac{m_X}{m_Y}S_2$$

$$\left(\frac{r_{10}r_{20}}{r_{10} + r_{20}}\right)\delta = -qS_1$$

where

$$q = 1 + \frac{m_Y}{m_X}\left(\frac{r_{20}}{r_{10} + r_{20}}\right)^2 + \frac{m_Y}{m_Z}\left(\frac{r_{10}}{r_{10} + r_{20}}\right)^2$$

(e) Are S_1, S_2, and S_3 normal-mode coordinates?

7.43 The molecule N_2O has a microwave spectrum consisting of lines uniformly spaced 0.838 cm^{-1} apart. Its three vibrational frequencies, 2223.76, 588.78, and 1284.91 cm^{-1} are each ir active and Raman active. What structures of N_2O are consistent with these data?

7.44 Devise an experiment to determine whether an unknown molecule has a center of symmetry.

7.45 Figure 7.27 shows the normal-mode vibrations of an octahedral MX_6 molecule.

(a) List the frequencies that correspond to g vibrations and those that correspond to u vibrations.

(b) Only one of the vibrations is both ir and Raman inactive. Which one must it be?

(c) List the ir- and Raman-active vibrations.

> *Ans.* (a) g: ν_1, ν_2, ν_5; u: ν_3, ν_4, ν_6
> (b) ν_6
> (c) ir: ν_3, ν_4; Raman: ν_1, ν_2, ν_5

7.46 Show that when shielded Slater orbitals are used for obtaining $<r_i^{-1})$ in Eq. 7.181 for σ_{at}

$$\sigma_{at} = 17.754 \sum_i \left(\frac{\zeta_i}{n_i}\right) \times 10^6$$

7.47   Calculate σ_{at} for Al, Si, and P from Eq. 7.181 (see Problem 7.46 and Table 4.9).

7.48   Calculate the hyperfine constant a for a ground-state H atom (Eq. 7.202).

7.49   Given the hyperfine structure constants $a_\alpha = 4.92$ G and $a_\beta = 1.80$ G in the napthalene anion, calculate the value of a_α if the α hydrogens are replaced by deuterium atoms (see Table 7.12). Sketch the ESR spectrum expected for the napthalene-1-d_1 anion.

7.50   Apply the linear variational principle (see Section 5.64 and Problem 5.29) to the problem of a ground-state linear rigid-rotor molecule, with a permanent electric dipole moment μ_0, in a weak static electric field ε. Take the trial wave function to be of the form

$$\chi(\theta, \phi) = aS_{00}(\theta, \phi) + bS_{10}(\theta, \phi)$$

where

$$S_{00}(\theta, \phi) = \frac{1}{(4\pi)^{1/2}}$$

$$S_{10}(\theta, \phi) = \left(\frac{3}{4\pi}\right)^{1/2} \cos \theta$$

Find the value of b/a that minimizes the energy. Expand the expression for the energy through the second power in ε^2 [using the approximation $(y^2 + x^2)^{1/2} \simeq y + x^2/2y$ for small x]. Compare your result with Eq. 7.29b. What is the electric polarizability of the molecule? *Hint:* The Hamiltonian has the form

$$H = H_0 - \mu_0\, \varepsilon \cos \theta$$

since the direction of ε can be arbitrarily taken to be the z direction. Since S_{00} and S_{10} are eigenfunctions of the zero-field Hamiltonian H_0, they satisfy the equations

$$H_0 S_{00} = 0 \quad H_0 S_{10} = \left(\frac{\hbar^2}{I_e}\right) S_{10}$$

Ans. $b/a \simeq [I_e/(\sqrt{3}\,\hbar)^2]\mu_0\varepsilon$ for small values of $(\mu_0\varepsilon)$

7.51   Use Eq. 7.182 to diagram the energy levels of HD in a magnetic field of 10,000 G. Show that the allowed transitions correspond to those indicated in Figs. 7.42a and 7.42b, which treat the H and D levels independently for simplicity.

7.52   Calculate ω_e for HCl from Eq. 7.62 and the vibrational absorption data given in Table 7.5.

7.53   Show that if ψ_{vib} has the factored form given in Eq. 7.123, the Schrödinger equation for vibration separates into the set of equations given in Eq. 7.124.

7.54   According to Eq. 1.39, the angular momentum of a particle of mass m moving in the xy plane is given by

$$M_z = p_\phi = mr^2\omega$$

where $\omega = \dot\phi$ and

$$x = r\cos\phi \qquad y = r\sin\phi$$

Show that

$$x\dot{y} - y\dot{x} = r^2\omega$$

and thus that

$$M_z = m(x\dot{y} - y\dot{x})$$

7.55   Determine whether the electric dipole transitions $2b_2 \leftrightarrow 2b_1$ ($n \leftrightarrow \pi^*$) and $1b_1 \leftrightarrow 2b_1$ ($\pi \leftrightarrow \pi^*$) in formaldehyde are allowed or forbidden; that is, determine whether the integrals

$$\int 2b_2 \begin{Bmatrix} x \\ y \\ z \end{Bmatrix} 2b_1 \, dv \qquad \int 1b_1 \begin{Bmatrix} x \\ y \\ z \end{Bmatrix} 2b_1 \, dv$$

are nonzero or zero. (*Hint:* From Fig. 7.29b, find out whether the integrands change sign or not under 180° rotation about the z axis and under reflection in the yz plane; then use the fact that the integrals themselves must remain unchanged by these operations.)

7.56   (a) Using Eq. 7.189, derive expressions for the three absorption transition frequencies in which $\Delta M_{Ia} = -1$, ignoring the coupling term involving J_{ab}.

(b) Repeat the derivation of part (a), but include the coupling term in Eq. 7.189, taking into account the possible combinations of M_{Ia} and M_{Ib}. Compare your results with Eqs. 7.190.

7.57   Write down the expression for the electrostatic potential υ at the point P of Fig. 7.54a. Expand $\upsilon(R, r, \theta)$ in a Maclaurin series in (r/R). Show that if only the first nonzero term containing (r/R) is retained (i.e., if $R \gg r$), the potential reduces to that of Eq. 7.205.

7.58   Derive Eq. 7.207 in the following steps:

(a) Taking the value at the origin to be zero, expand the general electric potential $\upsilon(x, y, z)$ in a Maclaurin series to give

$$\upsilon(x, y, z) = x\left(\frac{\partial \upsilon}{\partial x}\right)_0 + y\left(\frac{\partial \upsilon}{\partial y}\right)_0 + z\left(\frac{\partial \upsilon}{\partial z}\right)_0$$

$$+ \frac{1}{2}x^2\left(\frac{\partial^2 \upsilon}{\partial x^2}\right)_0 + \frac{1}{2}y^2\left(\frac{\partial^2 \upsilon}{\partial y^2}\right)_0 + \frac{1}{2}z^2\left(\frac{\partial^2 \upsilon}{\partial z^2}\right)_0$$

$$+ xy\left(\frac{\partial^2 \upsilon}{\partial x \partial y}\right)_0 + yz\left(\frac{\partial^2 \upsilon}{\partial y \partial z}\right)_0 + zx\left(\frac{\partial^2 \upsilon}{\partial z \partial x}\right)_0 + \cdots$$

(b) If $\upsilon(x, y, z)$ is cylindrically symmetric about the z axis, then

$$\upsilon(-x, y, z) = \upsilon(x, -y, z) = \upsilon(y, x, z) = \upsilon(x, y, z)$$

Furthermore, if there is no source of the potential $\mathfrak{v}$ at the origin, electromagnetic theory (Poisson's equation) requires that

$$\left(\frac{\partial^2 \mathfrak{v}}{\partial x^2}\right)_0 + \left(\frac{\partial^2 \mathfrak{v}}{\partial y^2}\right)_0 + \left(\frac{\partial^2 \mathfrak{v}}{\partial x^2}\right)_0 = 0$$

so that for cylindrical symmetry

$$\left(\frac{\partial^2 \mathfrak{v}}{\partial x^2}\right)_0 = \left(\frac{\partial^2 \mathfrak{v}}{\partial y^2}\right)_0 = -\frac{1}{2}\left(\frac{\partial^2 \mathfrak{v}}{\partial z^2}\right)_0$$

For this case show that

$$\mathfrak{v}(x, y, z) = z\left(\frac{\partial \mathfrak{v}}{\partial z}\right)_0 + \tfrac{1}{4}(3z^2 - r^2)\left(\frac{\partial^2 \mathfrak{v}}{\partial z^2}\right)_0 + \cdots$$

where $r^2 = x^2 + y^2 + z^2$

(c) Show that the energy of interaction of $\mathfrak{v}$ with a set of charges e located at the points (x_i, y_i, z_i) is

$$\Delta E = e \sum_i (x_i, y_i, z_i) = \frac{e}{4}\sum_i (3z_i^2 - r_i^2)\left(\frac{\partial^2 \mathfrak{v}}{\partial z^2}\right)_0 + \cdots$$

provided the z component of the dipole moment for the set of charges is zero.

(d) If the positions of the charges of part (c) are determined in a coordinate system $\bar{x}$, $\bar{y}$, $\bar{z}$ obtained by rotating the x, y, z coordinate system through the angle θ about the x axis (Fig. 7.54d), then the positions in the x, y, z system are given by

$$x_i = \bar{x}_i$$
$$y_i = \bar{z}_i \sin\theta - \bar{y}_i \cos\theta$$
$$z_i = \bar{z}_i \cos\theta + \bar{y}_i \sin\theta$$

Show that if the charge distribution is cylindrically symmetric about the $\bar{z}$ axis

$$\left[\text{i.e., if } \sum_i \bar{x}_i^2 = \sum_i \bar{y}_i^2 = \frac{1}{2}\sum_i (r_i^2 - \bar{z}_i^2)\right],$$

the interaction energy is

$$\Delta E = \frac{e}{4}\sum_i (3\bar{z}_i^2 - r_i^2)\left(\frac{\partial^2 \mathfrak{v}}{\partial z^2}\right)_0 [\tfrac{1}{2}(3\cos^2\theta - 1)] + \cdots$$

$$= -\frac{e}{4}\sum_i (3\bar{z}_i^2 - r_i^2)\left(\frac{\partial \mathcal{E}_z}{\partial z}\right)_0 [\tfrac{1}{2}(3\cos^2\theta - 1)] + \cdots$$

$$= -\frac{e^2 qQ}{4} [\tfrac{1}{2}(3\cos^2\theta - 1)] + \cdots$$

7.59   For the charge distribution of Figs. 7.54a and 7.54c, show that eQ is equal to $2er^2$. For a charge e at point P in Fig. 7.54a, show that if $\theta = 0$, the field gradient eq at the center of the two dipoles is $-2e/R^3$. Evaluate ΔE_Q from Eq. 7.207 for this case and compare the result with Eq. 7.206. Note that if the particle at point P is an electron its charge is $-e$, so that a minus sign is introduced on the right-hand side of Eq. 7.206.

For the following problems a programmable desk calculator or digital computer will be useful.

C7.1   Calculate the zero-point energy in eV from ω_e and $\omega_e x_e$ for H_2, HCl, and Cl_2 listed in Table 7.3, and obtain D_e for each of these molecules. Compare plots of a Morse function fitted to D_e, ω_e, and R_e with those of the parabolas plotted for these molecules in Problem 7.12.

C7.2   (a) Derive Eq. 7.86.

(b) For the O_2 molecule the spectroscopic constants (in cm^{-1}) for the $X^3\Sigma_g^-$ and $B^3\Sigma^-$ states are

State	ω_e	$\omega_e x_e$	$\tilde{B}_e$
$X^3\Sigma_g^-$	1580.361	12.0730	1.44567
$B^3\Sigma_u^-$	700.36	8.002	0.819

The $\tilde{\nu}_{00}$ transition is at 49357.6 cm^{-1}. Assuming the rigid-rotor, harmonic-oscillator model, use Eq. 7.86 to calculate the first-order rotational lines in the $v' = 0$, $v'' = 0$ transition and plot the Fortrat parabola. Draw appropriately spaced spectral lines, using full lines for the R branch and dotted lines for the P branch. Is $R_e(B^3\Sigma_u^-)$ greater or smaller than $R_e(X^3\Sigma_g^-)$? Compare the answer obtained from the qualitative appearance of your Fortrat parabola with that obtained from Fig. 7.15.

C7.3   Using a potential-energy function for H_2O and D_2O of the form

$$2V = k_1[(r_1 - r_0)^2 + (r_2 - r_0)^2] + 2k_{12}(r_1 - r_{10})(r_2 - r_{20}) + k_\delta r_0^2 \delta^2$$

where r_1, r_2 are the two O—H bond lengths, r_0 is the equilibrium bond length, and δ is the change in the bond angle from the equilibrium value, find values of k_1, k_{12}, and k that give the best least-squares fit to $\lambda_1 + \lambda_2$, $\lambda_1\lambda_2$, and λ_3 for both H_2O and D_2O. The observed frequencies are

	H_2O	D_2O
$\tilde{\nu}_1$ (cm^{-1})	3652	2666
$\tilde{\nu}_2$	1595	1179
$\tilde{\nu}_3$	3756	2789

The equilibrium bond angle is $105°$.

Compare your results with the values reported for H_2O by G. Herzberg [*Infrared and Raman Spectra* (Van Nostrand, Princeton, N.J., 1945)]: $k_1 = 7.66 \times 10^5$ dyn cm^{-1}, $k_{12} = -0.097 \times 10^5$ dyn cm^{-1}, and $k_\delta = 0.703 \times 10^5$ dyn cm^{-1}.

C7.4 The following frequency shifts (in cm^{-1}) are observed in the pure rotational Raman spectrum of N_2 [B. P. Stoicheff, *Can. J. Phys.* **32**, 630 (1954)]:

19.908 (in cm^{-1})	75.566	123.189
27.857	83.504	131.093
35.812	91.455	139.024
43.762	99.406	146.901
51.721	107.327	154.849
59.662	115.243	162.752
67.629		

(a) Assuming that all of the molecules are initially in the ground vibrational state, find the values of $\tilde{B}_0$ and $\tilde{D}_0$ that give the best least-squares fit to the data.

(b) From your value for $\tilde{B}_0$ determine the internuclear distance and compare with the value of R_e given for N_2 in Table 7.3.

(c) Compare your value for $\tilde{D}_0$ with the value for $\tilde{D}_e$ obtained by assuming the Morse relation

$$\tilde{D}_e = \frac{4\tilde{B}_e^3}{\omega_e^2}$$

(See Table 7.3 for the value of ω_e for N_2.)

Appendix

ELECTROSTATIC UNITS

The electrostatic unit of charge is defined by Coulomb's law for a vacuum, namely,

$$F = \frac{q_1 q_2}{4\pi \epsilon_0 r^2} \tag{A1}$$

where q_1 and q_2 are the charges on two particles, r is the distance between the charges, F is the force, and ϵ_0 is a constant. Since F has the units of mass (M) times length (L) divided by the time squared (T^2), we write the dimensional equation

$$[F] = [MLT^{-2}] \tag{A2}$$

From Eqs. (A1) and (A2) we have

$$[q] = [\epsilon_0^{1/2} M^{1/2} L^{3/2} T^{-1}] \tag{A3}$$

In cgs units (i.e., with the gram, centimeter, and second taken as the units of M, L, and T, respectively, and letting $4\pi\epsilon_0 = 1$) the unit of q is

$$[q] = [M^{1/2} L^{3/2} T^{-1}] = 1 \text{ esu or statcoulomb} \tag{A4}$$

The unit of current is the amount of current that passes unit charge in unit time; thus in cgs units the current i is

$$[i] = [M^{1/2}L^{3/2}T^{-2}] = 1 \text{ esu or statampere} \tag{A5}$$

The unit of electric field $\mathcal{E}$ obtained from the defining equation

$$F = q\mathcal{E} \tag{A6}$$

From Eqs. (A2), (A3), and (A6), we have

$$[\mathcal{E}] = [M^{1/2}L^{-1/2}T^{-1}] = 1 \text{ esu or statvolt cm}^{-1} \tag{A7}$$

in cgs units. The electric potential $\mathcal{V}$ is defined by

$$\mathcal{E}_x = -\frac{\partial \mathcal{V}}{\partial x} \tag{A8}$$

and similarly from the other components of the field; in cgs units we have from Eqs. (A7) and (A8)

$$[\mathcal{V}] = [M^{1/2}L^{1/2}T^{-1}] = 1 \text{ esu, or statvolt} \tag{A9}$$

ELECTROMAGNETIC UNITS

The electromagnetic unit of charge is defined by Ampere's law of magnetic forces, namely, that the force per unit length between two infinitely long parallel conductors a distance d apart is given by

$$\frac{dF}{dl} = \frac{\mu_0}{2\pi} \frac{i_1 i_2}{d} \tag{A10}$$

where i_1 and i_2 are the currents in the conductors and μ_0 is a constant. In electromagnetic units, the cgs system for M, L, and T are used, but one chooses $\mu_0/4\pi = 1$. The electromagnetic unit of current is, therefore,

$$[i] = [M^{1/2}L^{1/2}T^{-1}] = 1 \text{ emu or abampere} \tag{A11}$$

From Eq. (A11) we obtain

$$[q] = [M^{1/2}L^{1/2}] = 1 \text{ emu or abcoulomb} \tag{A12}$$

The unit of magnetic induction $\mathcal{B}$ is the gauss (G) defined by

$$F = qv\mathcal{B} \sin (\mathbf{v}, \mathbf{\mathcal{B}}) \tag{A13}$$

where $\mathbf{v}$ is the velocity of the charge q; this gives

$$[\mathcal{B}] = [M^{1/2}L^{-1/2}T^{-1}] = 1 \text{ G} \tag{A14}$$

In a vacuum, the magnetic field $\mathcal{H}$ is related to the magnetic induction by

$$\mathcal{B} = \mu_0 \mathcal{H} \tag{A15}$$

where μ_0 is the permeability of free space. Thus, since μ_0 is dimensionless in the electromagnetic system, $\mathcal{H}$ and $\mathcal{B}$ have the same dimension. The unit of $\mathcal{H}$ is called the oersted in cgs units.

$$[\mathfrak{K}] = [M^{1/2}L^{-1/2}T^{-1}] = 1 \text{ oersted} \tag{A16}$$

The combination of constants $(\epsilon_0\mu_0)^{-1/2}$ has the dimensions LT^{-1} and from electromagnetic theory,

$$(\epsilon_0\mu_0)^{-1/2} = c \text{ (the velocity of light)} \tag{A17}$$

Thus, with $\mu_0 = 4\pi$,

$$\epsilon_0 = (4\pi)^{-1}c^{-2} \tag{A18}$$

From Eqs. (A18), (A1), and (A3), we obtain the relation

$$1 \text{ abcoulomb} = c \text{ statcoulomb} \tag{A19}$$

GAUSSIAN-MIXED UNITS

Although all electric and magnetic quantities can be expressed in either esu or emu, the most convenient system of units for our purposes is the Gaussian-mixed system in which q and $\mathcal{E}$ are expressed in esu and i, $\mathfrak{B}$, and $\mathfrak{K}$ are expressed in emu. Unless otherwise stated (for example, occasional use of esu for current), this system is followed throughout the text.

RATIONALIZED MKS UNITS

By using the kilogram, meter, and second as the units of M, L, and T, respectively, and choosing

$$\frac{4\pi}{\mu_0} = 10^{-7}$$
$$4\pi\epsilon_0 = 10^{-7}c^{-2}$$

the units of the electric and magnetic quantities coincide with the "practical" units. In this system

$$q : 1 \text{ coulomb} = \frac{c}{10} \text{ statcoulomb}$$

$$= \frac{1}{10} \text{ abcoulomb}$$

$$\Phi : 1 \text{ volt} = \frac{10^8}{c} \text{ statvolt}$$

$$\mathcal{E} : 1 \text{ volt m}^{-1} = \frac{10^6}{c} \text{ statvolt cm}^{-1}$$

$$\mathfrak{B} : 1 \text{ weber m}^{-2} = 10^4 \text{ G}$$

$$\mathfrak{K} : 1 \text{ ampere-turn m}^{-1} = 4\pi \times 10^{-3} \text{ oersted}$$

ATOMIC UNITS

The quantum-mechanical equations describing electronic properties of atoms and molecules are considerably simplified if Hartree's atomic units (a.u.) are adopted. In these units, the units of charge and mass are taken to be the charge and rest mass of the electron, the unit of length is the Bohr radius, and the unit of time is chosen so that $\hbar$ has the numerical value of unity. Table A.1 lists the a.u., based upon the SI values of the physical

Table A.1   *Atomic units*[a]

Quantity	Unit	Physical significance	Value in cgs units
Charge	e	Electron charge	4.80298×10^{-10} esu
Mass	m or m_e	Electron mass	9.1091×10^{-28} gm
Length	$a_0 = \hbar^2/me^2$	Bohr radius	0.529167×10^{-8} cm
Velocity	$v_0 = e^2/\hbar$	Electron velocity in first Bohr orbit	2.18765×10^8 cm sec^{-1}
Time	a_0/v_0	Time required for electron in first Bohr orbit to travel one Bohr radius	2.41888×10^{-17} sec
Energy	e^2/a_0	Twice the IP for hydrogen (infinite nuclear mass)	4.35942×10^{-11} erg

[a] From H. A. Bethe and E. E. Salpeter, *Quantum Mechanics of One- and Two-Electron Atoms* (Springer-Verlag, Berlin, 1957), p. 3; revised in accordance with SI values of the physical constants.

constants (see below). Note that the Hartree a.u. of energy is twice the energy corresponding to the Rydberg frequency. Although several authors use the Rydberg itself as the a.u. of energy, throughout this text the Hartree a.u. of energy is used.

In atomic units the Schrödinger equation for an N-electron atom with a nucleus of charge Z is

$$\left[-\sum_{i=1}^{N} \left(\frac{1}{2} \nabla_i^2 + \frac{Z}{r_i} \right) + \sum_{i>j=1}^{N} \frac{1}{r_{ij}} \right] \psi = E\psi$$

PHYSICAL CONSTANTS AND CONVERSION FACTORS

The values of the physical constants and the conversion factors derived from them used in this text and given in Table A2 are those of the Système International d'Unités (SI). For more complete tables, see *Handbook of*

Mathematical Functions, edited by M. Abramowitz and I. A. Stegun (U.S. Department of Commerce, National Bureau of Standards, Washington, D.C., 1968), seventh printing, pp. 5–8. Recent remeasurements of e/h, if adopted as standard, would change values of α, e, h, m_e, N_A, and the constants derived from them by amounts up to about 1 part in 10,000. A discussion may be found in B. Taylor, W. Parker, and D. Langenberg, *Rev. Mod. Phys.* **41**, 375 (1969).

As an illustration of the conversion of a quantity from cgs units to a.u., we determine the velocity of light to be

$$c \text{ cm sec}^{-1} = \frac{c}{v_0} \text{ a.u.} = \frac{\hbar c}{e^2} \text{ a.u.} = \frac{1}{\alpha} \text{ a.u.} = 137.039 \text{ a.u.}$$

Factors for converting energy units based upon the SI values of the physical constants are given in Table A.3.

Table A.2 *Selected values of the physical constants*

Constant	Symbol	Value	Unit
Speed of light in vacuum	c	2.997925×10^{10}	cm sec^{-1}
Electronic charge	e	4.80298×10^{-10}	esu
Avogadro number	N_A	6.02252×10^{23}	molecules mole^{-1}
Electron mass	m_e	5.48597×10^{-4}	amu[a]
Proton mass	m_p	1.00727663	amu[a]
Neutron mass	m_n	1.0086654	amu[a]
Faraday constant	$\mathfrak{F}$	$96,487.0$	coulomb mole^{-1}
Planck constant	h	6.6256×10^{-27}	erg sec
	$\hbar$	1.05450×10^{-27}	erg sec
Fine-structure constant[b]	α	7.29720×10^{-3}	
	α^{-1}	137.0388	
Rydberg constant	R_∞	1.0973731×10^5	cm^{-1}
Bohr radius	a_0	5.29167×10^{-9}	cm
Bohr magneton	μ_B	9.2732×10^{-21}	erg gauss^{-1}
Nuclear magneton	μ_N	5.0505×10^{-24}	erg gauss^{-1}
Proton magnetic moment	μ_p	1.41049×10^{-23}	erg gauss^{-1}
Gas constant	R	8.3143×10^7	erg °K^{-1} mole^{-1}
Boltzmann constant	k	1.38054×10^{-16}	erg °K^{-1}
Gravitational constant	G	6.670×10^{-8}	dyne cm^2 g^{-2}

[a] 1 amu $= N_A^{-1}$ g $= 1.66043 \times 10^{-24}$ g   [b] $\alpha = e^2/\hbar c$

Table A.3 *Energy conversion factors*

	eV	erg	kcal/mole	Hz	cm⁻¹	°K	a.u.
eV	1	1.60210×10^{-12}	23.0609	2.41804×10^{14}	8.06573×10^{3}	1.16049×10^{4}	3.67502×10^{-2}
erg	6.24181×10^{11}	1	1.43942×10^{13}	1.50929×10^{26}	5.03448×10^{15}	7.24356×10^{15}	2.29388×10^{10}
kcal/mole	4.33634×10^{-2}	6.94725×10^{-14}	1	1.04854×10^{13}	3.49757×10^{2}	5.03228×10^{2}	1.59362×10^{-3}
Hz	4.13558×10^{-15}	6.62561×10^{-27}	9.53702×10^{-14}	1	3.33565×10^{-11}	4.79930×10^{-11}	1.51983×10^{-16}
cm⁻¹	1.23981×10^{-4}	1.98630×10^{-16}	2.85911×10^{-3}	2.99793×10^{10}	1	1.43879	4.55633×10^{-6}
°K	8.61705×10^{-5}	1.38054×10^{-16}	1.98717×10^{-3}	2.08364×10^{10}	6.95028×10^{-1}	1	3.16678×10^{-6}
a.u.	27.2107	4.35943×10^{-11}	6.27503×10^{2}	6.57966×10^{15}	2.19474×10^{5}	3.15777×10^{5}	1

NOTE: To convert a quantity expressed in the unit given in the left-hand column to a quantity expressed in a unit given in the top row, multiply by the factor appearing at the intersection of the column and row. For example, to convert from eV to erg, multiply 1.60210×10^{-12}.

INDEX

ATOMS & MOLECULES

**Martin Karplus and
Richard N. Porter**

This textbook is intended for the part of a beginning undergraduate course in physical chemistry concerned with basic principles of quantum theory and their application to atoms and molecules. It is based on lectures given originally by Martin Karplus at Columbia University and has been supplemented with material developed and taught by the authors over a period of about eight years.

The prerequisites for this book are introductory courses in calculus, chemistry, and physics. The authors manage to build an integrated development on such an elementary background by using a style that often forgoes the succinctness of purely mathematical exposition to explain *in words* the meaning and utility of a principle before it is formulated mathematically. Examples of applications include many from the recent literature of physical chemistry and chemical physics. Problems are chosen to reinforce and extend the textual material and are of sufficient number and variety to allow the individual instructor flexibility in his emphasis. Thus it is hoped that the student will be left with a realistic picture of the physical chemistry of atoms and molecules and an appreciation of the significance of experimental and theoretical techniques for discoveries in this field.

The discussion of atomic and molecular structure goes beyond any available elementary physical chemistry text, but these topics are treated in a style which places them within the context of the elementary course rather than in a more advanced course in quantum theory. In addition, the systematic and integrated treatment of van de Waals interactions, ionic bonds, transition metal complexes, and covalent bonds is unique in its organization with respect to open and closed shell interactions of atoms and ions.